Springer Series on
SIGNALS AND COMMUNICATION TECHNOLOGY

Signals and Communication Technology

Passive Eye Monitoring
Algorithms, Applications and Experiments
R.I. Hammoud (Ed.)   ISBN 978-3-540-75411-4

Multimodal User Interfaces
From Signals to Interaction
D. Tzovaras   ISBN 978-3-540-78344-2

Human Factors and Voice Interactive Systems
D. Gardner-Bonneau, H.E. Blanchard (Eds.)
ISBN: 978-0-387-25482-1

Wireless Communications
2007 CNIT Thyrrenian Symposium
S. Pupolin (Ed.)   ISBN: 978-0-387-73824-6

**Satellite Communications
and Navigation Systems**
E. Del Re, M Ruggieri (Eds.)
ISBN: 978-0-387-47522-6

Digital Signal Processing
An Experimental Approach
S. Engelberg   ISBN: 978-1-84800-118-3

**Digital Video and Audio
Broadcasting Technology**
A Practical Engineering Guide
W. Fischer   ISBN: 978-3-540-76357-4

Three-Dimensional Television
Capture, Transmission, Display
H.M. Ozaktas, L. Onural (Eds.)
ISBN 978-3-540-72531-2

**Foundations and Applications
of Sensor Management**
A.O. Hero, D. Castañón, D. Cochran,
K. Kastella (Eds.)   ISBN 978-0-387-27892-6

**Digital Signal Processing
with Field Programmable Gate Arrays**
U. Meyer-Baese   ISBN 978-3-540-72612-8

Adaptive Nonlinear System Identification
The Volterra and Wiener Model Approaches
T. Ogunfunmi   ISBN 978-0-387-26328-1

Continuous-Time Systems
Y.S. Shmaliy   ISBN 978-1-4020-6271-1

Blind Speech Separation
S. Makino, T.-W. Lee, H. Sawada (Eds.)
ISBN 978-1-4020-6478-4

**Cognitive Radio, Software Defined Radio,
and Adaptive Wireless Systems**
H. Arslan (Ed.)   ISBN 978-1-4020-5541-6

Wireless Network Security
Y. Xiao, D.-Z. Du, X. Shen
ISBN 978-0-387-28040-0

Terrestrial Trunked Radio – TETRA
A Global Security Tool
P. Stavroulakis   ISBN 978-3-540-71190-2

Multirate Statistical Signal Processing
O.S. Jahromi   ISBN 978-1-4020-5316-0

Wireless Ad Hoc and Sensor Networks
A Cross-Layer Design Perspective
R. Jurdak   ISBN 978-0-387-39022-2

**Positive Trigonometric Polynomials
and Signal Processing Applications**
B. Dumitrescu   ISBN 978-1-4020-5124-1

Face Biometrics for Personal Identification
Multi-Sensory Multi-Modal Systems
R.I. Hammoud, B.R. Abidi, M.A. Abidi (Eds.)
ISBN 978-3-540-49344-0

**Cryptographic Algorithms
on Reconfigurable Hardware**
F. Rodríguez-Henríquez
ISBN 978-0-387-33883-5

**Ad-Hoc Networking
Towards Seamless Communications**
L. Gavrilovska   ISBN 978-1-4020-5065-7

Multimedia Database Retrieval
A Human-Centered Approach
P. Muneesawang, L. Guan
ISBN 978-0-387-25627-6

Broadband Fixed Wireless Access
A System Perspective
M. Engels; F. Petre
ISBN 978-0-387-33956-6

Acoustic MIMO Signal Processing
Y. Huang, J. Benesty, J. Chen
ISBN 978-3-540-37630-9

Algorithmic Information Theory
Mathematics of Digital Information
Processing
P. Seibt   ISBN 978-3-540-33218-3

Continuous-Time Signals
Y.S. Shmaliy   ISBN 978-1-4020-4817-3

Interactive Video
Algorithms and Technologies
R.I. Hammoud (Ed.)   ISBN 978-3-540-33214-5

Handover in DVB-H
Investigation and Analysis
X. Yang   ISBN 978-3-540-78629-0

IP Traffic Theory and Performance
C. Grimm, G. Schlüchtermann
ISBN 978-3-540-70603-8

Christian Grimm · Georg Schlüchtermann

IP Traffic Theory and Performance

Christian Grimm
Gottfried Wilhelm Leibniz
Universität Hannover
Germany

Georg Schlüchtermann
Ludwig-Maximilians-Universität Munich
and University of Applied Sciences
Munich
Germany

ISBN 978-3-540-70603-8   e-ISBN 978-3-540-70605-2
DOI 10.1007/978-3-540-70605-2

Springer Series on Signals and Communication Technology ISSN 1860-4862

Library of Congress Control Number: 2008931483

© 2008 Springer-Verlag Berlin Heidelberg
This work is subject to copyright. All rights are reserved, whether the whole or part of the material is concerned, specifically the rights of translation, reprinting, reuse of illustrations, recitation, broadcasting, reproduction on microfilm or in any other way, and storage in data banks. Duplication of this publication or parts thereof is permitted only under the provisions of the German Copyright Law of September 9, 1965, in its current version, and permission for use must always be obtained from Springer. Violations are liable for prosecution under the German Copyright Law.

The use of general descriptive names, registered names, trademarks, etc. in this publication does not imply, even in the absence of a specific statement, that such names are exempt from the relevant protective laws and regulations and therefore free for general use.

Cover design: WMXDesign GmbH, Heidelberg

Printed on acid-free paper

9 8 7 6 5 4 3 2 1

springer.com

To the memory of my mother, Georg

Preface

> *Reading without meditation is sterile;*
> *meditation without reading is liable to error;*
> *prayer without meditation is lukewarm;*
> *meditation without prayer is unfruitful;*
> *prayer, when it is fervent, wins contemplation,*
> *but to obtain contemplation without*
> *prayer would be rare, even miraculous.*
>
> Bernhard de Clairvaux (12th century)

Nobody can deny that IP-based traffic has invaded our daily life in many ways and no one can escape from its different forms of appearance. However, most people are not aware of this fact. From the usage of mobile phones – either as simple telephone or for data transmissions – over the new form of telephone service Voice over IP (VoIP), up to the widely used Internet at the users own PC, in all instances the transmission of the information, encoded in a digital form, relies on the Internet Protocol (IP). So, we should take a brief glimpse at this protocol and its constant companions such as TCP and UDP, which have revolutionized the communication system over the past 20 years.

The communication network has experienced a fundamental change, which was dominated up to end of the eighties of the last century by voice application. But from the middle of the nineties we have observed a decisive migration in the data transmission.

If the devoted reader of this monograph reads the title 'IP traffic theory and performance', she/he may ask, why do we have to be concerned with modeling IP traffic, and why do we have to consider and get to know new concepts. She/he may argue that on the one hand, since the early days of Erlang and his fundamental view on the traffic description in the emerging communication world, formulas and tables have contributed to the building of powerful communication networks. On the other hand she/he may be guided by the argument that, even if we do not meet the correct model, i.e. if our standard knowledge does not suffice and fails, there is enough technical potential in the classical telecommunication network, in terms of equipment, to overcome any bottleneck.

In some respect, we will disprove this misleading attitude. But before going into details, we can already argue that on the one side, and this is done in several parts of the monograph, IP-based traffic does not fit into the classical framework of the Erlang theory. Since the network connections is no longer end-to end built, but it is chaotic at the first glance, and that is the strength of the IP-based networks, since the Internet is *self-organized*, i.e. deciding more less at each router or node, which route it will take. This introduces the stochastic aspect, which runs through the IP modeling as well through our book as a dominant feature.

On the other side making server or router as powerful as possible so as to make any modeling superfluous, has its decisive drawback. The network is not clearly structured, so that at each possible node, the capacity and service rate is large enough to encounter any traffic load. Only a few bottlenecks diminish the performance and would especially influence extensively the time sensitive traffic, as Voice over IP or video streaming, with its strong quality of service (QoS) requirement. The customer would avoid any of these services as a consequence. This in fact has occurred already in reality and is not a fiction. Incorrect design of networks according to IP traffic requirements lead to a rejection or at least an unexpected delay of new services like VoIP or video on demand.

In addition, a variety of technical possibilities raises more expectations: if we have the traffic capacity, we want, and we will use it without restriction – and in turn the network will meet its limits soon. Avoiding these difficulties and being prepared for future challenges, we present models, indicate consequences and outline major key aspects, like queueing for judging performances of the network.

After the discovery of the Bellcore group in the early 1990's it was Ilkka Norros who developed a first approach in describing the IP-based traffic. He used the fractional Brownian motion as stochastic perturbation to incorporate the basic phenomena of self-similarity and long-range dependence of the connectionless traffic. As already mentioned this is in contrast to the classical circuit switched traffic, where averaging over all scales is leveling the bursty character – this bursty character does not change over all time scales in the IP case.

In fact, this approach was not new, since in the sixties Benoît Mandelbrot and John Winslow Van Ness used the fractional Brownian motion and its self-similarity for the description of stock pricing in financial markets. The race for the most appropriate model was opened! A significant variety of models was introduced, which is a consequence of the fast growing number of applications accompanied by different protocols, especially with the TCP/IP. Some tried to describe the large scales and the more Gaussian traffic, other the more bursty traffic and again others the small scales influenced by the control cycle triggered by the TCP/IP. Here, especially the multifractal models and cascades entered the scene, with some of course relatively complicated and with the lack of a suitable chance for application. Up to now a unified theory is missing, which of course may not be near to fulfillment, since the variety of modern networks counteracts these efforts.

Hence, we will present the major models and indicate their advantages and limitation. It is clear that it would be beyond the scope of the book to give a full list of all approaches. It is our aim on the one hand to give a certain feeling, why the IP traffic differs from the classical Erlang description, to give insight into the different approaches, where of course we will to some extent only introduce them and will not go into details. This is left to the reader for further study. We will show, how one can map the traffic to the models using standard statistical methods and we will finally try to answer the question, as to what are the decisive key values, like queueing, waiting time and QoS thresholds. These factors enter finally into the major question of optimization – from the network point of view as well as from the economical standpoint. Summarizing we do not pursue completeness or even a full and profound description of all existing models for IP traffic. In fact, we would like the monograph to be considered as a 'window not mirror' as Rainer Maria Rilke in his 'testamony' once put it [79]. Writing the book was for us like the metaphor of Rilke, with only a small 'window hole' compared to the huge building of nature.

The book is organized as follows: We start with the fundamental ingredients and properties of the IP-based traffic and its resulting concepts and models. The second chapter is, what we call the classical traffic theory. We build up the foundations from the well-known Erlang formulas and the resulting basic ideas of telecommunication traffic in the circuit switched context to the new development of incorporating phase distributed approaches for inter arrival and service times. In addition, deterministic approaches mixed with probability aspects applied to the control cycles and stochastic influences of the chaotic structure of the network, will be considered as well.

As illustrated in the first chapter, the basic phenomena in IP traffic can be described as *self-similarity* and *long-range dependence*. Already outlined in chapter 2, the traffic is determined by stochastic perturbation. Thus, we have to deal with the basic stochastic processes as fractional Brownian motion (FBM), the FARIMA time series and fractional α-stable processes and its key value, the *Hurst exponent*. Their influence and significance in the IP traffic will be outlined in the sequel, one using the approach with stochastic differential equation (like the Norros approaches) and the other from a more physical point of view. Finally, we incorporate the protocol influence which leads to the multifractal Brownian motion (mBm) or the pure multifractal models.

To implement the models for a given observed network, we have to collect different samples of data and analyze them using standard statistical methods. This is especially important for estimating one key value, the Hurst parameter. Methods like the absolute value, the Whittle and wavelet estimator are to be mentioned. But, as the entire monograph demonstrates, no method is the dominant and fundamental one. The preferred model as well as the estimation methods depends on the particular application.

Finally, in the fifth chapter, we consider the major field of application for each of the various models – the performance and optimization aspect. Within the narrative, we present key models, like the Norros and multifractal approaches, and its influence on the performance key values. In conclusion, we apply this to the optimization under some selected aspects. The last section follows more or less the idea of fair prices introduced by Kelly, Maulhoo and Tan [138] and we apply this together with the special perturbation of the FBM. Here, we select the toolbox for optimal control, developed in mathematical economy. Since most approaches require mathematics for a suitable description, evitable, the reader should be familiar with some key concepts, such as differential equations, probability theory, and stochastic analysis. It should be mentioned that we have tried to keep most techniques as simple as possible, to avoid burdening the reader with details. These details and a more profound understanding is left to the interested reader in the specific monographs or the original literature. But the mathematical toolbox is essential and it seems suitable to cite Galileo Galilei:

> *The book of nature can only be understood, if we have learned its language and letters, in which it is written. And it is written in mathematical language and the letters are triangle, circles and geometrical figures; without those auxiliaries it is not possible for mankind to understand a single word.*

Even though geometry does play only a minor role for us, we can transfer the idea behind Galilei's words to the analytical field in mathematics, used for this monograph.

Hannover and München
May 2008

Christian Grimm
Georg Schlüchtermann

Contents

1 Introduction to IP Traffic 1
  1.1 TCP/IP Architecture Model 1
      1.1.1 Physical Layer 3
      1.1.2 Data Link Layer 4
      1.1.3 Network Layer 5
      1.1.4 Transport Layer 5
      1.1.5 Application Layer 8
  1.2 Aspects of IP Modeling 9
      1.2.1 Levels of Modeling 10
      1.2.2 Traffic Relations 13
      1.2.3 Asymmetry in IP Traffic 17
      1.2.4 Temporal Behavior 17
      1.2.5 Network Topology 18
  1.3 Quality of Service 19
      1.3.1 Best Effort Traffic 19
      1.3.2 Time Sensitive Data Traffic 20
      1.3.3 Overprovisioning 20
      1.3.4 Prioritization 21
  1.4 Why Traditional Models Fail 22

2 Classical Traffic Theory 29
  2.1 Introduction to Traffic Theory 29
      2.1.1 Basic Examples 29
      2.1.2 Basic Processes and Kendall Notation 32
      2.1.3 Basic Properties of Exponential Distributions 33
  2.2 Kolmogorov Equation 34
      2.2.1 State Probability 36
      2.2.2 Stationary State Equation 37
  2.3 Transition Processes 38
  2.4 Pure Markov Systems $M/M/n$ 40
      2.4.1 Loss Systems $M/M/n$ 40

		2.4.2	Queueing Systems $M/M/n$	44
		2.4.3	Application to Teletraffic	52
	2.5	Special Traffic Models		58
		2.5.1	Loss Systems $M/M/\infty$	58
		2.5.2	Queueing Systems of Engset	58
		2.5.3	Queueing Loss Systems	59
	2.6	Renewal Processes		60
		2.6.1	Definitions and Concepts	60
		2.6.2	Bounds for the Renewal Function	65
		2.6.3	Recurrence Time	67
		2.6.4	Asymptotic Behavior	68
		2.6.5	Stationary Renewal Processes	71
		2.6.6	Random Sum Processes	71
	2.7	General Poisson Arrival and Serving Systems $M/G/n$		73
		2.7.1	Markov Chains and Embedded Systems	73
		2.7.2	General Loss Systems $M/G/n$	74
		2.7.3	Queueing Systems $M/G/n$	74
		2.7.4	Heavy-Tail Serving Time Distribution	81
		2.7.5	Application of $M/G/1$ Models to IP Traffic	95
		2.7.6	Markov Serving Times Models $GI/M/1$	102
	2.8	General Serving Systems $GI/G/n$		107
		2.8.1	Loss Systems	107
		2.8.2	The Time-Discrete Queueing System $GI/G/1$	109
		2.8.3	$GI/G/1$ Time Discrete Queueing System with Limitation	117
	2.9	Network Models		122
		2.9.1	Jackson's Network	122
		2.9.2	Systems with Priorities	130
		2.9.3	Systems with Impatient Demands	131
		2.9.4	Conservation Laws Model	133
		2.9.5	Packet Loss and Velocity Functions on Transmission Lines	134
		2.9.6	Riemann Solvers	141
		2.9.7	Stochastic Velocities and Density Functions	146
	2.10	Matrix-Analytical Methods		148
		2.10.1	Phase Distribution	148
		2.10.2	Examples for Different Phase Distributions	153
		2.10.3	Markovian Arrival Processes	156
		2.10.4	Queueing Systems $MAP/G/1$	161
		2.10.5	Application to IP Traffic	173
3	**Mathematical Modeling of IP-based Traffic**			**181**
	3.1	Scalefree Traffic Observation		181
		3.1.1	Motivation and Concept	181
		3.1.2	Self-Similarity	183
	3.2	Self-Similar Processes		184

		3.2.1	Definition and Basic Properties 184

- 3.2.1 Definition and Basic Properties 184
- 3.2.2 Fractional Brownian Motion 190
- 3.2.3 α-stable Processes 194
- 3.3 Long-Range Dependence 202
  - 3.3.1 Definition and Concepts........................... 203
  - 3.3.2 Fractional Brownian Motion and Fractional Brownian Noise 207
  - 3.3.3 Farima Time Series 211
  - 3.3.4 Fractional Brownian Motion and IP Traffic – the Norros Approach 218
- 3.4 Influence of Heavy-Tail Distributions on Long-Range Dependence... 226
  - 3.4.1 General Central Limit Theorem 226
  - 3.4.2 Heavy-Tail Distributions in M/G/∞ Models 233
  - 3.4.3 Heavy-Tail Distributions in On-Off Models 235
  - 3.4.4 Aggregated Traffic................................ 240
- 3.5 Models for Time Sensitive Traffic 245
  - 3.5.1 Multiscale Fractional Brownian Motion 245
  - 3.5.2 Norros Models for Differentiating Traffic 249
- 3.6 Fractional Lévy Motion in IP-based Network Traffic 259
  - 3.6.1 Description of the Model 259
  - 3.6.2 Calibration of a Fractional Lévy Motion Model........ 260
- 3.7 Fractional Ornstein-Uhlenbeck Processes and Telecom Processes 261
  - 3.7.1 Description of the Model 261
  - 3.7.2 Fractional Ornstein-Uhlenbeck Gaussian Processes..... 262
  - 3.7.3 Telecom Processes 263
  - 3.7.4 Representations of Telecom Processes 263
  - 3.7.5 Application of Telecom Processes 265
- 3.8 Multifractal Models and the Influence of Small Scales 267
  - 3.8.1 Multifractal Brownian Motion 267
  - 3.8.2 Wavelet-Based Multifractal Models 270
  - 3.8.3 Characteristics of Multifractal Models................ 280
  - 3.8.4 Multifractal Formalism............................. 292
  - 3.8.5 Construction of Cascades 296
  - 3.8.6 Multifractals, Self-Similarity and Long-Range Dependence 308
- 3.9 Summary of Models for IP Traffic 316

4 Statistical Estimators 321
- 4.1 Parameter Estimation.................................... 321
  - 4.1.1 Unbiased Estimators............................... 322
  - 4.1.2 Linear Regression 329
  - 4.1.3 Estimation of the Heavy-Tail Exponent α 335
  - 4.1.4 Maximum Likelihood Method 344

4.2 Estimators of Hurst Exponent in IP Traffic 349
    4.2.1 Absolute Value Method (AVM) 349
    4.2.2 Variance Method 352
    4.2.3 Variance of Residuals 354
    4.2.4 R/S Method 356
    4.2.5 Log Periodogram – Local and Global 359
    4.2.6 Maximum Likelihood and Whittle Estimator 363
    4.2.7 Wavelet Analysis 368
    4.2.8 Quadratic Variation 379
    4.2.9 Remarks on Estimators 380

5 Performance of IP: Waiting Queues and Optimization 383
5.1 Queueing of IP Traffic for Perturbation with Long-Range Dependence Processes 383
    5.1.1 Waiting Queues for Models with Fractional Brownian Motion 384
    5.1.2 Queueing in Multiscaling FBM 392
    5.1.3 Fractional Lévy Motion and Queueing in IP Traffic Modeling 395
    5.1.4 Queueing Theory and Performance for Multifractal Brownian Motion 405
5.2 Queueing in Multifractal Traffic 411
    5.2.1 Queueing in Multifractal Tree Models 411
    5.2.2 Queueing Formula 417
5.3 Traffic Optimization 423
    5.3.1 Mixed Traffic 423
    5.3.2 Optimization of Network Flows 424
    5.3.3 Rate Control: Shadow Prices and Proportional Fairness 436
    5.3.4 Optimization for Stochastic Perturbation 442
    5.3.5 Optimization of Network Flows Using an Utility Approach 449

References 465

Index 479

1
Introduction to IP Traffic

> *He that will not apply new remedies must expect new evils: for time is the greatest innovator.*
>
> Francis Bacon (16th and 17th century)

Before starting with the mathematical analysis of IP traffic, we give an introduction to the characteristics and special properties of the Internet, so as to get a better understanding in the different models and their justification. In this first chapter, we will explain the structure of the protocols and networks as well as the applications in the Internet with regard to their impact on mathematical modeling. We will take a birds eye position, without in depth explanations of technical details. For a mathematical treatment of selected aspects and discussions of further details we refer to the remainder of this book.

1.1 TCP/IP Architecture Model

The Internet is a highly developed and complex system. Its infrastructure and usage is marked by heterogeneity and it is subject to both ongoing changes and permanent growth. Today, several hundred thousand local networks and wide area networks connect via millions of routers and links some ten millions end systems and hundreds of million users. All end systems and intermediate components exchange data with each other by help of a few common rules for their communication: the so-called TCP/IP protocols.

In addition, these joint protocol packet switching is the other main characteristic of the Internet. Each messages that is exchanged between end systems is divided by the sender into small units. Generally, the transmitting units of a protocol are denoted as *Protocol Data Units* (PDU). Every PDU is provided with control information for addressing of target processes, protocol-inherent mechanisms or checksums to identify transmission errors. With this additional information each PDU is forwarded by routers to the receiver, in principle completely independent of any preceding or succeeding PDU. The size, quantity and timely correlation of PDUs sent depends on different parameters in each protocol.

Packet switching builds the robust and efficient network infrastructure for the Internet. The failure of a single link merely affects the PDUs being transferred at that moment. If necessary at all, only these PDUs must be transferred again. With this, the packet switching seems more efficient compared to the circuit switched networks: the available resources are completely assigned to any arbitrary single PDU and not a priori to dedicated connections. Waiting or inactive periods which occur in every communication process can immediately be saturated with PDUs that are sent by concurrent users or processes. However, this results in a permanent competition of all PDUs around the available resources. The transmission quality, that is the guaranteed temporal behavior of PDUs and a constant data rate, as it is natural for circuit switched networks is possible in packet switched networks only with considerable additional overhead.

Figure 1.1 illustrates the traditional TCP/IP communication architecture with five distinct layers on the end systems. A further subdivision of the topmost application layer in three layers leads to the well known ISO/OSI (International Organization for Standardization/Open Systems Interconnection) model with seven layers in total. On every end system all layers have to be implemented to exchange data among each others. In routers which serve as forwarding units for the PDUs or IP packets between the end systems, merely the three lower layers have to be implemented. The protocols illustrated as horizontal arrows in figure 1.1 represent the logical communication between the same layers on different systems.

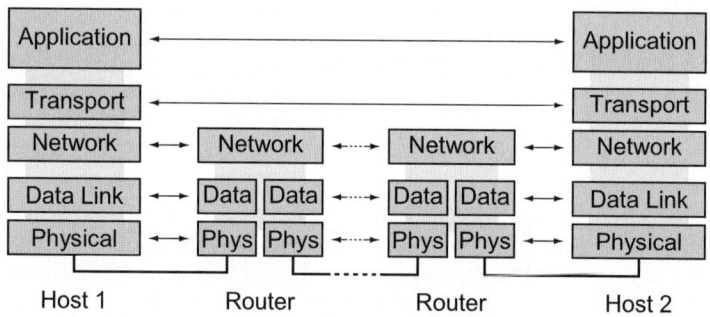

Fig. 1.1. TCP/IP Model by Department of Defense (DoD)

In addition to figure 1.1 figure 1.2 illustrates the typical implementation of all layers on the end systems as well as the addressing used in the respective layer. Typically, the two lower layers are build in hardware on the network interface card and must be implemented strictly according to the respective standards released by e.g. IEEE or ISO, since additional modifications are hardly possible.

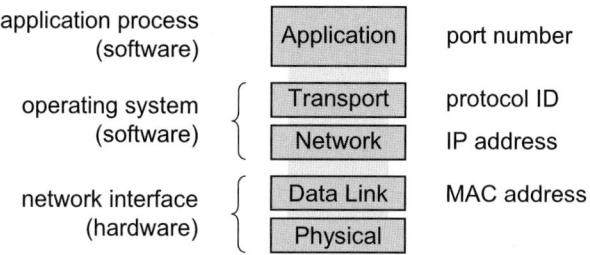

Fig. 1.2. Implementation and Addressing in TCP/IP

The protocols on network and transmission layer are implemented as part of the operating systems. The specifications are mainly published by the *Internet Engineering Task Force* (IETF). In principle, only a low degree of freedom is allowed with the implementations, since otherwise different end systems would have incompatible operating systems which could not communicate with each other. However, the behavior of single protocols can be varied in certain limits. We will show later that e.g. different adaptive algorithms in TCP are possible without violating the interoperability of end systems.

On application layer, an individual design of protocols is possible with the implementation of the respective application. Depending on the purpose of the application the programmer can freely define 'his own' protocols. By providing executables only the details of the protocols used in the application may even remain unpublished.

From these simple facts, important criteria follow for our models. The protocols on lower layers are described by models that were already build during the development of the respective technology. They were used as a necessary proof for the expected success of the technology, before a decision about the production of the respective hardware was made. We may regard these models as static because the protocols must be strictly compliant to their specifications. On the other hand, protocols on higher layers can be described only with more complex models, since they are subject to a variety of parameters and degrees of freedom and therefore have a higher variability. Even for unpublished protocols, a mathematical modeling is possible merely based on measured traffic.

Before we go into the details of IP traffic characteristics, we will briefly explain the typical properties of the five layers in the TCP/IP protocol architecture from figure 1.1 and their impact on modeling.

1.1.1 Physical Layer

The physical layer describes the transmission of signals over different types of media like copper cables, glass fibers or the air. Parameters like signal level and codings, modulation methods and frequency domains, but also the properties of the transmission media, antennas and cable connectors are defined here.

From this it follows that the physical layer is of limited influence for the modeling of IP traffic. However, modeling of new technologies in the physical layer is essential today, especially for mobile communication.

1.1.2 Data Link Layer

The protocols on data link layer determine the rules by which the transmission medium is accessed. The typical procedures differ considerably in local and wide area networks. During the second half of the nineties Ethernet emerged as the predominant technology in local area networks to transmit best effort traffic. Today, the Ethernet protocol family is used with data rates from – historically – 10 MBit/second to currently 10 GBit/second. Ethernet represents, like the typical access mode in Wireless LAN, a non deterministic behavior. By the robust decentralized approach for media access, little technical requirements and simple scalability of the data rate over several orders of magnitude, Ethernet proves to be the ideal protocol on data link layer below IP.

In the original Ethernet, several stations were connected in bus topology to a *shared* media. Due to the decentralized algorithm for media access more than a single station could send data at the same time. Competing end systems caused overlays of the signals on the link (often denoted as colliding PDUs or collisions) which made all data sent erroneous. Today, by switching technology with an underlying star topology and full duplex transmission between at least the end system and the next switch port, Ethernet provides full channel capacity at every interface. Significant impact with regard to modeling will only arise if accompanying protocols for packet prioritization or resource reservation with explicit demands for a given quality of service are established to support real time applications.

In wide area networks the circuit switched SONET/SDH technologies are used besides the cell switched ATM. The invention of *Synchronous Optical NETwork* (SONET) in the US and the almost compatible *Synchronous Digital Hierarchy* (SDH) in Europe and Japan at the end of the eighties made homogeneous, world wide data networks possible. SONET/SDH provides data rates of currently up to 160 GBit/s (OC-3072/STM-1024). Small data rates can be easily multiplexed with low overhead to high-rate streams and vice versa. An effective management of the network resources leads to an almost failsafe operation with high availability. The introduction of failover circuits and centralized management provides short bridging times and a robust behavior in case of failures of single components. With *Dense Wavelength Division Multiplexing* (DWDM) even multiple SDH channels can be transferred on different carriers over a single optical fiber. Depending on the applied techniques, at present, an immediate multiplication of the data rate to up to a total of 160 GBit/s is possible.

For many protocols in the data link layer excellent mathematical models exist. This applies both to the protocols in wide area networks as well as to Ethernet or WLAN in local area networks.

1.1.3 Network Layer

The most prominent protocol on network layer is the *Internet Protocol* (IP). From figure 1.1 the significance of IP in the Internet is immediately obvious because the network layer is the topmost layer on all components involved between the end systems. On network layer the necessary information is exchanged and evaluated for the appropriate routing of the so-called IP packets. Based on the calculated routing information, every router determines the next optimal link for every IP packet on the path to the receiver. In general, each IP packet is handled independent of all preceding and succeeding packets. This means that each IP packet carries the full information that is needed for routing decisions. Additionally, subsequent IP packets may even follow different paths through the network.

Each router may fragment IP packets, if the next link allows the transmission of smaller units only. The corresponding defragmentation of IP packets is done only by the receiver. By determining the maximum common packet size on all links between sender and receiver this fragmentation may be avoided. With the so-called *Maximum Transmission Unit* (MTU) the best possible IP packet size us used for a connection. Today, the typical MTU size for most connections is slightly below 1,500 bytes which corresponds directly with the maximum frame size of Ethernet.

For a single IP packet, a router needs only bounded information for its routing decision. In most cases only the receiver address of the IP packet as well as the information about the status of the links and the directly connected next routers are required. The *decentralized structure* arising from this hop-by-hop principle is the foundation for bypassing single components in case of failures and the resulting robustness of the Internet.

The behavior of routers is of high significance for our models. On one hand, with their decisions about the adequate next links on the path, routers can substantially influence the delay of IP packets and the achievable data rate. On the other hand, routers buffer IP packets in queues if a suitable path to the receiver is not immediately available. The mathematical treatment of these queues is crucial for modeling. Considering additional prioritization of IP packets leads to approaches which correspond to a withdrawal from the simple FIFO procedure towards more complex models.

1.1.4 Transport Layer

The protocols on transport layer represent the lowest layer of the so called end-to-end communication on the Internet between sender and receiver. The transport protocols operate exclusively between the end systems and are not interrupted by routers or other network components. In the TCP/IP protocol stack, two different protocols are available on transport layer. The *User Datagram Protocol* (UDP) can be considered as a rudimentary protocol which merely puts a header in front of the payload sent by the application. The goal of UDP is to address the application processes on the communicating hosts and to identify transmission errors. The PDUs in UDP are described as UDP datagrams.

Unlike UDP the *Transmission Control Protocol* (TCP) represents a complex protocol which provides a reliable communication between sender and receiver. Here, the term 'reliable communication' preferably means complete and correct transmission of data but also guaranteed identification of occurring errors to the applications with respective signaling.

For this, TCP introduces acknowledgments and timeouts as well as explicit phases for connection set-up (three way handshake) and connection tear-down (four way close). The duration of timeouts is not given by a fixed value but continuously calculated dependent on the current and possibly varying *Roundtrip Time* (RTT) between sender and receiver. The PDUs transferred with TCP are described as *TCP segments*. As an example, figure 1.3 illustrates the download of a World Wide Web object via TCP with its three phases.

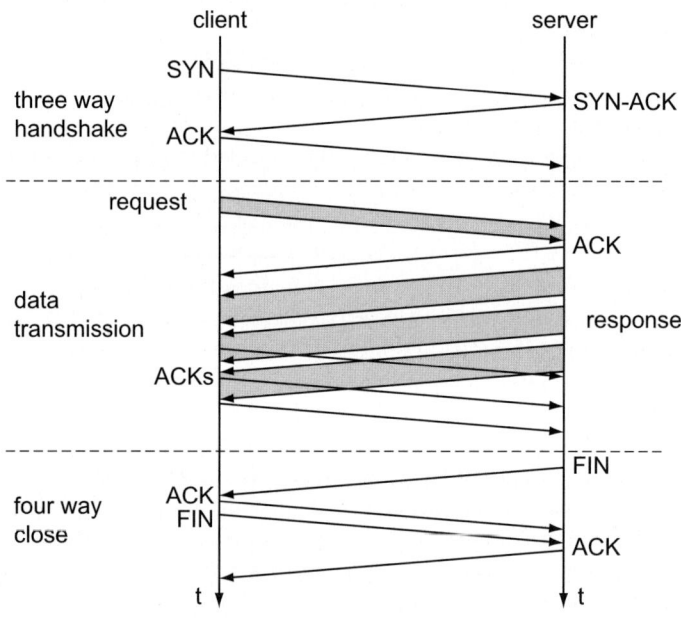

Fig. 1.3. Phases of data transmission in TCP

After the initial three way handshake the transmission of the data starts with a TCP segment containing the HTTP request of the client. The server acknowledges the request first and then sends the object in three separate TCP segments which are in turn acknowledged by the client. The reliable connection tear-down is done with the four way close in the last phase. In our example, the server sends the first segment of the four way close but it may also be initiated by the client.

The three way handshake in TCP corresponds to a check of accessibility and ready-to-receive state of the receiver before the payload or application data is transmitted. Therefore, TCP is also described as connection-oriented transport protocol. Correspondingly UDP in which comparable mechanisms are missing, is often called a connectionless transport protocol.

In TCP, we find two adaptive algorithms for the regulation of the data rate:

- flow control: avoid flooding of a possibly slow or temporarily overloaded receiver and
- congestion control: avoid flooding of network components such as routers with slow links.

For flow control, the receiver informs the sender with every header of a TCP segment about the free memory in its receive buffer (*RecvWindow*). Consequently the sender is allowed to send only less than RecvWindow byte with the next TCP segment. With this simple algorithm it is guaranteed that the sender will not overflow the receiver.

In most cases, packet loss is caused by congestion of routers or links in the network. The sender identifies these situations as an incomplete sequence of acknowledgments (duplicate ACKs) or as completely missing acknowledgments (timeouts) and treats both situations differently. To avoid flooding of the network already at the beginning of the transmission, TCP starts with a low transmission rate, increases the rate at first exponentially to a certain value (slow start phase) and then linearly (congestion avoidance phase). At congestion events, the sender halves (at duplicate ACKs) the current transmission rate or reduces it to a minimum (at timeouts). Afterwards, TCP increases the data rate again according to slow start or congestion avoidance phase. The corresponding sawtooth of the data rate as transmitted by the sender is illustrated in figure 1.4.

The described behavior was proposed with TCP Reno which was the typical implementation of TCP in most modern operation systems for several years. Fairness and robustness of the Internet are significantly based on this algorithm, since at congestion events all senders immediately reduce their transmission rate by, at least, a half. Today, new TCP implementations such as TCP BIC, TCP CuBIC or Highspeed TCP arise. For example, we find TCP BIC as the default TCP algorithm in current Linux kernels since version 2.16.9. In most cases these variants modify the behavior in congestion avoidance phase to gain better performance. Of course, by the use of a different algorithm the traditional fairness in the Internet is no longer guaranteed.

The dynamic behavior of the transmission rate in a single TCP stream with its partial exponential increase is of high importance for the modeling and leads to complex approaches. With multifractal models, we will explain possible solutions more precisely in sections 3.8.2 and 5.2.

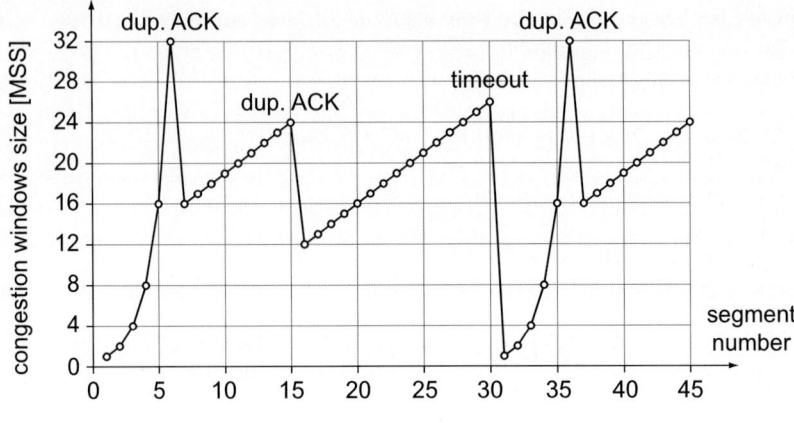

Fig. 1.4. TCP Slow Start

1.1.5 Application Layer

The application layer contains all protocols that are specified with the implementation of the respective applications. Compared to lower layers the term 'protocol' must be interpreted more broadly here, since also transformation or representation of data must be taken into account. By the implementation and the choice of the underlying transport protocol (TCP or UDP), essential characteristics are defined with regard to the sending behavior of an application. Typical best effort applications like World Wide Web or Email are ideally based on the connection-oriented and reliable transport protocol TCP. The application merely submits data to the TCP process on transport layer which provides the complete and reliable transmission. Transient problems concerning the transmission of the data are therefore handled and solved by TCP (that is the operating system) and remain hidden to the user. Even the programmer does not have to worry about the detection – or correction – of transmission errors. Only a corresponding error message to the user in case of failed connection set-up or complete failure of transmission should be taken into account. Correspondingly, the characteristics of the transmitted data stream depends on the application itself and the user behavior, as well as on the implementation of TCP on both communicating system.

Due to adaptive algorithms and possible retransmits in TCP neither the temporal behavior of PDUs nor the effective data rate can be reliably controlled by the application. Therefore, real time applications such as live streaming, video conferencing, Voice over IP or online games largely use UDP instead of TCP as transport protocol. With UDP, the temporal behavior of PDUs as well as the effective data rate are under direct control of the application.

In the previous paragraphs, we considered protocols that are used for the transmission of application data only. However, we must point out that almost hidden to both, the programmer and the user, further information has to be exchanged between services involved in order to operate the whole infrastructure of the Internet in a reliable way. Among others these services contain the following protocols:

- resolution or mapping of alphanumerical or numerical addresses (e.g. DNS and ARP)
- routing (e.g. BGP, OSPF, RIP, ISIS)
- network management (e.g. SNMP)
- signalling (e.g. ICMP, RSVP)
- authentication (e.g. RADIUS)

Table 1.1 summarizes some selected applications and infrastructure services with their corresponding protocols on application and transport layer.

Table 1.1. Corresponding protocols on application and transport layer

application	protocol on application layer	protocol on transport layer
World Wide Web	HTTP	TCP
Email	SMTP, IMAP, POP	TCP
File Transfer	FTP	TCP
TELNET, Secure Shell	TELNET, SSH	TCP
Peer to Peer	diverse	TCP
Network File System	NFS, SMB	UDP
Network Management	SNMP	UDP
Domain Name Service	DNS	UDP
Authentication	RADIUS, Diameter	TCP
Voice over IP	RTP	UDP
Streaming	RTP, RTCP	UDP
Control of Streams	RTSP	TCP

This table does not represent a complete or even definite picture. For example, apart from UDP the domain name service also uses TCP for so-called zone transfers, that is the exchange of address tables between DNS servers. Current versions of NFS can alternatively be used over TCP as well.

1.2 Aspects of IP Modeling

We explained in the previous section that data traffic in the Internet is substantially affected by both, the protocols used in distinct layers with their respective implementation and the dependences between different layers. Measurements of e.g. the temporal behavior of IP packets or the flow of TCP segments contain a large amount of relevant information. The description of single parameters like interarrival times of IP packets or transmission duration of single files are of limited use and may not be viewed in isolation for our models. For a comprehensive analysis particularly with regard to modeling the complete context of several layers must be considered carefully.

The measurement of sample data or a path Ω of IP traffic and the separation into single data streams does not lead to distinct series of random variables, that is a discrete stochastic process or a time series of PDUs. With our illustrations presented in the previous section it is obvious that data traffic occurs in different forms in different layers. This leads to two relevant considerations for our models:

- we have to apply specific models for each layer and
- we may not neglect dependencies between different layers.

1.2.1 Levels of Modeling

In accordance with the TCP/IP architecture model in figure 1.1 figure 1.5 illustrates different levels at which the user behavior and the resulting data traffic can be measured and modeled (we remark that we use the term 'level' and not 'layer' here).

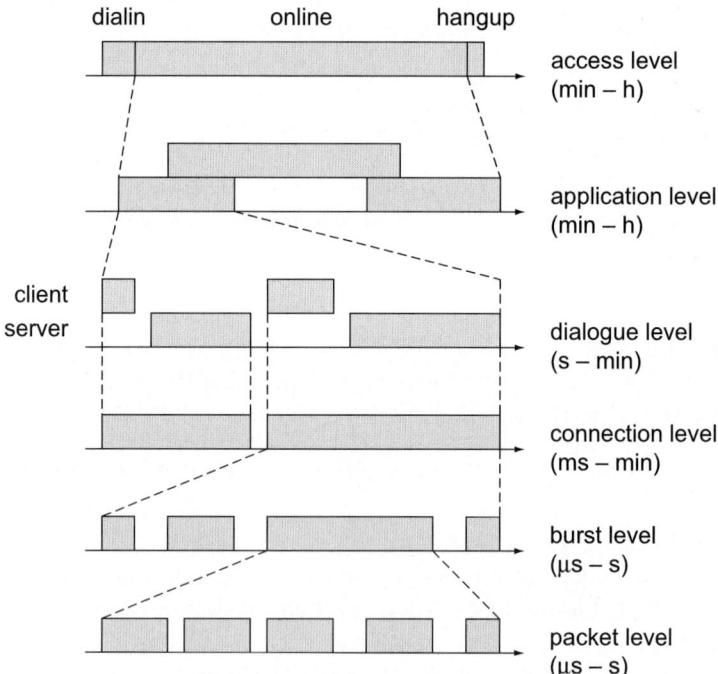

Fig. 1.5. Usage profiles at different levels

We assume that on each level the begin and end of events can be determined and the data volume transferred during this period can be measured. The necessary information is provided by logfiles in the operating system and network measurements of the data traffic. We have already noted that different approaches are available for the modeling of the characteristics observed on the respective level. Models which take the characteristics of different levels into account are denoted as *multi-level models*. We will apply a multi-level model to World Wide Web traffic as an example at the end of this section.

Access Level

At the topmost layer we find the time at which the user is connected to the Internet. For subscribers with dial-in lines this period corresponds to the duration between connection set-up and tear-down of their modem or respective equipment. For subscribers with permanent connection to the Internet this period correspond to the uptime of their end system. The basic structure in this level is significantly determined by different tariffs.

Application Level

The applications are represented by the next level. Several applications can be executed in parallel on clients as well as on servers. The execution and termination of an applications is initiated by the user, the operating system, other applications or even the application itself. Similar to the access level events, in the application level, events are mainly characterized by on and off periods. However, we have to consider the parallel execution of different or even the same application. With this, we get different usage times for the individual applications which are typically correlated with each other.

Dialogue Level

The next level represents the interaction between users and the applications. By the dialogue with the application, the user typically initiates data transmissions. Here, the application itself as well as the response times of other hosts and the network substantially influence the behavior of the user. As an example we illustrate three common phases of a user dialogue in the World Wide Web:

- Click: the user initiates the dialogue by clicking on a link.
- Wait: the download of the webpage starts, including all embedded images. The waiting phase is finished when all objects are retrieved from potentially different servers and the whole webpage is displayed in the browser.
- Think: depending on the content of the webpage and the interpretation by the user the webpage is viewed more or less carefully. During this time no further data is transferred and an inactive phase occurs. By clicking a new link this pause is finished and another dialogue is initiated.

Connection Level

Applied to the previous example of a dialogue in the World Wide Web with HTTP on application layer and TCP on transport layer the connection level represents the time period between connection set-up (three way handshake) and connection tear-down (four way close). In particular for HTTP we have to remember that not all objects of a web page have to be retrieved from the same web server and that not all objects have to be transferred over a single TCP connection (see mathematical modeling in the sections 2.8.1 and 3.8). Thus, the behavior on the connection level during the transmission highly depends on the version of HTTP agreed between the browser and each distinct web server.

If an application uses UDP instead of TCP as the transport protocol, no definite time period can be determined from the measured traffic, since UDP contains no explicit information about connection set-up and tear-down. For an in-depth analysis of UDP traffic we have to either consider further information of higher protocols or define empirical time periods for the typical duration of entire data flows.

Burst Level

IP packets contain TCP segments or UDP datagrams that are already measured on connection level. However, an important class of models does not consider individual but narrow series of IP packets. These so called *Bursts* of IP packets form the basis of on-off models which distinguish merely between active and passive phases of data transmission. We give a mathematical analysis of on-off models in section 3.4.3 (see also section 2.10).

Packet Level

With the analysis of IP packets on packet level, we close our illustration of different levels of modeling. At this level we consider all parameters such as interarrival or interreceive times between single IP packets. For the complex modeling of this layer, multifractal models are frequently consulted. For the mathematical treatment of multifractal models, we refer to section 3.8.

We remark that in networks with high data rates (above 1 GBit/s) an appropriate analysis of these parameters requires a resolution in the area of nanoseconds. It it obvious that the measurement equipment must guarantee the corresponding precision. Current PC-based equipment provides timestamps with a precision of only around a few microseconds and it is thus not suitable for such measurements without special hardware support.

Multilevel Models for World Wide Web Traffic

We summarize our considerations about multilevel models with an application to World Wide Web traffic. Several multilevel models for World Wide Web traffic exist, including those published by Mah [171], Feldmann et al. [84] or Choi and Limb [53]. All three models start from an application perspective at the client or server. Although the investigations are focused on application, dialogue and connection level, the target is to identify reasons for the characteristics on burst and packet level. In figure 1.6, we illustrate some of the investigated parameters. We see different sessions on application layer, number of pages from the same web server and different objects on connection layer. As already pointed out with our description of the connection level above it depends on the version of HTTP whether all objects are transferred over a single or over multiple TCP connections.

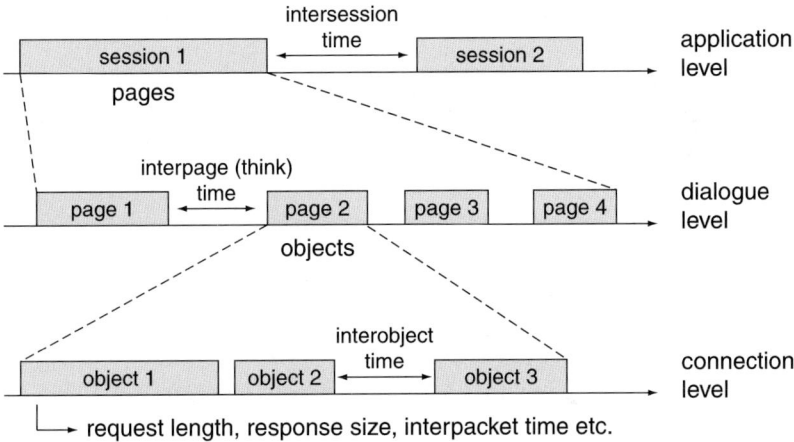

Fig. 1.6. Multilevel model for World Wide Web traffic

We omit an in depth comparison about the exact findings of the cited models. For our introduction, we just remark that all models agree that most parameters, even on higher levels, are described best with heavy-tail distributions like Pareto or Weibull.

Example 1.1. To give an early example for heavy-tail distributions we measured the size of video files in world wide web traffic. Figure 1.7 shows the empirical complementary distribution function with logarithmic scales on both axis. Starting at a certain scale (around 10^6) we see an almost straight line which clearly indicates a heavy-tail distribution. The line represents a Pareto distribution with adjusted parameters. We will illustrate the estimation of such distributions in the section 4.1.3.

1.2.2 Traffic Relations

In our explanations about different levels of modeling, we referred essentially to the communication between two end systems or even between two processes

14   1 Introduction to IP Traffic

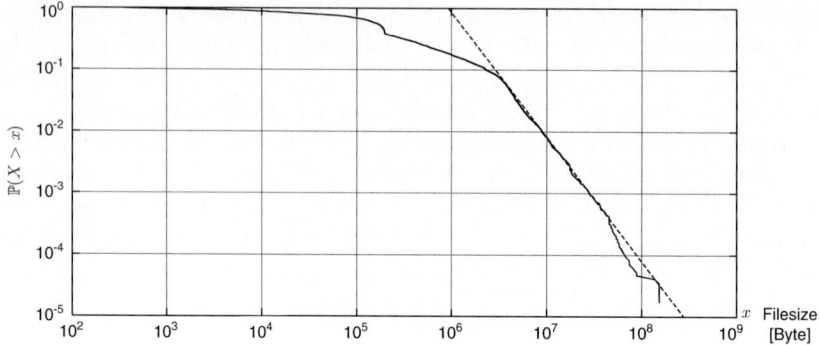

Fig. 1.7. CCDF for volume of video files in world wide web traffic

only. In contrast, data traffic in particular in wide area networks consists of a huge number of concurrent data streams between many distributed partners. To analyze this traffic the superposition can be split up after different communication partners at first. We may consider routers and end systems as well as subnetworks as communication partners. In a next step, the analysis of the separated streams and an appropriate superposition leads to the so-called *composite traffic* models. For a mathematical analysis of these models we refer to section 3.4.4.

In the following paragraphs we illustrate several criteria after which data traffic can be split up into different traffic relations. We refer to the addressing in the individual layers in accordance with figure 1.1.

Port-to-Port

A directed data stream between two processes can be represented by a so-called flow. A flow is defined as a tuple of the five elements {*source IP address, source port number, destination IP address, destination port number, protocol ID*}. The IP addresses indicate the two communicating systems involved, the port numbers the respective processes on both systems and the protocol ID the transport protocol used in this flow.

Table 1.2 represents as an example six different flows which were measured during the transmission of a simple webpage that consists of one HTML file and 15 embedded images of different size. The objects were retrieved by the client (131.34.62.123) from two different servers (195.34.129.243 and 187.250.45.136). Our choice of WWW as application implies HTTP as application protocol and TCP as underlying transport protocol. Therefore, we omit the protocol ID 19 for TCP in the remainder of our example.

At first we recognize that the application on both servers is addressed by the destination port 80. This value is recommended by IETF as port number for WWW services. Furthermore we see that the client initiates connections to the server 195.34.129.243 from two different source ports 2267 and 2268.

Table 1.2. Unidirectional port-to-port flows during transmission of a webpage

Source IP address	Port	Destination IP address	Port	IP packets	Volume [Byte]
131.34.62.123	2267	195.34.129.243	80	76	11,929
131.34.62.123	2268	195.34.129.243	80	27	8,078
131.34.62.123	3265	187.250.45.136	80	5	1,078
195.34.129.243	80	131.34.62.123	2267	125	151,322
195.34.129.243	80	131.34.62.123	2268	35	26,583
187.250.45.136	80	131.34.62.123	3265	8	4,397

Table 1.2 represents a direction-dependent, that is a unidirectional analysis of the transferred data traffic. However, in some cases we are less interested in data flows per direction than in bidirectional communication between two application processes. With a simple pair-wise exchange of the elements {source IP, destination IP} and {source port, destination port} in the tuple we obtain this direction-independent analysis. By this reduction, we get for our example the result in table 1.3.

Table 1.3. Bidirectional port-to-port flows during transmission of a webpage

IP address	Port	IP address	Port	IP packets	Volume [Byte]
131.34.62.123	2267	195.34.129.243	80	201	163,251
131.34.62.123	2268	195.34.129.243	80	62	34,661
131.34.62.123	3265	187.250.45.136	80	13	5,475

Host-to-Host

An analysis based on hosts offers an alternative view on our example. Here, we consider merely the IP addresses of the communicating systems and the transport protocol and not the port numbers of the communicating processes on each system. That is our tuple is reduced to {Source IP, Destination IP, Protocol ID}. The analysis of our example is made straightforward by the addition of the lines with identical source and destination IP addresses. Again, a unidirectional as well as a bidirectional variant is possible.

We show the host-to-host analysis merely with the direction-dependent view in table 1.4 as a proof for the strong asymmetry in WWW traffic. A comparison of the transferred data volume shows that the relationship between the directions Client→Server and Server→Client is approximately 1:9. Another explanation for this is found in the next section.

Table 1.4. Unidirectional host-to-host flows during transmission of a webpage

Source IP address	Destination IP address	IP packets	Volume [Byte]
131.34.62.123	195.34.129.243	103	20,007
195.34.129.243	131.34.62.123	160	177,905
131.34.62.123	187.250.45.136	5	1,078
187.250.45.136	131.34.62.123	8	4,397

End System

The analysis represented above were strongly focused on data traffic between communicating systems. With two further possible relations, we show results from the perspective of single entities. At first, data traffic can be summarized for single systems only, that is independent of their communicating partners. If this analysis is done direction-dependent, the difference between a client (typical data sink, high degree of downstream traffic) and a server (typical data source, high degree of upstream traffic) will be emphasized again. Table 1.5 shows the corresponding results for our example.

Table 1.5. Up- and downstream flows of single end systems during transmission of a webpage

Host	Upstream		Downstream	
	IP packets	Volume	IP packets	Volume
131.34.62.123	108	20.085	168	182.302
195.34.129.243	160	177.905	103	20.007
187.250.45.136	8	4.397	5	1.078

Net-to-Net and Net

If we consider whole subnetworks instead of single end systems we will obtain another rough structure of the data traffic. Again, both unidirectional and bidirectional variants are possible as well as the analysis for a single network independent of the communicating partners (Net) or between selected subnetworks (Net-to-Net). The criteria by which subnetworks can be defined in a suitable simple manner follow from the known structure of IP addressing and from routing information. We remark that in both cases, dependent of the selected point of measurement, the internal traffic is also measured, i.e. between two hosts within the same subnetwork.

1.2.3 Asymmetry in IP Traffic

A characteristic property of legacy TCP based services, such as World Wide Web, FTP or Email, is the asymmetry of the volume exchanged between clients and servers. For the description of this asymmetry we simply build the ratio of upstream load up_{data} to downstream load $down_{data}$. Because we can assume the connection times in both directions are almost equal, the ratio of the mean bit rates in upstream and downstream is approximately equal to the ratio of the respective volumes.

It was shown in [51] that the relation $\frac{up_{data}}{down_{data}}$ of the volumes of transfer in World Wide Web or FTP traffic can be described by lognormal distributions. The reason for this behavior is straightforward, since the size of acknowledgments in TCP can be neglected and thus the relation is substantially determined by the size of the transferred objects.

In contrast, asymmetry with regard to the number of IP packets varies with the number of acknowledgments sent by the client. If each received TCP segment is confirmed with an acknowledgment, the number of IP packets in upstream and downstream will be almost equal – independent of the application used. As already explained in section 1.1 this behavior of TCP is defined by the implementation in the operating system.

For Email, we have to remember that asymmetry is already given by different protocols for upload and download between clients and servers. Emails are downloaded by the clients from the mailboxes stored on mail servers with the Internet Message Access Protocol (IMAP) or the Post Office Protocol (POP). The upload of Emails, that is the sending to the outgoing mail relay (mail transfer agents, MTA), is done with the Simple Mail Transfer Protocol (SMTP). This protocol is also used for forwarding emails between different mail relays. From this it follows that we may expect a strong asymmetry for IMAP and POP as well as for SMTP between clients and mail relays. However, SMTP traffic between mail relays is almost symmetric.

With the examples above, we simply illustrate the differences between asymmetric and symmetric applications. Most other applications can be assigned to one of these categories. For example, video streaming is strictly asymmetric while Voice over IP, video conferencing or peer-to-peer applications show a clear symmetric behavior.

1.2.4 Temporal Behavior

Besides the mainly technical driven factors we have explained so far, we now discuss further parameters which we have to consider for measurements as well as for the selection of suitable models.

Like most systems that are affected by human behavior, the activity in the Internet strongly depends on the particular day of the week and the time of day. Figure 1.8 shows the data rate measured at the transition router between a local network and the providers backbone. The traffic was measured in a public facility where most of the employees access the Internet continuously during their daily work. To eliminate momentary peaks, as well as distinct temporal activity, we consider the total of incoming and outgoing traffic and plot averages over intervals of 20 minutes. To demonstrate the similar characteristics of daily courses, we also give the results for two successive weeks.

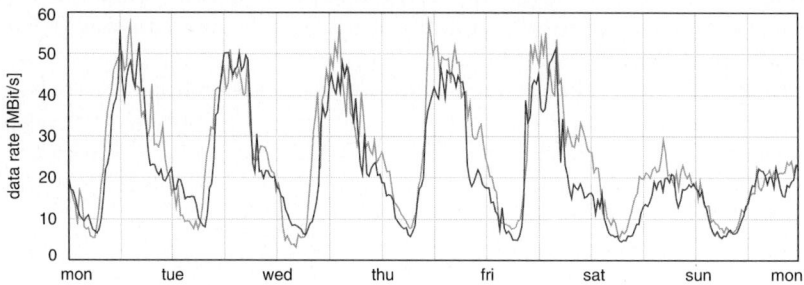

Fig. 1.8. Temporal behavior of IP traffic over two weeks

From the diagram, a periodic course can be recognized within each day as well as within a week. The factor between phases of intense activity (on working days from 1:00 to 2:00 pm) and lowest traffic (all days from 5:00 to 6:00 am) is approximately 5. We remark that this factor is considerably higher for measurements over shorter intervals and with distinct directions.

Corresponding curves of data traffic for e.g. private households reveal slightly shifted characteristics. Here, the main traffic hours fundamentally depend on the tariffs offered by the provider but also on the customers behavior during their leisure activities.

Consequently, the temporal behavior must be taken into account dependent of the environment in which the traffic flows are measured. A measurement over several hours will likely contain phases of different intensity. The dimensioning of network components derived from this inappropriate measurement may lead to weak results. Similar to the traditional modeling of telephony networks, it is therefore essential also for IP traffic to carefully identify main traffic hours.

1.2.5 Network Topology

Since the Internet is a global system of many distributed components, the topology of the network between sender and receiver plays an important rôle. Rather than the geographical distance, the 'network distance', which can be indicated by the number of routers passed through, is of special interest. Because of competing traffic in every router, it is likely that the average transmission time as well as its variance increases with the network distance between both entities. Unfortunately, it is difficult to take the network distance and its characteristics into account for a comprehensive analysis of IP traffic.

The number of routers which are passed through between sender and receiver, the load of the queues and processors in the routers, as well as the available capacity on the links cannot be determined with the common end-to-end measurement. Further investigation of these parameters could be done solely with concurrent hop-by-hop measurements. The measurement with suitable equipment at every router together with an environment for a centralized data processing can be conducted, but only at a very high cost.

Still a crude consideration of topographical aspects arises from the classification of net-to-net traffic relation as represented in section 1.2.2. A possible distinction is offered by dividing traffic into:

- data traffic to all nodes in the local network, that is without traversal of a router, or
- data traffic to all nodes that are connected to the same router, or
- data traffic to all nodes that are reachable over the same network provider, that is over the same backbone network, or
- data traffic to all other nodes.

The necessary information as to which IP addresses can be found in the regarded networks is at least known to the operator of the local network or the network provider.

Again, we remark that the uncertainty of this analysis still lies particularly in the unknown capacity and load of all network segments involved even for a known topology. For example, the achievable data rate to a host that is connected to the same, but overloaded router might be considerably lower than to a host that is connected via unloaded, large backbone routers and high capacity links.

1.3 Quality of Service

Here, network engineering results that are obtained from observation and modeling of IP traffic, are applied to the validation and dimensioning of network components as well as network topology. As part of these efforts, the compliance with *quality of service* (QoS) parameters such as data rate, delay, jitter or packet loss is of major interest. Before we give a more detailed introduction into QoS, we will first differentiate between two fundamental kinds of data traffic.

1.3.1 Best Effort Traffic

Best effort traffic is characterized by the use of TCP as protocol on transport layer. The receiver is reached over paths that are a priori unknown and may vary for each IP packet. The queues in the routers along these paths show different characteristics, depending on the load on other directly connected links. As already described in section 1.1, TCP implements an error control, which automatically initiates retransmission of lost segments after dynamically calculated timeouts. Additionally, TCP features adaptive algorithms for flow control (avoid flooding of the receiver) as well as congestion control (avoid flooding of the network). From this it follows that the transferred TCP segments and consequently the underlying IP packets show no constant size or interarrival times, even already at the sender side. The data is just sent out *as fast as possible*, the focus of TCP is clearly on an entire and error free transmission. Best effort traffic merely saturates the available bandwidth in the present of giving long lasting, error free transmissions.

Traditional queueing models are not applicable to best effort traffic. Suitable models and corresponding dimensioning of networks are focused on minimizing blocking and loss probability. In section 5.3.2 we give further details of these models.

1.3.2 Time Sensitive Data Traffic

The transmission of applications with real time requirements, like live streaming, Voice over IP or video conferencing is not possible with best effort traffic. These applications relay on time-critical parameters like maximum delay of packets, maximum oscillation of delays (that is jitter) as well as clear limits for an acceptable packet loss and a minimum bound of available data rate. A guarantee of these four so-called QoS parameters must be fulfilled for the transmission of time sensitive traffic and thus represent special demands for the dimensioning of the IP networks. Furthermore, it is obvious that TCP is inappropriate for these applications and UDP has to be used as transport protocol instead.

In sections 3.5.2 and 3.5.1, we will describe the characteristics of time sensitive traffic and the implications for the dimensioning of IP networks. Generally packet delay and packet loss must be kept under control to fulfill real time requirements, data rate and jitter may be regarded as a result. The demand to hold a predefined data rate within narrow bounds corresponds to a low variance of the delay. Together, we find the first reasons here why modeling of time sensitive traffic is done by means of Gaussian marginal distributions, i.e. the fractional Brownian motion. Furthermore, a low blocking probability is an essential model prerequisite because of the strong time sensitivity. We will point out in sections 3.5.1 and 3.8.1 that multifractal Brownian motion and general multifractals with Gaussian marginal distributions are suitable approaches for mathematical models of time sensitive traffic.

In principle two possible attempts exist to meet the requirements of time sensitive traffic.

1.3.3 Overprovisioning

Overprovisioning is a simple, straightforward solution for dimensioning of data networks, since resources are provided *more than enough*. From a technical perspective, the main challenge lies in an appropriate load sharing between the components that are operated partly in parallel. A value of 3–5 compared to the mean load in the main traffic hour is typically accepted as a factor for the necessary overdimensioning of IP based networks. With this factor the high costs for components that are rarely completely utilized emerge as an obvious disadvantage of overprovisioning.

However, overprovisioning does not guarantee compliance with predefined QoS parameters. Not all routers on the path may treat IP packets of time sensitive applications and best effort applications in the same manner. In other words, there is no guarantee that all routers prioritize real time application in their queues. Because of missing mechanisms like flow control or congestion control in TCP, applications based on UDP may flood a network. This may suppress TCP based applications almost completely and arbitrarily disrupt competing flows. After all, the dynamicity of IP traffic, which is easily proved by the high variance of the QoS parameters, cannot be completely compensated by an overprovisioning of resources. We go deeper into this finding by means of mathematical models in sections 3.5.1, 5.1 and 5.2.

1.3.4 Prioritization

With *Integrated Services* (IntServ) and *Differentiated Services* (DiffServ), the IETF tried to establish two different architectures on the network layer for assigning distinct priorities to data flows as an alternative to overprovisioning. While IntServ allows for almost arbitrary, fine granular requirements of QoS parameters, DiffServ is restricted to a few prioritized classes with predefined QoS parameters. Both approaches need additional signaling mechanisms, especially for allocation and deallocation of resources, as well as for confirmation or rejection of reservation requests. Because of packet switching or the hop-by-hop delivery of IP packets, the necessary implementations for IntServ or DiffServ cover both routers in the network and operating systems at the sender and receiver.

Apart from the technical challenges for reservation mechanisms authorization (who is allowed to reserve what resources) and accounting (which data has to be collected to write the bills for prioritized IP traffic) are just two more aspects that have prevented the deployment of a comprehensive architecture for prioritization in the Internet so far. However, we see a progressive use of *Multiprotocol Label Switching* (MPLS) today. MPLS adapts certain features of IntServ and DiffServ between data link and network layer. With this approach, QoS parameters can be managed at least in the backbones of individual providers.

In section 3.5.2, we will introduce a rather complex and not well suited model of Norros et al. to analyze the use of priorities and their impact on IP traffic. In section 3.5.1, which is about the two scale FBM and also in sections 3.8.2 and 3.8, which are about multifractals, we will consider alternative approaches which offer a more favorable access to prioritization.

1.4 Why Traditional Models Fail

In the previous sections, we have explained the main characteristics of IP traffic and have discussed possible solutions for respective models. These models were not as yet based on strict mathematical derivations but clearly revealed the challenges and provided first insight towards a dimensioning. We now want to answer the developing question as to why traditional traffic models fail with the paradigm shift in telecommunications networks from circuit switched telephone networks (*Public Switched Telephone Networks*, PSTN) to packet switched (or IP based) data networks.

With traditional telecommunication systems, the following four steps were defined as the state-of-the-art framework for building analytical models:

- measurement and statistical description of sample data
- preparation of mathematical models
- analysis and
- optimization of queueing models and performance.

Models built with this framework for telecommunication networks were also applied to the communication processes in the Internet as recently as 15 years ago. At that time, the planning of Internet backbones was just a marginal aspect in dimensioning of conventional telecommunication networks. But the observed properties of the heavy IP traffic soon cast doubts on this procedure. We give a short example: If we ask how often telephone connections fail we will get the answers 'once a month' or 'once a year'. In case of unsuccessful connections to arbitrary servers in the Internet the question is possibly about 'once a day' or 'once a week'.

Why are telephony networks operated with such astonishing reliability? The conventional PSTN is a static system to a large extent. It is both well structured and internationally standardized and shows only a low degree of variability. The applications are limited and well-defined. Additionally, the behavior of the users is also static and hardly subject to changes. This situation leads to well-established models with Poisson distributed arrival and service processes for the dimensioning of telephony networks. A further advantage of these models is the limited number of parameters and thus the few degrees of freedom. The simplicity and proved validity of the models led to the perspective that the world of telecommunication is strictly 'Poisson' minded.

Today, the migration from circuit to packet switched networks is realized with great efforts by all providers and carriers. It is widely accepted that packet switched networks provide higher efficiency, since free capacities lead to higher data rates of current flows or can be immediately utilized by new traffic.

The challenge for a comprehensive modeling is that single IP packets and not only entire flows compete for the available resources. At low load, the full capacity of the resources is used and every IP packet is forwarded almost without additional delay. With increasing load, the delay of every IP packet is significantly determined by queueing and scheduling algorithms in the routers.

Another important finding about IP traffic is that already the offered load of the sender is of high variability and thus almost unpredictable. Especially with TCP, the sender may transmit *bursts* of IP packets, there is no steady flow. This behavior is an essential difference which we have to consider accurately for the mathematical modeling.

If we now ask for the main characteristics of the Internet, we may give the following answers:

- The Internet is a heterogeneous system that depends on many processes on different layers. A clear pattern of applications or even of user behavior can not be recognized. We may define several characteristic profiles, but a clearly definable structure, as is well known in the conventional telecommunication networks, does not exist.
- The Internet is not exactly describable, it is even 'chaotic' in its behavior. Clear circuit switching structures are not recognizable, that is a circuit set-up between sender and receiver does not exist.
- The Internet is unpredictable in its growth. Even if an end of growth was occasionally predicted, new applications like e.g. Peer to Peer networks, Voice over IP, Live Video Streaming, Grid Computing or Storage Area Networks (SAN) frequently arise which cause a continuous increase of demands for more resources. Even today, after 15 years of intensive growth, an assured prediction about the development of the Internet within the next few years is almost impossible.

In circuit switched traffic, we find Poisson processes and consequently the exponential distribution which leads to an Erlang distribution for the summation of service times. The derivation requires the same distribution for both the arrival and the service processes (with possibly different rates). This is e.g. described by the arrival rate λ and service rate μ. If we assume e.g. $\lambda = 100$ 1/s, then the scale of bursts is around 10 ms. Variations of this scale, which means greater or smaller gaps in the interarrival times resp. service time, appear only with rapidly changing probability (decreasing resp. increasing).

IP traffic is characterized by frequent changes of passive and active phases in different time scales. An adequate modeling with Poisson processes and exponential distributions is not feasible, since variations would lead to large active or passive phases because of the rapidly decreasing probability.

A significant finding for IP traffic modeling was published by the group of W. E. Leland at Bellcore in 1992 [160]. The sample data was measured during an hour on a router between the local network and the providers backbone. The comparison of measured data rate over different time scales in figure 1.9 clearly illustrates the difference between Poisson (left column) and IP traffic (right column). Each black shaded area corresponds to the entire time range in the diagram above. With increasing time scales the Poisson process shows a clear smooth pattern while the measured traffic is still of high variability.

Poisson traffic is easily treated with regard to dimensioning of the network. As of a certain time scale no 'surprises' occur and we may avoid measurements with larger scales. Then the corresponding process gets stationary and the transition probabilities are therefore *time-independent*. Questions about the dimensioning of queues in routers or switches as well as the design of backbones and even the explicit analysis of QoS would get superfluous under these conditions.

However, according to the right column of figure 1.9, IP traffic exhibits a different behavior. From this follows immediately the need for larger queues along with separate consideration about QoS.

In summary we can state some early results about Poisson-based models applied to packet switched networks for IP traffic:

- Poisson processes cannot cover all parameters of IP traffic.
- Poisson processes cannot reflect the dynamics in IP traffic.
- Poisson processes cannot represent the multi-layer complexity of characteristic patterns in IP traffic.

We clearly see that Poisson processes are of limited relevance for IP traffic models. According to these results, we have to develop new suitable models for IP traffic. We can statistically describe the high variability with the so-called *Long-Range Dependence* processes (LRD). In contrast to Poisson processes the autocorrelation of LRD processes does not rapidly decrease. This is expressed e.g. by the fact that the autocorrelation for a sequence of increasing time spots is not summable (in contrast to the Poisson process). The variance of LRD processes decrease with a rate of a power bein greater than -1 and not linear as the Poisson process reveals.

The service times or file sizes can be described by heavy-tailed or subexponential distributions instead of exponential distributions. We will give more details about these distributions in section 2.7.4, for example with the Pareto distribution and its complementary distribution function $Cx^{-\alpha}$ with $\alpha \in]0,2[$. In chapter 2, we will extensively treat models of the form M/G/n, GI/G/n or M/GI/∞. We will mathematically derive, why long-range dependence or *fractal* processes appear for heavy-tail distributed service processes. We will also illustrate that fractal processes behave similarly in large and small time scales.

To give a simple graphical proof of these assumptions in figure 1.10 we replaced the Poisson process by a fractal process, the *fractional Gaussian Noise*. We introduce the Gaussian Noise as a process $(X_t)_{t \in [0,\infty[}$ with the autocorrelation

$$\mathbb{C}\mathrm{or}(X_t, X_{t+s}) = \frac{1}{2}\left((s+t)^{2H} - 2s^{2H} + |s-t|^{2H}\right)$$

The finite dimensional marginal distributions of this process are normal distributed. The parameter $H \in [\frac{1}{2}, 1[$ is denoted as the *Hurst exponent* (see section 3.1.2). We remark that we only need a single parameter to determine the strength of the fractional scales and that this process is different from a Brownian motion with the difference increasing as H increases. We may

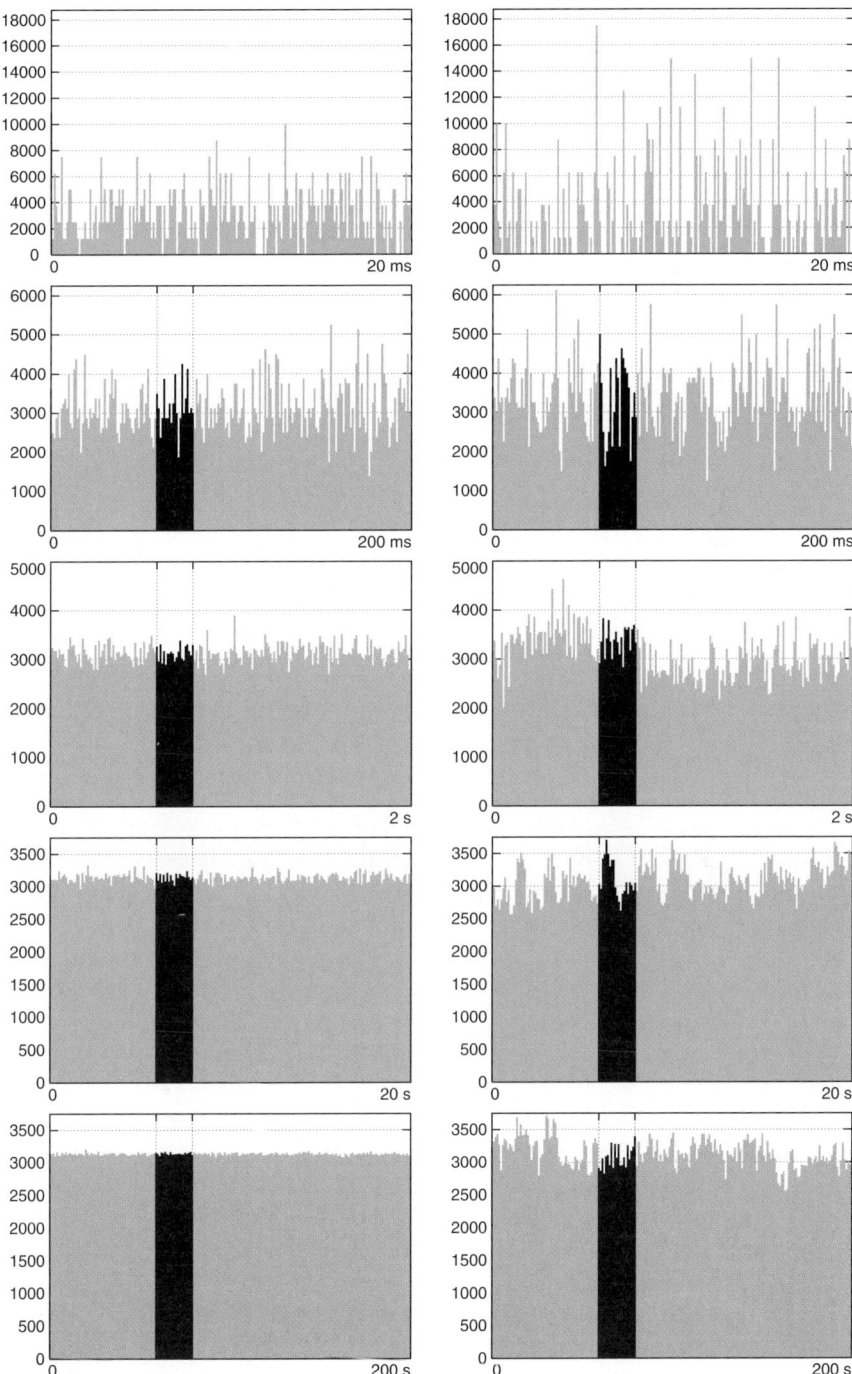

Fig. 1.9. Comparison of Poisson process (left) and measured IP traffic (right)

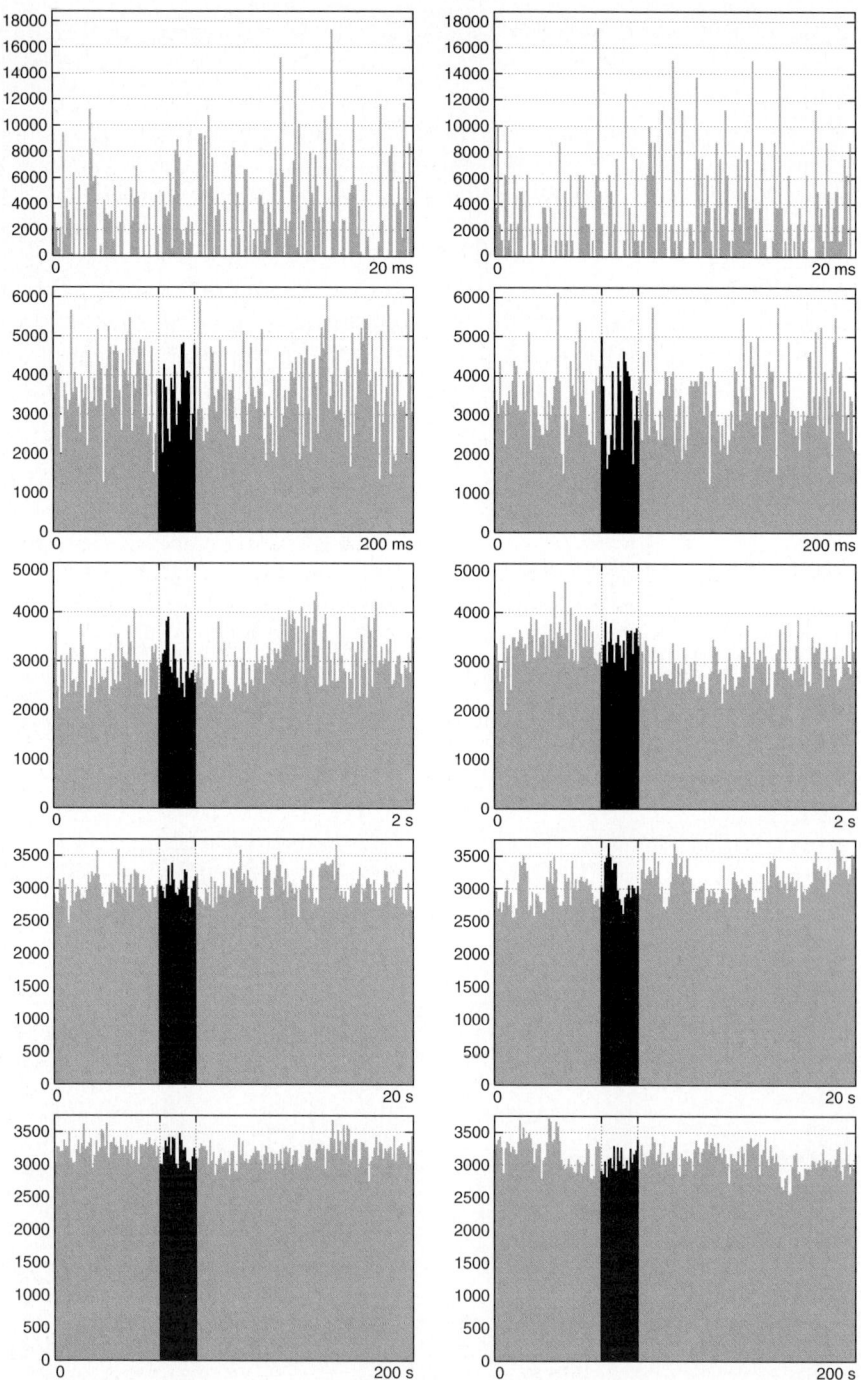

Fig. 1.10. Comparison of fractal process (left) and measured IP traffic (right)

already assume that models based on such processes can be suitably applied to IP traffic.

As in figure 1.9, the black shaded areas in figure 1.10 represent the entire time range of the diagram immediately above. The right column contains the measured data again and the left columns shows the fractal process (in this case a fractional Gaussian noise). It is obvious that in contrast to the Poisson process illustrated in figure 1.9 the structures in both columns remain similar for all time scales.

We will intensively treat fractal processes in sections 3.2 and 3.3 but we want to briefly introduce this topic in our introduction. We start with a time-discrete process $(X_k)_{k \in \mathbb{N}}$ with mean 0 and stationary covariance, that is the process depends on increases only and not on time. We denote the process *exactly self-similar* or *fractal* if there is a parameter $H \in [\frac{1}{2}, 1[$ with

$$X^m \stackrel{d}{\sim} m^{H-1} X$$

X is the multidimensional random vector and X^m is the mean of block wise formed discrete time values

$$X^{(m)}(k) = \frac{1}{m} \left(X_{(m-1)k+1} + X_{(m-1)k+2} + \ldots + X_{mk} \right)$$

The fractional Gaussian noise is such a process with an appropriate Hurst parameter $H \in [\frac{1}{2}, 1[$. For $H = \frac{1}{2}$ we obtain the conventional *white noise*. For an exactly self-similar process we can show that

$$\mathrm{Var}(X^m) \sim \kappa m^{2H-2}$$

This reflects a decrease of the autocorrelation which is no longer summable.

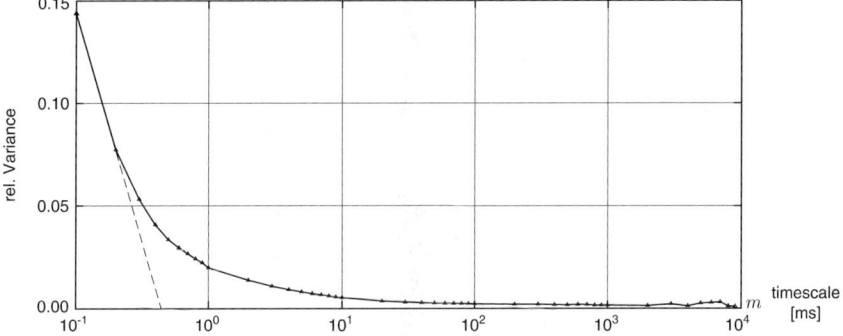

Fig. 1.11. Variance-Time Plot. The steeply falling line represents a Poisson process

Figure 1.11 gives another graphical comparison of Poisson and fractal behavior. Here, a log-log diagram of the variance of measured traffic depending on the scaling factor or time interval m is shown. Instead of absolute values we selected the relative variance on the y-axis compared to the value at $m = 10^{-2}$ ms which is not shown. The dotted line represents the case of a Poisson distribution and clearly illustrates the difference against long-range dependent (LRD) processes, depicted by the hyperbole of the solid line.

When using self-similar processes, we always have to consider the absence of stationarity. This may occur with strongly varying sample data which leads to stationary self-similar processes. To identify this behavior we measure e.g. service times, interarrival times or data volume over a period of two minutes and compare the values with a succeeding period of two minutes and another period of four minutes. If the results correspond to a stationary self-similar process, we have a suitable assumption for our model.

With all these findings we conclude that there is no easy recipe with which one can dimension arbitrary IP networks. Nevertheless, we already identified some important invariants:

- Poisson processes are still suitable for model connections initiated by the users.
- For the modeling of transferred data volume, file sizes or service times, we use heavy-tail distributions with a complementary distribution function $F^c(x) = x^{-\alpha}L(x)$, whereas L is a slowly varying function and $\alpha < 2$ or even $\alpha < 1$.
- If we sum up measured data we see a fractal structure which is not visible for individual connections. The question arises as to how we can consider an infinite variance as invariant for e.g. individual connection durations. Again, this is an asymptotic result which we obtain as mean over a large number of connections.
- The multi-layer architecture of protocols in the Internet significantly determines the characteristics of appropriate fractal models.

Because an exact description or a common model cannot be found for IP traffic (e.g. which α-stable process is suitable or which Hurst estimator is best), broader models like *multifractals* were developed. In these models the Hurst parameter is regarded, time-dependent, that is it depends on the time scale. We will go into the details of these models in section 3.8.2.

2
Classical Traffic Theory

> *There is nothing more practical than a good theory.*
>
> Immanuel Kant (18th century)

2.1 Introduction to Traffic Theory

2.1.1 Basic Examples

To illustrate the complexity of traffic theory we begin this chapter with three basic examples. Each example refers to certain aspects of traffic theory briefly and motivates the broad variety of models we will introduce in later sections of this chapter.

PAR Protocols

We already noted in section 1.1.4 that protocols following positive acknowledgment with retransmit (PAR protocols) are basic for reliable data transfers as implemented with TCP. A sender is transmitting data packets to a receiver. If the packets are received without error according to the underlying protocol, those packets are acknowledged by a signal, called the ACK (positive acknowledgment). If the ACKs are not received within a certain timeout interval, the sender retransmits the respective packets. This procedure is repeated until the sender receives an ACK for each transmitted packet or a given number of retransmits is reached and an error message is delivered to the respective application.

We denote by η_N the transmission time and by τ the delay, caused by a single retransmission cycle. We want to compute the virtual transmission time η_ν. This is the actual time for the successful transmission. We have to impose a certain probability for the failure transmission, which we denote by p_η. So we have by assuming the independence of the repeated transmission:

- with a probability of $1 - p_\eta$ the virtual transmission time $\eta_\nu = \eta$ and
- with probability p_η an additional transmission time of $\eta + \tau$ for η_ν.

Considering the packet arrival time, we get a classical traffic model, where the serving time is the virtual transmission time. As we will see in section 2.8.2, this is a so called one level GI/G/1 system, assuming general distributed arrival and serving time. As expected the analysis will provide e.g. queueing time and traffic load depending on the error probability p_η.

Data Transmission

To give another example for the application of the classical traffic theory, we consider the basic concept of data transmission using the packet switched technique. As described in section 3.4.3 we apply the most basic model, given in the figure 2.1.

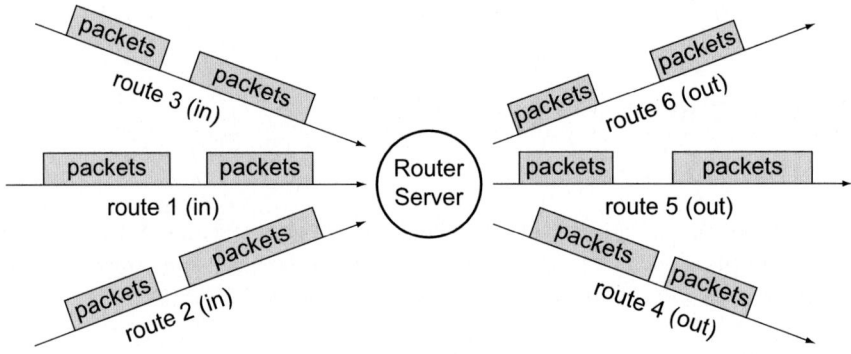

Fig. 2.1. Scheme of incoming and outgoing data transmission at a router on the packet level

It corresponds to the packet or burst level, which we discussed in section 1.2.1 and explained by the figure 1.5. We can consider each of the packets as *demands*, which have to directed in the router resp. server to further outgoing lines. These packets arrive with a certain rate, depending e.g. on the user behavior. This is determined e.g. by the dial in of all users into the Internet or intranet and indicates the interarrival times of the packets. Thus, the incoming stream is modeled as the *arrival process*. The router resp. server plays the rôle of the *serving part* of the traffic model. Here, depending of the future structure of the net and the destination of each of the transmitted data, these packets are distributed or served. Hence, we will call this the *serving process*. Since the incoming streams may surpass the outgoing capacity, we will observe an storage of data prior to the further transmission. Thus, these packets have to wait and will go into the queueing. We will denote this as the *waiting room* and the corresponding process the *queueing process*.

Call Attempts in a Cellular Mobile System

Let's consider a cell in a mobile communication system, more precisely in the most prominent global system for mobile communication, the *Global System*

2.1 Introduction to Traffic Theory

for Mobile Communications (GSM). This system is the most successful cellular communication system which was introduced in 1991, mainly in Europe and with adjustments in the USA and Japan. It has been based from the outset on the already existing large variety of different cellular systems. In GSM several new technologies entered the mobile communication, where, next to the unique standardization, we encounter the first time a digital transmission technique on the air interface. For our present purpose we have to know that the area covered by the mobile communication is split into a beehive-like structure, i.e. into several cells, each equipped with one base station or *Base Transceiver Station* (BTS). Every mobile station can get in contact with this fixed station, where data (mostly the encoded speech) is transferred via fixed line, finally to the core network for further handling.

The basic transmission technique on the air interface is a combination of time and frequency multiplex, since each user has to be separated from the other. To avoid interference between neighbouring cells, network management has to apply a careful frequency planning for each cell, a fact which can be neglected for the *Universal Mobile Telecommunications System* (UMTS) technique. Guard bands have to be implied as well.

- Time multiplex: Each frequency band is splitted in 8 time slots, where for each second frequency multiplex channel one channel is used for service transmission.
- Frequency multiplex: For each user there exits a pair of frequencies – the upper frequency for download the lower one for upload. In GSM900 the first and mostly used standard the band 935 to 960 MHz are receiver frequency and 890 to 915 MHz is the upload frequency band. With a distant of 200 kHz a maximum of 124 channel pairs can be provided.

The basic quality of service in GSM is that a certain blocking of incoming call attempts are not allowed to be surpassed by a fixed given probability. For the modeling of the GSM cell we use:

- Serving process: In each cell we have n given frequency pairs for the voice transmission. They build the serving units. The duration of a call is describe as serving time by a random variable η. An accepted call will occupy a pair of channel, i.e. a serving unit. If all serving units resp. channel pairs are occupied, the next incoming call is blocked and thus, rejected. We encounter a pure loss system.
- Arrival process: In each cell we find a finite number of m users, which are 'silent', 'active' or 'idle'. We can determine three kinds of arrival processes.
  - Arrival process with finite sources: we have m users, i.e. a finite number. The aggregated traffic is the compound of all users in the state 'silent' (see section 2.4.3).
  - Arrival process with infinite sources: suppose the number m of users is sufficient large. Then we can approximate the system by a model with infinite sources. This model is easier to handle and leads to the well known Erlang formula (see section 2.4.1).

– Model with call repetition: in reality an overload situation can be detected, which emerges faster, since usually rejected call in a blocking situation try to repeat the call in decreasing time intervals (so called snow ball effect) (see section 2.4.3).

2.1.2 Basic Processes and Kendall Notation

The classical model in the traffic theory can be divided into three basic processes: arrival, serving and state processes. We briefly summarize the main characteristics of these three models.

Arrival Process

The incoming demands are modeled as discrete time spots. They can include call attempts in the telecommunication, data packets or data units as observed in high speed networks like ATM. We can divide this class of arrival processes into single or group arrivals. The time between two succeeding arrivals is called interarrival time and represents a central issue in modeling (fig. 2.2).

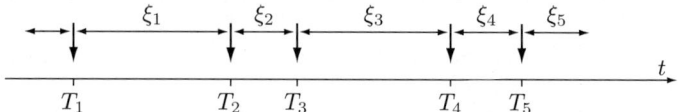

Fig. 2.2. Arrival process: arrival times T_i and interarrival times ξ_i

Serving Process

After its arrival the demand is handed over to the serving process. This leads to a number of demands in the serving process (a discrete process) and to the serving time (a time continuous process). The interarrival times as the serving time (serving process) are often modeled as Markov processes, i.e. we consider both processes as memoryless. It is assumed that the remaining time has the same distribution as the whole interarrival or serving time. Here, the remaining time is the time left in the serving process (fig. 2.3).

State Process

Often two kinds of state processes are considered:
- Number of demands in the system: Here, we have a discrete process. For example every incoming call increases the number of demands, where each of these calls could be served or fail.

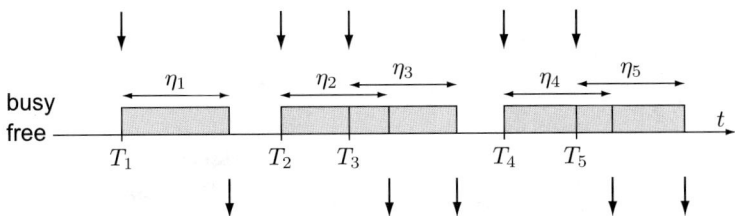

Fig. 2.3. Serving process: arrival times T_i and service times η_i

- Residual work in the system: Every incoming demand increases the remaining amount of work. The residual work is not identical with the serving time, since we have to consider the waiting time in addition. The residual time will decrease continuously and thus, we deal with a continuous process (fig. 2.4).

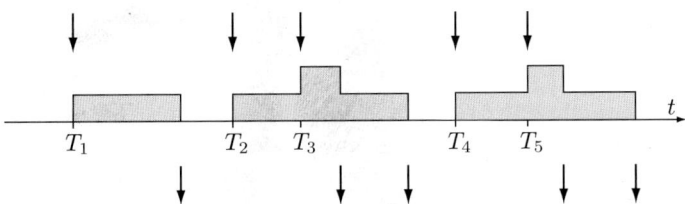

Fig. 2.4. State process: upper arrows mark arrivals, lower arrows departures of demands; gray shaded areas the number of demands in the system

We can represent arrival and serving processes in a short notation, the so called *Kendall notation*, e.g. M/M/n − S. Here, the first symbol reflects the arrival process, the second one the serving process, with n we indicate the number of possible simultaneous serving processes and S is the number of waiting places. In the above example the symbol M means that we have chosen a 'Markov' process as arrival and serving process. If we do not assume a Markov process – and this happens especially in the IP serving processes – then we use the notation M/G/n − S, GI/M/n − S or GI/G/n − S. Here, the symbol G means 'generally' distributed resp. GI 'generally independent' distributed. Hence, the arrivals (resp. calls or initializations of Internet connections) are stochasticly independent. The distribution of arrival resp. serving processes are arbitrary.

2.1.3 Basic Properties of Exponential Distributions

We call a distribution $F_\xi(t) = 1 - e^{-\alpha t}$ with density $f_\alpha(x) = \alpha e^{-\alpha x}$, the *exponential distribution* for α, where the parameter α is called *intensity*. If we choose independent RV $\xi_1, \ldots, \xi_k$ for a $k \in \mathbb{N}$ exponential distributed with the same intensity λ, then we call the distribution of the sum $\xi = \sum_{i=1}^{k} \xi_i$ the *Erlang-k distribution*.

We want to cite an important property of exponential distributions. To do so, let ξ be a random variable, which e.g. represents the serving time (like the duration of a call). We say that ξ is *memoryless*, if the conditional probability that the serving time lasts further s time units is independent of the already time passed t, i.e.

$$\mathbb{P}(\xi > s + t \,|\, \xi > t) = \mathbb{P}(\xi > s \,|\, \xi > 0), \quad s, t \geq 0$$

Now, we assume in addition that $\mathbb{P}(\xi > t) > 0$ (a call could last arbitrary long). We have according to the definition of conditional probability

$$\mathbb{P}(\xi > s + t \,|\, \xi > t) = \frac{\mathbb{P}(\xi > s + t)}{\mathbb{P}(\xi > t)}$$

If ξ is exponential distributed for an intensity λ, then we conclude

$$\frac{\mathbb{P}(\xi > s + t)}{\mathbb{P}(\xi > t)} = \frac{e^{-\lambda(s+t)}}{e^{-\lambda t}} = e^{-\lambda s} = \mathbb{P}(\xi > s \,|\, \xi > 0)$$

where we remark that $\mathbb{P}(\xi > 0) = 1$, since calls always last a certain time. Thus, the exponential distribution is memoryless. But we have even more.

Theorem 2.1. *Let ξ be a random variable, which is memoryless. In addition we assume $\mathbb{P}(\xi \leq 0) = 0$ and $\mathbb{P}(\xi > t) > 0$ for all $t \geq 0$. Then there is a $\lambda > 0$, such that ξ is exponential distributed according to λ.*

This theorem is of great importance for our purposes of IP traffic modeling, since we can also assume in the case of IP traffic $\mathbb{P}(\xi \leq 0) = 0$ and $\mathbb{P}(\xi > t) > 0$ (for the time of data transmission). Thus, we can conclude: If the statistic data will reveal that the connection time is not exponential distributed, then according to the above theorem the random variable *cannot* be memoryless. For modeling with stochastic processes this is of decisive importance.

2.2 Kolmogorov Equation

We begin with a discrete process $(X_t)_{t \in \mathbb{N}}$, i.e. a process with a discrete parameter space and select a sequence of succeeding time spots $t_0, t_1, \ldots, t_n$. We raise the question for a Markov process: If the state depends at time t_{n+1} only on the preceding t_n, can we fix a process only by knowing its *transition probabilities* (i.e. the conditional probabilities)

$$p_{ij}(t_n, t_{n+1}) = \mathbb{P}(X_{t_{n+1}} = j \,|\, X_{t_n} = i)$$

If the transition behavior is identical for all time spots, then we call the process homogeneous. Then, we have for all n

2.2 Kolmogorov Equation

$$p_{ij}(t_n, t_{n+1}) = p_{ij}(t_{n+1} - t_n) = p_{ij}(\Delta t)$$

Since each demand can either stay in state i or reach a new one, we have the completeness relation

$$\sum_j p_{ij}(\Delta t) = 1, \text{ for all } \Delta t > 0$$

Accordingly, we define the transitions from i to j of the transition probability of a homogeneous process in dependence of time t and represent all transition probabilities in a matrix form. Here, the i-th row and the j-th column show the transition probability from state i to j

$$\mathcal{P}(t) = \begin{bmatrix} p_{00}(t) & p_{01}(t) & \cdots & p_{0j}(t) & \cdots \\ p_{10}(t) & p_{11}(t) & \cdots & p_{1j}(t) & \cdots \\ \vdots & \vdots & \ddots & \vdots & \ddots \\ p_{i0}(t) & p_{i1}(t) & \cdots & p_{ij}(t) & \cdots \\ \vdots & \vdots & \ddots & \cdots & \ddots \end{bmatrix}$$

With this representation we can formulate the *Chapman-Kolmogorov equation*. It describes the relation of the transition probability between the time spots t and $t + \Delta t$, $\Delta t > 0$

$$\mathcal{P}(t + \Delta t) = \mathcal{P}(t)\mathcal{P}(\Delta t)$$

or if we represent it for every point (i, j) in the left matrix

$$p_{ij}(t + \Delta t) = \sum_k p_{ik}(t) p_{kj}(\Delta t) \tag{2.1}$$

We reformulate this expression and divide by Δt

$$\frac{p_{ij}(t + \Delta t) - p_{ij}(t)}{\Delta t} = \frac{\sum_{k \neq j} p_{ik}(t) p_{kj}(\Delta t)}{\Delta t} - \frac{p_{ij}(t)(1 - p_{jj}(t))}{\Delta t}$$

We assume the existence of all appearing limits and investigate, what comes out, if the time difference Δt tends to 0. By this, we obtain different intensities

$$\lim_{\Delta t \downarrow 0} \frac{p_{ij}(t + \Delta t) - p_{ij}(t)}{\Delta t} = \frac{d}{dt} p_{ij}(t)$$

(representing the first derivative of the transition probabilities $p_{ij}(t)$ at time t)

$$\lim_{\Delta t \downarrow 0} \frac{p_{kj}(\Delta t)}{\Delta t} = q_{kj}, \; k \neq j$$

(this is the transition probability density for the transition $k \to j$)

$$\lim_{\Delta t \downarrow 0} \frac{(1 - p_{jj}(t))}{\Delta t} = q_j = \sum_{k \neq j} q_{jk}$$

(we consider this as the transition probability density for leaving the state j). These values describe the tendency in changing the transition probability. The expression $\lim_{\Delta t \downarrow 0}$ means that Δt tends to 0 from positive values, thus, from 'above'. The Chapman-Kolmogorov equation can be restated after a limiting process to

$$\frac{d}{dt} p_{ij}(t) = \sum_{k \neq j} q_{kj} p_{ik}(t) - q_j p_{ij}(t)$$

Writing the values q_{kj} into a matrix, we obtain

$$\mathcal{Q}(t) = \begin{bmatrix} q_{00}(t) & q_{01}(t) & \cdots & q_{0j}(t) & \cdots \\ q_{10}(t) & q_{11}(t) & \cdots & q_{1j}(t) & \cdots \\ \vdots & \vdots & \ddots & \vdots & \ddots \\ q_{j0}(t) & q_{j1}(t) & \cdots & q_{jj}(t) & \cdots \\ \vdots & \vdots & \ddots & \vdots & \ddots \end{bmatrix}$$

Since $\sum_k q_{jk} = 0$, we have for the probability for remaining in state j

$$q_{jj} = \sum_{k \neq j} q_{jk} = -q_j$$

This leads to the *Kolmogorov forward equation* for Markov processes, a system of differential equations formulated in matrix notation

$$\frac{d\mathcal{P}(t)}{dt} = \mathcal{P}(t) \mathcal{Q}$$

2.2.1 State Probability

We denote by $x(j, t) = \mathbb{P}(X_t = j)$ the probability that the process is in the state j at time t. We can express this by applying conditional probability and get

$$x(j, t) = \sum_i \mathbb{P}(X_t = j \mid X_0 = i) \mathbb{P}(X_0 = i) = \sum_i x(i, 0) p_{ij}(t)$$

Multiplying the Kolmogorov forward equation with $x(i, 0)$ and summing over all i, this implies

$$\sum_i \frac{d}{dt} p_{ij}(t) x(i, 0) = \left(\sum_{k \neq j} q_{kj} \left(\sum_i x(i, 0) p_{ik}(t) \right) \right) - \left(\sum_i x(i, 0) p_{ij}(t) q_j \right)$$

Finally we obtain a system of partial differential equations, the so called *Kolmogorov forward equations for state probabilities*

$$\frac{\partial}{\partial t}x(i,t) = \sum_{k\neq j} q_{kj}x(k,t) - q_j x(j,t) \qquad (2.2)$$

and the completeness relation

$$\sum_j x(j,t) = 1$$

With this, one describes the development of the state process X_t, who is in state j at time t. With the help of differential equation we can compute the time dependent state probabilities for the general case of an instationary Markov process. An instationary view is e.g. necessary for the overload behavior of server and communication systems. But for stationary processes we can simplify this problem to ordinary differential equation.

2.2.2 Stationary State Equation

In this case we have that the process is not dependent of time, i.e.

$$\frac{d}{dt}\mathbb{P}(X_t = j) = \frac{\partial}{\partial t}x(j,t) = 0 \qquad (2.3)$$

With this, we obtain for the state probability

$$x(j) = \lim_{t\to\infty} \mathbb{P}(X_t = j)$$

The system of partial differential equations (2.2) will turn with the help of (2.3) into a more simple system of linear equations

$$q_j x(j) = \sum_{k\neq j} q_{kj}x(k) \quad \text{and} \quad \sum_j x(j) = 1$$

Example 2.2. Transition probabilities for a Poisson process: We consider a state process with Poisson arrivals. For the distribution of the interarrival time ξ as for the recurrence time R we have (see also (2.23))

$$F_\xi(t) = F_R(t) = 1 - e^{-\lambda t}$$

We denote by X_t the state process at time t, e.g. the numbers of call attempts or the numbers of dial-ins to the network at time t. Let $X_t = i$. The probability density for the transition of i to $i+1$ can be computed according to

$$q_{i,i+1} = \lim_{\Delta t \to 0} \frac{p_{i,i+1}(\Delta t)}{\Delta t} = \lim_{\Delta t \to 0} \frac{\mathbb{P}(R \leq \Delta t)}{\Delta t} = \lim_{\Delta t \to 0} \frac{1 - e^{-\lambda \Delta t}}{\Delta t}$$

$$= \lim_{\Delta t \to 0} \frac{1 - \left(1 - \frac{\lambda \Delta t}{t} + \frac{(\lambda \Delta 1!)^2}{2!} + \ldots\right)}{\Delta t} = \lambda$$

Example 2.3. Transition probabilities with exponential distributed serving time (call duration): This time we consider the call duration. For that we start more generally and denote by X_t the number of all enduring servings at time t. Let k be active and let the serving time be exponential distributed. For the distribution of the serving time η and the recurrence time R it holds

$$F_\eta(t) = F_R(t) = 1 - e^{-\mu t}$$

The time until the next end of serving (or the end of a call) is the minimum of all recurrence times of the independent calls

$$R^* = \min(R_1, \ldots, R_k)$$

Thus, we get for the distribution of R^*

$$F_{R^*} = 1 - \prod_{i=1}^{k}(1 - F_R(t)) = 1 - e^{-k\mu t}$$

For computing the probability of the transition of k to $k-1$, i.e. the probability for the end of a call and the resulting serving of the remaining $k-1$ calls, we obtain

$$q_{k,k-1} = \lim_{\Delta t \to 0} \frac{p_{k,k-1}(\Delta t)}{\Delta t} = \lim_{\Delta t \to 0} \frac{\mathbb{P}(R^* \leq \Delta t)}{\Delta t} = \lim_{\Delta t \to 0} \frac{1 - e^{-k\mu \Delta t}}{\Delta t}$$
$$= \lim_{\Delta t \to 0} \frac{1 - \left(1 - \frac{k\mu\Delta t}{t} + \frac{(k\mu\Delta 1!)^2}{2!} + \cdots\right)}{\Delta t} = k\mu$$

2.3 Transition Processes

Birth and death processes are fundamental for the description in telecommunication. If we want to describe Markov processes, we need the list of the transitions probability rates q_{ij}, to determine the transition probability in the matrix according to the Kolmogorov forward equation.

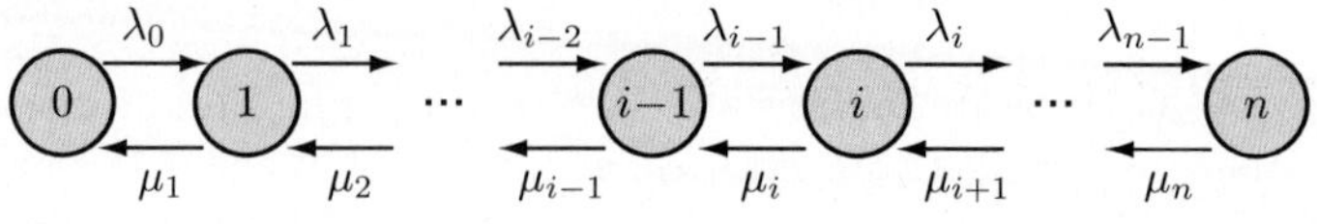

Fig. 2.5. State transition diagram for birth and death processes

We interpret the state transition diagram from figure 2.5. The single birth rates are λ_i, i.e. the transition from i to $i+1$. The death rate is given by μ_i and indicates the transition from i to $i-1$. Thus, we get

2.3 Transition Processes

$$q_{ij} = \begin{cases} \lambda_i & \text{for } i = 0, 1, \ldots, n-1, \; j = i+1 \\ \mu_i & \text{for } i = 1, 2, \ldots, n, \; j = i-1 \\ 0 & \text{else} \end{cases}$$

The next steps consist in splitting into instationary and stationary birth and death processes, as well as the formulation of the respective Kolmogorov forward equation for $i = 1, \ldots, n-1$

$$\frac{\partial}{\partial t} x(0, t) = -\lambda_0 x(0, t) + \mu_1 x(1, t)$$

$$\frac{\partial}{\partial t} x(i, t) = -(\lambda_i + \mu_i) x(i, t) + \lambda_{i-1} x(i-1, t) + \mu_{i+1} x(i+1, t)$$

$$\frac{\partial}{\partial t} x(n, t) = -\mu_n x(n, t) + \lambda_{n-1} x(n-1, t)$$

In the above problem we solve the partial differential equation using the example of Poisson processes as pure birth process. For this let X_t denote the number of incidents during the time period t, e.g. the active telephone connection. Then we get for the state probability

$$x(i, t) = \mathbb{P}(X_t = i)$$

We start at time $t = 0$ from an empty system, i.e. we choose $X_0 = 0$. As example we pick a pure birth process, i.e we have a birth rate λ and consequently a death rate $\mu = 0$. Thus,

$$q_{ij} = \begin{cases} \lambda & \text{for } i = 0, 1, \ldots, n-1, \; j = i+1 \\ 0 & \text{else} \end{cases}$$

The differential equations reads as

$$\frac{\partial}{\partial t} x(0, t) = -\lambda x(0, t)$$

$$\frac{\partial}{\partial t} x(i, t) = -\lambda x(i, t) + \lambda x(i-1, t), \; i = 1, 2, \ldots$$

With the Laplace transform Φ_X (see e.g. [74]) we deduce

$$x(i, t) \overset{LT}{\leftrightarrow} \Phi_X(i, s)$$

$$\frac{\partial}{\partial t} x(i, t) \overset{LT}{\leftrightarrow} s\Phi_X(i, s) - x(i, 0)$$

The system of differential equations turns into a system of linear equations

$$s\Phi_X(0, s) - 1 = -\lambda \Phi_X(0, s)$$
$$s\Phi_X(i, s) - 0 = -\lambda \Phi_X(i, s) + \lambda \Phi_X(i-1, s), \; i = 1, 2, \ldots$$

With successive insertions we obtain

$$\Phi_X(i,s) = \frac{\lambda^i}{(s+\lambda)^{i+1}}$$

The inverse Laplace transform gives

$$x(i,t) = \mathbb{P}(X_t = i) = e^{-\lambda t}\frac{(\lambda t)^i}{i!}, \text{ with } i = 0,1,\ldots, t \geq 0.$$

Hence, (X_t) is a Poisson process, derived as pure birth process.

2.4 Pure Markov Systems M/M/n

2.4.1 Loss Systems M/M/n

Description of the Model

We treat the special case of a loss system M/M/n as first important birth and death process. Figure 2.6 represents the components of the system. For the

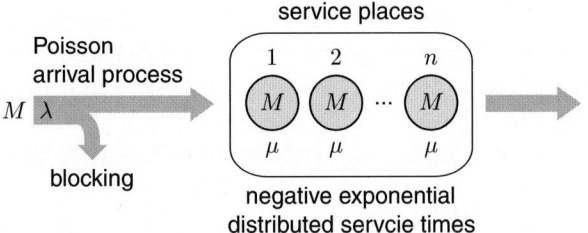

Fig. 2.6. Loss System M/M/n

model we choose the following assumptions:

- Poisson arrival process (call attempts): arrival rate λ (birth rate), i.e. the interarrival time ξ is exponential distributed w.r.t. λ. The distribution of the interarrival times reads as

$$F_\xi(t) = 1 - e^{-\lambda t} \quad \text{and} \quad \mathbb{E}(\xi) = \frac{1}{\lambda}$$

- Exponential distributed serving time η (call duration): serving rate μ, i.e. the serving time η is exponential distributed w.r.t. μ

$$F_\eta(t) = 1 - e^{-\mu t} \quad \text{and} \quad \mathbb{E}(\eta) = \frac{1}{\mu}$$

- Occupation of n places and blocking, if all places are occupied.

Erlang Loss Formula

We assume that the initial value of the state process runs at the beginning through an instationary phase and ends up in a stationary state. Since the process will then be time independent, we will describe it by the random variable X

$$x(i) = \mathbb{P}(X_t = i) = \mathbb{P}(X = i)$$

We consider two fundamental events:

- Arrival event: The transition $X = i$ to $X = i+1$ happens according to a Poisson arrival process with rate λ, if the demands (calls) are all served. If $X = n$, then all further demands are rejected. Thus, we have a pure loss system.
- Serving end: If $X = i$, we have i demands in the serving phase (i calls). For an exponential distributed serving time we know that the transition from $X = i$ to $X = i - 1$ happens with a rate of $i\mu$.

Thus, we obtain for the transition probability rates

$$q_{ij} = \begin{cases} \lambda & \text{for } i = 0, 1, \ldots, n-1, \ j = i+1 \\ i\mu & \text{for } i = 1, 2, \ldots, n, \ j = i-1 \\ 0 & \text{else} \end{cases}$$

A state transition diagram of a loss system M/M/n can be found in figure 2.7.

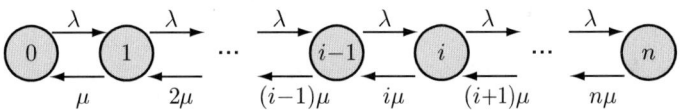

Fig. 2.7. State transition diagram for loss system M/M/n

The linear system of equations for the solution of the state probabilities is

$$\lambda x(i-1) = i\mu x(i), \ i = 1, 2, \ldots, n \quad \text{and} \quad \sum_{i=0}^{n} x(i) = 1$$

Setting $\alpha = \frac{\lambda}{\mu}$, called the *offer*, we obtain the respective state probabilities by successive insertion

$$x(i) = \frac{\frac{\alpha^i}{i!}}{\sum_{k=0}^{n} \frac{\alpha^k}{k!}} \qquad (2.4)$$

This leads for the blocking probability to the *Erlang-B formula* or *Erlang formula for a loss system*.

Blocking Probabilities

An important key value of the loss system is the blocking probability. The system is in a blocked state, if we have $X = n$. That means the blocking probability or Erlang-B formula can be computed according to

$$p_B = x(n) = \frac{\frac{\alpha^n}{n!}}{\sum_{k=0}^{n} \frac{\alpha^k}{k!}} \quad \text{Erlang-B formula} \tag{2.5}$$

The *traffic value* Y is the average number of occupied serving units, which is given by the product of the average serving time and the mean rate of accepted demands. The mean rate of accepted demands is gained as the product of arrival rate and the non-blocking probability, thus, $\lambda(1 - p_B)$. This can be expressed in the formula

$$Y = \lambda(1 - p_B)\frac{1}{\mu} = \alpha(1 - p_B) \quad \text{Unit: Erlang}$$

We consider a bundle of lines in a telephone network and deduce the following facts:

- Arrival process: As arrival process we have the call attempts for occupation of the bundle of lines. For sufficient large number of telephone users we can assume a Poisson arrival process. This leads to a simple analysis, whose assumption is proved by empirical study.
- Serving process: If the call is accepted the line is busy. The line corresponds to a serving unit, thus, the serving time to the call duration. If all n lines are busy, a call attempt is rejected.
- Multiplexing gain: In telephone networks line bundles are built according the principle that large bundles are more efficient than smaller ones. This gain can be extracted from the Erlang formula. If we denote by n the number of lines and with Y the traffic value as indicated above, then we have by Y/n the average traffic value per line

From the relation

$$\frac{Y}{n} = \lambda(1 - p_B)\frac{1}{n\mu} = \frac{\alpha(1 - p_B)}{n}$$

we deduce the following:

- If λ and p_B are fixed, then a growing n gives an increasing coefficient and hence, an increasing α a larger rate as n. Thus, the fraction $\frac{Y}{n}$ increases. Each line can serve more traffic (calls).
- The value $\frac{Y}{n}$ can be regarded as multiplexing gain. But the curve $\frac{Y}{n}$ gets more horizontal for constant p_B if n grows and approaches a saturation. The important interpretation of this evidence is that larger networks do not provide necessarily a bigger gain!

Multiplexing Gain

We start with a figure for the multiplexing of several lines in a telecommunication network. We consider up to 150 lines. In fact, at the beginning all lines are separated, while we bundle up to 150 bundles in the final system. The gain will be expressed by Y/N, where N expresses the number of lines combined together.

For demonstrating the multiplexing gain, we just consider for a given n, $\lambda > 0$ and μ two exactly identical loss systems, say $1a$ and $1b$. For the system $1a$ and $1b$ we can compute the blocking probability $p_{1,B}$ according to the Erlang formula (2.4). Now we multiplex both systems and obtain a new loss system, called system 2, with $2n$ serving units, 2λ as arrival intensity and the same μ. Hence, the new load $\alpha_2 = 2\alpha$. The major observation is (see figure 2.8)

$$p_{2,B} < p_{1,B}$$

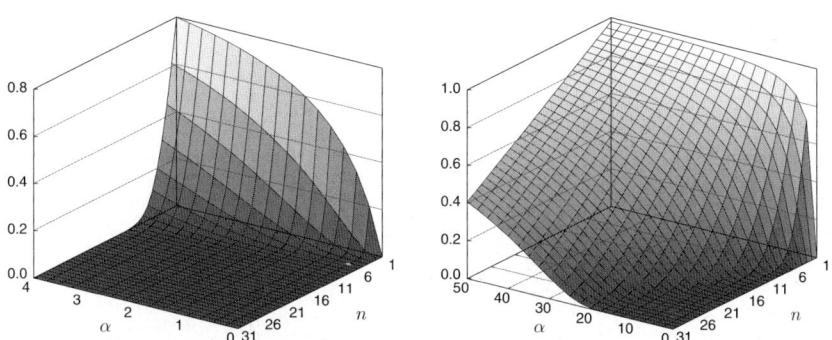

Fig. 2.8. Blocking probability (Erlang-B formula) depending on the number of lines n and the traffic load α

This phenomena can be explained by the fact that in two separated systems one demand can be rejected, since its subsystem is occupied, while the other subsystem has still free capacity. In the compounded system 2 the free capacity is incorporated in the whole system and the new demand is not rejected. For illustrating this we look at the following example.

Example 2.4. Consider a telephone phone system between location A and B. We have an arrival rate of $2\lambda = 0.8$ calls per second. The duration of a call η in B is exponential distributed with mean $\mathbb{E}(\eta) = 100$ seconds. Thus the serving rate $\mu = \frac{1}{\mathbb{E}(\eta)} = 0.01$ calls per second. An incoming call meeting an occupied line is rejected. Thus we have a loss system M/M/n, where the results can be transformed to the general case of M/GI/n. We consider two alternative ways of installations:

- System 1: First we consider two subsystems with each $n = 50$ switching servers, each having an arrival rate of $\lambda_1 = 0.4$ transactions per second. The load in each system is $\alpha_1 = \lambda_1 \cdot \mathbb{E}(\eta) = 40$ Erl. The load per server is $\rho_1 = \frac{\lambda_1 \cdot \mathbb{E}(\eta)}{50} = \frac{4}{5}$ Erl. Thus, each line is loaded up to 80%. The loss probability reads as $p_{1,B} = 0.0187$ or 1.87%. For the traffic value we have each $Y_1 = 39.25$. Thus, over all $Y = 78.5$.
- System 2: Now we combine the two subsystems to a whole of $n = 100$ switching servers, each having, thus, an arrival rate of $\lambda_2 = 2\lambda_1 = 0.8$ transactions per second. The load in each system is $a_2 = \lambda_2 \cdot \mathbb{E}(\eta) = 120$ Erl. The load per server is $\rho_2 = \frac{\lambda_2 \cdot \mathbb{E}(\eta)}{100} = \frac{4}{5}$ Erl. Thus, each server has a traffic value of 80%. The loss probability turns out to be $p_{2,B} = 0.004$ or 0.4%. The traffic value over all in the system II reads as $Y_2 = 79.68$.

The example reveals impressively that the traffic load increases from 78.5 to 79.68, and even more the loss probability decreases drastically from 1.87% to 0.4%. Figure 2.9 depicts that with an increase of lines the multiplex gain increases rapidly – the free traffic capacity per line increases up to 1, but the gain for growing numbers of n is fast getting slowlier, for all traffic amounts α.

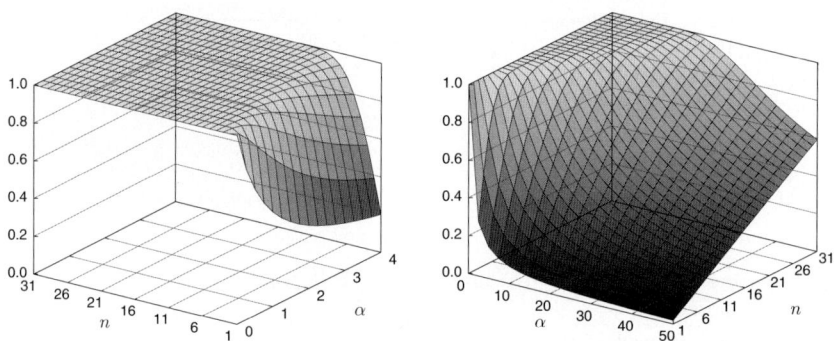

Fig. 2.9. Multiplexing gain M/M/n: Depending on the traffic load α and the number of coumpounded lines. We see that with increasing n the gain grows to 1, though its increase gets significant slowlier starting from 40. This indicates that more lines need not increase the gain in the same scale

2.4.2 Queueing Systems M/M/n

Description of the Model

Similar to the loss system we choose for the queueing system M/M/n the following assumptions:

- Poisson arrival process (call attempts): arrival rate λ (birth rate), i.e. the interarrival time ξ is exponential distributed

$$F_\xi(t) = 1 - e^{-\lambda t} \quad \text{and} \quad \mathbb{E}(\xi) = \frac{1}{\lambda}$$

- Exponential distributed serving time η (call duration): service rate μ, i.e. the service time μ is exponential distributed

$$F_\eta(t) = 1 - e^{-\mu t} \quad \text{and} \quad \mathbb{E}(\eta) = \frac{1}{\mu}$$

- Occupation of n places and an infinite queue.

Figure 2.10 shows a representation of a queueing system M/M/n. For a pure

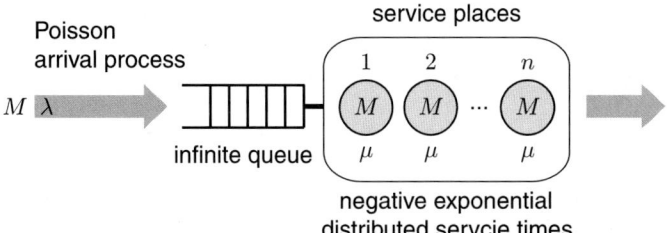

Fig. 2.10. Queueing System M/M/n

queueing system the load is identical to the offer and thus,

$$\alpha = \frac{\lambda}{\mu} = \frac{\lambda}{\mathbb{E}(\eta)}$$

For the load of a unit we obtain the loading coefficient

$$\nu = \frac{\alpha}{n}$$

Now we have $\lambda \mathbb{E}(\eta)$ as average number of arrivals during a serving time. If $\lambda \mathbb{E}(\eta)$ grows more than n, then the system will become unstable. Hence, we have to introduce a stability criteria

$$\nu = \frac{\alpha}{n} < 1 \quad \text{or} \quad \alpha < n$$

Erlang-C Formula

Since beside the demands $X_\eta(t)$ in the serving system there exist demands in the queue, we introduce another process of demands in the queue $X_W(t)$. We have due to the restriction of n serving places $X_\eta(t) \leq n$. If $X_\eta(t) < n$, so we deduce $X_W(t) = 0$.

The state process is time continuous, state discrete and Markovian, since the arrival and serving process are Markovian. If the process turns to be stationary, then we denote the demand process by X_η resp. X_W. According to

$$x(i) = \mathbb{P}(X(t) = i) = \mathbb{P}(X = i)$$

we can determine the transition probabilities as in the loss system:

- Arrival event: $X = i \longrightarrow X = i+1$ with rate λ and $i = 0, 1, 2, \ldots$
- Serving end:
  - If $X \leq n$, then we assume $X = i \leq n$. As in the loss system we have the transition $X = i \longrightarrow X = i-1$ with rate (death rate) $i\mu$.
  - If $X > n$, then all serving places are occupied and $X_W = X - n$. Hence, it follows as for $X \leq n$ the transition $X = i \longrightarrow X = i-1$ with rate $n\mu$ and $i = n+1, n+2, \ldots$

The state transition diagram for the queueing system M/M/n can be seen in figure 2.11.

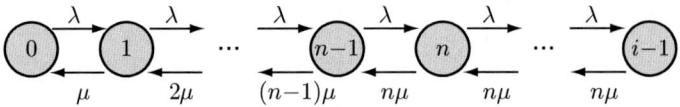

Fig. 2.11. State transition diagram for queueing system M/M/n

Thus, we obtain the Kolmogorov equations

$$\lambda x(i-1) = i\mu x(i), \quad i = 1, 2, \ldots, n$$
$$\lambda x(i-1) = n\mu x(i), \quad i = n+1, \ldots$$
$$\sum_{i=0}^{\infty} x(i) = 1 \qquad (2.6)$$

This simple system of linear equations can successively be solved, and we get

$$x(i) = \begin{cases} x(0)\frac{\alpha^i}{i!} & \text{for } i = 0, 1, 2, \ldots, n \\ x(0)\frac{\alpha^n}{n!}\left(\frac{\alpha}{n}\right)^{i-n} = x(n)\nu^{i-n} & \text{for } i = n+1, n+2, \ldots \end{cases}$$

Hence, we only need to determine $x(0)$ and obtain from the normalization equation (2.6) with $\nu < 1$

$$x(0)^{-1} = \sum_{k=0}^{n-1} \frac{\alpha^k}{k!} + \frac{\alpha^n}{n!} \sum_{k=0}^{\infty} \nu^k = \sum_{k=0}^{n-1} \frac{\alpha^k}{k!} + \frac{\alpha^n}{n!} \frac{1}{1-\nu}$$

For $\nu \geq 1$ the series $\sum_{k=0}^{\infty} \nu^k$ does not converge. The reasoning can immediately be deduced. If $\nu \geq 1$ holds, then we know that the mean offer is always bigger than the number of serving units. The system gets unstable.

As in the system of the loss system, we determine some system characteristics. Before we continue with considering the queueing probability in the next subsection, we want to cite a standard result in the classical traffic theory. We formulate the *theorem of Little*.

Proposition 2.5. *If $\lambda > 0$ is the intensity of the arrival process and if $\eta > 0$ is the serving time of a demand, then we have for the average number $\mathbb{E}(\Psi)$ of all demands in the system*

$$Y = \mathbb{E}(\Psi) = \lambda \cdot (\eta)$$

For a proof see [253, p. 12] resp. [11, p. 240].

Queueing Probability

According to the model a unit has to wait before being served, if $X = n$ holds. Thus, we get for the queueing probability the sum of all state probabilities is larger or equal than n, thus,

$$p_W = \sum_{i=n}^{\infty} x(i)$$

Expressing this with the help of α, we find the *Erlang queueing formula* or *Erlang-C formula*

$$p_W = \frac{\frac{\alpha^n}{n!(1-\nu)}}{\sum_{i=0}^{n-1} \frac{\alpha^i}{i!} + \frac{\alpha^n}{n!(1-\nu)}} \tag{2.7}$$

The figure 2.12 depicts the decrease for the waiting probability if two separated systems are compounded. We consider the traffic value as the mean number

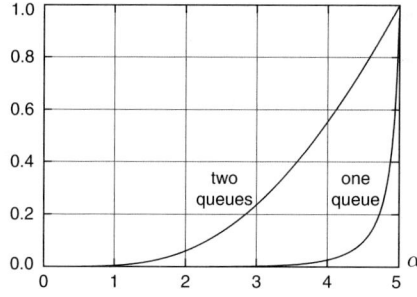

Fig. 2.12. Queueing probability p_W depending on the traffic load α with $n = 5$ for two separated systems, thus queues, and a compounded system with one queue

of occupied serving units $\varUpsilon$. With the same notation we get

$$Y = \mathbb{E}(\varUpsilon) = \sum_{i=0}^{n-1} ix(i) + \sum_{i=n}^{\infty} x(i) = \alpha \quad \text{Unit: Erlang}$$

According to the Little formula from proposition 2.5 we can express this directly

$$Y = \frac{\lambda}{\mu} = \alpha$$

The figure 2.13 shows, depending on the traffic load ρ and for different multiplexed lines, the logarithmic queueing time probability.

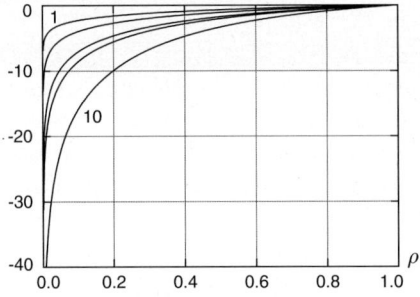

Fig. 2.13. Logarithmic queueing probability (Erlang formula) for different multiplexed lines from $n = 1, 2, 4, 5$ to 10

Queueing Distribution

According to the definition it coincides with the expectation of the numbers of units in the queue, thus, we get the mean queueing length $\varPsi$

$$\varPsi = \mathbb{E}(X_W) = \sum_{i=n}^{\infty}(i-n)x(i) = \sum_{i=n}^{\infty}(i-n)x(n)\nu^{i-n} = x(n)\sum_{i=n}^{\infty} i\nu^i$$

$$= x(n)\frac{\nu}{(1-\nu)^2} = x(0)\frac{\alpha^n}{n!}\frac{\nu}{(1-\nu)^2}$$

According to our formula for the queueing probability we deduce

$$\varPsi = p_W \frac{\nu}{1-\nu}$$

With this we can now state the mean queueing or waiting time. For this we split our system into two kinds of waiting times, hence, two subsystems I and II:

2.4 Pure Markov Systems M/M/n

- I: the mean waiting time of *all* demands $\mathbb{E}(W)$ and
- II: the mean waiting time of the *queueing* elements $\mathbb{E}(W_1)$.

We start with the investigation of the subsystem I, the waiting time of all demands:

- The mean arrival rate is $\lambda_\mathrm{I} = \lambda$.
- The average number of demands equals the sum of all demands in the waiting queue and of the serving units

$$\mathbb{E}(X_\mathrm{I}) = \mathbb{E}(X_W) + \mathbb{E}(X_\eta) = \Psi + Y$$

- The average time spent in the system is the sum of the waiting time according to all demands and the serving time

$$\mathbb{E}(T_\mathrm{I}) = \mathbb{E}(W) + \mathbb{E}(\eta)$$

The mean waiting time decreases, if two systems of each n lines are compounded. The multiplexing gain is depicted in the figure 2.14 for $n = 5$. It also reveals the intense increase approaching the value $\alpha = \to 5$, which is equivalent to the relative traffic $\rho \to 1$ in this situation. We selected two disjoint regions of α for better understanding and used $\lambda = \frac{1}{200}$.

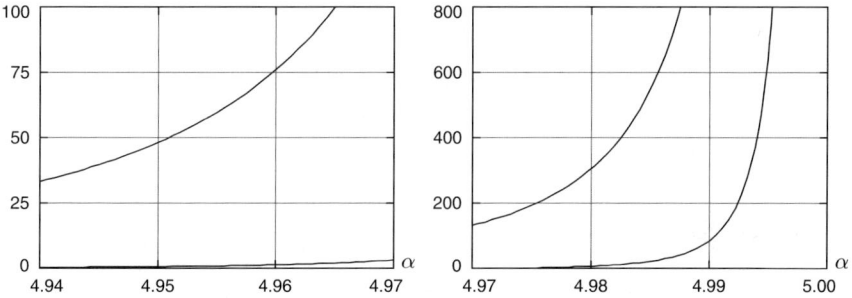

Fig. 2.14. Mean waiting time $\mathbb{E}(W)$ with respect to all demands, depending on the traffic load α for three systems with each $n = 5$, the upper line represents three separated systems, thus queues, the lower one compounded with only one queue

Using the Little formula, we deduce

$$\lambda_\mathrm{I} \mathbb{E}(T_\mathrm{I}) = \mathbb{E}(X_\mathrm{I})$$

and thus,

$$\mathbb{E}(W) = \frac{\Psi}{\lambda} \tag{2.8}$$

Now we turn to the subsystem II. Here, we have a pure waiting queue:

- The mean arrival rate is $\lambda_\mathrm{II} = p_W \lambda$. This is the arrival rate of all demands weighted by the waiting probability.

- The average number of all demands in the system $\mathbb{E}(X_{\mathrm{II}})$ equals the mean queueing length, thus, $\mathbb{E}(X_{\mathrm{II}}) = \nu$.
- The average time spent in the system $\mathbb{E}(T_{\mathrm{II}})$ equals the mean waiting time according to all waiting elements, thus, $E(W_1)$.

With the Little formula we can deduce $\lambda_{\mathrm{II}} \mathbb{E}(T_{\mathrm{II}}) = \mathbb{E}(X_{\mathrm{II}})$ and hence,

$$\mathbb{E}(W_1) = \frac{\nu}{p_W \lambda} = \frac{1}{\lambda} \frac{\nu}{1-\nu}$$

The figure 2.15 shows the mean queueing length $\mathbb{E}(W_1)$ for different number of lines, indicating that the length decreases for growing n for fixed ρ, while the increase gets more intense with growing number of lines for $\rho \to 1$.

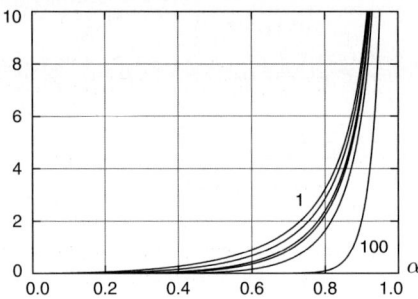

Fig. 2.15. Mean queueing length for multiplexed lines $n = 1, 2, 4, 5, 10$ and $n = 100$ (from above to below)

Important for the consideration of the waiting time is the *queueing distribution*, which is finally considered in this section of queueing systems. We start with the queueing system using the First-In-First-Out (FIFO) principle. We consider a sample demands arriving for the derivation of the distribution and we want to compute the probability that this unit has a positive waiting time. For this, we use the concept of the conditional probability, thus for $t > 0$

$$\mathbb{P}(W > t \,|\, W > 0) = \frac{\mathbb{P}(W > t, W > 0)}{\mathbb{P}(W > 0)} = \frac{\mathbb{P}(W > t)}{\mathbb{P}(W > 0)}$$

According to the derivation of the queueing probability we know

$$\mathbb{P}(W > 0) = p_W = x(n) \frac{1}{1-\nu}$$

We use an arbitrary sample demand and want to investigate its way through the system. At the arrival time it meets i elements in the system. A positive waiting time gives

$$\mathbb{P}(W > t \,|\, W > 0) = \sum_{i=0}^{\infty} \mathbb{P}(W > t \,|\, \Phi = i+n) \mathbb{P}(\Phi = i+n)$$

2.4 Pure Markov Systems M/M/n

For this we have
$$\mathbb{P}(\Phi = i + n) = x(i+n) = x(n)\nu^i$$

As first result we deduce
$$\mathbb{P}(W > t \mid W > 0) = \sum_{i=0}^{\infty} \mathbb{P}(W > t \mid \Phi = i + n)(1 - \nu)\nu^i$$

For computing the probability distribution $\mathbb{P}(W > t)$ we need a formula for $\mathbb{P}(W > t \mid \Phi = i + n)$.

At the arrival time our sample meets $\Phi = i + n$ demands. Thus, all serving places are occupied and i elements are waiting in the queue. The period η^* between two serving ends in this state is exponential distributed according to
$$F_{\eta^*} = \mathbb{P}(\eta^* \leq t) = 1 - e^{-n\mu t}$$

The waiting time of the sample is built up as follows. First, we have to consider the time between the arrival of the sample and the first serving end. This is the forward recurrence time of η^*, which has, as derived later, the same distribution as η^* itself. Next, we compute the time from the first serving end until the time spot, for which all $i = \Psi - n$ are switched over to the serving process. These are i intervals, each distributed as η^*. Hence, the waiting time consists of $i + 1$ intervals of the type η^*. The distribution of the waiting time resembles an Erlang distribution of order $(i + 1)$

$$\mathbb{P}(W > t \mid \Phi = i + n) = e^{-n\mu t} \sum_{k=0}^{i} \frac{(n\mu t)^k}{k!}$$

Inserted in the above equation we deduce
$$\mathbb{P}(W > t \mid W > 0) = e^{-n\mu t}(1 - \nu) \sum_{i=0}^{\infty} \sum_{k=0}^{i} \nu^i \frac{(n\mu t)^k}{k!}$$
$$= e^{-n\mu t}(1 - \nu) \sum_{i=0}^{\infty} \frac{(n\mu t)^k}{k!} \sum_{k=0}^{i} \nu^i$$
$$= e^{-(1-\nu)n\mu t}$$

Thus, it follows
$$\mathbb{P}(W > t) = \mathbb{P}(W > t \mid W > 0)\mathbb{P}(W > 0)$$
$$= e^{-(1-\nu)n\mu t} p_W$$
$$= 1 - F_W(t)$$

and the waiting distribution is finally computed according to
$$F_W(t) = 1 - e^{-(1-\nu)n\mu t} p_W$$

The figure 2.16 depicts the complementary waiting time distribution each for two different number of multiplexed lines and three different values of relative traffic load ρ in a logarithmic scale.

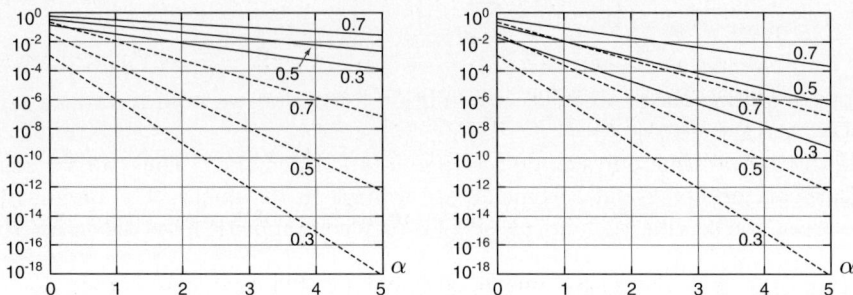

Fig. 2.16. Complementary queueing time distribution for $n = 10$ (dotted lines, left and right), $n = 2$ (solid lines, left) and $n = 5$ (solid lines, right), each for $\rho = 0.3$, 0.5, and 0.7 and $\mu = 1$

The figure 2.17 demonstrates the relative load per line depending of traffic load α and the multiplexed lines n.

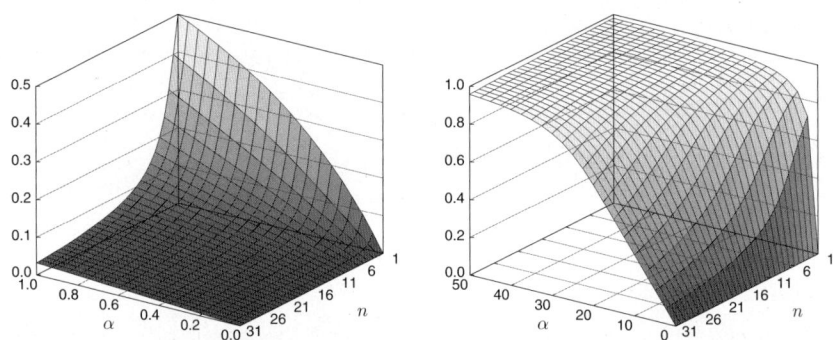

Fig. 2.17. Proportional occupancy of traffic amount per line depending on the traffic load α and number of lines n.

2.4.3 Application to Teletraffic

Multiplexing Gain in Queueing Systems

Already in section 2.4.1 on loss systems we considered the multiplexing gain in connection with telecommunication systems using the Erlang loss formula. The basic result was that multiplexed systems reduce the blocking probability compared to single systems. The compound system is working more efficiently.

2.4 Pure Markov Systems M/M/n

We want to follow that line in queueing systems as well. For this we consider a simple configuration, which of course can be generalized, though the main idea are already revealed in this situation:

- System I
  - Consisting of two subsystems Ia and Ib of type M/M/n queueing systems.
  - Subsystems Ia and Ib have identical serving places n and service rates. The arrival process has the same rate $\lambda = \lambda_1$.
  - The waiting probability in system I is p_{W_1}.
  - The average waiting time of all demands is W_1.
- System II
  - All demands are served by a common system of $2 \cdot n$ units (multiplexing of system I).
  - We have one central queue, e.g. implemented as shared memory.
  - The queueing system has $2 \cdot n$ units and one arrival process with rate $\lambda_2 = 2 \cdot \lambda$.
  - The waiting probability of system II is denoted by p_{W_2}.
  - Let W_2 be the waiting time of all demands.

We can easily detect (see also figures 2.12 and 2.14) using formulas (2.7) and (2.8) that

$$p_{W_2} < p_{W_1} \quad \text{and} \quad W_2 < W_1$$

How can we explain this phenomena? Suppose a demand in subsystem Ia arrives and encounters all service units occupied. Then it has to wait, though there maybe a vacant spot in subsystem Ib. Since both systems are strictly separated, the test demand in system Ia has to wait. In contrast in system II this demand would be served immediately, which implies a reduction of the waiting queue and mean waiting time.

Example 2.6. Consider a server system with arrival rate of $2\lambda_1 = 1,600$ transactions per second. The duration of one transaction η is exponential distributed with mean $\mathbb{E}(\eta) = 10$ ms. Thus the serving rate $\mu = \frac{1}{\mathbb{E}(\eta)} = 100$ transactions per second.

- System I: First we consider two subsystems with each $n = 12$ switching servers, each having, thus, an arrival rate of $\lambda_1 = 800$ transactions per second. The load in each system is $a_1 = \lambda_1 \cdot \mathbb{E}(\eta) = 8$ Erl. The load per server is $\rho_1 = \frac{\lambda_1 \cdot \mathbb{E}(\eta)}{n} = \frac{2}{3}$ Erl. Thus, each server is loaded up to 66.6%. The waiting time probability according to (2.8) is $p_{W_1} = 0.14$ or 14%, the mean waiting time of all demands is $\mathbb{E}(W_1) = 0.35$ ms.
- System II: Now we combine the two subsystems to a whole of $n = 24$ switching servers, having thus, an arrival rate of $\lambda_2 = 2\lambda_1 = 1,600$ transactions per second. The load in each system is $a_2 = \lambda_2 \cdot \mathbb{E}(\eta) = 16$ Erl. The load per server is $\rho_2 = \frac{\lambda_2 \cdot \mathbb{E}(\eta)}{24} = \frac{2}{3}$ Erl. Thus, each server is loaded up to 66.6%. The waiting time probability according to (2.8) is $p_{W_2} = 0.043$ or 4.3%, the mean waiting time of all demands is $\mathbb{E}(W_2) = 0.05$ ms.

The numerical example shows that multiplexing decreases the waiting time probability significant from 14% to 4.3% and the mean waiting time falls from 0.35 ms to 0.05 ms, the seventh part.

Loss Systems with Finitely Many Sources

We consider the following model:
- There are n service units (e.g. ongoing calls) with exponential distributed service time (call duration) $F_\eta(t) = 1 - e^{-\mu t}$, $\mathbb{E}(\eta) = \frac{1}{\mu}$.
- We have $m > n$ incoming clients. Thus, the arrival process is not a Poisson process, since we do not have infinite many sources.

The incoming user can be divided into two classes - the *active* and the *silent* one.

- The active clients are just served. The sojourn time in the active phase coincides with the serving time.
- The silent client stays in this *idle* mode until the next attempt is started. Either the attempt is successful or the client is again blocked the client returns into the idle mode. The duration of this idle phase I is exponential distributed with parameter a, i.e. $F_I(t) = \mathbb{P}(I \leq t) = 1 - e^{-at}$.

State probabilities: From the macro state M, which are combined by several micro states $X = 0, 1, \ldots, i - 1$, we deduce the Kolmogorov equations:

$$(m - i + 1)\tau x(i - 1) = i\mu x(i), \; i = 1, 2, \ldots, n \quad \text{and} \quad \sum_{i=0}^{n} x(i) = 1$$

Inserting successively reveals with $\tilde{a} = \frac{\tau}{\mu}$

$$x(i) = \frac{\binom{m}{i}\tilde{a}^i}{\sum_{k=0}^{n} \binom{m}{k}\tilde{a}^k}$$

When do we observe a blocking? Here is the blocking probability

$$p_B = x_A(n) = \frac{\binom{m-1}{n}\tilde{a}^n}{\sum_{k=0}^{n} \binom{m-1}{k}\tilde{a}^k} \quad \text{Engset formula}$$

The radio-based network ALOHA was an early example of such a loss system with finitely many sources, working with the phase model according to Erlang (see e.g. [260]). With ALOHA, university campuses between Hawaiian islands were connected over a shared radio channel. The basic idea of this network was to omit a centralized approach granting access to the channel. In contrast, a distributed algorithm was implemented at the cost of possible simultaneous transmissions of several stations causing collisions on the shared channel. With the so called Aloha protocol basic algorithms on the data link layer for channel access and retransmits in case of failures resp. collisions were implemented on each station of the entire ALOHA network.

Started in late 1970, ALOHA is regarded today as the world's first wireless packet-switched network. The most prominent successor of the Aloha protocol was Ethernet, which employed with *Carrier Sense Multiple Access/Collision Detection* (CSMA/CD) a 'polite' version of the Aloha protocol. In CSMA/CD, the stations are listening to the channel and waiting for a free period before transmitting to avoid collisions with an already ongoing transmission (see e.g. [143]). We remark that in the current Gigabit or 10 Gigabit Ethernet standard different access protocols are implemented.

Example 2.7. We consider a mobile cell with finite sources in more detail now, since we have already looked at the cellular system GSM in the introductory examples. There we sketched how a mobile system works in principle. It is a loss system with finitely many resources. This fact and together with the Engset formula will be used for dimensioning the GSM cells. We present the necessary ingredients:

- $\mathbb{E}(\eta) = \frac{1}{\mu}$.
- α is the arrival rate of a user in the state 'idle'. The load of a user in the 'idle' or silent state is $Y = \alpha \cdot \mathbb{E}(\eta) = \frac{\alpha}{\mu}$ Erl. Thus, the overall load in the cell is $a = m \cdot Y = m \cdot \alpha \cdot \mathbb{E}(\eta)$ Erl. If $m \to \infty$ we get approximately a Poisson process with $\lambda = \alpha \cdot m$. The number n of usable channels (which implies the required frequencies) has to dimensioned in such a way that the total offer should not surpass a given certain QoS threshold, expressed by the blocking probability.

We consider certain further directions and specifications of the above Markovian systems, which we shortly consider. For a detailed investigation the reader should consult the literature.

Call Repetition with Finitely Many Resources

Though the above model may somehow describe the traffic within a mobile cell, in a lot of situation rejected call attempts lead to a call repetition. In case of a model with finitely many sources this has a decisive influence on the over all traffic. Hence, this model is an extension of the previous one. We shortly indicate the model description and note that a major application consists in the traffic analysis of a mobile cell with call repetition. For further information please consult [255, 257].

Instead of describing the model abstract, we change immediately to the situation of the telephone communication. We start with the user view. Thus a first call attempt will be just denoted by 'call', while a repeated call after blocking is denoted by 'subsequent call'. In figure 2.18 we have depicted the model from the user's point of view. All parts are Markovian. Let's characterize the used RV:

- Idle time: Time, where the source resp. telephone is not calling. We characterize it by a exponential distributed RV

$$F_\iota(t) = \mathbb{P}(\iota \le t) = 1 - e^{-\alpha t} \quad \text{and} \quad \mathbb{E}(\iota) = \frac{1}{\alpha}$$

- Serving time resp. connection time: this is the duration of the active call, again exponential distributed

$$F_\eta(t) = \mathbb{P}(\eta \le t) = 1 - e^{-\mu t} \quad \text{and} \quad \mathbb{E}(\eta) = \frac{1}{\mu}$$

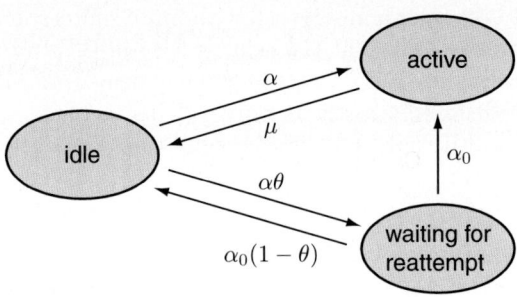

Fig. 2.18. Call repetition with finitely many resources from the user's point of view

- Inter-reattempt time: the time between the first call and the first repeated call attempt resp. between the succeeding call attempts. The repeating intervals are independent of each other and exponential distributed assumed

$$F_\varpi(t) = \mathbb{P}(\varpi \le t) = 1 - e^{-\alpha_0 t} \quad \text{and} \quad \mathbb{E}(\varpi) = \frac{1}{\alpha_0}$$

In general the mean 'interreattempt' time is decisive smaller than the mean idle time, i.e. $\mathbb{E}.(\varphi) < \mathbb{E}.(\iota)$. In figure 2.19 the transitions of all states are depicted. We included the factor θ to indicate the probability how impatient the user is. Usually the parameter decreases with the number of call attempts, but we assume it constant.

We incorporate all parts into the general model, which is sketched in the figure 2.19 and indicate its basic components:

- We have a finite number m of users, forming the arrival space. The traffic intensity depends on the number of users in the 'idle' phase.
- The number n of serving spots (e.g. channels in a cell of a mobile communication system) characterizing the serving space. As done for the users, it is exponential distributed with intensity μ. We have a loss system, provided an incoming call meets n occupied serving units. By a probability of $\theta \in\,]0,1[$ the call is repeated or with $1 - \theta$ stopped.
- The repetition space, where at most $m - n$ calls wait for reattempts, has a leaving rate of α_0 as indicated in figure 2.19.

2.4 Pure Markov Systems M/M/n

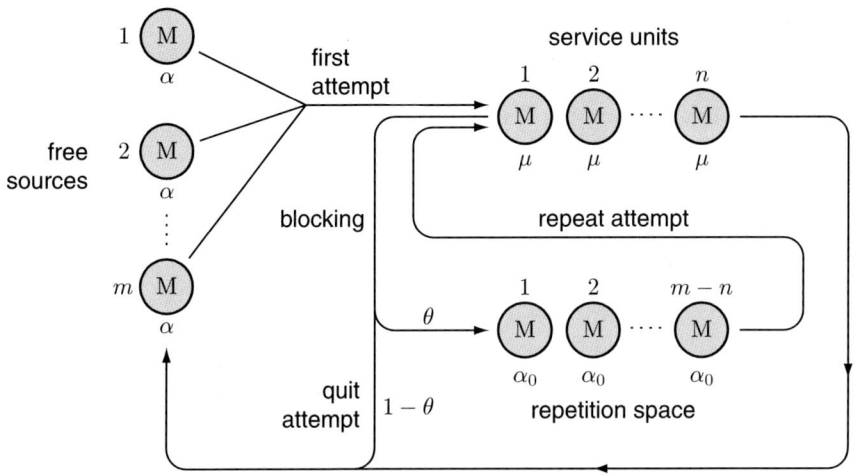

Fig. 2.19. Call reattempts with finite number of users

Reduction Methods for Markovian Systems

We shortly summarize this area and refer for a detailed description to the literature [255]. In many communication system, we do not have a one dimensional state space, as seen in the classical telephone system: everyone just calls, there is no quality or service difference. Especially in mobile communication but also the use of the Internet often offers different resource classes to the user. This leads to a higher dimensional state process. Since one can allow combination between classes, the increase of resource classes leads to an exponential increase in the states. This again makes numerical implementations more difficult.

Since we stick to the simple Markovian case, for each class one has to analyze a M/M/n loss system. So higher dimension for the state space implies an exponential increase of numerical blocking analysis. The major method is introduced by Kaufman and Roberts [135].

We briefly describe the model. There are l resources, which may e.g. be the number of available orthogonal codes in the CDMA used in UMTS, the capacity of a CPU or the capacity of a plain line. We assume c different classes of users requiring different applications or services. A user in class c is part of a Poisson process with the arrival rate λ_c. The serving time is described by the RV η_c and with rate μ_c. Each user of class c demands an amount a_c. Assuming at the arrival time spot we do have a_s resources available, the new user occupies all resources. After the serving time η_c the user leaves the system.

As remarked above, since we do not require the detailed description for the further study, we do not go deeper in the material. The basic idea consists in determination of the state probability using the reduction method due to

Kaufman and Roberts with the help of recursion. It is based on the reduction of the higher dimensional state space to a one dimensional one, whose states corresponds to the occupied resources. In accordance, in one particular state all higher dimensional states are compounded in occupying an equal number of resources. Instead of computing all state equation for all classes it suffices in determination of the blocking probability and the computation of the distribution of occupied resources.

2.5 Special Traffic Models

In this section we will sketch some special traffic models and discuss some of the main results. A detailed description can be found in monographs as [21, 30].

2.5.1 Loss Systems M/M/∞

The loss system M/M/∞ is neither a queueing nor a pure loss system. Since each demand gets on a serving spot, we obtain a pure birth and death process with the transition probability rates

$$q_{i,i+1} = \lambda, \ i = 0, 1, \ldots \quad \text{and} \quad q_{i,i-1} = i\mu, \ i = 1, 2, \ldots$$

As state probabilities in the stationary case we get (provided we have $n \to \infty$ in the system M/M/$n - 0$)

$$x_i = \frac{\rho^i}{i!} e^{-\rho}$$

with $\rho = \frac{\lambda}{\mu}$. Thus, the mean number of serving elements equals the traffic load

$$\mathbb{E}(X) = \rho$$

We will consider a generalization of M/G/∞ in chapter 3, because it was one of the first models in IP traffic theory, explaining the so called asymptotic self-similarity.

2.5.2 Queueing Systems of Engset

We give a short description of the queueing system of Engset. From s sources demands arrive independently according to a Poisson process with intensity λ and are distributed on $n \leq m$ serving channels. Again the serving procedures are stochasticly independent and exponential distributed with parameter μ. In contrast to the loss system of Engset rejected units are stored in queueing space with $m - n$ spots. Each unit receive at least one waiting place. The source cannot send new demands to the system, if there are units in the serving procedure or in the waiting space. E.g. under the constraint $X_t = i$, only $m-i$ sources can be active. We obtain for the stationary state probabilities

$$x_i = \begin{cases} \binom{m}{i} \rho^i x_0 & \text{for } 1 \leq i \leq n \\ \dfrac{m!}{n^{i-n} n! (m-i)!} \rho^i x_0 & \text{for } n \leq i \leq m \end{cases}$$

and for the empty probability with $\rho = \frac{\lambda}{\mu}$ as traffic intensity

$$x_0 = \left(\sum_{i=1}^{n} \binom{m}{i} \rho^i + \sum_{i=n+1}^{m} \frac{m!}{n^{i-n} n! (m-i)!} \rho^i \right)^{-1}$$

If we allow $n > m$, every demand finds a serving channel and the system is no longer a queueing system. It is binomial distributed with parameters m and $\frac{\lambda}{\lambda + \mu}$.

2.5.3 Queueing Loss Systems

Here, we have a system of the form $M/M/n-s$. We start with the description of the model:

- The arrival process is a Poisson process with interarrival times exponential distributed according to the rate λ.
- We have n serving spots. The serving time is again exponential distributed according to μ.
- There are s queueing spots.
- Each demand to serve, which neither finds a waiting spot nor a serving channel, is lost.
- As above we have $\rho = \frac{\lambda}{\mu} < n$, i.e. we assume $a = \frac{\rho}{n} < 1$, to keep the system stable.

First we obtain the state space $\mathbb{Z} = \{0, 1, \ldots, n+s\}$. The transition probabilities read as

$$q_{i,i+1} = \lambda, \quad i = 0, \ldots, s+m-1$$

and

$$q_{i,i-1} = \begin{cases} i\mu & \text{for } i = 1, \ldots, s \\ n\mu & \text{for } i = n+1, \ldots, n+s \end{cases}$$

We get for the state probabilities in the stationary case

$$x_i = \begin{cases} \dfrac{1}{i!} \rho^i x_0 & \text{for } i = 1, \ldots, n-1 \\ \dfrac{1}{n! n^{i-n}} \rho^i x_0 & \text{for } i = n, \ldots, n+s \end{cases}$$

$$x_0 = \left(\sum_{i=0}^{n-1} \frac{1}{i!} \rho^i + \sum_{i=n}^{n+s} \frac{1}{n! n^{i-n}} \rho^i \right)^{-1}$$

2 Classical Traffic Theory

We can sum up the second term (geometric series!) and conclude

$$x_0 = \begin{cases} \left(\sum_{i=0}^{n-1}\frac{1}{i!}\rho^i + \frac{1}{n!}\frac{1-(\frac{\rho}{n})^{s+1}}{1-\frac{\rho}{n}}\rho^n\right)^{-1} & \text{for } \rho \neq n \\ \left(\sum_{i=0}^{n-1}\frac{1}{i!}\rho^i + (s+1)\frac{n^n}{n!}\right)^{-1} & \text{for } \rho = n \end{cases}$$

We assumed arbitrary values for ρ, since we do not have a pure loss system. Let us summarize some important probabilities:

- x_0 is the *void probability*.
- x_{n+s} is the *loss probability*.
- $p_f = \sum_{i=0}^{n-1} x_i$ is the probability that a serving channel is not occupied.
- $p_W = \sum_{i=n}^{n+s} x_i$ is the queueing probability.

As in a pure loss system we can describe the traffic load and the mean number of occupied serving spots

$$Y = \rho(1 - x_{n+m}) \quad \text{mean number of occupied serving spots}$$
$$\nu = \frac{\rho}{n}(1 - x_{s+n}) \quad \text{traffic load}$$

2.6 Renewal Processes

The area of renewal processes is fundamental and widely explored in the traffic theory. Thus, we only selected a small angle and treat in particular the following items:

- basic examples (Poisson process),
- elementary renewal theorem and the theorem of Blackwell,
- limit theorems,
- stationary processes.

2.6.1 Definitions and Concepts

A classical renewal process can be found in telecommunication in the form of call attempts. Here, the interarrival times of the call attempts at times $t_0, t_1, \ldots, t_n, \ldots$ build a point process ξ_i. The time between the time spots t_i and t_{i+1} is denoted as interarrival time and has the distribution function $F_{\xi_i}(t)$. It is clear that the particular interarrival times are independent, since the single user does not know of the other calls. Simultaneously, we model the interarrival time for each user homogeneously, i.e. identical distributed. Only the time up to the first call is often differently assumed. Hence, the following definition is motivated.

2.6 Renewal Processes

Definition 2.8. *A non negative point process* $(\xi_i)_{i\geq 1}$, *which is independent (for $i \geq 1$) and identical distributed (for $i \geq 2$), is called* renewal process. *Thus, $F_{\xi_i}(t)$ turns to $F_\xi(t)$ for all $i = 2, 3, \ldots$ The value ξ_i is called* life time *or in traffic theory* interarrival time. *If $F_\xi(t) \neq F_{\xi_1}(t)$, then the process is called* modified *or* delayed. *Otherwise we call it* ordinary *or* simple.

Important random variables are the *renewal times* or *arrival times*

$$T_n = \sum_{i=1}^{n} \xi_i$$

and the *renewal counting process*

$$N(t) = \sup\{n \in \mathbb{N}; T_n \leq t\}, \text{ with } N(t) = 0, \text{ for } t < T_1$$

We have $N(t) \geq n \Leftrightarrow T_n \leq t$, thus,

$$F_{T_n}(t) = \mathbb{P}(T_n \leq t) = \mathbb{P}(N(t) \geq n) \tag{2.9}$$

For computing a probability density of $N(t)$, we proceed recursively as follows:

$$F_{T_n}(t) = F_{\xi_1} \star F_\xi^{*(n-1)}, \text{ with } F^{*(0)}(t) = 1, \ n = 1, 2, \ldots \tag{2.10}$$

Suppose $f_1 = F'_{\xi_1}$ and $f = F'_\xi$ are the corresponding densities of the distributions of F_{ξ_1} and F_{ξ_i}, $i \geq 2$, then it follows

$$f_{T_n}(t) = f_1 \star f^{*(n-1)}, \text{ with } f^{*(0)}(t) = 1, \ n = 1, 2, \ldots$$

Since we have $\mathbb{P}(N(t) \geq n) = \mathbb{P}(N(t) = n) + \mathbb{P}(N(t) \geq n+1)$, we obtain with (2.9)

$$\mathbb{P}(N(t) = n) = F_{T_n}(t) - F_{T_{n+1}}(t), \text{ with } F_{T_0}(t) = 1, \ n = 0, 1, \ldots$$

With this we can tackle several problems, e.g. the question, how much capacity a system still has. For an answer we compute the additional load, which a system can allow not to surpass a certain security line $1 - \alpha$, so that a renewal process (i.e. new arrival events) does not stop. This means to solve the following inequality

$$1 - F_{T_n}(t) = F_{T_n}^c \geq 1 - \alpha$$

In this connection we introduce the next notion of the *random walk* as generalization of the renewal process. Let S_0 and $(\zeta_n)_{n\geq 1}$ be a sequence of independent random variables, where the $(\zeta_n)_{n\geq 1}$ are identical distributed. Then we call the (discrete) process

$$S_n = S_0 + \zeta_1 + \ldots + \zeta_n, \ n \geq 0 \tag{2.11}$$

a *random walk* for the initial value S_0. A random variable ζ is called *generator* of the random walk, if it is distributed as ξ_1 (and thus, as all ξ_n). Since all random variables (ζ_n) are identical distributed, we will often skip the index and only use the generator. A random walk is called *discrete*, if it assumes values in $\mathbb{Z}$ (or generally in a discrete set), and if ζ is Binom(n,p)-distributed. We call a discrete random walk *symmetric*, if $\mathbb{P}(\zeta = -1) = \mathbb{P}(\zeta = 1) = \frac{1}{2}$. Further properties of random walks can be found in monographs like those of Schürger [232], Alsmeyer [11] and Feller [86].

We proceed to the important notion of the *renewal function*.

Definition 2.9. *We call*
$$H(t) = \mathbb{E}(N(t))$$
a subordinated renewal function. *It indicates the expected numbers of arrival events in the interval* $[0, t]$.

We see that
$$H(t) = \sum_{n=0}^{\infty} \mathbb{P}(N(t) \geq n)$$
and with (2.9) and (2.10)
$$H(t) = \sum_{n=1}^{\infty} F_{\xi_1} \star F_{\xi}^{*(n-1)}(t) \tag{2.12}$$

We insert the definition of the convolution and obtain
$$H(t) = \sum_{n=1}^{\infty} F_{\xi_1} \star F_{\xi}^{*(n-1)}(t) = F_{\xi_1}(t) + \sum_{n=1}^{\infty} \int_0^t F_{\xi_1} \star F_{\xi}^{*(n-1)}(t-x) dF_{\xi}(x)$$
$$= F_{\xi_1}(t) + \int_0^t \sum_{n=1}^{\infty} \left(F_{\xi_1} \star F_{\xi}^{*(n-1)}(t-x) \right) dF_{\xi}(x)$$

According to (2.12) we can rewrite the integrand as $H(t-x)$. Thus, we get
$$H(t) = F_{\xi_1}(t) + \int_0^t H(t-x) dF_{\xi}(x) \tag{2.13}$$

Analogously we conclude
$$H(t) = F_{\xi_1}(t) + \sum_{n=1}^{\infty} \int_0^t F_{\xi}^{*(n)}(t-x) dF_{\xi_1}(x)$$
$$= F_{\xi_1}(t) + \int_0^t \sum_{n=1}^{\infty} F_{\xi}^{*(n)}(t-x) dF_{\xi_1}(x)$$

Thus, it follows

$$H(t) = F_{\xi_1}(t) + \int_0^t H_g(t-x)dF_{\xi_1}(x)$$

where H_g is the renewal function of the ordinary renewal process. Because of $F_{\xi_1} = F_\xi$ for an ordinary renewal process, we obtain the integral equation

$$H_g(t) = F_\xi(t) + \int_0^t H_g(t-x)dF_\xi(x) \tag{2.14}$$

The equations (2.13) and (2.14) are called *renewal equations* and according to Feller can be solved uniquely [88].

Up to now we determined uniquely the distributions of the renewal function using an integral equation. In fact, in most cases we cannot derive an explicit solution. The problem consists in the convolution of the distribution functions, which can be transformed by the Laplace transform into an algebraic equation (see e.g. [74]). For this we first differentiate (2.12) and get

$$\frac{dH}{dt}(t) = h(t) = \sum_{n=1}^\infty f_{\xi_1} \star f_\xi^{*(n-1)}(t)$$

We call $h(t)$ the *renewal density*. An analogous definition can be obtained for the renewal density h_g of an ordinary renewal process. Thus, we rewrite the integral equation using simple differentiation

$$h(x) = f_{\xi_1}(x) + \int_0^x h(x-s)f_\xi(s)ds \tag{2.15}$$

$$h(x) = f_{\xi_1}(x) + \int_0^x h_g(x-s)f_{\xi_1}(s)ds \tag{2.16}$$

$$h_g(x) = f_\xi(x) + \int_0^x h_g(x-s)f_\xi(s)ds \tag{2.17}$$

For solving the equations (2.15), we apply the Laplace transform and obtain

$$\mathcal{L}h(s) = \mathcal{L}f_{\xi_1}(s) + \mathcal{L}h(s) \cdot \mathcal{L}f_\xi(s) \quad \text{resp.} \quad \mathcal{L}h_g(s) = \mathcal{L}f_\xi(s) + \mathcal{L}h_g(s) \cdot \mathcal{L}f_\xi(s)$$

Solving the above equation, we get

$$\mathcal{L}h(s) = \frac{\mathcal{L}f_{\xi_1}(s)}{1 - \mathcal{L}f_\xi(s)} \quad \text{resp.} \quad \mathcal{L}h_g(s) = \frac{\mathcal{L}f_\xi(s)}{1 - \mathcal{L}f_\xi(s)}$$

Considering the relation between Laplace transform and the Laplace transform of the derivative, we can deduce

$$\mathcal{L}H(s) = \frac{\mathcal{L}f_{\xi_1}(s)}{s(1 - \mathcal{L}f_\xi(s))} \quad \text{resp.} \quad \mathcal{L}H_g(s) = \frac{\mathcal{L}f_\xi(s)}{s(1 - \mathcal{L}f_\xi(s))} \tag{2.18}$$

Using the Laplace transform we can determine the renewal function with the help of the inverse Laplace transform. Now we present some examples, which are important for the traffic theory, as the exponential distribution.

Example 2.10. Exponential distribution: Taking the density $f_\xi(t) = \lambda e^{-\lambda t}$, we compute for the Laplace transform

$$\mathcal{L}f_\xi(s) = \frac{\lambda}{s+\lambda}$$

and get according to (2.18)

$$\mathcal{L}H_g(s) = \frac{\frac{\lambda}{s+\lambda}}{s - \frac{\lambda s}{s+\lambda}} = \frac{\lambda}{s^2}$$

With the help of the inverse we can determine the renewal function

$$H_g(t) = \lambda t$$

Because of the uniqueness of the solution we can say that the renewal process is exactly exponential distributed, if the renewal function is linear.

Example 2.11. Erlang-k distribution: We consider interarrival times of serving times which are Erlang-k distributed for a parameter $\lambda > 0$. The renewal function can be computed again with the help of the Laplace transform. The Laplace transform of the Erlang-k distribution is

$$\mathcal{L}f_\xi(s) = \left(\frac{\lambda}{\lambda+s}\right)^k$$

Thus, it follows by (2.18):

$$\mathcal{L}H_g(s) = \frac{\mathcal{L}f_\xi(s)}{s(1-\mathcal{L}f_\xi(s))} = \frac{\left(\frac{\lambda}{\lambda+s}\right)^k}{s\left(1-\left(\frac{\lambda}{\lambda+s}\right)^k\right)}$$

Hence we get for the renewal function (see e.g. [43])

$$H_g(t) = e^{\lambda t} \sum_{n=1}^{\infty} \sum_{i=kn}^{\infty} \frac{(\lambda t)^i}{i!}$$

As example, we get for $k = 2$

$$H_g(t) = \frac{1}{2}\left(\lambda t - \frac{1}{2} + \frac{1}{2}e^{-2\lambda t}\right)$$

and for $k = 4$

$$H_g(t) = \frac{1}{4}\left(\lambda t - \frac{3}{2} + \frac{1}{2}e^{-\lambda t} + \sqrt{2}e^{-2\lambda t}\sin(\lambda t + \frac{\pi}{4})\right)$$

Example 2.12. Normal or Gaussian distribution: We cite the renewal function for the normal distribution $\mathcal{N}(\mu, \sigma^2)$ without proof

$$H_g(t) = \sum_{n=1}^{\infty} \Phi\left(\frac{t - n\mu}{\sigma\sqrt{n}}\right)$$

where Φ represents the distribution function of $\mathcal{N}(0,1)$.

Provided the first moment exists we want to cite an important formula for the renewal function. For this we set

$$f_I(t) = \frac{1}{\mu}(1 - F_\xi(t)) = \frac{1}{\mu} F_\xi^c(t)$$

resp.

$$F_I(t) = \frac{1}{\mu} \int_0^t F_\xi^c(s)\,ds \qquad (2.19)$$

This definition reflects the integrated complementary distribution function F_I, which wil be discussed in detail in section 2.7.4.

Theorem 2.13. *Let* $\mu = \mathbb{E}(\xi) = \int_0^\infty F_\xi^c(s)\,ds$ *be the expectation value. Then the renewal function H has exactly the form*

$$H(t) = \frac{t}{\mu}$$

if $f_{\xi_1}(s) = f_I(s)$ *resp.* $F_{\xi_1}(t) = F_I(t)$.

We state finally a useful relationship for the generating random variable $\hat{\xi}$, having F_I as distribution function. It is possible to derive that

$$\mathbb{E}(\hat{\xi}) = \frac{\mu^2 + \sigma^2}{2\mu} \quad \text{resp.} \quad \mathbb{E}(\hat{\xi}^2) = \frac{\mathbb{E}(\xi^3)}{3\mu} \qquad (2.20)$$

2.6.2 Bounds for the Renewal Function

Considering the theorem of Little, it is evident that one is interested in the knowledge of the expected value $\mathbb{E}(N(t))$, thus, in the computation of the renewal function. But as already described it is difficult or even impossible to indicate the exact form of the renewal function in full generality. Hence, one is often interested in estimations or asymptotic behavior. Let us begin with some elementary estimations for ordinary renewal processes (thus, we write $H(t)$ instead of $H_g(t)$).

Simple Approach

To determine $N(t)$, we have to estimate T_n

$$\sup_{1 \leq i \leq n} \xi_i \leq \sum_{i=1}^{n} \xi_i = T_n$$

Hence, by the iid-property

$$F_\xi^{*(n)}(t) = \mathbb{P}(T_n \leq t) \leq \mathbb{P}\left(\sup_{1 \leq i \leq n} \xi_i \leq t\right) = (F_\xi(t))^n$$

Summing up both sides over n and assuming that $F_\xi(t) < 1$ holds (geometric series on the right side!), we get

$$F_\xi(t) \leq H(t) \leq \frac{F_\xi(t)}{1 - F_\xi(t)} = \frac{F_\xi(t)}{F_\xi^c(t)}$$

Linear Approach

As exercise the reader may derive the following estimation: Defining F_I as in (2.19), we get with $\mu = \mathbb{E}(\xi)$, $\mathcal{F} = \{t \geq 0; F_\xi(t) < 1\}$ and $L_u = \inf_{t \in \mathcal{F}} \frac{F_\xi(t) - F_I(t)}{F_\xi^c(t)}$, $L_o = \sup_{t \in \mathcal{F}} \frac{F_\xi(t) - F_I(t)}{F_\xi^c(t)}$

$$\frac{t}{\mu} + L_u \leq H(t) \leq \frac{t}{\mu} + L_o$$

As we will learn in the section on subexponential distributions (see section 2.7.4), F_I is again a distribution function. Thus, we can build the corresponding complementary distribution function F_I^c. Defining

$$L(t) = \frac{f_I(t)}{F_I^c(t)}, \quad t \geq 0$$

it follows

$$L(t) = \frac{F_\xi^c(t)}{\int_t^\infty F_\xi^c(s) ds}$$

Hence, this gives the representations

$$L_u = \frac{1}{\mu} \inf_{t \in \mathcal{F}} \frac{1}{L(t)} - 1 \quad \text{and} \quad L_o = \frac{1}{\mu} \sup_{t \in \mathcal{F}} \frac{1}{L(t)} - 1$$

Finally, we have

$$\frac{t}{\mu} + \frac{1}{\mu} \inf_{t \in \mathcal{F}} \frac{1}{L(t)} - 1 \leq H(t) \leq \frac{t}{\mu} + \frac{1}{\mu} \sup_{t \in \mathcal{F}} \frac{1}{L(t)} - 1$$

2.6.3 Recurrence Time

We define two new processes, based on a particular renewal process. They represent a virtual and independent observer watching at two renewal time spots at the renewal process. We call

$$R_r(t) = t - T_{N(t)}$$

the *backward recurrence time* and

$$R_v(t) = T_{N(t)+1} - t$$

the *forward recurrence time*, each taken for $t \geq 0$. We indicate with R_r the time intervals as backward recurrence time, that is the time from the last event to the observing time and R_v the forward recurrence time, that is the time from the observer time until the next renewal event. With this we compute the particular distributions

$$F_{R_r(t)}(x) = \mathbb{P}(t - T_{N(t)} \leq x) = \mathbb{P}(t - x \leq T_{N(t)})$$
$$= \sum_{n=1}^{\infty} \mathbb{P}(t - x \leq T_n, N(t) = n)$$
$$= \sum_{n=1}^{\infty} \mathbb{P}(t - x \leq T_n \leq t < T_{n+1}) = \sum_{n=1}^{\infty} \int_{t-x}^{t} F_\xi^c(t-s) dF_{T_n}(s)$$
$$= \sum_{n=1}^{\infty} \int_{t-x}^{t} F_\xi^c(t-s) d\left(F_{\xi_1} \star F_\xi^{*(n-1)}\right)(s)$$
$$= \int_{t-x}^{t} F_\xi^c(t-s) \sum_{n=1}^{\infty} d\left(F_{\xi_1} \star F_\xi^{*(n-1)}\right)(s) = \int_{t-x}^{t} F_\xi^c(t-s) dH(s)$$

Thus, we get

$$F_{R_r(t)}(x) = \begin{cases} \int_{t-x}^{t} F_\xi^c(t-s) dH(s) & \text{for } 0 \leq x \leq t \\ 1 & \text{for } x > t \end{cases} \quad (2.21)$$

For the derivative and hence for the density follows

$$f_{R_r(t)}(x) = \begin{cases} F_\xi^c(x) h(t-x) & \text{for } 0 \leq x \leq t \\ 0 & \text{for } x > t \end{cases}$$

Analogously one can derive a representation of the distribution for the forward recurrence time

$$F_{R_v(t)}(x) = F_{\xi_1}(t+x) - \int_0^t F_\xi^c(t+x-s) dH(s) \quad (2.22)$$

$$f_{R_v(t)}(x) = f_{\xi_1} - \int_0^t f_\xi(t+x-s) h(s) ds$$

Example 2.14. We consider the Poisson process as renewal process. For a Poisson process the interarrival times are exponential distributed. Thus, we have for the density

$$f_\xi(x) = \lambda e^{-\lambda x}$$

and for the distribution function

$$F_\xi(t) = 1 - e^{-\lambda t}$$

As distribution of the recurrence time we get the density

$$f_R(t) = \lambda \int_t^\infty f_\xi(x)dx = \lambda(1 - F_\xi(t)) = \lambda e{-\lambda t} = f_\xi(t)$$

With this the identity in distribution

$$F_R(t) = F_\xi(t) \tag{2.23}$$

follows. This means that the remaining time period R until the next event, watched by the observer, possesses the same distribution as if she/he would watch the process at the renewal time spots. The process develops itself independently from the observed time spot, which again indicates the memoryless of the Poisson process. It possesses the Markov property.

As exercise the reader may explicitly compute the corresponding distributions $F_{R_r(t)}$ and $F_{R_v(t)}$. Again easily done for exercise, one can show that $N(t)+1$ is a stopping time (see for the notion of stopping time e.g. [208]) for the renewal process but not $N(t)$. We can apply the Wald identity (see e.g. [208]) and get

$$\mathbb{E}(R_v(t)) = \mathbb{E}(\xi_1) + \mu H(t) - t$$

If $H = H_g$, i.e. if (ξ_i) is an ordinary renewal process, we can simplify this to

$$\mathbb{E}(R_v(t)) = \mu(H_g(t) + 1) - t$$

In both cases and for further derivation let $\mu = \mathbb{E}(\xi)$. A representation for $\mathbb{E}(R_r(t))$ cannot be derived using the same arguments. The reader is invited to give a reasoning for this as exercise.

2.6.4 Asymptotic Behavior

As already mentioned above there is in general no satisfying representation of the renewal function, even using the Laplace transform. But, which is useful in a lot of cases, one can derive an asymptotic behavior. A first result in this direction is the so called *elementary renewal theorem*.

Theorem 2.15. *Let (ξ_i) be a renewal process. Then we have the asymptotic*

$$\lim_{t \to \infty} \frac{H(t)}{t} = \frac{1}{\mu}$$

where $\mu = \mathbb{E}(\xi)$, if the first moment exists. Otherwise set $\mu = \infty$.

2.6 Renewal Processes

The elementary renewal theorem states nothing else than that the expectation value of the arrival process, averaged over large times, behaves like the arrival rate, thus, $\frac{1}{\mathbb{E}(\xi)} = \frac{1}{\mu}$. To formulate the *fundamental renewal theorem*, we have to introduce the notion of an *arithmetic* renewal process.

Definition 2.16. *A renewal process (ξ_i) resp. its distribution function F_ξ is called* arithmetic, *if*

$$\sum_{i=0}^{\infty} \mathbb{P}(\xi_i = i \cdot a) = 1$$

for a suitable $a > 0$.

Expressed in other words, this means that a non-arithmetic renewal process does not live on a fixed grain (with distance $a > 0$). Equivalently, an arithmetic renewal process assumes values in a subset of $\mathbb{R}$, whose elements have a distance of an integer multiple of a fixed $a > 0$. Thus, we can formulate the fundamental renewal theorem.

Theorem 2.17. *Suppose F_ξ is not arithmetic and $g : [0, \infty[\longrightarrow \mathbb{R}$ is integrable. Then we have for an arbitrary initial distribution F_{ξ_1}*

$$\lim_{t \to \infty} \int_0^t g(t-s)dH(s) = \frac{1}{\mu} \int_0^\infty g(s)ds$$

As example we set for $u > 0$

$$g(s) = \begin{cases} 1 & \text{for } 0 \leq s \leq u \\ 0 & \text{else} \end{cases}$$

With theorem 2.17 we can deduce the following well-known renewal theorem of Blackwell.

Theorem 2.18. *If $h > 0$ and F_ξ not arithmetic, then independently of the initial distribution F_{ξ_1} we have*

$$\lim_{t \to \infty} (H(t+h) - H(t)) = \frac{h}{\mu}$$

The elementary renewal theorem shows an asymptotic 'swing in' into a stationary state. The result of Blackwell describes on the other hand a local behavior, which depends on the step width $h > 0$.
Assuming the existence of the variance of the renewal theorem, we can state another asymptotic.

Proposition 2.19. *Let F_ξ be not arithmetic and $\sigma^2 = \mathbb{V}ar(\xi) < \infty$. then it follows*

$$\lim_{t \to \infty} \left(H(t) - \frac{t}{\mu} \right) = \frac{\sigma^2}{2\mu^2} - \frac{\mathbb{E}(\xi_1)}{\mu} + \frac{1}{2}$$

Let us remember the distribution of the forward and backward recurrence times (2.22) and (2.21). Then we can e.g. deduce with (2.22) and the fundamental renewal theorem (for fixed $x \geq 0$ and $g(t) = F_\xi^c(t+x)$)

$$\lim_{t \to \infty} F_{R_v(t)}(x) = \lim_{t \to \infty} F_1(t+x) - \frac{1}{\mu} \int_0^\infty F_\xi^c(x+s)ds = 1 - \frac{1}{\mu} \int_x^\infty F_\xi^c(t)dt$$

We have because of $F_{I,\xi}^c(x) = \frac{1}{\mu} \int_x^\infty F_\xi^c(t)dt$

$$\lim_{t \to \infty} F_{R_v(t)}(x) = F_{I,\xi}(x) \tag{2.24}$$

Analogously it follows

$$\lim_{t \to \infty} F_{R_r(t)}(x) = F_{I,\xi}(x)$$

One would expect because of the symmetry of the forward resp. backward recurrence times that building the expected value we obtain

$$\lim_{t \to \infty} \mathbb{E}(R_v(t)) = \frac{\mu}{2}$$

This could be interpreted in the way that in the mean the observation of an arrival process (e.g. the call attempts) lies exactly in the middle of the interarrival time interval. In fact, one has because of (2.24) and (2.20)

$$\lim_{t \to \infty} \mathbb{E}(R_v(t)) = \mathbb{E}(\hat{\xi}) = \frac{\mu^2 + \sigma^2}{2\mu} > \frac{\mu}{2}$$

if $\sigma^2 > 0$. This is called the *paradox of the renewal theory*. A reasoning for this can be find in the fact that it is more probable, if an observation time spot lies in an interarrival time of longer distance. We can deduce similar to the renewal theorem resp. proposition 2.19

$$\lim_{t \to \infty} h(t) = \frac{t}{\mu}$$

Assuming the existence of the first moments $\mathbb{E}(\xi_1)$ and $\mu = \mathbb{E}(\xi)$, we can conclude the 'central limit theorem' for renewal count processes.

Theorem 2.20. *Let $N(t)$ be asymptotically Gaussian distributed with an expected value of $\frac{t}{\mu}$ and the variance $\frac{\sigma^2 t}{\mu^2}$. Thus, we have*

$$\lim_{t \to \infty} \mathbb{P}\left(\frac{N(t) - \frac{t}{\mu}}{\sigma\sqrt{t\mu^{-3}}} \leq x\right) = \Phi(x)$$

where Φ is the Gaussian distribution $\mathcal{N}(0,1)$.

2.6.5 Stationary Renewal Processes

The notion of stationarity is, as in the case of stochastic processes, central for the treatment of the renewal theory. For the definition we pick a renewal process (ξ_i) and switch to the equivalent process of the forward recurrence times $(R_v(t))$.

Definition 2.21. *A renewal process (ξ_i) is called* stationary, *if the process of the forward recurrence times $(R_v(t))$ is stationary in the stricter sense.*

This notion could be equivalently introduced using the backward recurrence times or the corresponding count process $N(t)$. We can define the stationarity, since the forward recurrence times are Markovian, using

$$F_{R_v(t)}(x) = F_v(x), \text{ for all } x \geq 0$$

The distribution of the forward recurrence time is independent of the time spot t. We remember that the expected value $\mu = \mathbb{E}(\xi)$ can be written in the form $\mu = \int_0^\infty F_\xi^c(x)dx$ according to (2.34).

Theorem 2.22. *If F_ξ is not arithmetic and if ξ has a first moment μ, then the renewal process is stationary if and only if*

$$H(t) = \frac{t}{\mu}$$

Remark 2.23. According to theorem 2.13 the random variable (ξ_i) is exactly stationary if

$$F_1(x) = F_I(x) = \frac{1}{\mu}\int_0^x F^c(t)dt$$

2.6.6 Random Sum Processes

We will return to random sums in section 2.7.4 on subexponential distributions. Thus, we give a short introduction for renewal processes. For this we define first an aggregated process.

Definition 2.24. *Let $N(t)$ be a counting process with $T_i = \inf\{t \geq 0, N(t) = i\}$ and (η_j) a sequence of random variables. We call*

$$S(t) = \sum_{i=1}^{N(t)} \eta_i, \ t \geq 0$$

an aggregated process *or* sum process.

This definition will be applied to renewal processes resp. to traffic theory. For this we pick for an ordinary renewal process (ξ_i) the corresponding counting process $N(t)$ and a sequence of iid random variables (η_j). Here, the ξ_i and η_j are assumed to be stochastic independent for $i \neq j$. Simultaneously, we assume for ξ, as for η, the existence of the first and second moments.

Example 2.25. As known, one can consider the (ξ_i) as interarrival times, e.g. the time between two succeeding calls $i-1$ and i. Then the η_i represents the serving time (or the i-th call duration) to the i-th demand and thus, cannot be assumed independently to the ξ_i, but to the other interarrival times. We can consider the amount $S(t) = \sum_{j=1}^{N(t)} \eta_j$ as overall load in the system, where the $N(t)$ represents the corresponding counting process to the (ξ_i).

We want to investigate the expectation value. A simple computation shows

$$S(t) = \sum_{j=1}^{N(t)} \eta_j = \sum_{j=1}^{N(t)+1} \eta_j - \eta_{N(t)+1}$$

Applying the expectation operator we can use the Wald identity on the right side (see e.g. [208] for the Wald identity), since $N(t)+1$ is a stopping time. Thus it follows

$$\mathbb{E}(S(t)) = \mathbb{E}(\eta)(\mathbb{E}(N(t)) + 1) - \mathbb{E}(\eta_{N(t)+1}) = \mathbb{E}(\eta)(H_g(t) + 1) - \mathbb{E}(\eta)$$
$$= \mathbb{E}(\eta) H_g(t)$$

Defining $m(t) = \mathbb{E}(X(t))$ as the expected value of the stochastic process at time t, we deduce the following proposition.

Proposition 2.26. *Let (ξ_i) be an ordinary renewal process, $N(t)$ the corresponding counting process and (η_j) a sequence of iid random variables according to our model. Then we have*

$$m(t) = \mathbb{E}(\eta) H(t)$$

The elementary renewal theorem allows to formulate the following asymptotic fact.

Proposition 2.27. *Let (ξ_i) be an ordinary renewal process, $N(t)$ be the corresponding counting process and (η_j) be a sequence of iid random variables with an expectation value $\nu = \mathbb{E}(\eta)$. Then we have*

$$\lim_{t \to \infty} \frac{m(t)}{t} = \frac{\nu}{\mu} \qquad (2.25)$$

We can consider (2.25) as load in the system, whereas $\frac{1}{\mu}$ is the arrival rate and ν the serving rate. As we will see later, when considering the aggregated traffic for subexponential or heavy-tail distributions, we can investigate in particular a convergence in distribution, i.e. the derivation of the classical central limit theorem.

Theorem 2.28. *If*

$$\sigma^2 = \mathbb{V}ar(\mu\eta - \nu\xi) > 0 \qquad (2.26)$$

then it follows

$$\frac{X(t) - \frac{\nu}{\mu}t}{\mu^{-\frac{3}{2}}\sigma\sqrt{t}} \xrightarrow{d} \Phi$$

i.e., the convergence in distribution to the Gaussian distribution.

We can reformulate the equation (2.26) in the form

$$\sigma^2 = \mu^2 \mathrm{Var}(\eta) + \nu^2 \mathrm{Var}(\eta)$$

If the relation (2.26) is satisfied, then the convergence is expressed in the form

$$X(t) \xrightarrow{d} \mathcal{N}\left(\frac{\nu}{\mu}t, \mu^{-3}\sigma^2 t\right)$$

2.7 General Poisson Arrival and Serving Systems M/G/n

2.7.1 Markov Chains and Embedded Systems

If we assume, as necessary for modeling the IP-based traffic, that the inter-arrival times or serving times are no longer Markovian, we have to use more general methods to derive the key values of the traffic. This leads to the method of the *embedded Markov chains*.

We start with a state discrete process $(X(t))_{t \in I}$. For certain time spots (t_n), $n = 0, 1, \ldots$ we assume that the process enjoys the Markov property. Hence

$$\mathbb{P}(X_{t_{n+1}} = x_{n+1} \mid X_{t_n} = x_n, \ldots, X_{t_0} = x_0) = \mathbb{P}(X_{t_{n+1}} = x_{n+1} \mid X_{t_n} = x_n)$$

for the time spots $t_0 < t_1 < \ldots < t_n < t_{n+1}$. The whole development after the time spot t_n is described by the state X_{t_n} and can be computed. The sequence (X_{t_n}) is called a *Markov chain*. The method considering just the sequence (X_{t_n}) and not the whole process (X_t) is called the method of the *embedded Markov chain*. We define as above

$$x(i, n) = \mathbb{P}(X_{t_n} = i)$$

and $\mathcal{X}_n = (x(i, n); i = 0, 1, \ldots)$. The transition matrix $P = (p_{ij})$ with

$$p_{ij} = \mathbb{P}(X_{t_{n+1}} = j \mid X_{t_n} = i), \quad i, j = 0, 1, 2, \ldots$$

gives us

$$\mathcal{X}_{n+1} = \mathcal{X}_n \mathcal{P}$$

The matrix $\mathcal{P}$ is called a *stochastic matrix*, i.e., $\mathcal{P}e = e$ where $e = (1, 1, \ldots)$. $\mathcal{P}$ is different to the transition probability matrix, which was used for the treatment of time continuous processes.

2.7.2 General Loss Systems M/G/n

Description of the Model

Since we are interested in waiting queue systems, we just indicate the assumption of the pure loss model. Here we assume:

- Poisson arrival process: Rate λ (birth rate), i.e., the interarrival times ξ are exponential distributed. Thus, the distribution and the expectation value of the interarrival times are $F_\xi(t) = 1 - e^{-\lambda t}$, and $\mathbb{E}(\xi) = \frac{1}{\lambda}$.
- There are n serving places.
- The serving time is assumed general distributed.
- We have for the mean serving time $\mathbb{E}(\eta) = \mu$.
- The load of the system turns out to be $\lambda \cdot \mu$ according to the Little theorem.
- The relative load per serving place is $\frac{\lambda \mu}{n}$.

Erlang Formula

If we define $a = \frac{\lambda}{\mu}$, then the formula (2.4) and Erlang formula (2.5) hold respectively. This is shown e.g. in [238].

Consequences

Since this type of traffic model does not have any influence on the self-similar and long-range dependent processes (see e.g. section 3.3), which are needed for modeling the IP-based traffic, we only cite [20] for the interested reader as reference for further details. The key values of the M/G/n − 0 system are widely influenced by the Poisson arrival process but not the general serving distributions, as e.g. the heavy-tail distributions used in the IP-based data transfer. The Erlang-B formula is valid, respectively (here we have $\alpha = \lambda \cdot \mu$).

2.7.3 Queueing Systems M/G/n

Description of the Model

The waiting queue model M/G/1 is characterized by the following assumption:

- Poisson arrival process (call attempts): Rate λ (birth rate), i.e., the interarrival times ξ are exponential distributed

$$F_\xi(t) = 1 - e^{-\lambda t} \quad \text{and} \quad \mathbb{E}(\xi) = \frac{1}{\lambda}$$

- We assume one serving place.
- The serving time η is assumed to be general distributed with expected value $\mathbb{E}(\eta) = \mu$.

2.7 General Poisson Arrival and Serving Systems M/G/n

- We have an infinite waiting room with queueing principles FIFO, FCFS, LIFO or RANDOM.
- The load of the system is according to Little formula $\lambda \cdot \mu$.

We denote by λ the arrival rate and by μ the mean serving time. Thus, $\rho = \lambda\mu = \frac{E(\eta)}{E(\xi)}$ indicates the mean load.

Embedded Time Spots

Here, we use the principle of the embedded Markov times. Since the serving time process is the only process, which does not enjoy the Markov property, the process is memoryless at the end of each serving time. Thus, we choose for the embedded time spots the times of the single serving times. Hence, we denote by $(X(t_0)X(t_1), \ldots, X(t_n), X(t_{n+1}), \ldots)$ the system states, whereas t_n is the time of the n-th serving end (fig. 2.20).

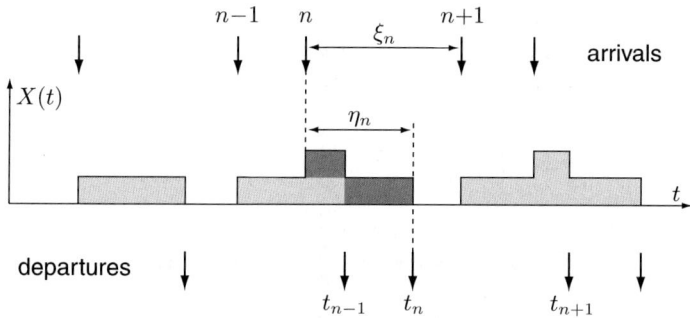

Fig. 2.20. State process for the queueing system M/G/1.

We introduce the random variable ζ. It describes the amount of arrival events during a single serving time η. The distribution is

$$F_\zeta(i) = \mathbb{P}(\zeta = i)$$

As generating function we get

$$\zeta_{\mathrm{EF}}(z) = \sum_{i=0}^{\infty} F_\zeta(i) z^i$$

Hence, it follows

$$\mathbb{E}(\zeta) = \left.\frac{d\zeta_{\mathrm{EF}}(z)}{dz}\right|_{z=1} = \lambda \mathbb{E}(\eta) = \rho$$

State Transitions

For analyzing the state transitions we need the transition probabilities

$$p_{ij} = \mathbb{P}(X_{t_{n+1}} = j \mid X_{t_n} = i)$$

We have to consider several cases for the investigation of the time period between two succeeding serving endings, i.e the interval $]t_n, t_{n+1}[$:

- $i = 0$: the system is empty at time t_n. If a demand enters the system, the serving process starts immediately. To be able to observe j demands in the system at time t_{n+1}, in the meantime, j demands have to enter the system. Hence,

$$p_{0j} = F_\zeta(j), \; j = 0, \ldots$$

- $i > 0$: the system is not empty at time t_n. The process of the state is given according to: At t_n we have i demands in the system. Immediately afterwards the serving process starts. At t_{n+1} we still observe j demands in the system. This means that during one serving period $(j-i+1)$ demands entered the system. Thus,

$$p_{ij} = F_\zeta(j - i + 1), \; i = 1, \ldots, \; j = i-1, i, \ldots$$

The state transition matrix $\mathcal{P}$ looks as follows:

$$\mathcal{P} = \begin{bmatrix} F_\zeta(0) & F_\zeta(1) & F_\zeta(2) & F_\zeta(3) & \cdots \\ F_\zeta(0) & F_\zeta(1) & F_\zeta(2) & F_\zeta(3) & \cdots \\ 0 & F_\zeta(0) & F_\zeta(1) & F_\zeta(2) & \cdots \\ 0 & 0 & F_\zeta(0) & F_\zeta(1) & \cdots \\ \vdots & \vdots & \vdots & \vdots & \ddots \end{bmatrix}$$

State Equations

To start with the analysis, we fix the state transition probabilities

$$x(j, n) = \mathbb{P}(X(t_n) = j), \; j = 0, 1, \ldots$$

We write this infinite vector in the form

$$\mathcal{X}_n = (x(0, n), x(1, n), \ldots, x(j, n), \ldots)$$

and thus, we can formulate the general state transition equation in a matrix writing

$$\mathcal{X}_n \mathcal{P} = \mathcal{X}_{n+1}$$

From the start vector $\mathcal{X}_n$ we can successively compute all state vectors $\mathcal{X}_m$, $m > n$. This is of importance for the analysis of instationary processes, as observed in communication networks while overload or swing in phases. Here as well, we can consider stationary states and obtain them according to

2.7 General Poisson Arrival and Serving Systems M/G/n

$$\mathcal{X}_n = \mathcal{X}_{n+1} = \ldots = \mathcal{X}$$

where $\mathcal{X} = (x(0), x(1), \ldots)$, and the stationary state transition equation reads as

$$\mathcal{X}P = \mathcal{X}$$

We see that a stationary state is obtained by using e.g. the estimation of the form

$$|\mathbb{E}(X(t_{n+1})) - \mathbb{E}(X(t_n))| < \epsilon = 10^{-8}$$

We want to gain an expression for the state probability. For this we use the method of the generating function. With the help of the state transition equation

$$x(j) = x(0)F_\zeta(j) + \sum_{i=1}^{j+1} x(i)F_\zeta(j-i+1), \; j = 0, 1, \ldots \quad (2.27)$$

we get using the concept of generating function

$$X_{\text{EF}}(z) = \sum_{j=0}^{\infty} x(j)z^j = x(0)\sum_{j=0}^{\infty} F_\zeta(j)z^j + \sum_{j=0}^{\infty}\sum_{i=1}^{j+1} x(i)F_\zeta(j-i+1)z^j \quad (2.28)$$

With the help of the identity

$$\zeta_{\text{EF}}(z) = \sum_{j=0}^{\infty} F_\zeta(j)z^j$$

and $\zeta_{\text{EF}}(z) = \sum_{j^*=0}^{\infty} F_\zeta(j^*)z^{j^*}$ we can rearrange the double sum into

$$\sum_{j=0}^{\infty}\sum_{i=1}^{j+1} x(i)F_\zeta(j-i+1)z^j = \sum_{i=1}^{\infty} x(i) \sum_{j=0}^{\infty} F_\zeta(j-i+1)z^j$$

$$\stackrel{j^*=j-i+1}{=} \sum_{i=1}^{\infty} x(i) \sum_{j^*=0}^{\infty} F_\zeta(j^*)z^{j^*} z^{i-1}$$

$$= \sum_{i=1}^{\infty} x(i)\zeta_{\text{EF}}(z)z^{i-1}$$

$$= \frac{1}{z}\zeta_{\text{EF}}(z)(X_{\text{EF}}(z) - x(0))$$

Inserting this expression into the double sum of (2.28), it finally follows

$$X_{\text{EF}}(z) = x(0)\zeta_{\text{EF}}(z) + \frac{\zeta_{\text{EF}}(z)}{z}(X_{\text{EF}}(z) - x(0))$$

resp.

$$X_{\text{EF}}(z) = x(0)\frac{\zeta_{\text{EF}}(z)(1-z)}{\zeta_{\text{EF}}(z) - z}$$

For a determination of $x(0)$ we choose the limit $z \to 1$ in the last equation. In case of $\lim_{z \to 1} X_{\mathrm{EF}}(z) = \frac{0}{0}$ we use the rule of l'Hospital

$$1 = \sum_{j=0}^{\infty} x(j) = \lim_{z \to 1} X_{\mathrm{EF}}(z)$$

$$= x(0) \lim_{z \to 1} \left(\frac{\frac{d}{dz}\zeta_{\mathrm{EF}}(z)(1-z) - \zeta_{\mathrm{EF}}(z)}{\frac{d}{dz}\zeta_{\mathrm{EF}}(z) - 1} \right)$$

$$= x(0) \frac{-1}{\lim_{z \to 1} \frac{d}{dz}\zeta_{\mathrm{EF}}(z) - 1}$$

Here, we applied

$$\lim_{z \to 1} \frac{d}{dz}\zeta_{\mathrm{EF}}(z) = \lim_{z \to 1} \sum_{j=0}^{\infty} F_\zeta(j) j z^{j-1} = \mathbb{E}(\zeta) = \lambda \mathbb{E}(\eta) = \rho$$

thus,

$$x(0) = 1 - \rho \quad \text{and} \quad X_{\mathrm{EF}}(z) = (1-\rho)(1-z)\frac{\zeta_{\mathrm{EF}}(z)}{\zeta_{\mathrm{EF}}(z) - z}$$

The generating function $\zeta_{\mathrm{EF}}(z)$ represents the Poisson arrivals during one particular serving period, thus, during an interval with distribution function $F(t)$ and corresponding Laplace transform $\Phi_\eta(s)$. Hence, it follows for the generating function

$$\zeta_{\mathrm{EF}}(z) = \sum_{j=0}^{\infty} F_\zeta(j) z^j = \sum_{j=0}^{\infty} \left(\int_0^\infty \frac{(\lambda t)^j}{j!} e^{-\lambda t} f_\eta(t) dt \right) z^j$$

$$= \Phi_\eta(s)|_{s=\lambda(1-z)} = \Phi_\eta(\lambda(1-z))$$

For the generating function of X we obtain

$$X_{\mathrm{EF}}(z) = (1-\rho)(1-z)\frac{\Phi_\eta(\lambda(1-z))}{\Phi_\eta(\lambda(1-z)) - z}$$

Thus, we derived the *Pollaczek-Khintchine formula* for state probabilities.

Queueing Distribution Functions

Demands could consist e.g. in 'clicks' for loading of certain contents of World Wide Web pages. The demands are directed into a virtual waiting room according to a queueing principle of types as FIFO or FCFS. As in the system M/M/n we start with a test demand and denote by D the random variable describing the time, used by this unit to be guided through the system. With f_D we mean its density function, and for the distribution function we write F_D. Then, we have

2.7 General Poisson Arrival and Serving Systems M/G/n

$$D = W + \eta$$
$$f_D = w(t) * f_\eta(t)$$

and for the Laplace transform accordingly

$$\Phi_D(s) = \Phi_W(s) \cdot \Phi_\eta(s)$$

We can represent the sojourn time in the system with the help of figure 2.21.

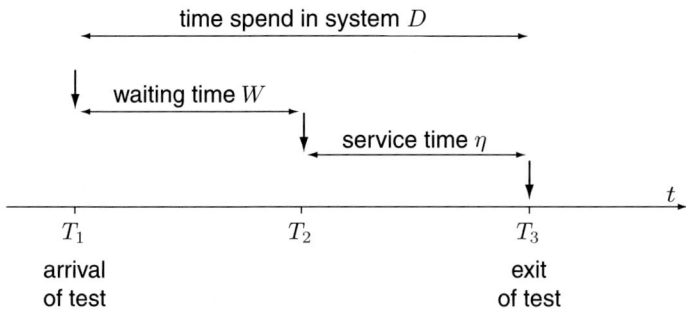

Fig. 2.21. Sojourn time in the system

We assume that at the time, when the test unit exits the system, there are still $X = m$ demands in the system left. Hence, it is $X = m$ the system state at the time of a Markov chain spot. Because of the FIFO-principle these demands entered the system during the time, the test unit spend in the system D. Thus,

$$x(m) = \mathbb{P}(\text{test unit leaves } m \text{ demands in the system})$$
$$= \mathbb{P}(m \text{ arrival during the time spend in the system } D)$$

Hence, it follows $X = m$ as the amount of Poisson arrivals during D, a time interval with distribution function $F_D(t)$

$$X_{EF}(z) = \sum_{j=0}^{\infty} x(j) z^j = \sum_{j=0}^{\infty} \left(\int_0^\infty \frac{(\lambda t)^j}{j!} e^{-\lambda t} f_D(t) dt \right) z^j$$
$$= \Phi_D(s)|_{s=\lambda(1-z)} = \Phi_D(\lambda(1-z))$$

Inserting this into the Pollaczek-Khintchine formula for state probabilities, it follows

$$\Phi_D(\lambda(1-z)) = (1-\rho)(1-z) \frac{\Phi_\eta(\lambda(1-z))}{\Phi_\eta(\lambda(1-z)) - z}$$

or, if we set $s = \lambda(1-z)$,

$$\Phi_D(s) = \Phi_\eta(s) \frac{s(1-\rho)}{s - \lambda + \lambda \Phi_\eta(s)}$$

Using the product representation of the Laplace transform of $\Phi_D(s)$ and solving for $\Phi_W(s)$, we finally obtain the *Pollaczek-Khintchine formula for waiting time distribution functions*

$$\Phi_W(s) = \frac{s(1-\rho)}{s - \lambda + \lambda \Phi_\eta(s)}$$

To gain the desired distribution function, we apply the inverse Laplace transform. In tables one can, depending on the case, find the corresponding distribution function.

Waiting Time Probabilities

The distribution of the waiting time will be computed using Laplace transform

$$F_W(t) \longleftrightarrow \frac{\Phi_W(s)}{s}$$

The probability not to wait, thus, $\mathbb{P}(W = 0)$, can be written according to the right side continuity of the distribution function

$$\mathbb{P}(W = 0) = \lim_{t \to 0} F_W(t) = \lim_{s \to \infty} \frac{s\Phi_W(s)}{s} = 1 - \rho \qquad (2.29)$$

For the above representation in the form of Laplace transform we applied the limit theorem for the Laplace transforms and, in the last equation, the Pollaczek-Khintchine formula for waiting time distribution functions.
Thus, we obtain with the help of the complementary distribution function the queueing probability

$$p_W = \mathbb{P}(W > 0) = 1 - \mathbb{P}(W = 0) = \lim_{t \to 0}(1 - F_W(t)) = \rho$$

Mean Waiting Time

As in the pure Markov model we divide into the mean waiting time $\mathbb{E}(W)$ of all demands and the mean waiting time of the demands in the waiting queue $\mathbb{E}(W_1)$. They will be again computed with the help of the Pollaczek-Khintchine formula for waiting time distribution functions

$$\mathbb{E}(W) = \frac{\mathbb{E}(\eta)\rho(1 + c_\eta^2)}{2(1-\rho)} = \frac{\lambda \mathbb{E}(\eta^2)}{2(1-\rho)}$$

and for mean waiting time of the demands in the queue

$$\mathbb{E}(W_1) = \frac{\mathbb{E}(W)}{p_W} = \mathbb{E}(\eta)\frac{1 + c_\eta^2}{2(1-\rho)}$$

Here, the amount $c_\eta = \frac{\sqrt{\text{Var}(\eta)}}{\mu}$ represents the variational coefficient of η. But in several cases this formula will not be very helpful for determine the mean waiting time in the Internet traffic, since the serving resp. the connection time do not have a second moment, will say, a finite variance and we will have to refer to other methods (see section 3). A relationship between the second moment of the serving time and the first moment of the waiting time will be established e.g. in section 3.3.4.

According to the above mentioned formulas, we realize that the mean waiting time is only depending on the first two moments of the serving time. The mean waiting time depends on the load of the system.

For an interpretation we remark that the expected waiting time is proportional to the factor $(1 + c_\eta^2)$. This means that a high variance of the serving time in the server implies a higher waiting time.

2.7.4 Heavy-Tail Serving Time Distribution

Introduction to Subexponential Distributions

For further results and investigations we need the introduction to the class of subexponential and in particular heavy-tail distributions. They enjoy properties, which stand mainly in contrast to those of the exponential distribution, which is, as we know, widely used in the classical traffic theory, especially for the description of the circuit switched networks. As we will point out below, the subexponential distributions fit perfectly in modeling the IP-based traffic. Let F_ξ be a distribution function of a random variable ZV ξ. Let $\xi_1, \ldots, \xi_n$ be further stochastic independent random variables, identical distributed as ξ. We denote by F^{c*n} the complementary distribution function of the sum variable $S_n = \xi_1 + \ldots + \xi_n$. This means

$$F^{c*n}(t) = P(S_n > t)$$

The relation between the sum and the maximum of a sequence of iid random variables gives rise to the following definition.

Definition 2.29. *The distribution F is called subexponential (denoted by $F \in S$) if*

$$\lim_{x \to \infty} \frac{F^{c*n}(x)}{F^c(x)} = n \qquad (2.30)$$

The following result is remarkable in this context and gives a good interpretation for the behavior of IP-based traffic:

> For the maximum $M_n = \max(\xi_1, \ldots, \xi_n)$ (which is again a random variable) of n identical, stochastic independent and subexponential distributed random variables holds
>
> $$\mathbb{P}(M_n > x) \sim \mathbb{P}(S_n > x)$$
>
> asymptotically for large x, if $t \to \infty$.

Exactly this result shows that for subexponential distributed events not each single event, but the *extreme* event determines the behavior. Suppose the connection times in the IP-based traffic are subexponential distributed, then not the major part of the connections determines the behavior of the network, but those of long durations.

Many of the known distributions F_ξ used in the classical theory satisfy the property

$$\limsup_{t\to\infty} e^{\nu t} F_\xi^c(t) < \infty \tag{2.31}$$

Here, ξ is a random variable distributed according to F_ξ. This means that if the expected value exists then

$$\text{there is a } \nu > 0, \text{ with } F_\xi^c(t) \leq e^{-\nu t}\mathbb{E}(\nu\xi),\ t \geq 0$$

This shows again that for those distributions extreme values are not very likely. In (2.31) we used the 'limes superior' of a function $f(t)$

$$\limsup_{t\to\infty} f(t) = \lim_{t\to\infty} \sup\{f(s); s \geq t\}$$

Equivalently one may introduce the 'limes inferior' $\liminf_{t\to\infty} f(t)$. We give some examples of the fast decaying distributions.

Example 2.30. Exponential distribution (fig. 2.22): $F_\xi^c(t) = e^{-\lambda t}$, $\lambda > 0$. This distribution characterizes, as we know, the positive random variables, which are memoryless, i.e.,

$$\mathbb{P}(\xi > t + s) = \mathbb{P}(\xi > s)\mathbb{P}(\xi > t)$$

If one detects by statistical methods that the sample does not behave like a exponential distribution (resp. no Poissson process), then consequently one does not have a memoryless random variable (resp. a Markov process) according to theorem 2.1.

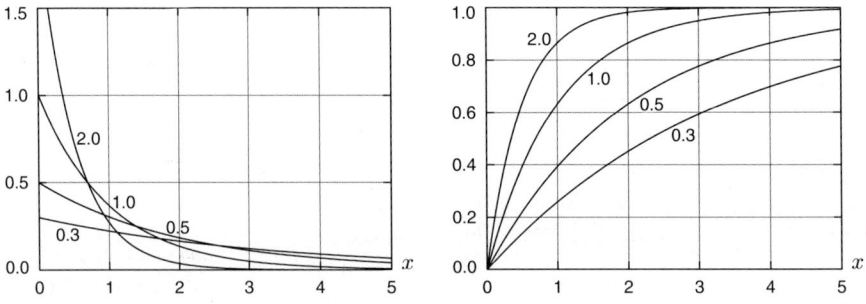

Fig. 2.22. Probability density function (pdf, left) and cumulative distribution function (cdf, right) of the exponential distribution for different λ

2.7 General Poisson Arrival and Serving Systems M/G/n

Example 2.31. Beta distribution with $a > 0$ and $b > 0$ (fig. 2.23):

$$f(x) = \begin{cases} \frac{x^{a-1}(1-x)^{b-1}}{B(a,b)} & \text{for } 0 < x < 1 \\ 0 & \text{else} \end{cases}$$

Here $B(a,b)$ is the so called *beta function*

$$B(a,b) = \frac{\Gamma(a)\Gamma(b)}{\Gamma(a+b)}$$

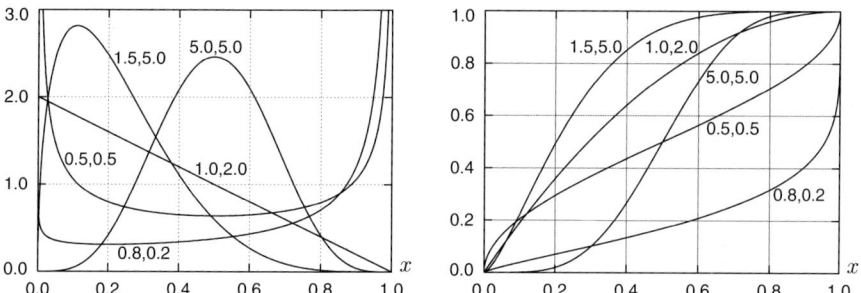

Fig. 2.23. Probability density function (pdf, left) and cumulative distribution function (cdf, right) of the beta distribution for different a and b

Example 2.32. Weibull distribution with $c > 0$ and $\tau \geq 1$ (fig. 2.24):

$$F_\xi^c(x) = e^{cx^\tau}$$

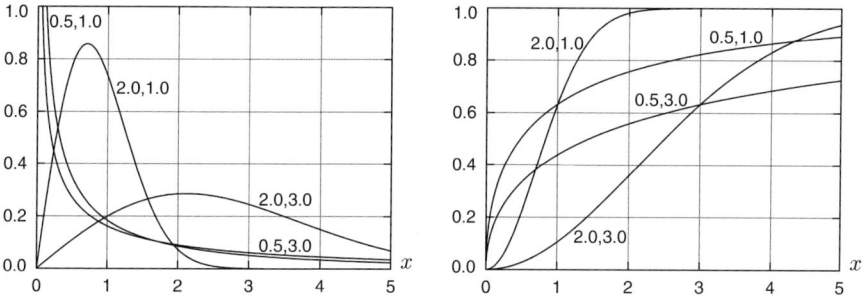

Fig. 2.24. Probability density function (pdf, left) and cumulative distribution function (cdf, right) of the Weibull distribution for different τ and β

Example 2.33. Normal distribution on $[0, 1[$ (fig. 2.25):

$$f_\xi(x) = \sqrt{\frac{2}{\pi}} e^{-\frac{x^2}{2}}$$

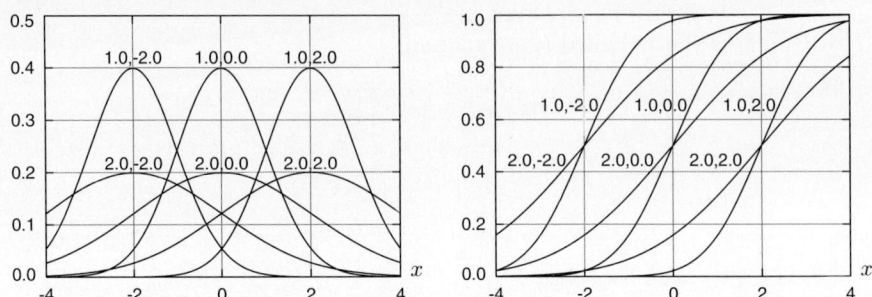

Fig. 2.25. Probability density function (pdf, left) and cumulative distribution function (cdf, right) of the normal distribution for different μ and σ^2

Example 2.34. Every distribution with compact support.

In contrast to the above distributions, we encounter certain different properties of the subexponential distributions or its subclass the *heavy-tail distributions*. The reasoning for the notion of subexponential is given by the following fact:

If a distribution F is subexponential, i.e., $F \in \mathcal{S}$, then we have for all $\epsilon > 0$
$$e^{\epsilon x} F^c(x) \to \infty, \text{ for } x \to \infty$$

This implies that the complementary distribution function (also called 'tail') does not decay as fast as every positive power of the exponential function grows. Thus, often distribution functions with a special behavior are used.

Definition 2.35. *We say that the distribution function F is* heavy-tailed *if*
$$F_\xi^c(x) \sim x^{-\alpha} L(x), \text{ for } x \to \infty, \ \alpha \in \ ]0,2] \tag{2.32}$$

where $L \in \mathcal{R}_0$ is a slowly varying function, *that means*
$$\lim_{x \to \infty} \frac{L(tx)}{L(x)} = 1, \text{ for all } t > 0$$

Those slowly varying functions are e.g. the constant functions or $\log(x)$. We can say that how much the scaling is changed, larger or smaller, in the limit of infinity, they all behave the same. This is central for the behavior of the complementary distribution functions. Substituting the number '1' in the above limit by the function 't^p', we can analogously define the class $\mathcal{R}_p$:

2.7 General Poisson Arrival and Serving Systems M/G/n

A function L on $[0,\infty[$ belongs to the class $\mathcal{R}_p$, $p \in \mathbb{R}$, if we have for all $t > 0$
$$\lim_{x \to \infty} \frac{L(tx)}{L(x)} = t^p$$

Example 2.36. We give some examples:

- Every constant functions belongs to the class $\mathcal{R}_0$.
- For all $p \in \mathbb{R}$ we find the following functions in the class $\mathcal{R}_p$
$$x^p, \ x^p \log(1+x), \ (x \log(1+x))^p, \ x^p \log(\log(e+x))$$
  The functions
  $$2 + \cos x, \ \sin(\log(1+x))$$
  do not belong to any class $\mathcal{R}_p$.
- It can happen that a function in these classes oscillates heavily
$$\liminf_{x \to \infty} L(x) = 0 \quad \text{and} \quad \limsup_{x \to \infty} L(x) = \infty$$
  An example is
  $$L(x) = \exp\left(\log(1+x)^{\frac{1}{2}} \cos\left(\log(1+x)^{\frac{1}{2}}\right)\right)$$

The so called 'heavy-tail' distributions play a central rôle in the IP traffic theory, we will treat them later more intensively. First, we will sketch some interesting properties and relationships of the heavy-tail distributions, while examples of distributions being heavy-tail, will be given below later.

Property 2.37. We indicate two remarkable properties of distributions $F \in \mathcal{S}$:

- If $F \in \mathcal{S}$, then we have
$$\lim_{x \to \infty} \frac{F^c(x-y)}{F^c(x)} = 1$$
  uniformly for y in bounded subsets of $]0, \infty[$.
- By the property of subexponential distributions $F \in \mathcal{S}$, we can obtain for every $\epsilon > 0$ and $y \geq 0$:
$$\int_y^\infty e^{\epsilon x} f(x) dx \geq e^{\epsilon y} F^c(y), \text{ with } f \text{ density of } F$$

We deduce briefly the inequality
$$\int_y^\infty e^{\epsilon x} f(x) dx \geq e^{\epsilon y} \int_y^\infty f(x) dx$$
$$= e^{\epsilon y}\left(1 - \int_{-\infty}^y f(x)dx\right)$$
$$= e^{\epsilon y}(1 - F(y)) = e^{\epsilon y} F^c(y)$$

This means for the Laplace transform of f:

If $F \in \mathcal{S}$, then we have $\mathcal{L}(f)(-\epsilon) = \infty$, for all $\epsilon > 0$ \hfill (2.33)

In other words the Laplace transform has a *singularity of first order* at 0.

We want to sketch a first relationship to the arrival processes in the IP-based telecommunication. For this we describe shortly the model. We denote by ξ_k the independent and identical distributed data amounts (ξ_k for the k-th connection). The distribution of the until time t summed up data amount $S(t) = \sum_{k=1}^{N(t)} \xi_k$ is

$$G_t(x) = \mathbb{P}(S(t) \le x) = \sum_{k=0}^{\infty} e^{-\mu t} \frac{(\mu t)^k}{k!} F_{\xi_1}^{k*}(x)$$

Here, the arrival process is a homogeneous Poisson process with intensity $\mu > 0$, i.e.,

$$\mathbb{P}(N(t) = k) = e^{-\mu t} \frac{(\mu t)^k}{k!}, \quad k \ge 0$$

Assuming an arbitrary arrival process, then we have

$$G_t(x) = \sum_{k=0}^{\infty} p_t(k) F_{\xi_1}^{k*}(x), \quad x \ge 0$$

where we set

$$p_t(k) = \mathbb{P}(N(t) = k), \quad k \ge 0$$

as the distribution of the arrival process (compare this with the usual Poisson arrival process). Since we assume a subexponential distribution F for the data amount of a particular connection, we can deduce the following theorem for the distribution of the aggregated data amounts.

Theorem 2.38. *Let $F_{\xi_1} \in \mathcal{S}$ and $t > 0$. Furthermore, the distribution of the arrival process (p_t) satisfies the inequality*

$$\sum_{k=0}^{\infty} (1+\epsilon)^k p_t(k) < \infty$$

for an $\epsilon > 0$. Then we have $G_t \in \mathcal{S}$ and

$$G_t^c(x) \sim \mathbb{E}(N(t)) F_{\xi_1}^c, \text{ for } x \to \infty$$

Remark 2.39. We give some remarks on the preceding theorem.

- The condition

$$\sum_{k=0}^{\infty} (1+\epsilon)^k p_t(k) < \infty$$

2.7 General Poisson Arrival and Serving Systems M/G/n

tells us that $p_t(k)$ decays relatively fast, while in contrast $(1+\epsilon)^k$ grows for $k \to \infty$ exponential towards ∞ ($\epsilon > 0$). With this for example we try to overweight the probability of the first connection attempts. This incorporate the fact that a server capacity cannot be increased arbitrary, i.e. that we cannot have arbitrary many call or connection attempts.
- We can deduce that the aggregated data amount are subexponential or heavy-tail distributed at time t.
- The theorem indicates in addition an approximative formula for large data amounts and hence, it can be used for simulations resp. computing the overall data amount.

For the treatment of the single distribution functions, to establish a table and for a better investigation of the particular properties we define four further classes of distributions (fig. 2.26):

$$\mathcal{D} = \{F \text{ distribution on }]0,\infty[; \limsup_{x \to \infty} \frac{F^c(\frac{x}{2})}{F^c(x)} < \infty\}$$

$$\mathcal{E} = \{F \text{ distribution on }]0,\infty[; \lim_{x \to \infty} \frac{F^c(x-y)}{F^c(x)}, \text{ for all } y > 0\}$$

$$\mathcal{K} = \{F \text{ distribution on }]0,\infty[; \mathcal{L}(f)(-\epsilon) = \int_0^\infty e^{\epsilon x} f(x) dx = \infty, \text{ for all } \epsilon > 0\}$$

$$\mathcal{R} = \{F \text{ distribution on }]0,\infty[; F^c \in R_{-\alpha}, \text{ for a } \alpha \geq 0\}$$

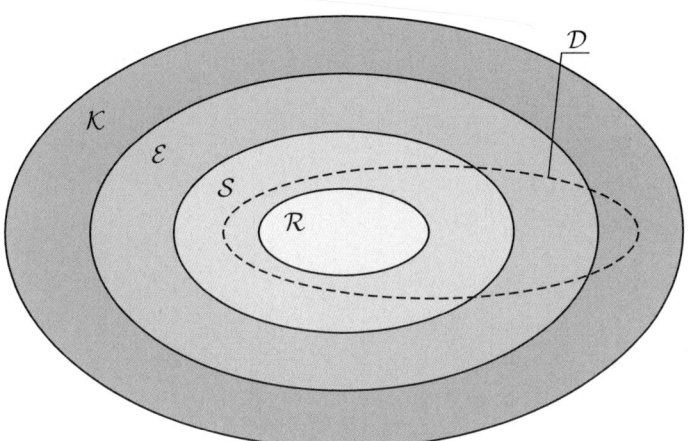

Fig. 2.26. Scheme of subclasses of subexponential distributions

As we see from the figure the class $\mathcal{R}$ is contained in all the other and thus, possesses all the characteristic properties. On the other side the representation of the tail of these distribution is very useful for applications (especially often, if a closed expression of the distribution is not possible). These are the facts, why one prefers choosing for modeling the data amount or connection time distributions from the class $\mathcal{R}$. We will describe further reasons in section 3.2.3.

We introduce for further purposes the important notion of the *integrated complementary) distribution*.

Definition 2.40. *If F_ξ is a distribution function of a random variable ξ with mean $\mu = \mathbb{E}(\xi)$, then we define by*

$$F_{I,\xi}(t) = \frac{1}{\mu}\int_0^t F_\xi^c(x)dx,\ t \geq 0,$$

the integrated (complementary) distribution function to ξ.

We integrated the complementary distribution function. This method is often used, to level out jumps or abrupt behavior. The new function $F_{I,\xi}$ is 'smoother'. Because of the existence of the expected value we could divide by μ. Thus, the resulting function reveals the properties, necessary for a distribution function of positive random variables ($F_{I,\xi}(0) = 0$, $F_{I,\xi}(t)$ is monotone increasing, since F_ξ^c is positive, $F_{I,\xi}(t)$ is continuous, because of the integral and $\lim_{t\to\infty} F_{I,\xi}(t) = 1$). As exercise one can deduce the fact that

$$\int_0^\infty F_\xi^c(x)dx = \mu \tag{2.34}$$

From this follows the property $\lim_{t\to\infty} F_{I,\xi}(t) = 1$

$$\int_0^t F^c(x)dx = \int_0^t (1 - F(x))dx = t - \int_0^t 1 \cdot F(x)dx$$

$$= t - xF(x)|_0^t + \int_0^t xf(x)dx$$

$$= (1 - F(t))t + \int_0^t xf(x)dx$$

$$= t\int_t^\infty f(x)dx + \int_0^t xf(x)dx$$

The last integral $\int_0^t xf(x)dx$ converges for $t \to \infty$ towards the expected value μ. The first integral can be estimated by (note that the density is positive)

$$0 \leq t\int_t^\infty f(x)dx \leq \int_t^\infty xf(x)dx$$

Since the integral $\int_0^\infty xf(x)dx = \mu$ exists (note that the random variable is positive), the expression $\int_t^\infty xf(x)dx$ converges towards 0 for $t \to \infty$. Hence, the integral $t\int_t^\infty f(x)dx$ does it also, and we overall conclude

2.7 General Poisson Arrival and Serving Systems M/G/n

$$\int_0^t F^c(x)dx \to \int_0^\infty xf(x)dx = \mu$$

Before establishing the connection between subexponential and subexponential integrated distribution, we give some examples of subexponential distributions. We indicate only the complementary distribution function, since we have $F_\xi(t) = 1 - F_\xi^c(t)$ and for the density f_ξ of F_ξ.

Example 2.41. Weibull distribution (fig. 2.27):

$$f_\xi(x) = a\tau t^{\tau-1} e^{-at^\tau}, \quad F_\xi^c(t) = e^{-at^\tau}, \quad \text{for } a > 0,\ 0 < \tau < 1$$

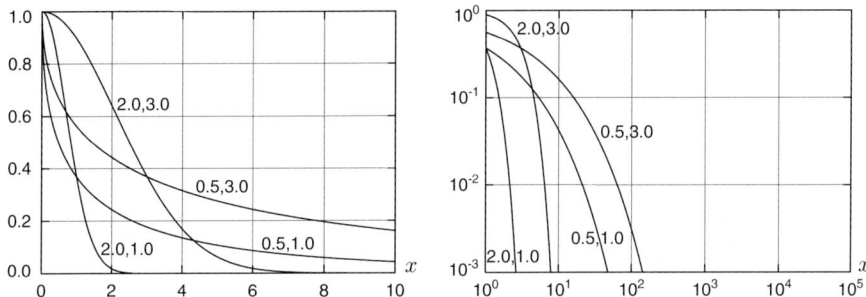

Fig. 2.27. Complementary cumulative distribution function (ccdf) of the Weibull distribution, linear (left) and logarithmic (right) scale on x-axis

Example 2.42. Pareto distribution (fig. 2.28):

$$f_\xi(x) = \alpha\kappa^\alpha \left(\frac{1}{\kappa+x}\right)^{\alpha+1}, \quad F_\xi^c(t) = \left(\frac{\kappa}{\kappa+t}\right)^\alpha, \quad \text{for } \alpha, \kappa > 0$$

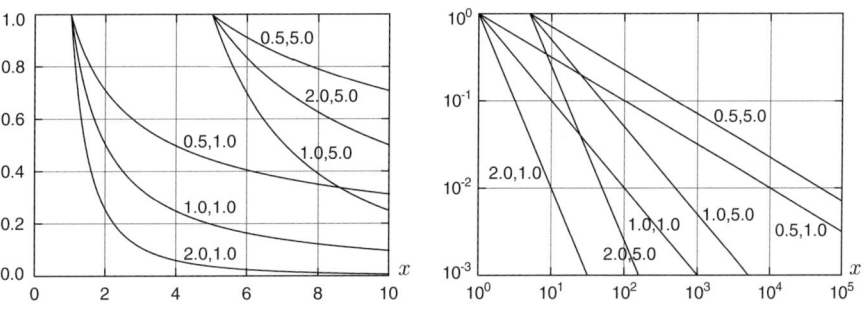

Fig. 2.28. Complementary cumulative distribution function (ccdf) of the Pareto distribution, linear (left) and logarithmic (right) scale on x-axis

90   2 Classical Traffic Theory

Example 2.43. Burr distribution (fig. 2.29):

$$f_\xi(x) = \kappa^\alpha \tau x^{\tau-1} \left(\frac{1}{\kappa + x^\tau}\right)^{\alpha+1}, \quad F_\xi^c(t) = \left(\frac{\kappa}{\kappa + t^\tau}\right)^\alpha, \quad \text{for } \alpha, \kappa, \tau > 0$$

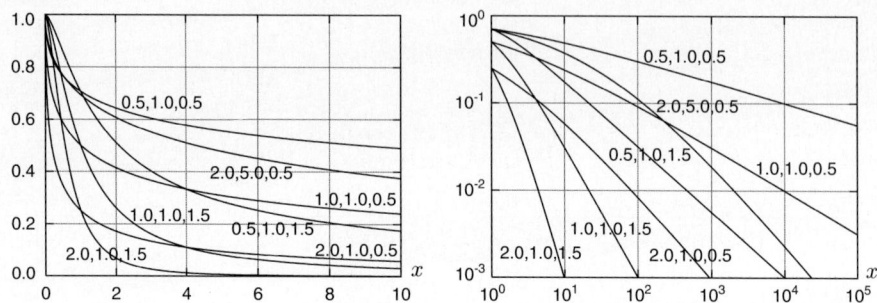

Fig. 2.29. Complementary cumulative distribution function (ccdf) of the Burr distribution, linear (left) and logarithmic (right) scale on x-axis

Example 2.44. Benktander distribution type I (fig. 2.30):

$$F_\xi^c(t) = \left(1 + 2\left(\frac{\beta}{\alpha}\right)\log t\right) e^{-\beta(\log t)^2 - (\alpha+1)\log t}, \quad \text{for } \alpha, \beta > 0$$

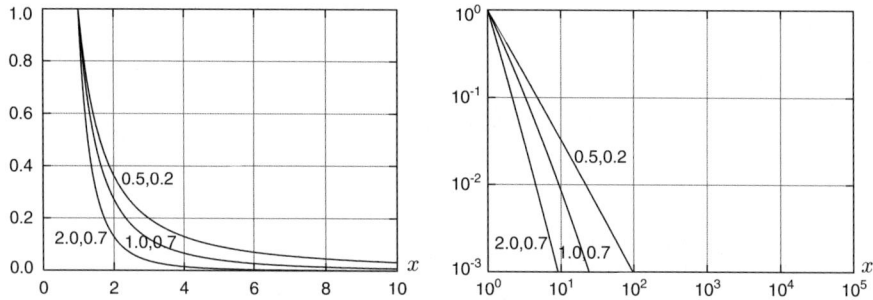

Fig. 2.30. Complementary cumulative distribution function (ccdf) of the Benktander I distribution, linear (left) and logarithmic (right) scale on x-axis

Example 2.45. Benktander distribution type II (fig. 2.31):

$$F_\xi^c(t) = e^{\frac{\alpha}{\beta}} t^{-(1-\beta)} e^{-\alpha \frac{t^\beta}{\beta}}, \quad \text{for } \alpha > 0,\ 0 < \beta < 1$$

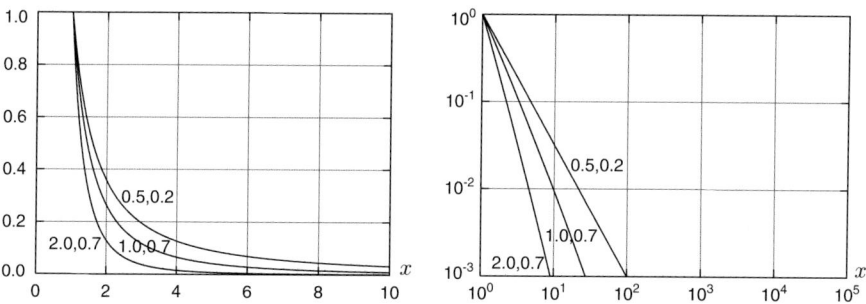

Fig. 2.31. Complementary cumulative distribution function (ccdf) of the Benktander II distribution, linear (left) and logarithmic (right) scale on x-axis

Example 2.46. Lognormal distribution (fig. 2.32):

$$f_\xi(x) = \frac{1}{\sqrt{2\pi}\sigma x} e^{-\frac{(\log x - \mu)^2}{2\sigma^2}}, \text{ for } \mu \geq 0,\ \sigma > 0$$

(F_ξ^c cannot be given in closed form.)

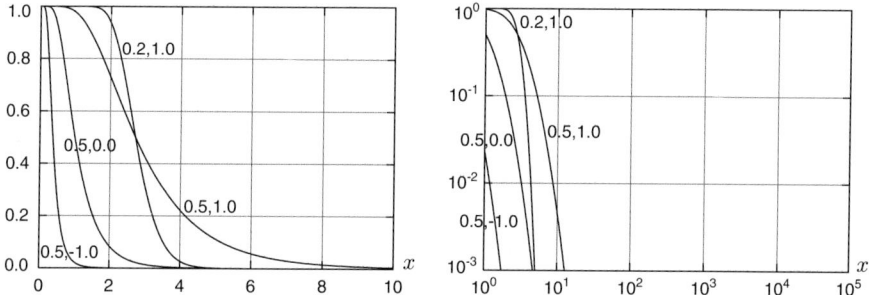

Fig. 2.32. Complementary cumulative distribution function (ccdf) of the Lognormal distribution, linear (left) and logarithmic (right) scale on x-axis

Example 2.47. Log-gamma distribution:

$$f_\xi(x) = \frac{\alpha^\beta}{\Gamma(\beta)} (\log x)^{\beta-1} x^{-\alpha-1}, \text{ for } \alpha > 0,\ \beta > 0$$

where Γ is the gamma function (F_ξ^c cannot be given in a closed form).

Next we state the relationship between the complementary and integrated (complementary) distribution function. Unfortunately neither $F \in \mathcal{S} \Rightarrow F_I \in \mathcal{S}$ or $F_I \in \mathcal{S} \Rightarrow F \in \mathcal{S}$ does hold. But there is for most applications some hope, since for distributions interesting for modeling the IP-based traffic performance the equivalence hold: they fulfill $F \in \mathcal{S}$ as $F_I \in \mathcal{S}$. Here is a list of samples, in fact identical with the former list of examples:

- Pareto distribution,
- Weibull distribution for $\tau < 1$,
- Lognormal distribution,
- Benktander type I and Benktander type II distribution,
- Burr distribution,
- Log-gamma distribution.

An important rôle for the estimation of subexponential resp. heavy-tail distributions plays the so called *mean excess function* or *excess distribution*.

Definition 2.48. *Let ξ be a random variable:*

- *For a random variable ξ with distribution function F_ξ and a threshold x we define*
$$F_{x,\xi}(t) = \mathbb{P}(\xi - x \leq t \,|\, \xi > x), \ x \geq 0$$
as the excess distribution *of ξ above the threshold x.*
- *The function*
$$CME_\xi(x) = \mathbb{E}(\xi - x \,|\, \xi > x)$$
for a random variable ξ will be called excess function.

If the distribution F is continuous, then the function CME_ξ provides a useful representation. We have

$$CME_\xi(x) = \mathbb{E}(\xi) \frac{F_{I,\xi}(x)^c}{F^c(x)}$$

if ξ has a first moment. We can express it in other words

$$F^c(x) = \frac{CME_\xi(0)}{CME_\xi(x)} \exp\left(-\int_0^x \frac{1}{CME_\xi(t)} dt\right), \ x > 0$$

For characterizing the heavy-tail distributions, we cite the theorem of Karamata [133] (resp. [76, A.3.6]), to which we will come back, when modeling the so called on-off models.

Theorem 2.49. *Let $L \in \mathcal{R}_0$ locally bounded on the interval $[x_0, \infty[$, $x_0 > 0$. We have:*

- *If $\alpha > -1$, then it follows*
$$\int_{x_0}^x t^\alpha L(t) dt \sim (\alpha + 1)^{-1} x^{\alpha+1} L(x), \ x \to \infty$$

- *If $\alpha < -1$, then it holds*
$$\int_{x_0}^x t^\alpha L(t) dt \sim -(\alpha + 1)^{-1} x^{\alpha+1} L(x), \ x \to \infty$$

2.7 General Poisson Arrival and Serving Systems M/G/n

- For the case $\alpha = -1$ we get

$$\frac{1}{L(x)} \int_{x_0}^{x} \frac{L(t)}{t} dt \longrightarrow \infty, \quad x \to \infty$$

and $x \longmapsto \int_{x_0}^{x} \frac{L(t)}{t} dt \in \mathcal{R}_0$.
- If $\alpha = -1$ and suppose $\int_{x_0}^{x} \frac{L(t)}{t} dt < \infty$, then we deduce

$$\frac{1}{L(x)} \int_{x}^{\infty} \frac{L(t)}{t} dt \longrightarrow \infty, \quad x \to \infty$$

and $x \longmapsto \int_{x}^{\infty} \frac{L(t)}{t} dt \in \mathcal{R}_0$.

With these facts we obtain, provided $F^c \in \mathcal{R}_{-\alpha}$, for $\alpha > 1$

$$CME_\xi(x) \sim \frac{x}{\alpha - 1}, \quad \text{for } x \to \infty \tag{2.35}$$

Since the mean excess function plays the central rôle for the estimations, we give for certain important distributions their corresponding excess functions in the table 2.1.

Table 2.1. Excess functions

Exponential	$CME(x) = \frac{1}{\lambda}$
Lognormal	$CME(x) = \frac{\sigma^2 x}{\log x - \mu}(1 + o(1))$
Pareto	$CME(x) = \frac{x_0 + x}{\alpha - 1}, \, \alpha > 1$
Weibull	$CME(x) = \frac{x^{1-\tau}}{c\tau}(1 + o(1))$
Normal	$CME(x) = \frac{1}{x}(1 + o(1))$

In the figures 2.33 to 2.33e we depict the excess functions from table 2.1, where we omitted the expression $o(1)$, i.e. setting it 0. Thus, we have to mention that the excess function for 'lognormal', 'Weibull' and 'standard normal' are asymptotic.

As we will see, the Pareto and lognormal distributions exhibit a prominent rôle within the class of subexponential distributions. Thus, we will deduce some properties of them. At first we consider a Pareto distribution $F_\xi(x) = 1 - \left(\frac{\kappa}{\kappa + x}\right)^\alpha$, where a random variable ξ is distributed according to F_ξ. Then it follows

$$\mathbb{P}(\xi > 2x) = c\mathbb{P}(\xi > x), \quad x > 0$$

where c is a constant independent of x. Furthermore the Pareto distribution is the only distribution, which is invariant for cutting from below, i.e., suppose $y \geq x_0 > 0$, then we have

$$\mathbb{P}(\xi > y \mid \xi > x_0) = \mathbb{P}\left(\left(\frac{x_0}{\kappa}\right)\xi > y\right)$$

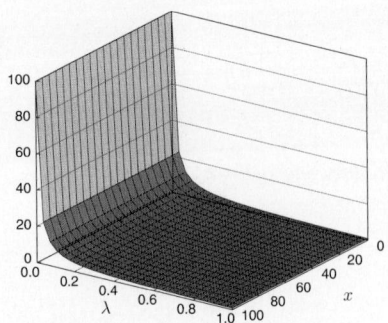

Fig. 2.33. Excess function for the exponential distribution

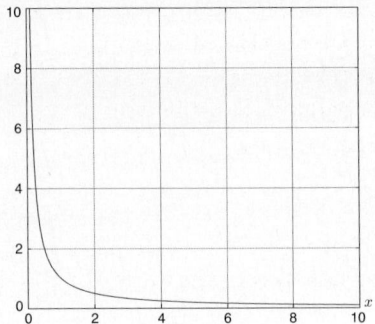

Fig. 2.33a. Excess function for the standard normal distribution

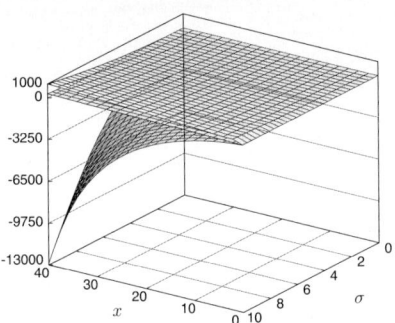

Fig. 2.33b. (Asymptotic) excess function for the lognormal distribution with mean $\mu = 4, 0, -9$ from above

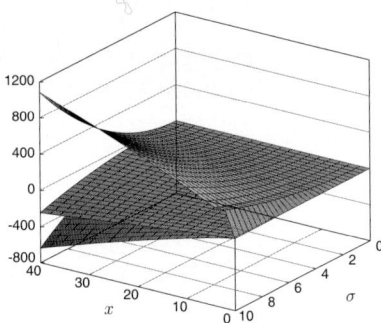

Fig. 2.33c. (Asymptotic) excess function for the lognormal distribution with mean $\mu = 0, 10, 20$ from above

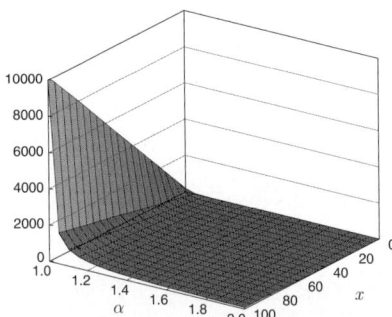

Fig. 2.33d. Excess function for the Pareto distribution with $x_0 = 0, 1, 2$. As seen, they are all equal

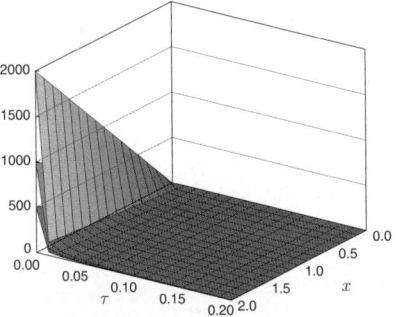

Fig. 2.33e. Excess function for the Weibull distribution with $c = 1, 2, 4$ from above

Thus the conditional distribution is again a Pareto distribution, just with different parameters.

For the lognormal distribution we estimate the complementary distribution function first, since it is not available in closed form. We have as density for the parameter $\mu = 0$ and $\sigma = 1$, a standard lognormal distribution

$$f_\xi(t) = \frac{1}{\sqrt{2\pi}\sigma t} e^{-\frac{(\log t)^2}{2}}$$

This means the random variable $\log \xi$ is Gaussian. For the standard normal distributed random variable ZV η it holds

$$\mathbb{P}(\eta > y) \sim \frac{1}{\sqrt{2\pi} y} e^{-\frac{y^2}{2}}$$

This implies

$$\mathbb{P}(\xi > x) \sim \frac{1}{\sqrt{2\pi} \log x} e^{-\frac{(\log x)^2}{2}}$$

Thus, there is a constant c with

$$\mathbb{P}(\xi > x) \sim c \frac{e^{-\frac{(\log x)^2}{2}}}{\log x}$$

We have for $n \in \mathbb{N}$

$$\log x \cdot e^{\frac{(\log x)^2}{2}} > x^n \qquad (2.36)$$

for sufficient large x. We can deduce (2.36) in the following way:

Find for $n \in \mathbb{N}$ a x_0, such that it holds

$$\log x > n, \text{ for all } x \geq x_0$$

Then, we obtain

$$(\log x)^2 > n \log x, \text{ thus } e^{(\log x)^2} > x^n, \text{ for } x \geq x_0$$

The lognormal distribution is subexponential but according to our definition not heavy-tail, since the complementary distribution function does **not** enjoy the behavior

$$F^c(x) \sim x^{-\alpha} L(x)$$

for large x with a suitable $0 < \alpha$ and a slowly varying function $L \in \mathcal{R}_0$.

2.7.5 Application of M/G/1 Models to IP Traffic

Motivation for the following section are the measurements done by the Leland group at Bellcore [160], to which we will return in connection with the Norros model and the introduction of the FBM into the traffic modeling. These measurements show the decisive differences in local networks, when e.g. considering transmission times of FTP transfers. The main difference consists in the great variability of the average serving rates μ, what clearly indicates that the Poisson process is no longer applicable.

In addition, the measurements revealed that the CPU times as well as the amount of data follow the rules of heavy-tail distributions. The time series of data in IP traffic show the so called long-range dependence, which we will discuss in the next chapter intensively (e.g. in the section on α-stable processes). This implies that the present state is strongly depending on the former one. We keep to the standard definition that for a discrete process $(X_n)_{n \in \mathbb{N}}$ the correlation function $\mathrm{Cor}(X_0, X_n)$ decays more slowly than exponential and is not summable (see theorem 3.37). The classical ARMA processes or time series have a correlation function decaying exponentially (see e.g. section 3.3.3).

We start with a brief summary on the classical queueing model M/G/1:

- The arrival process is Poisson with rate λ, i.e., the interarrival time ξ_n is Markovian resp. exponential. The arrival times T_n, with $\xi_1 = T_1$ and $\xi_n = T_n - T_{n-1}$, $n \geq 2$ are Poisson with rate λ.
- The serving rate is general distributed according to the function F with mean (expected value) μ.
- We assume a stable system, i.e. the average load is $\rho = \frac{\mathbb{E}(\eta)}{\mathbb{E}(\xi)} = \lambda\mu < 1$.

The aim is to obtain results on the waiting time distribution, residual work and queueing length under the assumption of a subexponential distributed serving time. Differently to section 2.7.3, we will formulate only qualitative results. No detailed proofs will be presented, which is left to the reader in the given literature. The following results are mainly based on the article [104]. We denote by W_n the waiting time of the n-th user, by η_n (set $\eta_0 = 0$) the serving time of the n-th user and with $\xi_{n+1} = T_{n+1} - T_n$ the interarrival time from the n-th to the $(n+1)$-th user. We have the relation

$$W_0 = 0, \quad W_{n+1} = (W_n + \eta_n - \xi_n)^+$$

For a fixed $n \in \mathbb{N} \cup \{0\}$ we derive from [11, Lemma 11.1.1] the identity in distribution

$$W_n = \max_{0 \leq k \leq n} \sum_{i=0}^{k} (\eta_i - \xi_{i+1})$$

where $(S_n = \sum_{n=0}^{n} (\eta_i - \xi_{i+1}))_{n \geq 0}$, with $S_0 = 0$, is a stochastic random walk with generator

$$\zeta = \eta - \xi$$

(see (2.11) from section 2.6). For the expected value of the generator we get

$$\mathbb{E}(\zeta) = \mathbb{E}(\eta - \xi) = \mu - \frac{1}{\lambda} = \frac{1}{\lambda}(\lambda\mu - 1) < 0$$

2.7 General Poisson Arrival and Serving Systems M/G/n

This indicates that (W_n) is distributed like the maximum of a random walk with negative drift and converges almost surely towards a RV W_∞, for $n \to \infty$ (see [11, Satz 11.1.3a)]). The random walk W_∞ is finite (because of convergence) with a distribution function $\mathcal{W}(t)$, $t > 0$. According to a result of [11, S. 238] resp. the waiting time distribution due to Pollaczek-Khintchine (see section 2.7.3), we deduce

$$\mathcal{W}(t) = (1-\rho) \sum_{n=0}^{\infty} \rho^n F_I^{n*}(t), \ t \geq 0 \tag{2.37}$$

where $F_I^{0*} = \delta_{[0,\infty[}$ is the Dirac distribution with measure 1 at 0. According to our notation, we have

$$F_I^{n*}(t) = \frac{1}{\mu} \int_0^t F^{n*,c}(y) dy$$

where $F^{n*,c}(y)$ is the complementary distribution function of the n-times convolution of the distribution function F, i.e. the complementary distribution function of the random variable $\eta_1 + \ldots + \eta_n$. Changing to the complementary distribution of $\mathcal{W}(t)$, it follows

$$\mathcal{W}^c(t) = (1-\rho) \sum_{n=0}^{\infty} \rho^n (F_I^{n*,c}(t)), \ t \geq 0$$

It is obvious to divide both sides by $1 - F_I^c(t)$. This implies

$$\frac{\mathcal{W}^c(t)}{F_I^c(t)} = (1-\rho) \sum_{n=0}^{\infty} \rho^n \frac{1 - F_I^{n*}(t)}{1 - F_I^c(t)}$$

Results on the asymptotic behavior of $\mathcal{W}^c(t)$ can be derived, if interchanging limit and summation is possible. According to [233, Lemma 2.10] and the dominated convergence theorem (see e.g. [98, Satz 1.6.10]), this can be done. Using the characterization of subexponential distributions in (2.30) and the convergence of the geometric series $\sum_{n\geq 0} \rho^n$ (note that $\rho < 1$), we can derive the following results, which serves even as another characterization of subexponential distributions, because of the stationary distribution of $\mathcal{W}$ and the integrated distribution of F_I.

Theorem 2.50. *In a M/G/1 model with $\rho < 1$ the following assertions are equivalent*

$$\lim_{t \to \infty} \frac{\mathcal{W}^c(t)}{F_I^c(t)} = \frac{\rho}{1-\rho} \Leftrightarrow F_I \text{ is subexponential} \Leftrightarrow \mathcal{W}(t) \text{ is subexponential}$$

The limit exists, if and only if one has $\frac{\rho}{1-\rho}$ as asymptotic value of $\frac{\mathcal{W}^c(t)}{F_I^c(t)}$. Modeling serving times subexponential, we obtain a subexponential equilibrium distribution F_I. This is again equivalent to the fact that the waiting time distribution is subexponential, since the complementary distribution of the waiting time differs only by a constant factor, namely $\frac{\rho}{1-\rho}$. For a rigorous proof of this heuristic justification see [78].

This result is valid with the restriction, if one changes to a general arrival process, since the necessary result from [11] is also valid in the GI/G/1 case – only the first moments of ξ and η have to exist.

Theorem 2.51. *Assume a GI/G/1 waiting system. Then the following implications hold*

$$F_I \text{ is subexponential} \Leftrightarrow \mathcal{W}(t) \text{ is subexponential} \Rightarrow \lim_{t \to \infty} \frac{\mathcal{W}^c(t)}{F_I^c(t)} = \frac{\rho}{1-\rho}$$

When does greater load occurs in the systems? For this let

$$\tau = \inf\{n > 0; W_n = 0\} = \inf\{n > 0; S_n < 0\}$$

(see e.g. [11, Kor. 11.1.2]), where $S_n = \sum_{i=1}^{n}(\eta_i - \xi_i) = \sum_{i=1}^{n} \eta_i - T_n$ is a random walk with $S_0 = 0$ and generator $\zeta = \eta - \xi$. S_n represents the aggregated n serving times of the n first interarrival times. The value τ is denoted as weak decreasing laddar index (LI) (see e.g. [11, S. 38]) and indicates the first demand, which does not have to wait. We assume that $F \in \mathcal{L}$, (see the definition form section 2.7.4) i.e. it holds $\lim_{x \to \infty} \frac{F^c(x+y)}{F^c(x)} = 1$ for $y \in \mathbb{R}$. This tells us that the distribution function is subexponential with $F^c(x) = L(x)x^{-\alpha}$ and $L \in \mathcal{R}_0$ slowly varying – we know the notion from section 2.7.4. Then the complementary distribution function G^c of ζ looks like

$$G^c(x) = \mathbb{P}(\zeta > x) = \int_0^\infty \mathbb{P}(\eta > x + y) dF_\xi(y)$$
$$\sim \mathbb{P}(\eta > x) = F^c(x), \text{ for } x \to \infty$$

The complementary distribution $\mathcal{W}^c$ is thus equivalent to the complementary distribution function of F, that means

$$\mathcal{W}^c(x) \sim \frac{1}{\nu} \int_x^\infty F^c(y) dy = \frac{\mu}{\nu}(1 - F_I(x)), \text{ for } x \to \infty \qquad (2.38)$$

where $\nu = \mathbb{E}(\zeta - \xi)$. Since we assumed $\rho < 1$, this gives $\tau < \infty$ a.s., but we have even more. It holds $\mathbb{E}(\tau) < \infty$, and with the Wald identity (see [98, Lemma 6.3.18])

$$\mathbb{E}(\tau) = \lambda \mathbb{E}(\xi_1)^{\leq}$$

where $\xi_1^{\leq} = T_\tau$ (see [11, S. 234, (11.2.2)]). The distribution of the (T_n) is Poisson.

We define the 'stopped' maximum

$$M_\tau = \sup_{1 \leq n \leq \tau} S_n$$

2.7 General Poisson Arrival and Serving Systems M/G/n

This reflects the greatest values of S_n up to that demand, which is the first one sent without waiting. It is natural to ask, when the first moment (or the first demand) occurs, where the random walk (S_n) steps over a fixed threshold x, i.e.

$$\tau(x) = \inf\{n > 0; S_n > x\}$$

Then it is immediately clear, that $\mathbb{P}(M_\tau > x) = \mathbb{P}(\tau(x) < \tau)$. The paper [104] examines, how and when such a cyclic maximum M_τ occurs. Thus, we introduce for $0 < x_0 < x < \infty$ three values

$$N(x, x_0) = \operatorname{card}\{n \geq 0; n < \tau, S_n \leq x_0, S_{n+1} > x\}$$
$$p_1(x, x_0) = \mathbb{P}(S_{n+1} > x, \text{ for a } 0 \leq n < \tau \text{ with } S_n \leq x_0)$$
$$p_2(x, x_0) = \mathbb{P}(\tau(x) < \tau \text{ and } x_0 \leq S_{\tau(x)-1} \leq x)$$

Before we formulate the main result for the relationship of these value, we shortly look closer to them. The value $N(x, x_0)$ indicates the amount of demands n, which are in the first cycle (i.e. not to wait until the first demand), for which the first serving time S_n is smaller than x_0 and the next demand of S_{n+1} steps over the threshold x. Correspondingly $p_1(x, x_0)$ measures the probability, that this happens at least once within a cycle, and $p_2(x, x_0)$ the probability that for given $x_0 < x$ before the incident $S_{\tau(x)} > x$ the demand $S_{\tau(x)-1}$ surpasses the value x_0. We have now the following theorem.

Theorem 2.52. *(Greiner/Jobman/Klüppelberg) Let $m = \mathbb{E}(\tau)$. Then it follows:*

a) $p_1(x, x_0) \leq \mathbb{P}(M_\tau > x) \leq p_1(x, x_0) + p_2(x, x_0)$ *for $x > x_0$.*
b) $\mathbb{E}(N(x, x_0)) \sim p_1(x, x_0) \sim mW^c(x_0)F^c(x)$ *for $x_0 > 0$ and asymptotic $x \to \infty$.*
c) *If $\mathcal{W}$ possesses a density w, such that $w(x) \sim \frac{F^c(x)}{\nu}$ for $x \to \infty$, then it follows*

$$\lim_{x_0 \to \infty} \limsup_{x \to \infty} \frac{p_2(x, x_0)}{F^c(x)} = 0$$

d) $\mathbb{P}(M_\tau > x) \sim mF^c(x)$, $x \to \infty$.

The proof is given in [104].

Remark 2.53. We give some remarks on theorem 2.52:

- By comparing c) of the result in theorem 2.52 with relation (2.38), we realize that the assumption of c) in the theorem is fulfilled, if we choose $F \in \mathcal{L}$. Examples and further explanation to the class $\mathcal{L}$ can be found in section 2.7.4.
- The relevance of the value τ and $\tau(x)$ lies in the judgment, when the buffer (or server, router) is again empty or at least works without delay (in the case τ) and when the traffic grows over the threshold (case $\tau(x)$). The theorem reveals, in which form this occurs in dependence of the serving distribution.

- How can we interpret the result? At first it shows that the traffic (i.e. the random walk S_n) develops with negative drift of the generator $\zeta = \eta - \xi$. Is the serving time F strongly heavy-tailed, then the surpass of a high threshold occurs at x (see point c) in the theorem). After that we have a system similar to the beginning with the help of τ, and the negative drift occurs again. But on the way down an overflow can occur (see point b) in the theorem), which is justified by the heavy-tail of F.
- A weak point of the M/G/1 model consists in the consideration of only one serving unit. Thus, mostly so called processor sharing systems are considered, as we do partly in the section 3.5.2. Another extreme case, the M/G/∞ models, are done in section 2.5.1.

A next value of interest is the remaining or queueing traffic load (V_t). It is also called the virtual waiting time and consists of the sum of all serving times of the waiting requests and handled loads up to time t, i.e. all waiting demands and just sent IP packets in a server or router. V_t is nothing else as the waiting time of an incoming data request at time t until its successful sending. This can be formally described as

$$V_t = \sum_{n=0}^{\infty} (T_n + W + \eta_n - t)^+ \mathbf{1}_{[T_n, T_n+1[}(t)$$

Suppose Θ is a p Bernoulli distributed random variable, which is independent to all appearing random variables, then under $\rho < 1$ the identity holds in distribution

$$V_t \stackrel{d}{=} (1 - \Theta) + \Theta(W_\infty + \eta^*) \stackrel{d}{=} (W_\infty + \eta - \xi) \stackrel{d}{=} (M + \eta - \xi)$$

where $M = \sup_{n \geq 0} S_n$ and η^* is a random variable, distributed according to F_I and independent to the already introduced random variables. A proof can be found e.g. in [11, prop. 11.3.2.]. It can be seen that the distribution of the remaining load or request is determined by the waiting time resp. serving time in case of exponential distributed interarrival times. This means that it is also subexponential distributed, provided F or F_I are subexponential.

We investigate another key quantity, the waiting queue Q_t in the system up to time t, as we already looked at in the chapter on the classical traffic theory under FIFO principle. Let $D_n = \eta_n + W_n$ the sojourn time of the n-th user, which should be independent to all later arrival times in the system. Let furthermore N_t a stationary version of the counting process, subordinated to the arrival times, i.e. w.l.o.g.

$$N_t = \sup\{n \geq 0; T_n \leq t\}$$

This corresponds to the amount of requests up to time t. Furthermore we define with $D = \eta + W$ the stationary sojourn time (consider in the section 'renewal process'the notion 'stationary'). A request n will be part of the waiting queue at time t, if $T_n \leq t$ holds for its arrival time, and it has not been

2.7 General Poisson Arrival and Serving Systems M/G/n

served yet until time $T_n + D_n$, i.e. $T_n + D_n > t$. Thus, we have the identity in distribution

$$Q_t = \sum_{n \geq 0} \mathbf{1}_{(T_n \leq t; T_n + D_n > t)}$$

With Q_∞ we denote the stationary waiting queue distribution, then in this case follows (see [11, p. 240])

$$Q_t \longrightarrow Q_\infty \text{ in distribution, for } t \to \infty$$

Let $(\hat{T}_n)$ be a copy of (T_n) independent to all appearing random variables. Then we have the following waiting queue distribution (see [11, prop. 14.4.2]).

Theorem 2.54. *For the asymptotic length of the waiting queue Q_∞ and all $n \geq 1$ we have*

$$\mathbb{P}(Q_\infty = 0) = 1 - \rho \quad \text{(compare with (2.29))}$$
$$\mathbb{P}(Q_\infty \geq n) = \mathbb{P}(W_\infty + \eta^* > \hat{T}_{n-1}) = \mathbb{P}(D_n > \xi^* + \hat{T}_{n-1})$$
$$= \mathbb{P}(V > \hat{T}_{n-1})$$

In particular

$$\mathbb{E}(Q_\infty) = \lambda \mathbb{E}(D) = \lambda \mathbb{E}(W_\infty) + \rho \qquad (2.39)$$

The equation (2.39) is called the *Theorem of Little* (see proposition 2.5). This identity can be described heuristically as follows: The expected value $\mathbb{E}(D)$ indicates the mean sojourn time of the data in the system in the stationary case. Then $\lambda \mathbb{E}(D)$ is the average amount of incoming requests during this time period. But this corresponds also to the mean number of waiting requests, if just one is severed and leaves the server or router. Hence, we can reformulate the theorem of Little

$$Q_\infty \stackrel{d}{=} N_D$$

If F_η is subexponential, then we have $F_\eta^c(x) = o(1 - F_{\eta,I}(x))$ according to the result in [233, Lemma 3.1.]. Hence, F_η^c decays faster than $\mathcal{W}^c$ (see e.g. the representation of the distribution $\mathcal{W}$ from (2.37)). Thus, W_∞ predominates η, the serving time in the sum of D, and it follows

$$\mathbb{P}(D > x) \sim \mathbb{P}(W_\infty > x), \text{ for } x \to \infty$$

If we assume a heavy-tail distributed serving time, as done in the most models, so we get the following result according to Assmussen et al. [22].

Theorem 2.55. *We consider a waiting time system M/G/1 with arrival rate λ and traffic identity $\rho = \mu\lambda < 1$. Assume that the serving time is distributed according to F_η, where the integrated (complementary) equilibrium distribution $F_{\eta,I} \in \mathcal{S}$, i.e. it is subexponential. Let W_∞ be the stationary waiting time. Furthermore it is assumed that*

$$\lim_{x \to \infty} \frac{F_\eta^c(xe^{\frac{y}{\sqrt{x}}})}{F_\eta^c(x)} = 1, \ y \in I \tag{2.40}$$

holds uniformly on bounded intervals $I \subset \mathbb{R}$. Then, for $k \to \infty$ the stationary waiting queue distribution $\mathcal{Q}$ fulfills the relationship

$$\mathcal{Q}^c(k) = \mathbb{P}(Q_\infty > k) \sim \mathbb{E}(\lambda W_\infty > k) \sim \frac{\rho}{1-\rho}(1 - F_{I\eta}(\frac{k}{\lambda}))$$

The assumption (2.40) guaranties that the complementary distribution decays more slowly than the complementary Weibull distribution with density $\exp(-\sqrt{x})$. Consequently, this results holds for the Pareto distribution, lognormal distribution and Weibull distribution with complementary distribution function of order $F_\eta^c(x) \sim \exp(-x^\beta)$, $\beta < 0.5$. In these cases the waiting queue will be large because of the long serving time. The Poisson-arrival does not contribute anything substantial. But, if the complementary serving time distribution $F_\eta^c(x) \sim \exp(-x^\beta)$ decays with $\beta > 0.5$, then one has to take the arrival process into account and both influence the waiting queue resp. its distribution. Some examples demonstrate this fact:

- If $F_\eta^c(x) \sim \exp(-\sqrt{x})$, then we have

$$\mathbb{P}(W > k) \sim e^{\frac{1}{8\lambda}} e^{-\sqrt{\frac{xd}{\lambda}}}, \text{ for } k \to \infty$$

- Assume $F_\eta^c(x) \sim \exp(-x^\beta)$, $\beta =]\frac{1}{2}, \frac{2}{3}[$, then it follows

$$\mathbb{P}(W > k) \sim e^{-\left(\frac{k}{8\lambda}\right)^\beta + \frac{(1-\beta)\beta^2}{\lambda}\left(\frac{k}{\lambda}\right)^{2\beta-1}}, \text{ for } k \to \infty$$

- If $\beta > \frac{2}{3}$ then we get an expression with terms of higher order and the expression gets rather complicated.

2.7.6 Markov Serving Times Models GI/M/1

Though the following traffic situation is not rarely observed in IP traffic, we present it for completeness. We initially describe the ingredients of the model.

Description of the Model

In accordance to the mixed system M/GI/1 we consider now:

- General distributed interarrival time ξ with expected value $\mathbb{E}(\xi)$.
- One serving place.
- Poisson distributed serving time $F_\eta(t) = \mathbb{P}(\eta \le t) = 1 - e^{-\mu t} \Rightarrow \mathbb{E}(\eta) = \frac{1}{\mu}$.
- For the load coefficient we get $\rho = \frac{\mathbb{E}(\eta)}{E(\xi)} = \frac{1}{\mu \mathbb{E}(\xi)}$.

2.7 General Poisson Arrival and Serving Systems M/G/n

Methods for Treatment of the Model

Since the interarrival times are general distributed we choose for the embedding time spots according to the memoryless the particular arrival times. If $X(t)$ denote the system state at time t, then we form the sequence $(X(t_0), X(t_1), \ldots, X(t_n), X(t_{n+1}), \ldots)$ as embedded Markov chain. As in the treatment of the waiting system M/G/1, we define the random variable ζ, as the amount of demands in the system during an interarrival time periods. Accordingly we denote the distribution function by

$$F_\zeta(i) = \mathbb{P}(\zeta = i)$$

The generating function can be written in the form

$$\zeta_{EF}(z) = \sum_{i=0}^{\infty} F_\zeta(i) z^i$$

whereas $\mathbb{E}(\zeta) = \left.\frac{d\zeta_{EF}(z)}{dz}\right|_{z=1} = \mu \mathbb{E}(\xi) = \frac{1}{\rho}$.

As already done in the treatment of the waiting system M/G/1 in section 2.7.3 we use for the computation of the transition probabilities

$$p_{ij} = \mathbb{P}(X_{t_{n+1}} = j \,|\, X_{t_n} = i)$$

a division into several cases. During the interval $]t_n, t_{n+1}[$ we have no arrival events. This implies a pure loss process:

- $j = 0$: the system is empty at time t_{n+1}. Immediately after the n-th arrival we find $i + 1$ demands in the system. The probability p_{i0} equals the one that at least $i + 1$ demands are served, i.e.,

$$p_{i0} = \sum_{k=i+1}^{\infty} F_\zeta(k) = 1 - \sum_{k=0}^{i} F_\eta(k), \quad i = 0, 1, 2, \ldots$$

- $j > 0$: the system is not empty at time t_{n+1}. Directly before the n-th arrival we have i demands in the system; shortly afterwards we have $i + 1$ demands. Just before the $(n+1)$-th arrival there are still j demands in the system. Hence, $(i+1-j)$ demands are served during the interval $]t_n, t_{n+1}[$. Hence

$$p_{ij} = F_\zeta(i + 1 - j), \quad i = 0, 1, \ldots, \quad j = 1, \ldots, i+1$$

The transition matrix is written in the compact form

$$\mathcal{P} = (p_{ij}) = \begin{bmatrix} 1 - F_\zeta(0) & F_\zeta(0) & 0 & 0 & \cdots \\ 1 - \sum_{k=0}^{1} F_\zeta(k) & F_\zeta(1) & F_\zeta(0) & 0 & \cdots \\ 1 - \sum_{k=0}^{2} F_\zeta(k) & F_\zeta(2) & F_\zeta(1) & F_\zeta(0) & \cdots \\ \vdots & \vdots & \vdots & \vdots & \ddots \end{bmatrix}$$

State Probabilities

We proceed as in the case of the waiting system M/GI/1. For this let $x(i,n) = \mathbb{P}(X(t_n) = j)$ and the state probability vector

$$\mathcal{X}_n = (x(0,n), x(1,n), \ldots, x(j,n), \ldots)$$

We conclude the state transition equation

$$\mathcal{X}_n P = \mathcal{X}_{n+1}$$

With the initial vector $\mathcal{X}_0$ we can successively compute all vectors $\mathcal{X}_n$ in the instationary case. Again we find a condition for the stationary state

$$\mathcal{X}_n = \mathcal{X}_{n+1} = \ldots = \mathcal{X} = (x(0), x(1), \ldots, x(j), \ldots)$$

We find for the state equation

$$\mathcal{X} P = \mathcal{X}$$

Again, we derive the state probabilities for the stationary case. For this we split the matrix notation into the components:

$$x(0) = \sum_{i=0}^{\infty} x(i)\left(1 - \sum_{k=0}^{i} F_\zeta(k)\right) = \sum_{i=0}^{\infty} x(i) \sum_{k=i+1}^{\infty} F_\zeta(k)$$

$$x(j) = \sum_{i=j-1}^{\infty} x(i) F_\zeta(i+1-j) = \sum_{i=0}^{\infty} x(i+j-1) F_\zeta(i), \; j = 1, 2, \ldots$$

We choose a so called geometric approach (see e.g. [141, 253])

$$x(j+1) = \varrho x(j), \; j = 0, 1, 2, \ldots \quad \text{resp.} \quad x(j+1) = \varrho^{j+1} x(0)$$

The aim is now to determine the particular parameter ϱ. We insert this approach in the above equation and deduce

$$x(j) - (x(j-1) F_\zeta(0) + x(j) F_\zeta(1) + x(j+1) F_\zeta(2) + \ldots) = 0$$

This gives

$$\varrho x(j-1) - x(j-1) F_\zeta(0) - \varrho x(j-1) F_\varrho(1) - \varrho^2 x(j-1) F_\zeta(2) - \ldots = 0$$

or

$$x(j-1) \left(\varrho - (F_\varrho(0) + \varrho F\zeta(1) + \varrho^2 F\zeta(2) + \ldots)\right)$$
$$= x(j-1) \left(\varrho - \sum_{i=0}^{\infty} F_\zeta(i) \varrho^i\right) = 0$$

2.7 General Poisson Arrival and Serving Systems M/G/n

Assuming that $x(j-1) > 0$, then the solution of the last equation must coincide with the one of the following equation (also called fix point equation)

$$z = \zeta_{\text{EF}}(z)$$

since this is exactly the content of the brackets. Excluding the trivial solution $\varrho = 1$, i.e., $x(i) = x(i+1)$ for all $i = 0, 1, \ldots$, we consider the area of convergence of the series $\sum_{i=0}^{\infty} F_\zeta(i) z^i$ for $|z| \leq 1$. For this we differentiate the generating function twice

$$\frac{d\zeta_{\text{EF}}(z)}{dz} = \sum_{i=1}^{\infty} F_\zeta(i) i z^{i-1} \geq 0$$

$$\frac{d^2\zeta_{\text{EF}}(z)}{dz^2} = \sum_{i=2}^{\infty} F_\zeta(i) i (i-1) z^{i-2} \geq 0$$

Thus, the generating function is monotone increasing and convex in the interval $]0, 1[$.

Depending on the value $\frac{d\zeta_{\text{EF}}(z)}{dz}$ at the spot $z = 1$, we find:

- an intersection point at $z = 1$, if $\left.\frac{d\zeta_{\text{EF}}(z)}{dz}\right|_{z=1} \leq 1$, or
- two intersection points, in particular one in $]0, 1[$, if $\left.\frac{d\zeta_{\text{EF}}(z)}{dz}\right|_{z=1} > 1$.

Since $\left.\frac{d\zeta_{\text{EF}}(z)}{dz}\right|_{z=1} = \mathbb{E}(\zeta) = \frac{1}{\rho}$, this implies that there exists exactly one non trivial solution, if $\frac{1}{\rho} > 1$ or $\rho < 1$. This is a fact, which we already observed for the stability of the system M/M/1. We have for the solution ϱ

$$x(j) = \varrho^j x(0) \quad \text{and} \quad \sum_{i=0}^{\infty} x(i) = 1$$

Hence, we obtain by inserting the state probabilities

$$x(j) = (1 - \varrho) \varrho^j$$

The solution for ϱ can be obtained in general using numerical techniques, as the approximation methods according to Newton.

Waiting Time Distributions

Here, we follow the same line as in the last case of the M/G/1 - case and consider the queueing principle FIFO. The approach uses the conditional probabilities

$$\mathbb{P}(W > t \mid W > 0) = \frac{\mathbb{P}(W > t, W > 0)}{\mathbb{P}(W > 0)} = \frac{\mathbb{P}(W > t)}{\mathbb{P}(W > 0)}$$

whereas

2 Classical Traffic Theory

$$\mathbb{P}(W > 0) = \sum_{i=0}^{\infty} x(i) = 1 = 1 - x(0) = \varrho$$

Again, we choose a test unit and analyze the demands in the system. Only if $X > 0$, we can observe a positive waiting probability. Thus,

$$\mathbb{P}(W > t) = \sum_{i=1}^{\infty} \mathbb{P}(W > t \mid X = i)\mathbb{P}(X = i) = \sum_{i=1}^{\infty} \mathbb{P}(W > t \mid X = i)(1 - \varrho)\varrho^i$$

From the representation of $\mathbb{P}(W > 0)$ we can divide the last equation by ρ and get

$$\mathbb{P}(W > t \mid W > 0) = \frac{\mathbb{P}(W > t)}{\mathbb{P}(W > 0)} = \sum_{i=1}^{\infty} \mathbb{P}(W > t \mid X = i)(1 - \varrho)\varrho^{i-1}$$

To obtain a final form of the waiting distribution, we have to express $\mathbb{P}(W > t \mid X = i)$. As assumption we choose $X = i$ in the system. The waiting time of a test unit is divided into two phases:

- The time period starting from the arrival of the test unit until the first end of the serving time. This coincides with the forward recurrence time of the serving time η and has, because of the Markov property, the same distribution as η.
- The time period from the first serving end until the time spot, when all demands are served. This means the time of $i - 1$ serving units.

Thus, the waiting time is combined of i serving periods η. Hence, we have that $\mathbb{P}(W > t \mid X = i)$ is an Erlang distribution of order i

$$\mathbb{P}(W > t \mid X = i) = e^{-\mu t} \sum_{k=0}^{i-1} \frac{(\mu t)^k}{k!}$$

Substituting into the equation for $\mathbb{P}(W > t \mid W > 0)$ we deduce

$$\mathbb{P}(W > t \mid W > 0) = e^{-\mu t} \sum_{i=1}^{\infty} \sum_{k=0}^{i-1} \frac{(\mu t)^k}{k!}(1 - \varrho)\varrho^{i-1}$$

$$= e^{-\mu t} \sum_{k=0}^{\infty} \frac{(\mu t)^k}{k!} \sum_{i=k+1}^{\infty} (1 - \varrho)\varrho^{i-1}$$

$$= e^{-\mu t} \sum_{k=0}^{\infty} \frac{(\varrho\mu t)^k}{k!} = e^{-(1-\varrho)\mu t}$$

Hence, we can solve for $\mathbb{P}(W > t)$ and realize

$$\mathbb{P}(W > t) = \mathbb{P}(W > t \mid W > 0)\mathbb{P}(W > 0) = \varrho e^{-(1-\varrho)\mu t} = 1 - W(t)$$

For the final waiting distribution we note

$$W(t) = 1 - \varrho e^{-(1-\varrho)\mu t}$$

Remark 2.56. The treatment of the general GI/G/1 loss resp. waiting system is still to do. We will come to it in more detail in section 2.8.1 (see also [11, p. 230]).

2.8 General Serving Systems GI/G/n

2.8.1 Loss Systems

We presented in section 2.7.3 queueing probabilities and further key dates, which were not representable in closed form. The only way consisted in using the mean generating function. With F_ξ we indicate, as usually, the distribution of the interarrival times. Let Φ_F be the Laplace transform of a function F. In a GI/G/1 − 0 loss system all new arriving demands will be blocked during the serving of this particular unit. We denote by N_A the amounts of arriving demands during the serving time. Then the blocking probability can be computed according to (see also section 2.7.3)

$$p_B = \frac{\mathbb{E}(N_A)}{1 + \mathbb{E}(N_A)}$$

in other words the relation of the incoming demands to all demands in the system. As in section 2.7.3 we want to determine the expected number $\mathbb{E}(N_A)$ via the conditional probability of η

$$\mathbb{P}(N_A = n \,|\, \eta = x) = F_\xi^{(n)}(x) - F_\xi^{(n-1)}(x) \tag{2.41}$$

where $F_\xi^{(n)}$ is the n times convolution of the distribution function F_ξ. Here, we use the fact that the single arrivals, thus, the incomings of the data in the server or router are independent. From the formula for the conditional expectation we obtain with (2.41)

$$\mathbb{E}(N_A \,|\, \eta = x) = \sum_{n=1}^{\infty} n \mathbb{P}(N_A = n \,|\, \eta = x) = \sum_{n=1}^{\infty} F_\xi^{(n)}(x)$$

With the help of the formula for the conditional expectation we deduce furthermore

$$\mathbb{E}(N_A) = \int_0^\infty \mathbb{E}(N_A \,|\, \eta = x) f_\eta(x) dx \tag{2.42}$$

where f_η is the density of the serving duration.

To gain a upmost explicit representation, we want to assume a hyperexponential distributed serving time. There are more possibilities for considering heavy-tail distributed serving times. For this let

$$f_\xi(x) = \sum_{j=1}^{m} \beta_j \mu_j e^{-\mu_j x} \tag{2.43}$$

Here the coefficients β_j and μ_j are suitable chosen. We can rewrite (2.42) with the help of (2.43) (assuming that the integrals are convergent absolutely)

$$\mathbb{E}(N_A) = \sum_{j=1}^{m} \beta_j \mu_j \int_0^\infty F_\xi^{(n)}(x) e^{-\mu_j x} dx \qquad (2.44)$$

As known from the results of sums over random variables that the n times convolution (thus, the distribution of the sum of n random variables) can be written as the product of the Laplace transform of the corresponding functions. Hence, we can represent the integrand with (1.20) and (1.19) from [102] as

$$\int_0^\infty F_\xi^{(n)}(x) e^{-\mu_j x} dx = \frac{1}{\mu_j} (\Phi_F(\mu_j))^n \qquad (2.45)$$

Here we pick $s = \mu_j$ (inserted into (1.18) of [102]). Putting into (2.44), we deduce

$$\mathbb{E}(N_A) = \sum_{j=1}^{m} \beta_j \frac{\Phi_F(\mu_j)}{1 - \Phi_F(\mu_j)}$$

With increasing serving time the blocking probability decreases – a surprising fact from result (2.45). What is the reason for this? For this we consider the probability that the interarrival items are larger than the serving times, i.e.

$$\Delta = \mathbb{P}(\xi > \eta) \qquad (2.46)$$

These amounts play also a rôle in the consideration of the stationarity. In a loss system GI/G/$r-0$ the state process is stationary, if $\mathbb{E}(\eta) < \infty$ and $\mathbb{P}(\eta_0 < r\xi_0)$, where η_0 and ξ_0 are the serving and interarrival time distributions of the first demand (see section 2.7.3). In contrast to the above example of the hyperexponential distributed serving time, we have chosen a general density f_η. With the complementary distribution function F_ξ^c of the interarrival times we represents (2.46) (consider e.g. section 2.7.4)

$$\Delta = \int_0^\infty F_\xi^c(x) f_\eta(x) dx \qquad (2.47)$$

Let's first consider the hyperexponential distributed interarrival and serving times

$$f_\chi(x) = \sum_{j=1}^{m_\chi} \beta_j^\chi e^{\mu_j^\chi x}, \ \chi \in \{\xi, \eta\}$$

Computing the integral in (2.47) this leads in the hyperexponential case

$$\Delta = \sum_{i=1}^{m_\xi} \sum_{j=1}^{m_\eta} \beta_i^\xi \beta_j^\eta \frac{\mu_j^\eta}{\mu_i^\xi + \mu_j^\eta}$$

On the other hand, choosing a Pareto distribution for the interarrival rep. serving times

$$F_\chi(x) = 1 - \left(\frac{x_\chi}{x}\right)^{\alpha_\chi}, \quad \chi \in \{\xi, \eta\}, \ x \geq x_\chi$$

we obtain for the complementary distribution function (see section 2.7.4)

$$F_\xi^c(x) = \mathbf{1}_{x \leq x_\xi} + \mathbf{1}_{x > x_\xi} \left(\frac{x_\xi}{x}\right)^{\alpha_\xi} \tag{2.48}$$

The density of the serving time reads as

$$f_\eta(x) = \frac{\alpha_\eta}{x}\left(\frac{x_\eta}{x}\right)^{\alpha_\eta}, \quad x > 0 \tag{2.49}$$

We can again insert (2.48) and (2.49) into (2.47) and get

$$\Delta_{Par} = \begin{cases} \frac{\alpha_\eta}{\alpha_\xi + \alpha_\eta}\left(\frac{x_\xi}{x_\eta}\right)^{\alpha_\xi} & \text{for } x_\xi \leq x_\eta \\ 1 - \frac{\alpha_\xi}{\alpha_\xi + \alpha_\eta}\left(\frac{x_\eta}{x_\xi}\right)^{\alpha_\eta} & \text{for } x_\xi > x_\eta \end{cases} \tag{2.50}$$

The Δ in (2.48) as in (2.50) depends heavily on the parameters, influencing the variance (e.g. the α in the expression of the Pareto distribution). A growing Δ means an increase in variance of the serving time distribution and this diminishes the blocking probability. But we cannot consider $1 - \Delta$ as an upper threshold for the blocking probability, since there is a correlation between the existing serving interval and the amount of blocked demands. In the Pareto situation a larger serving time variance is gained in the case of the lowest value of α_η. We know that there does not exist a finite variance for $\alpha_\eta < 2$. But also in this case the blocking probability $1 - \Delta$ decreases (as next blocking) with increasing serving time variance.

Even if the subexponential distributions, as we saw (and will furthermore deduce later) lead to the so called long-range dependence and self-similarity with all the undesirable side effects (as increased waiting time probability see section 3), there are interesting facts, which, as demonstrated in this section, with increasing variance of the serving time lead to decreasing blocking probability, provided the interarrival times have a sufficient large variance.

2.8.2 The Time-Discrete Queueing System GI/G/1

As we already saw in the above treatment of the loss system, in the general case of a GI/G/1 queueing system it will be difficult, mostly impossible, to establish formulas for queueing in closed form. But nevertheless, we can give, using the Z-resp. Fourier-transform, some formulas, which will enable a suitable analysis with the help of computer programmes.

Description of the Model

- The interarrival times are general distributed, but the events occur stochasticly independent.

- The serving time is general distributed, as well.
- We allow an infinite queue, where the entrance into the serving part will occur according to the principles FIFO, FCFS, LIFO or RANDOM.

Since we are interested in the discrete time model, we define

$$\xi(k) = \mathbb{P}(\xi = k \cdot \Delta t), \quad k = 0, 1, \ldots$$
$$\eta(k) = \mathbb{P}(\eta = k \cdot \Delta t), \quad k = 0, 1, \ldots$$

where ξ resp. η denote the interarrival resp. the serving time of each demand, and Δt is a fixed time length. The traffic load of one unit is given by

$$\rho = \frac{\mathbb{E}(\eta)}{\mathbb{E}(\xi)}$$

General Methods

As already mentioned an analytical solution to the queueing problem is not at hand. The fundamental tool for attacking the problem consists in the consideration of the state process of the remaining work. To solve this question the integral equation due to Lindley (see [165]) is a suitable approach, designed for the continuous case. For getting the idea, what is about the integral equation method, we briefly turn to the time continuous case. We denote, as usual, with F_ξ resp. F_η the distribution function of the interarrival resp. the serving time, while f_ξ resp. f_η denote its densities. If we write

$$f_\psi(t) = f_\xi(-t) * f_\eta(t)$$

then, according to Lindley [165], we have the stationary waiting time distribution given by

$$W(t) = \begin{cases} 0 & \text{for } t < 0 \\ W(t) * f_\psi(t) & \text{for } t \geq 0 \end{cases} \qquad (2.51)$$

which is an integral equation ($*$ represents the convolution). The function f_ψ is often called 'systems function', since it contains all random parameters necessary for the queueing system.

The above integral equation (2.51) is a slight modification of the Wiener-Hopf integral equation used in mathematical physics resp. stochastic analysis. If we differentiate (2.51), we obtain an integral equation for the density w

$$w(t) = \begin{cases} 0 & \text{for } t < 0 \\ \delta(t) \int_{-\infty}^{0+} (w(u) * f_\psi(u)) \, du & \text{for } t = 0 \\ w(t) * f_\psi(t) & \text{for } t > 0 \end{cases} \qquad (2.52)$$

Citing Kleinrock [141], we get the compact form

$$w(t) = \pi_0(w(t) * f_\psi(t))$$

where the operator π_0 is defined by

$$\pi_0(f(t)) = \begin{cases} 0 & \text{for } t < 0 \\ \delta(t) \int_{-\infty}^{0+} f(u)du & \text{for } t = 0 \\ f(t) & \text{for } t > 0 \end{cases}$$

Solutions for the Time Discrete Queueing System

The above introduced integral equation is now applied to the time discrete case. We can, in principle, derive the necessary expressions directly from the equation (2.51), but we will directly deduce them in the following.
In the figure 2.34 the arrival and serving process is depicted. The remaining workload $\mathcal{R}(t)$ at time t is defined as sum of the whole serving and residual serving time of all demands being in the system. Since we assume a discrete time, the remaining time process $\mathcal{R}$ is time discrete, too, consisting of integer valued amounts. If a serving unit is occupied, for each time step Δt exactly one working unit is done.

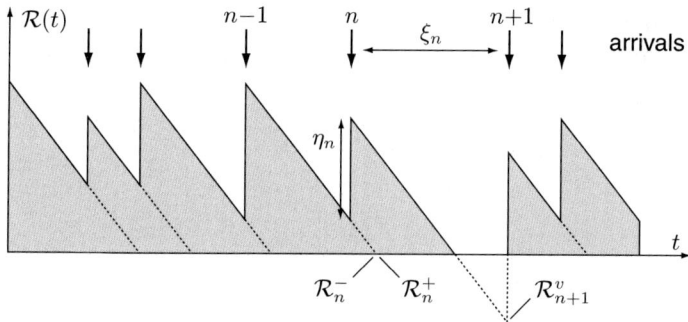

Fig. 2.34. Arrival and serving process for a discrete GI/G/1 system.

We consider the following random variables (RV):

- ξ_n: RV for the interarrival time between the n-th and $(n+1)$-th demand.
- η_n: RV for the serving time of the n-th demand.
- $\mathcal{R}_n^-$: RV for the residual work time in the system immediately before the arrival time of the n-th demand.
- $\mathcal{R}_n^+$: RV for the residual work time in the system immediately after the arrival time of the n-th demand.
- $\mathcal{R}_{n+1}^v$: RV for the *virtual residual work time* in the system immediately before the arrival time of the $(n+1)$-th demand.

The RV $\mathcal{R}_{n+1}^v$ just serves for the derivation of the results. For $\mathcal{R}_{n+1}^v$ it is assumed that the serving period continues, though there is no demand in the system. Hence, it can assume negative values.

We want to compute $\mathcal{R}_n^-$ with the help of $\mathcal{R}_{n+1}^-$. For this, we represent the expression in the figure 2.34

$$\mathcal{R}_n^- \longrightarrow \mathcal{R}_n^+ \longrightarrow \mathcal{R}_{n+1}^v \longrightarrow \mathcal{R}_{n+1}^-$$

- $\mathcal{R}_n^- \longrightarrow \mathcal{R}_n^+$: The residual workload increases after the arrival of the n-th demand by the serving time of this demand, i.e

$$\mathcal{R}_n^+ = \mathcal{R}_n^- + \eta_n$$
$$r_n^+(j) = r_n^-(j) * f_{\eta_n}(j)$$

Note that r_n^+, r_n^- and f_{η_n} are the discrete densities of the respective RV's, and that the density function of the sum of two independent RV is the convolution.

- $\mathcal{R}_n^+ \longrightarrow \mathcal{R}_{n+1}^v$: We can detect using the figure

$$\mathcal{R}_{n+1}^v = \mathcal{R}_n^+ - \xi_n$$
$$r_{n+1}^v(j) = r_n^+(j) * f_{\xi_n}(-j)$$

- $\mathcal{R}_n^v \longrightarrow \mathcal{R}_{n+1}^-$: The relation between the remaining time and virtual residual workload is given by the fact that in an empty system, the virtual time decreases, while the 'real' residual serving time stays at 0. Thus, the virtual residual workload may attain negative values.

$$\mathcal{R}_{n+1}^- = \max(0, \mathcal{R}_n^v)$$
$$r_{n+1}^-(j) = \pi_0(r_{n+1}^v(j))$$

where π_m is defined as an operator in the discrete time range, in analogy to (2.52)

$$\pi_m(x(j)) = \begin{cases} 0 & \text{for } j < m \\ \sum_{i=-\infty}^{m} x(i) & \text{for } j = m \\ x(j) & \text{for } j > m \end{cases}$$

We can deduce the general Lindley integral equation

$$r_{n+1}^-(j) = \pi_0(r_n^-(j) * f_{\xi_n}(-j) * f_{\eta_n}(j))$$
$$= \pi_0(r_n^-(j) * f_{\psi_n}(j))$$

where

$$f_{\psi_n}(j) = f_{\xi_n}(-j) * f_{\eta_n}(j) \tag{2.53}$$

denotes the *time discrete system function*. We assume FIFO regime and thus, $\mathcal{R}_n^-$ is identical with the waiting time W_n of the n-th demand. Hence, from (2.53) we conclude

$$w_{n+1}(j) = \pi_0\left(w_n(j) * f_\psi(j)\right) \tag{2.54}$$

The relation (2.54) indicates the queueing time distribution of two succeeding demands for general arrival and serving processes. The arrival resp. serving time can be demand dependent. We assume that the interarrival and serving time are iid distributed, i.e.

$$\xi = \xi_n, \ \eta = \eta_n, \ \text{for all } n \in \mathbb{N}$$

and the system is stationary, i.e.

$$W = \lim_{n \to \infty} W_n$$

Thus, we obtain a stationary probability

$$w(j) = \pi_0 \left(w(j) * f_\psi(j) \right), \ \text{with } f_\psi(j) = f_\xi(-j) * f_\eta(j) \qquad (2.55)$$

From equation (2.55) we can derive a formula for the queueing time distribution

$$F_W(j) = \sum_{k=0}^{j} f_W(k) = \sum_{k=-\infty}^{j} f_\psi(k) * f_W(k) = \sum_{k=-\infty}^{j} \sum_{i=-\infty}^{\infty} f_\psi(i) \cdot f_W(k-i)$$

$$= \sum_{i=-\infty}^{\infty} f_\psi(i) \sum_{k=-\infty}^{j} f_W(k-i) = \sum_{i=-\infty}^{\infty} f_\psi(i) \cdot F_W(k-i), \ j = 0, 1, \ldots$$

or shortly written

$$F_W(j) = \begin{cases} 0 & \text{for } k < 0 \\ f_\psi(j) * F_W(j) & \text{for } k \geq 0 \end{cases} \qquad (2.56)$$

Thus, it is obvious to state the queueing probability p_W in our system

$$p_W(j) = \mathbb{P}(\text{at least one demand has to wait until being served})$$
$$= \mathbb{P}(W > 0) = 1 - F_W(0) = 1 - f_W(0) \qquad (2.57)$$

Analysis by Z-Transformation

We can, using (2.56), deduce that F_W is the discrete convolution with $f_\psi(\cdot)$, provided we omit the negative time part. We introduce a correction term $F_{W-}(k)$ for $k < 0$ and obtain

$$F_{W-}(k) + F_W(k) = f_\psi(k) * F_W(k) \qquad (2.58)$$

We consider the Z-transform $W_Z(\cdot)$ of f_w and $\mathbb{W}_Z(\cdot)$ of the distribution F_W. Similar $\mathbb{W}_Z^-(\cdot)$ is the Z-transform of the correction term. We can formulate the relationships

$$f_w(j) \overset{\text{Z-trans.}}{\Leftrightarrow} \mathbb{W}_Z(z) = \sum_{j=0}^{\infty} f_w(j) z^{-j}$$

$$F_W(j) = \sum_{k=0}^{j} f_W(k), \; j = 0, 1, \ldots \overset{\text{Z-trans.}}{\Leftrightarrow} \mathbb{W}_Z(z) = \sum_{j=0}^{\infty} F_W(j) z^{-j}$$

$$F_{W^-}(j), \; j = 0, 1, \ldots \overset{\text{Z-trans.}}{\Leftrightarrow} \mathbb{W}_Z^-(z) = \sum_{j=-\infty}^{-1} F_{W^-}(j) z^{-j}$$

We have for the Z-transform of the F_W

$$\mathbb{W}_Z(z) = \sum_{j=0}^{\infty} F_W(j) z^{-j} = \sum_{j=0}^{\infty} \sum_{k=0}^{j} f_W(k) z^{-j} = \frac{W_Z(z)}{1 - z^{-1}} \quad (2.59)$$

By equation (2.58) and (2.59) we conclude with Z_{f_ψ} as the Z-transform of the density f_ψ

$$\mathbb{W}_Z^-(z) + \mathbb{W}_Z(z) = Z_{f_\psi}(z) \cdot \mathbb{W}_Z(z)$$

which can be transformed to the so called *characteristic equation*

$$\mathbb{W}_Z^-(z) \cdot \frac{1}{\mathbb{W}_Z(z)} = \frac{Z_{f_\psi}(z) - 1}{1 - z^{-1}} \quad (2.60)$$

We denote the right side of (2.60) as *characteristic function*

$$\mathbb{S}_Z(z) = \frac{Z_{f_\psi}(z) - 1}{1 - z^{-1}} \quad (2.61)$$

Remark 2.57. We give some further remarks on the results above.

- We can interpret the term $W_Z^-(z)$ as the Z-transform of a sequence of coefficients, which do not increase on the negative axis. According to the Eneström-Kakeya theorem all zero points lie outside of the unit circle (see [8]). Since F_{W^-} is a left hand side defined sequence with finite and non negative values, it can be shown according to [198] that the region of convergence of the function $W_Z^-(z)$ is within and on the unit circle. Thus, all singularities are outside. Hence, $W_Z^-(z)$ is a Z-transform of a maximal phase sequence.
- The function F_W is the Z-transform of a distribution. Hence, it converges within and on the unit circle and all singularity are outside.
- Because of the stability constraint $\rho < 1$, we conclude with the help of the limit theorem of Z-transforms

$$0 < \lim_{z \to \infty} W_Z(z) \leq 1$$

Example 2.58. The Geom(m)/Geom(m)/1 system: For some distributions concerning the interarrival times resp. serving times it is possible to derive

2.8 General Serving Systems GI/G/n

the characteristic equation directly. We choose the parameter m identical for both processes, the arrival as well as the serving one

$$f_\xi(j) = (1-\alpha)\alpha^{j-m}, \quad j \geq m, \quad \mathbb{E}(\xi) = m + \frac{\alpha}{1-\alpha}$$

$$f_\eta(j) = (1-\beta)\beta^{j-m}, \quad j \geq m, \quad \mathbb{E}(\xi) = m + \frac{\beta}{1-\beta}$$

with $\alpha, \beta \in [0,1]$. The Z-transforms read as

$$Z_{f_\xi}(z) = \frac{(1-\alpha)z^{-m}}{1-\alpha z^{-1}}$$

$$Z_{f_\eta}(z) = \frac{(1-\beta)z^{-m}}{1-\beta z^{-1}}$$

The stability constraint implies

$$\rho = \frac{\mathbb{E}(\eta)}{\mathbb{E}(\xi)} < 1 \Leftrightarrow \beta < \alpha$$

From equation in (2.53) we deduce the Z-transform of the system function

$$Z_{f_\psi}(z) = Z_{f_\xi}(z^{-1}) \cdot Z_{f_\eta}(z) = \frac{(1-\alpha)(1-\beta)}{(1-\alpha z)(1-\beta z^{-1})}$$

According to the definition of the characteristic function (2.60), we get

$$\mathcal{S}_Z = \frac{\alpha z - \beta}{(1-\alpha z)(1-\beta z^{-1})}$$

Since for further analysis the singularities and zero points are important, we give the useful alternative representation

$$\mathcal{S}_Z = \alpha \frac{z\left(1-\frac{\beta}{\alpha}z^{-1}\right)}{(1-\alpha z)(1-\beta z^{-1})} = \mathbb{W}_Z^-(z) \cdot \frac{1}{Z_W(z)} \qquad (2.62)$$

Thus, $\frac{1}{\alpha}$ is the only singularity outside the unit circle of $\mathcal{W}_Z^-(\cdot)$, which is the Y-transform of the maximal phase sequence $(F_{W^-}(j))$. We deduce by equation (2.62) the queueing time distribution

$$Z_W(z) = K_0 \frac{1}{\alpha} z^{K_1 - 1} \frac{1 - \beta z^{-1}}{1 - \frac{\beta}{\alpha} z^{-1}}$$

Because of the norming constraint $Z_W(1) = 1$, we get

$$K_0 = \frac{\alpha - \beta}{1 - \beta} \qquad (2.63)$$

Equation (2.61) tells us

$$\lim_{z \to \infty} Z_W(z) > 0 \tag{2.64}$$

and we obtain

$$K_1 = 1 \tag{2.65}$$

Finally, equations (2.63), (2.64) and (2.65) offer the queueing time distribution in the Z-transform

$$Z_W = \frac{\alpha - \beta}{1 - \beta} \cdot \frac{z - \beta}{z\alpha - \beta}$$

Analytical Cycles

From

$$f_{W_{n+1}}(j) = \pi_0 \left(f_{W_n}(j) * f_{\xi_n}(-j) * f_{\eta_n}(j) \right) = \pi_0 \left(f_{W_n}(j) * f_{\psi_n}(j) \right)$$

follows

$$f_{W_0}(j) = \delta(j) = \begin{cases} 1 & \text{for } j = 0 \\ 0 & \text{else} \end{cases}$$

Algorithm in the Transformed Range

The algorithm for determining the queueing probability for the general GI/G/1 queueing system are based on the Z-transform

$$\mathbb{W}_Z^-(z) \cdot \frac{1}{\mathbb{W}_Z(z)} = \frac{Z_{f_\psi}(z) - 1}{1 - z^{-1}}$$

$$\mathbb{W}_{\text{gf}}^-(z) \cdot \frac{1}{W_{\text{gf}}(z)} = \frac{\text{GF}_{f_\psi}(z) - 1}{1 - z^{-1}}$$

where $\mathbb{W}_{\text{gf}}^-$, W_{gf} and GF_{f_ψ} are the generating functions. To solve the above equations for the explicit queueing probability, there are two standard methods:

- Polynomial factorization: Separation method by explicit computation of the singularities and zero points using polynomial factorization (see [147, 252].
- Cepstrum separation: By considering the phase property terms the characteristic function will be separated by the Cepstrum concept (see [8, 256]).

We do not go into the details of both methods, since this topic is beyond the scope of this book. The reader is recommended to consult the given literature.

System Characteristics

Queueing time probability: As already analyzed in equation (2.57), the probability that an arriving demand has to wait, is identical with the probability that immediately before the arriving time spot the serving units are occupied, i.e.

$$p_W = \sum_{j=1}^{\infty} u^-(j) = 1 - u^-(0) = 1 - w(0) \quad \text{waiting time!probability}$$

Queueing Time of the Waiting Demands: The distribution $f_{w_1}(j)$ of the waiting time W_1 of the demands, which have to wait, can be computed from the general waiting time distribution $W(j)$ of an arbitrary demand

$$f_{w_1}(j) = \mathbb{P}(W_1 = j) = \begin{cases} 0 & \text{for } j \leq 0 \\ \frac{f_w(j)}{p_W} & \text{for } k > 0 \end{cases}$$

System load and mean queueing length: With the help of the Little formula and the mean waiting time we can determine the mean queueing length Ω and the system load $\mathbb{E}(X)$. For the entire system we get $\mathbb{E}(W) + \mathbb{E}(\eta)$ for the whole sojourn time, and with the mean interarrival time $\mathbb{E}(\xi)$, we conclude finally the system load $\mathbb{E}(X)$

$$\mathbb{E}(X) = \frac{\mathbb{E}(W) + \mathbb{E}(\eta)}{\mathbb{E}(\xi)} = \frac{\mathbb{E}(W)}{\mathbb{E}(\xi)} + \rho$$

with as usual $\rho = \frac{\mathbb{E}(\eta)}{\mathbb{E}(\xi)}$. Using the Little formula, we get for the mean queueing length

$$\Omega = \frac{\mathbb{E}(W_1) \cdot p_W}{\mathbb{E}(\xi)} = \frac{\mathbb{E}(W)}{\mathbb{E}(\xi)}$$

2.8.3 GI/G/1 Time Discrete Queueing System with Limitation

For the analysis in the preceding section an unlimited queueing time was assumed. In most communications systems we have to impose a limited waiting time.

Queueing in Systems with Limitation

In this system the waiting queue is not allowed to excced a certain threshold $W_{\max}$, which is given by a time interval Δt and a fixed length L. The time axis is divided in intervals of length Δ_n

$$W_{\max} = L \cdot \Delta t$$

This means that the residual workload does not exceed the threshold $W_{\max}$. We introduce the following random variables:

- ξ_n: RV for the interarrival time of the n-th to the $(n+1)$-th demand.
- η_n: RV for the serving time of the n-th demand.
- $\mathcal{R}_n$: RV for the residual time in the system, immediately before the arrival time spot of the n-th demand.

We treat the problem by splitting it into three cases. In the first case we assume that the residual time $\mathcal{R}_n$ is smaller than L, then that $\mathcal{R}_n$ exceeds L and finally we consider the general case:

- Case 1: Acceptance of the demand in case of $\mathcal{R}_n < L$. We consider the conditional RV $\mathcal{R}_{n,0} = \mathcal{R}_n | \mathcal{R}_n < L$, whose distribution reads as

$$r_{n,0}(j) = \frac{\sigma^{L-1}(r_j))}{\mathbb{P}(\mathcal{R}_n < L)} = \frac{\sigma^{L-1}(r_j))}{\sum_{k=0}^{L-1} r_n(k)} \qquad (2.66)$$

where $\sigma^m(x(j))$ is the operator mimicking the lower part ($k \leq m$) of the distribution $x(\cdot)$

$$\sigma^m(x(j)) = \begin{cases} x(j) & \text{for } j \leq m \\ 0 & \text{for } j > m \end{cases}$$

Dividing the equation (2.66) by $\mathbb{P}(\mathcal{R}_n < L)$ is nothing else than a kind of norming of the conditional distribution of the RV $\mathcal{R}_{n,0} = \mathcal{R}_n | \mathcal{R}_n < L$. It follows

$$\mathcal{R}_{n+1,0} = \mathcal{R}_{n,0} + \eta_n - \xi_n \qquad (2.67)$$
$$r_{n+1,0}(j) = \pi_0 \left(r_{n,0}(j) * f_{\eta_n}(j) * f_{\xi_n}(-j) \right)$$

- Case 2: Blocking of demands in case of $\mathcal{R}_n \geq L$. We analyze this case similar to the above one. Start with the conditional RV $\mathcal{R}_{n,1} = \mathcal{R}_n | \mathcal{R}_n \geq L$. For the distribution we get

$$r_{n,1}(j) = \frac{\sigma_L(r_{(j)})}{\mathbb{P}(\mathcal{R}_n \geq L)} = \frac{\sigma_L(r_{(j)})}{\sum_{k=L}^{\infty} r_n(k)} \qquad (2.68)$$

where $\sigma_m(x(j))$ is the operator passing the upper part ($k \leq m$) of the distribution $x(\cdot)$

$$\sigma_m(x(j)) = \begin{cases} x(j) & \text{for } j \geq m \\ 0 & \text{for } j < m \end{cases}$$

Again dividing the equation (2.68) by $\mathbb{P}(\mathcal{R}_n \geq L)$ is basically the norming of the conditional distribution of the RV $\mathcal{R}_{n,1} = \mathcal{R}_n | \mathcal{R}_n \geq L$. It follows

$$\mathcal{R}_{n+1,1} = \mathcal{R}_{n,1} - \xi_n \qquad (2.69)$$
$$r_{n+1,1}(j) = \pi_0 \left(r_{n,1}(j) * f_{\xi_n}(-j) \right)$$

- Case 3: Combination of Case 1 and Case 2. This case reflects both the acceptance and blocking of the residual workload from between the n-th and $(n+1)$-th demand. The distribution is a weighted combination of both cases (resp. equations (2.67) and (2.69))

$$\begin{aligned}r_{n+1}(j) &= \mathbb{P}(\mathcal{R}_n < L) r_{n+1,0}(j) + \mathbb{P}(\mathcal{R}_n \geq L) r_{n+1,1}(j) \\ &= \pi_0 \left(\sigma^{L-1}(r_{n,0}(j)) * f_{\eta_n}(j) * f_{\xi_n}(-j) \right) + \pi_0 \left(\sigma_L(r_{n,1}(j)) * f_{\xi_n}(-j) \right) \\ &= \pi_0 \left(\left(\sigma^{L-1}(r_{n,0}(j)) * f_{\eta_n}(j) + \sigma_L(r_{n,1}(j)) \right) * f_{\xi_n}(-j) \right) \end{aligned}$$

where the last equation is due to the fact that π_0 is piecewise linear

$$p_B = \sum_{j=L}^{\infty} r(j)$$

Analysis of Spacers in Communications Networks

Example 2.59. Spacer in communication networks: Before we apply the above theoretical framework to this particular model, we shortly give an overview on the concept of *spacers* in the communication networks. This concept is used for the traffic management and admission control at the *User Network Interface* (UNI). As we already mentioned in the previous chapter, usual IP traffic reveals a bursty character. The spacing technique is assigned to smooth the bursts. Intensively applying in ATM networks, it is introduced in the IP networks for the QoS solutions. Let us describe, how the concept works.

As simple idea, the spacing tries to separate two incoming demands up to interval of T time units. Thus, the arrival rate is at most $\frac{1}{T}$. This value will be agreed on by the users and providers in the so called service level agreements (SLA). The incoming traffic is smoothed up to this agreed SLA. Shortly speaking, the spacer is functioning as a device for keeping distances between the incoming demands, entering the network. The figure 2.35 depicts the spacing of two incoming demands for a parameter T. The arrival process $(\xi_1(t))$ is smoothed down, so that the outgoing process $(\xi_2(t))$ has a distance on T time units. In the right figure the arrivals $1, 2, 6$ are in time and are passing without delay the spacer, while $3, 4$ and 5 are too early and are stored. To avoid a too large storing, a maximal spacing time $T_{\max}$ is defined. If an incoming demand is predicted to surpass this maximal spacing time, it is rejected.

Example 2.60. Spacer model as GI/D/1 system with queueing time limit: We discretize the time axis in intervals Δt for the time discrete analysis:

- ξ_1: RV for the interarrival time of the incoming process resp. arrival process.
- ξ_2: RV for the 'interleaving' time of the outgoing resp. leaving process.

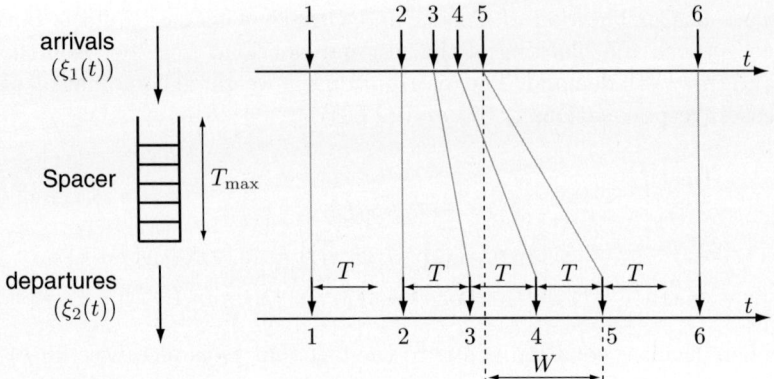

Fig. 2.35. Spacing of incoming demands

- Ξ: maximal interarrival distance (unit Δt), i.e the maximal allowed interarrival time is $\frac{1}{\Xi}$.
- $\tau_{\max}$: maximal delay in the spacer (unit Δt).

Our aim is to show that we can model the spacer as a GI/D/1 system. This is done, if we can demonstrate that the arrival process, the blocking probability and the outgoing process are identical. For this purpose we give evidence for:

- both models agree on the blocking behavior provided the same arrival process and
- the outgoing process coincides with a delay of T.

In the figure 2.36 we compare the state process in the spacer and the GI/D/1−$T_{\max}$ queueing system.
We now give an interpretation of the above examples. For this we investigate the remaining time $\mathcal{R}(t)$:

- Spacer: $\mathcal{R}(t)$ is the time until the arrival of the next demand will be accepted, since the minimal interarrival time is fulfilled. Demands 1 and 2 meet the spacer being empty and pass it (see figure 2.36), while demands 3, 4 and 5 see a positive residual work time $\mathcal{R}(t)$ and are delayed at about $\mathcal{R}(t)$, before leaving the spacer. Clearly, accepted demands increase the residual workload $\mathcal{R}(t)$ by the value T. Thus, incoming demand 6 would have to wait more than $T_{\max}$ and will be rejected, not increasing the remaining workload $\mathcal{R}(t)$.
- GI/D/1 system with queueing time limit: The remaining workload $\mathcal{R}(t)$ in the system is described similar to the general GI/G/1 queueing system.

The queueing time is identical for both systems. They only differ by the fact that the spacer does not need additional serving. Comparing the outgoing processes, we see that, because of the missing serving time T, these outgoing processes differ at the amount T. As the blocking behavior is identical in

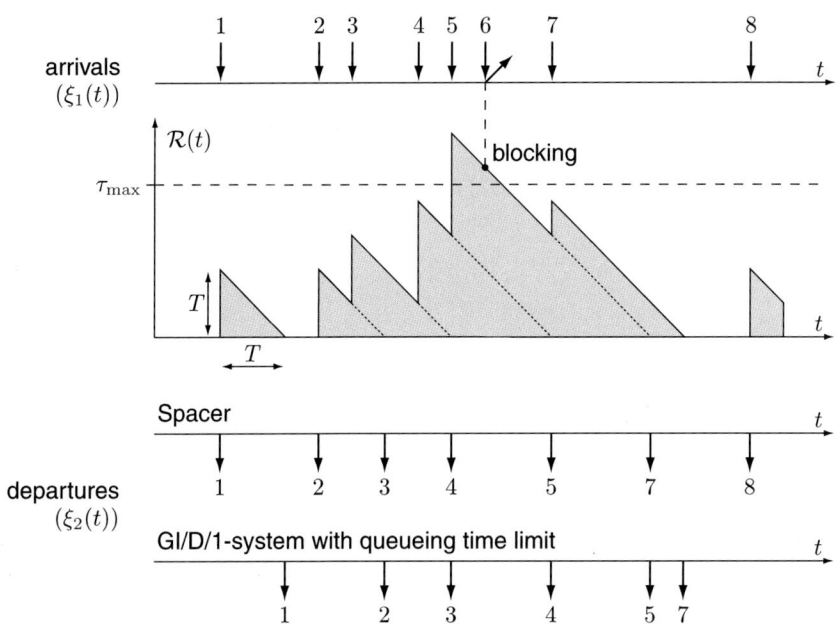

Fig. 2.36. Outgoing processes of spacer and GI/D/1 system with queueing time limit

both models, we can say that the GI/D/1 queueing system is suitable for the description of the spacer concept.

Serving Time in a Network

We start with the description, how the heavy-tail distribution exercises its influence on the waiting time probability. For this we denote with T_a the time already spend in the net, with T_R the remaining time and with T the present time of the flow in the net. Under the assumption that the flow in the network already durated T_a, we can compute the conditional probability that the time of the flow in the network will surpass the time value of $T_a + t$ and represent this in the form

$$\mathbb{P}(T > T_a + t \,|\, T > T_a) = \mathbb{P}(T_R > t \,|\, T > T_a) = \frac{F_T^c(T_a + t)}{F_T^c(T_a)}$$

Here, F_T^c describes the complementary distribution function of the duration of the flow in the net. With this equation we can determine the time, which remains to determine a shortcut for the flow. Here, we assume a probability that the flow will stay in the connection for at least t time units. It is clear that with growing t the probability will decrease and hence, we will determine an upper bound, before which the decision has to be taken.

The following consideration is crucial in connection with the heavy-tail distributions. With growing duration time T_a, the probability increases to stay further t time units in the net, which would not be the case in the classical exponential distributed duration time

$$\frac{F_T^c(T_a+t)}{F_T^c(T_a)} \sim \frac{L(T_a+t)}{L(T_a)} \frac{(T_a+t)^{-\alpha}}{T_a^{-\alpha}} \quad (L \in \mathcal{R}_0) \qquad (2.70)$$

Because of the property of slowly varying functions that $\frac{L(T_a+t)}{L(T_a)} \to 1$ for $T_a \to \infty$ holds, we get with (2.70)

$$\lim_{T_a \to \infty} \frac{F_T^c(T_a+t)}{F_T^c(T_a)} = \lim_{T_a \to \infty} \frac{(T_a+t)^{-\alpha}}{T_a^{-\alpha}} = \lim_{T_a \to \infty} \left(1 + \frac{t}{T_a}\right)^{-\alpha} > \left(1 + \frac{t}{T_{a'}}\right)^{-\alpha}$$

for all $T_{a'} > 0$. Thus, the conditional probability increases with growing T_a. This is in contrast to the exponential distributed serving times, which the reader may convince herself/himself easily as exercise.

2.9 Network Models

In this section we summarize a selection of well known network models. All models are extensively described in the respective literature and provide a foundation for numerous specific approaches.

2.9.1 Jackson's Network

In a lot of service systems a simple linear division in arrival part, service part and waiting room is not possible. The best examples can be found in communications and computer networks, where a large number of routers and hosts are connected to a network. We separate two larger systems, the open and the closed service networks and start with the open systems.

Since differing from the classical models, the representation will get more complex. Similar observations can be made by passing from the Poisson arrival as well as serving systems to the general distributed interarrival time resp. as well as serving times, as derived already in section 2.7.3. Thus, we will introduce the simple and often discussed service network due to Jackson (see e.g. [30]). First we give some necessary preliminaries:

- We start with n nodes $\mathbf{1}, \ldots, \mathbf{n}$.
- The node $\mathbf{j}$ has s_j channels ($1 \leq s_j \leq \infty$).
- In each node there is an infinite waiting room.
- Each node receives demands from 'outside'. These demands arrive independently at the node $\mathbf{j}$, and they are Poisson distributed with intensity λ_j.

- The serving time in each node **j** are independent and identically exponential distributed with parameter μ_j. In addition, between the nodes the serving times are independent.
- According to the probability α_j a certain demand leaves after serving in node **j** the network or enters the node **i** with transition probability p_{ji}, where it is served again. The transition distribution is independent to the preceding process. The serving process is also again a Markov process.
- Let $\mathcal{P} = (p_{ji})_{i,j=1,\ldots,n}$ be the transition matrix. If $\mathcal{E}$ is the unit matrix, then let $\mathcal{P}-\mathcal{E}$ be invertible, i.e. $(\mathcal{P}-\mathcal{E})^{-1}$ exists. It holds for all $j = 1,\ldots,n$

$$\alpha_j + \sum_{i=1}^{n} p_{ji} = 1 \tag{2.71}$$

We denote by σ_j the whole request rate in the node **j**, which consists of all demands within as well as outside the network. Since we assume stationarity in the whole system, we have to stress the equality of input and output rate. By $\sigma_j p_{ji}$ we have the part of the input rate of the node **i**, coming from **j**. Then $\sum_{j=1}^{n} \sigma_j p_{ji}$ reflects the whole input rate of the node **i**, and it follows the stationarity condition

$$\sigma_i = \lambda_i + \sum_{j=1}^{n} \sigma_j p_{ji}, \quad i = 1,\ldots,n \tag{2.72}$$

We can express this more elegantly

$$\sigma(\mathcal{P} - \mathcal{E}) = \lambda \tag{2.73}$$

where $\sigma = (\sigma_1,\ldots,\sigma_n)$ and $\lambda = (\lambda_1,\ldots,\lambda_n)$. If we know the transition probabilities and the arrival rates, then we can compute the specific input intensities

$$\sigma = \lambda(\mathcal{P} - \mathcal{E})^{-1}$$

The weighted sum of the Poisson streams is not necessarily again Poisson. As done in the other models we denote by $\mathcal{X}(t) = (X_1(t),\ldots,X_n(t))$ the state vector, where we indicate by

$$\mathbf{k} = (k_1,\ldots,k_n)$$

the vector of realizations. The stationary state probabilities read as

$$x_{\mathbf{k}} = \lim_{t \to \infty} \mathbb{P}(\mathcal{X}(t) = \mathbf{k})$$

By $\mathbf{e}_j = (0,\ldots,1,\ldots,0)$ we denote the j-th unit vector, with a 1 at the j-th place. We can represent each **k** as sum of different unit vectors $\mathbf{e}_1,\ldots,\mathbf{e}_n$. Let's start with the state **k**. Then we can treat the following transitions:

- If a demand enters the node **j** from outside, then the state $\mathcal{X}(t)$ goes into the state $\mathbf{k} + \mathbf{e}_j$.

- If a serving in node **j** is finished, then we have a transition into the state so $\mathbf{k} - \mathbf{e}_j$.
- If in the node **j** a serving unit is done and enters the node **i**, then we obtain $\mathbf{k} - \mathbf{e}_j + \mathbf{e}_i$.

Hence, we deduce the transition probability rates

$$q_{\mathbf{k},\mathbf{k}+\mathbf{e}_j} = \lambda_j$$
$$q_{\mathbf{k},\mathbf{k}-\mathbf{e}_j} = \min(k_j, s_j)\mu_j \alpha_j$$
$$q_{\mathbf{k},\mathbf{k}-\mathbf{e}_j+\mathbf{e}_i} = \min(k_j, s_j)\mu_j p_{ji}, \ j \neq i$$

Because of (2.71), we deduce $\sum_{i=1, j\neq i}^{n} p_{ji} = 1 - p_{jj} - \alpha_j$. The rate leaving the state **k** is given by

$$q_{\mathbf{k}} = -\sum_{j=1}^{n} \lambda_j - \sum_{j=1}^{n} \mu_j(1-p_{jj})\min(x_j, s_j)\mu_j$$

Since all states **k** fulfill the linear equations of the Chapman-Kolmogorov system, we obtain by (2.1)

$$q_{\mathbf{k}} x_{\mathbf{k}} = \sum_{j=1}^{n} \lambda_j x_{\mathbf{k}-\mathbf{e}_j} + \sum_{j=1}^{n} \alpha_j \mu_j \min(k_j+1, s_j) x_{\mathbf{k}+\mathbf{e}_j} \tag{2.74}$$

$$= \sum_{i=1}^{n} \sum_{j=1, j\neq i}^{n} \alpha_j \mu_j \min(k_j+1, s_j) p_{ji} x_{\mathbf{k}+\mathbf{e}_j-\mathbf{e}_i}$$

in conjunction with norming equation

$$1 = \sum_{\mathbf{k} \in \mathbb{Z}} x_{\mathbf{k}}$$

We remind the state probabilities of the waiting system $M/M/s_j - \infty$ from section 2.4.2. If all demands enter with intensity σ_j into the system, and if all demands are independent and identical exponential distributed on the s_j channels with traffic load $\rho_j = \frac{\sigma_j}{\mu_j}$, then we have under the condition $\sigma_j < s_j$

$$x_j(i) = \begin{cases} \frac{1}{i!}\rho_j^i x_j(0) & \text{for } i = 1, 2, \ldots, s_j - 1 \\ \frac{1}{s_j! s_j^{i-s_j}} \rho_j^i x_j(0) & \text{for } i = s_j, s_j+1, \ldots \end{cases}$$

where from the norming equation

$$x_j(0) = \left(\sum_{i=0}^{s_j-1} \frac{1}{i!}\rho_j^i + \frac{\rho_j^{s_j}}{(s_j-1)!(s_j-\rho_j)} \right)^{-1} \tag{2.75}$$

follows. Because of the independence, the channels are coupled multiplicatively and thus, we can formulate the following theorem for $j = 1, \ldots, n$.

Theorem 2.61. *If the intensity rate σ satisfies in addition to equation (2.72) and the relation*
$$\sigma_j < s_j \mu_j, \text{ for } j = 1, 2, \ldots, n$$
holds, then for $\mathbf{k} = (k_1, \ldots, k_n)$ the process X possesses the state probabilities
$$x_{\mathbf{k}} = \prod_{j=1}^{n} x_j(k_j)$$

We see that the system resembles a waiting system of the form $M/M/s - \infty$, but it is not of the same type. The $M/M/s - \infty$ waiting system was a special form of the birth and death process, i.e. (X_t) reflects a birth and death process. That is not the case here, since the input process is a compounded Poisson process, and this need not to be again Poissonian. The waiting queue length in the respective nodes (server or router) are mutually independent. An explicit computation for proving is rather sophisticated. For this, one has to compute (2.75) and consequently insert the result into (2.74). Then, with the help of (2.73) and (2.72) we establish an identity. We give two examples for special networks.

Example 2.62. For the most simple example we choose $n = 1$. But in addition we stress that a part of the served demand is not leaving the system, but appears again as demands. It could e.g. happen that a part of $(1-\alpha)100\%$ data are error transmitted and has to repeat the run through the router. Then, α denotes the probability leaving the system and we have $p_{11} = 1 - \alpha$. In the sequel we omit the indices. Because of (2.71), we get for the input rate
$$\sigma = \lambda + \sigma(1 - \sigma)$$
Then, we deduce
$$\sigma = \frac{\lambda}{\alpha}$$
By condition $\frac{\lambda}{\alpha} < s\mu$ (or equivalently $\rho < \alpha s$) we deduce the stationarity where $\rho = \frac{\lambda}{\mu}$). Thus, we can compute the stationary state probabilities
$$x_i = \begin{cases} \frac{1}{i!}(\frac{\rho}{\alpha})^i x_0 & \text{for } i = 1, 2, \ldots, s-1 \\ \frac{1}{s! s^{i-s}}(\frac{\rho}{\alpha})^i x_0 & \text{for } i = s, s+1, \ldots \end{cases}$$
where
$$x_0 = \left(\sum_{i=1}^{s-1} \frac{1}{i!}(\frac{\rho}{\alpha})^i + \frac{(\frac{\rho}{\alpha})^s}{(s-1)!(s - \frac{\rho}{\alpha})} \right)^{-1}$$
We have a $M/M/s - \infty$ waiting system without respond, whose demands have the intensity $\frac{\lambda}{\alpha}$.

Example 2.63. In this example we want to examine a sequential network. Demands resp. data from outside appear only at the node **1**, i.e. $\lambda_1 = \lambda$. The demands run through the network sequentially from **1** to **n**. Thus, we can state the following parameters for the system

$$\lambda_j = 0, \ j = 2, \ldots, n$$
$$p_{j,j+1} = 1, \ j = 1, \ldots, n$$
$$\alpha_j = 0, \ j = 1, \ldots, n-1$$
$$\alpha_n = 1$$

In the case of stationarity the input rates at all nodes have to match

$$\lambda = \sigma_1 = \sigma_2 = \ldots = \sigma_n$$

Assuming only one service channel for each node, we get

$$\rho_j = \frac{\lambda}{\mu_j} < 1, \ j = 1, \ldots, n$$

We deduce the estimation

$$\lambda < \min(\mu_1, \ldots, \mu_n)$$

Hence, the network is as efficient as its weakest member is. The state probability can be computed for $\mathbf{k} = (k_1, \ldots, k_n)$ according to

$$x_{\mathbf{k}} = \prod_{j=1}^{n} \rho_j^{k_j}(1 - \rho_j), \ \mathbf{k} \in \mathbb{Z}^n$$

State Dependent Parameter

In particular for communication networks, the input and service rates can be state dependent. Thus, the exterior arrival rates in the nodes **j** are state dependent according to the condition

$$\lambda_j(k) = r_j \lambda(k), \text{ with } r_1 + r_2 + \ldots + r_n = 1$$

if we have k demands in the system $(j = 1, \ldots)$. Assuming at time t an over all state in the single nodes of $X_1(t) + X_2(t) + \ldots + X_n(t) = k$, the over all arrival rate in the system is $\lambda(k)$. Similar we denote by $\mu_j(k_j)$ the service rate in the node **j** under the presumption that k_j demands are served there.

All other conditions of the Jackson service network are assumed to be still valid. In addition, we assume the existence of a number $N \in \mathbb{N}$, such that $\lambda(k) > 0$, if $k < N$ and $\lambda(k) = 0$ for $k \geq N$. Thus, we want to prevent that the system suffers an overflow, as soon as the system falls under the threshold N. Furthermore, let (we assume the unique solvability) be the vector $\sigma = (\sigma_1, \ldots, \sigma_n)$ as unique solution of the system of equations

$$\sigma_i = r_i + \sum_{j=1}^{n} \sigma_j p_{ji}, \ j = 1, \ldots, n$$

In addition, we start with an empty initial system, i.e. $x_i(0) = 1$ and

$$x_j(k) = \prod_{l=1}^{k} \frac{\sigma_j}{\mu_j(l)}, \ k = 1, 2, \ldots$$

We choose an arbitrary state vector $\mathbf{k} = (k_1, \ldots, k_n)$. The single nodes fulfill the boundary condition

$$|\mathbf{k}| = \sum_{i=1}^{n} k_i$$

This is the over all number of demands in the system. We can determine the state probability in dependence of the vector $\mathbf{k}$

$$x_{\mathbf{k}} = x_{\mathbf{k}_0} \prod_{l=1}^{|\mathbf{k}|} \lambda(l) \prod_{j=1}^{n} x_j(k_j), \ \mathbf{k} \in \mathbb{Z}^n, \ |\mathbf{k}| > 0$$

with $\mathbf{k}_0 = (0, 0, \ldots, 0)$ and

$$x_{\mathbf{k}_0} = \left(1 + \sum_{\{\mathbf{k}; |\mathbf{k}| > 0\}} \prod_{l=1}^{|\mathbf{k}|} \lambda(l) \prod_{j=1}^{n} x_j(k_j) \right)^{-1}$$

Though the state probabilities appear in product form, the particular states in the nodes are no longer mutually independent.

Closed Serving Networks

Here, we consider a network without demands entering from outside. That means that demands served in a node, are not leaving the system and are handled over to another serving node. Thus, we can state first that the number of demands do not change. The modeling reads as:

- The network consists of n nodes $\mathbf{1}, \ldots, \mathbf{n}$.
- After the serving in a node $\mathbf{j}$ the demand enters with probability p_{ji} the node $\mathbf{i}$. Hence, we have the presumption

$$\sum_{i=1}^{n} p_{ji} = 1, \ i = 1, \ldots, n \qquad (2.76)$$

- For the Markov process with state space $\mathbf{N} = \{1, 2, \ldots n\}$ and with transition matrix $\mathcal{P} = (p_{ji})$ the state probabilities fulfill for $x_1, \ldots, x_n$ the equations

$$x_i = \sum_{j=1}^{n} p_{ji} x_i, \ i = 1, \ldots, n, \ \sum_{i=1}^{n} x_i = 1 \qquad (2.77)$$

- We denote the serving rate in the node $\mathbf{j}$ by $\mu_j(k_j)$, provided k_j demands are in the node $\mathbf{j}$.

As usual we denote the state in the node $\mathbf{j}$ at time t by $X_j(t)$ and the over all state in the system by $\mathcal{X}(t) = (X_1(t), \ldots, X_n(t))$. Indicating by N the whole number of demands in the system, we deduce for the state space

$$\mathbf{Z} = \left\{ \mathbf{k} = (k_1, \ldots, k_n); \sum_{j=1}^{n} k_j = N, \ k_j = 0, 1, \ldots n \right\} \qquad (2.78)$$

The state space $\mathbf{Z}$ consist of $\binom{n+N-1}{N}$ elements. The transition probability rates for the state $\mathbf{k} = (k_1, \ldots, k_n)$ are indicated with the same notation

$$q_{\mathbf{k}-\mathbf{e}_j+\mathbf{e}_i, \mathbf{k}} = \mu_i(k_i+1) p_{ij}, \ j \neq i, \ \mathbf{k} - \mathbf{e}_j + \mathbf{e}_i \in \mathbf{Z}$$

and

$$q_{\mathbf{k}, \mathbf{k}-\mathbf{e}_j+\mathbf{e}_i} = \mu_j(k_j) p_{ji}, \ j \neq i, \ k_i \geq 1$$

Because of the relation (2.76) we can derive the rate leaving the state $\mathbf{k}$

$$q_{\mathbf{k}} = \sum_{j=1}^{n} \mu_j(k_j)(1 - p_{jj})$$

The state probabilities

$$x_{\mathbf{k}} = \lim_{t \to \infty} \mathbb{P}(X(t) = \mathbf{k}), \ \mathbf{k} \in \mathbf{Z}$$

of the stationary process $(X(t))_{t \geq 0}$ can be written as

$$\sum_{j=1}^{n} \mu_j(k_j)(1-p_{jj}) x_{\mathbf{k}} = \sum_{\substack{i,j=1 \\ j \neq i}}^{n} \mu_i(k_i+1) p_{ij} x_{\mathbf{k}-\mathbf{e}_j+\mathbf{e}_i} \qquad (2.79)$$

Suppose we have $\mathbf{k} - \mathbf{e}_j + \mathbf{e}_i \notin \mathbf{Z}$, then define $x_{\mathbf{k}-\mathbf{e}_j+\mathbf{e}_i} = 0$. Let $\varphi_j(0) = 1$ and for $j = 1, \ldots, n, \ i = 1, \ldots, N$

$$\varphi_j(i) = \prod_{m=1}^{i} \left(\frac{x_j}{\mu_j(m)} \right)$$

Then we are able formulating the main theorem for closed serving networks.

Theorem 2.64. *For a given state* $\mathbf{k} = (k_1, \ldots, k_n) \in \mathbf{Z}$ *the stationary state probabilities read as*

$$x_{\mathbf{k}} = c \prod_{j=1}^{n} \varphi_j(k_j)$$

with

$$c = \left(\sum_{\mathbf{z} \in \mathbf{Z}} \prod_{j=1}^{n} \varphi_j(z_j) \right)^{-1}$$

Using the probability in (2.79), we can prove e.g. the stationarity.

Example 2.65. (see [30]) We consider a simple computer network with central unit and two peripheric devices **1** and **2**. After the phase of computation in the central device **1** the programme changes with probability p over to **2**, and with probability $1 - p$ to **3**. From **2** the demand returns immediately back to the central unit with probability, the same holds for **3**, so that the demand in a queue builds up to the over all number in the system N. It is a *multi-programming* system with N level of multi-programming. We obtain a transition matrix

$$\mathcal{P} = \begin{bmatrix} 0 & 1-p & p \\ 1 & 0 & 0 \\ 1 & 0 & 0 \end{bmatrix}$$

Hence, we get as solution of (2.77)

$$x_1 = \frac{1}{2}, \quad x_2 = \frac{(1-p)}{2}, \quad x_3 = \frac{p}{2}$$

The service rates μ_1, μ_2 and μ_3 are assumed to be independent of the number of the demands. Then we have

$$\varphi_1(k_1) = \left(\frac{1}{2\mu_1} \right)^{k_1}, \quad \varphi_2(k_2) = \left(\frac{1-p}{2\mu_2} \right)^{k_2}, \quad \varphi_3(k_3) = \left(\frac{p}{2\mu_3} \right)^{k_3}$$

Let $\mathbf{k} = (k_1, k_2, k_3)$ with $k_1 + k_2 + k_3 = N$ be an admissible state. Then we conclude the following probabilities in the stationary state

$$x_{\mathbf{k}} = \frac{c}{2^N} \left(\frac{1}{\mu_1} \right)^{k_1} \left(\frac{1-p}{\mu_2} \right)^{k_2} \left(\frac{p}{\mu_3} \right)^{k_3}$$

where

$$c = \frac{2^N}{\sum_{\mathbf{z} \in \mathbf{Z}} \left(\frac{1}{\mu_1} \right)^{z_1} \left(\frac{1-p}{\mu_2} \right)^{z_2} \left(\frac{p}{\mu_3} \right)^{z_3}}$$

The state space is given by (2.78) with $N = 3$.

2.9.2 Systems with Priorities

We use as basic model the serving system with one channel and Poisson distributed arrival and serving processes, thus a system M/M/1 − 1. It can be derived that the model turns already very complex even with one channel and the extension is no longer sufficiently representable. Indeed, a transfer to many channels is canonical and does not contribute to further understanding. Beside the serving system we assume two types of demands for handling, **1** and **2**. Both types have exponential distributed interarrival times with parameters λ_1 resp. λ_2. Type **1** has absolute priority in the following sense:

- If there are demands of type **1** and **2** in the system, then the type **1** is served.
- An ongoing serving of type **2** will be interrupted immediately, if a demand of type **1** arrives. The interrupted demand of type **2** is lost, provided all waiting spots are occupied.
- A demand of **1** replaces a waiting demand of type **2** on the particular waiting spot. In this case a demand of type **1** is just served.
- Replaced demands are always lost.

The serving times are exponential distributed w.r.t. μ_1 resp. μ_2. The states in the systems are denoted by (i, j), where i is the number of demands of type **1**, j the number of demands of type **2**. We have $0 \le i, j \le 2$ and $i+j \le 2$. But, our system turns out not to be a birth and death process, since the representation is done in a triangle form and not linear. Nevertheless, the state process (X_t) consists of a Markov chain. From the stationary Kolmogorov equation we deduce the following system. We write in the stationary case

$$x_{i,j} = \mathbb{P}(X = (i,j))$$

This implies

$$(\lambda_1 + \lambda_2)x_{0,0} = \mu_1 x_{1,0} + \mu_2 x_{0,1}$$
$$(\lambda_1 + \lambda_2 + \mu_1)x_{1,0} = \lambda_1 x_{0,0} + \mu_1 x_{0,2}$$
$$(\lambda_1 + \lambda_2 + \mu_2)x_{0,1} = \lambda_2 x_{0,0} + \mu_1 x_{1,1} + \mu_2 x_{0,2}$$
$$(\lambda_1 + \mu_1)x_{1,1} = \lambda_2 x_{1,0} + \lambda_1 x_{0,1} + \lambda_1 x_{0,2}$$
$$\mu_1 x_{2,0} = \lambda_1 x_{1,0} + \lambda_1 x_{1,1}$$
$$(\lambda_1 + \mu_2)x_{0,2} = \lambda_2 x_{0,1}$$

Simultaneously, we have the norming condition

$$\sum_{i,j=0}^{2} x_{i,j} = 1$$

A general solution of all seven linear equation is complex and their representation does not provide any further insights. Thus, we skip their noting and keep to the fact of the unique solvability and to an example.

If the number of demands of type **1** is of interest, then we have the classical form of a loss and queueing systems M/M/1 − 1 (see section 2.4.2), since demands of type **2** do not influence those of type **1**.

Finally, we treat the case that there is no waiting spot, i.e. we are taking of a system M/M/1 − 0. Every demand is lost, which meets a serving spot, provided it is occupied by demands of type **1**. A demand of type **1** replaces one of type **2** on the serving channel. Thus, we obtain with the above notation the stationary probabilities

$$(\lambda_1 + \lambda_2)x_{0,0} = \mu_1 x_{1,0} + \mu_2 x_{0,1}$$
$$\mu_1 x_{1,0} = \lambda_1 x_{0,0} + \lambda_1 x_{0,1}$$

and the norming condition

$$x_{0,0} + x_{1,0} + x_{0,1} = 1$$

From this we deduce the state probabilities

$$x_{0,0} = \frac{\mu_1(\lambda_1 + \lambda_2)}{(\lambda_1 + \mu_1)(\lambda_1 + \lambda_2 + \mu_2)}$$
$$x_{0,1} = \frac{\mu_1 \lambda_2}{(\lambda_1 + \mu_1)(\lambda_1 + \lambda_2 + \mu_2)}$$
$$x_{1,0} = \frac{\lambda_1}{\lambda_1 + \mu_1}$$

Here, $x_{1,0}$ is the lost probability of the demand of type **1**. There is no information entering from type **2**, but only the probability is important that the random serving period of type **1** is larger than the interarrival time of type **1** in the serving system. At last, we determine the lost probability of a demand **2**, being in the serving channel. It coincides with the conditional probability under the constraint that a demand of **1** arrives, and reads as

$$\int_0^\infty e^{-\mu_2 t} \lambda_1 e^{-\lambda_1 t} = \lambda_1 \int_0^\infty e^{-(\lambda_1 + \mu_2)t} dt = \frac{\lambda_1}{\lambda_1 + \mu_1}$$

Thus, for the lost probability p_V of a demand of type **2** we have

$$p_V = \frac{\lambda_1}{\lambda_1 + \mu_1} x_{0,0} + x_{0,1} + x_{1,0}$$
$$p_V = 1 - \frac{\mu_1 \mu_2}{(\lambda_1 + \mu_1)(\lambda_1 + \lambda_2 + \mu_2)}$$

2.9.3 Systems with Impatient Demands

We cannot assume in general that in a system with infinite waiting space no incoming demand does not get lost. This e.g. happens, if we assume a limited waiting time. In computer networks the timeouts restrict the waiting time up to acknowledgments, where after the timeouts the data is sent again. Further examples for systems with limited waiting time are real-time steering units, booking and reservation systems, or computer based control centers. We realize by the number of examples that there is a variety of those model and select one exemplary model.

We start with a pure queueing system M/M/$n - \infty$, but introduce an *admissible waiting distribution*. The admissible waiting times of the serving units are independent and identical exponential distributed according to the rate ν. As above, we indicate the transition and state probabilities

$$q_{i,i+1} = \lambda, \ i = 0, \ldots$$

and

$$q_{i,i-1} = \begin{cases} i\mu & \text{for } i = 1, \ldots, n \\ n\mu + (i-n)\nu & \text{for } i = n+1, n+2, \ldots \end{cases}$$

With this we see that the death rate is practically unbounded. This means that independently of the traffic value $\rho = \frac{\lambda}{\mu}$, more demands leave a sufficiently large waiting queue than enter. In addition, the system swings into a stationary state with state probabilities according to

$$x_i = \begin{cases} \frac{1}{i!}\rho^i x_0 & \text{for } i = 1, \ldots, n \\ \frac{\rho^n}{n!} \frac{\lambda^{i-n}}{\prod_{j=1}^{i-n}(n\mu+j\nu)} x_0 & \text{for } i = n+1, n+2, \ldots \end{cases}$$

$$x_0 = \left(\sum_{i=0}^{n} \frac{1}{i!}\rho^i + \frac{\rho^n}{n!} \sum_{i=n+1}^{\infty} \frac{\lambda^{i-n}}{\prod_{j=1}^{i-n}(n\mu+j\nu)} \right)^{-1}$$

The mean waiting queue is

$$\mathbb{E}(W_L) = \sum_{i=n+1}^{\infty} x_i(i-n)$$

and reads after simple transformation

$$\mathbb{E}(W_L) = x_n \sum_{i=1}^{\infty} i\lambda^i \left(\prod_{j=1}^{i}(n\mu+j\nu) \right)^{-1}$$

Since we start from a pure waiting system and the lost possibilities are given by the surpassing of the waiting times, we have to proceed differently for computing the loss probability than in the pure loss system. We denote with p_V the loss probability. It represents the probability that a unit leaves the system without being served. Consequently $1 - p_V$ is the probability that a unit leaves the system after serving. This results in

$$p_V = \frac{\nu}{\lambda}\mathbb{E}(W_L)$$

Thus, the loss probability is directly proportional to the mean waiting queue length.

2.9.4 Conservation Laws Model

We start with a model due to [62, 204], based on deterministic traffic flow models, which are mainly used for describing road traffic. This approach requires a certain knowledge in the area of partial differential equation. Since most models are mainly centered around the stochastic setting, we do not go into details of the huge theory of partial differential equation. In addition, we briefly outline the approach and do not want to dive to deep into the models, since the original literature may serve as a good reference anyhow. Our aim is mainly to provide a certain insight into the idea and to embed this approach with the other stochastic models.

While we considered in most scenarios a single link network, here we change to a large number of nodes, using a simple routing algorithm, which is based on limiting procedure, obtained by partial differential equation for the packet density in the network. Each node consists on many incoming and outgoing lines. We start with formulating the underlying situation. We assume that each node (router) sends and receives packets. Three further assumptions are taken into account:

- Each packet travels with a fixed speed and final destination. Later we will incorporate some stochastic perturbation of the velocity.
- Each node handles the packets, i.e. sends and receives them.
- There exists a probability of data loss, which is resolved by the retransmission algorithm of TCP.

To get a first feeling, we focus on a straight line with consequent lines out of a router. As we consider data transfers via TCP between sender an receiver, the packets are sent, until they are acknowledged. So, on the macroscopic level we have all packets conserved. Thus, we conclude for the microscopic dynamic the conservation law

$$\frac{\partial}{\partial t}\rho(t,x) + \frac{\partial}{\partial x}f(\rho(t,x)) = 0 \qquad (2.80)$$

Here, $\rho(t,x)$ describes the packet density at time t and position x, while $f(\rho) = v\rho$ with a velocity function v, which is assumed constant in a first run. To incorporate a more complex structure of the network, We have to impose two different rules for the routing algorithm (according to [62]):

- RA1: Packets from incoming lines are sent to outgoing lines directly without taking into account any congestion or high load in lines.
- RA2: Packets are sent to outgoing lines to maximize the flux.

2.9.5 Packet Loss and Velocity Functions on Transmission Lines

Other network model consists of nodes N_k representing servers, routers or edges. Between consequent nodes a line is nothing else than a real interval I built up by the union of several edges. As mentioned above we are interested in the incorporation the packet loss phenomena: while the dynamic is given by the conservation law model (2.80), between the nodes N_k and N_{k+1} there may happen a loss, triggered by an overflow in the node N_{k+1} and resolved by the retransmission algorithm of TCP. We assume a constant velocity between nodes N_k and N_{k+1}. At discrete time t_i we have a packet amount of $R_k(t_i)$ at node N_k. As known from the elastic IP traffic, packets may got lost. This is represented by a function

$$p : [0, R_{\max}] \longrightarrow [0, 1]$$

making the loss probability depending on the amount of packets sent. We assign by δ the distance between the node N_k and N_{k+1} (here constant for different nodes). By Δt_0 we mean the transmission time between node N_k and N_{k+1}, provided the packet was sent successfully, and by Δt_{av} the average transmission time under the constraint that some packets are lost at the node N_{k+1}. At last by $\bar{v} = \frac{\delta}{\Delta t_0}$ resp. $v = \frac{\delta}{\Delta t_{\mathrm{av}}}$ we denote the packet velocity.

We assume a geometric distribution, i.e. we have a failure probability of $p \in]0, 1[$ and consequently a 'success' probability of $1 - p$. Hence, we conclude

$$\Delta t_{av} = \sum_{n=1}^{\infty} n \Delta t_0 (1-p) p^{n-1} \qquad (2.81)$$

This results into a average time, needed to travel from one node to the other, and consequently a mean velocity v.

$$v = \frac{\delta}{\Delta t_{av}} = \frac{\delta}{\Delta t_0}(1-p) = \bar{v}(1-p) \qquad (2.82)$$

Theorem 2.66. *If we assume a fixed initial input rate $R_k(t_0) = R$ for all nodes $k = 1, \ldots, n$, then the average transmission velocity is given by (2.81) and (2.82).*

The following examples are cited from [204]. As revealed by the above derivation the selection of the loss probability is fundamental.

Example 2.67. We choose

$$p(\rho) = \begin{cases} 0 & \text{for } 0 \leq \rho \leq \sigma \\ \frac{2(\rho-\sigma)}{\rho} & \text{for } \sigma \leq \rho \leq \rho_{\max} \end{cases}$$

where $\sigma \in]0, \frac{\rho}{2}[$. The function is depicted in figure 2.37. As average transmission velocity we obtain

$$v(\rho) = \overline{v}(1 - p(\rho)) = \begin{cases} \overline{v} & \text{for } 0 \leq \rho \leq \sigma \\ \overline{v}\frac{(2\sigma-\rho)}{\rho} & \text{for } \sigma \leq \rho \leq \rho_{\max} \end{cases}$$

If we suppose that $v(\rho_{\max}) = \overline{v}\frac{(2\sigma-\rho)}{\rho} = 0$ (i.e. for a maximum density there is flux anymore), we conclude that $\sigma = \frac{\rho_{\max}}{2}$. Because of the form of the flux $f(\rho) = v(\rho)\rho$ we get

$$f(\rho) = \begin{cases} \overline{v}\rho & \text{for } 0 \leq \rho \leq \sigma \\ \overline{v}(2\sigma - \rho) & \text{for } \sigma \leq \rho \leq \rho_{\max} \end{cases}$$

The results are depicted in the figure 2.37.

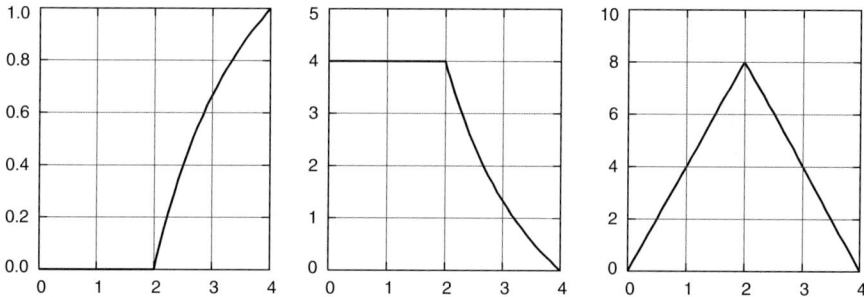

Fig. 2.37. The functions p for the probability selection, v as velocity and f as flow function

Example 2.68. We now choose

$$p(\rho) = \begin{cases} 0 & \text{for } 0 \leq \rho \leq \sigma \\ \frac{\rho-\sigma}{\sigma} & \text{for } \sigma \leq \rho \leq \rho_{\max} \end{cases}$$

where $\sigma \in \,]0, \rho_{\max}[$. The function is depicted in figure 2.39. As average transmission velocity we obtain

$$v(\rho) = \overline{v}(1 - p(\rho)) = \begin{cases} \overline{v} & \text{for } 0 \leq \rho \leq \sigma \\ \overline{v}\frac{(2\sigma-\rho)}{\sigma} & \text{for } \sigma \leq \rho \leq \rho_{\max} \end{cases}$$

If we suppose that $v(\rho_{\max}) = \overline{v}\frac{(2\sigma-\rho)}{\rho} = 0$, we deduce that $\sigma = \frac{\rho_{\max}}{2}$. Because of the form of the flux $f(\rho) = v(\rho)\rho$ we get

$$f(\rho) = \begin{cases} \overline{v}\rho & \text{for } 0 \leq \rho \leq \sigma \\ \overline{v}\rho\frac{2\sigma-\rho}{\sigma} & \text{for } \sigma \leq \rho \leq \rho_{\max} \end{cases}$$

The results are depicted in the figure 2.38.

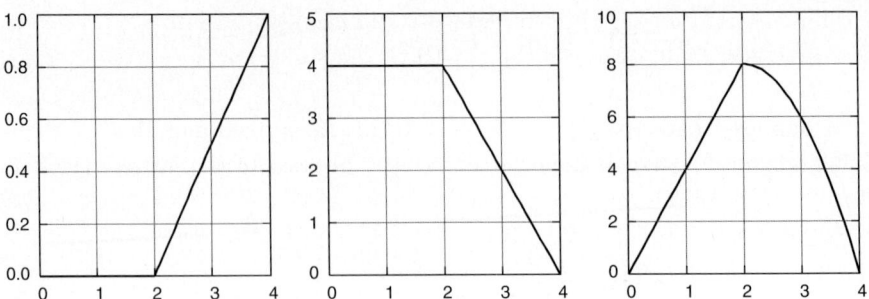

Fig. 2.38. The functions p for the probability selection, v as velocity and f as flow function

Example 2.69. Similar to the above example finally we proceed with an alternative loss probability

$$p(\rho) = \begin{cases} 0 & \text{for } 0 \le \rho \le \sigma \\ \frac{(\rho-\sigma)^2}{\sigma^2} & \text{for } \sigma \le \rho \le \rho_{\max} \end{cases}$$

where $\sigma \in {]}0, \rho_{\max}[$. The function is depicted in figure 2.38. As average transmission velocity we obtain

$$v(\rho) = \overline{v}(1-p(\rho)) = \begin{cases} \overline{v} & \text{for } 0 \le \rho \le \sigma \\ \overline{v}\rho\frac{(2\sigma-\rho)}{\sigma^2} & \text{for } \sigma \le \rho \le \rho_{\max} \end{cases}$$

If we suppose that $v(\rho_{\max}) = \overline{v}\frac{(2\sigma-\rho)}{\rho} = 0$, we conclude that $\sigma = \frac{\rho_{\max}}{2}$. Because of the form of the flux $f(\rho) = v(\rho)\rho$, we get

$$f(\rho) = \begin{cases} \overline{v}\rho & \text{for } 0 \le \rho \le \sigma \\ \overline{v}\rho^2\frac{(2\sigma-\rho)}{\sigma^2} & \text{for } \sigma \le \rho \le \rho_{\max} \end{cases}$$

The results are depicted in the figure 2.39.

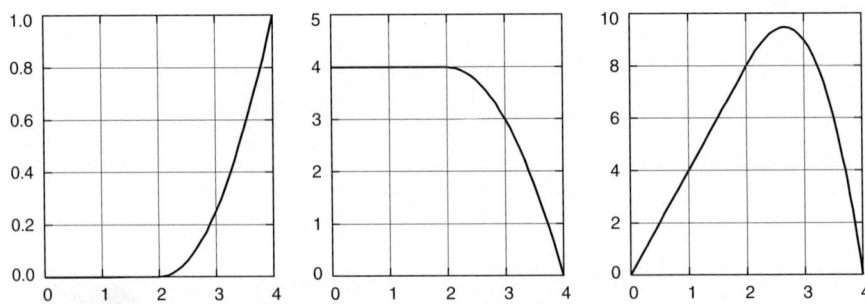

Fig. 2.39. The functions p for the probability selection, v as velocity and f as flow function

We have to realize that only in example 2.67 the flux satisfies at the point $\rho_0 \in \,]0, \rho_{\max}[$ the realizations

$$\lim_{\rho \to \rho_0+} f'(\rho) \neq 0 \quad \text{and} \quad \lim_{\rho \to \rho_0-} f'(\rho) \neq 0$$

So the density variation along discontinuities not crossing σ is equivalent to the flux ones. This motivates to consider in the sequel only this examples and gives rise by norming the packet density, i.e. assuming $\rho_{\max} = 1$, to impose on the flux the following condition

$$f : [0,1] \longrightarrow \mathbb{R}, \quad f(\rho) = \begin{cases} \overline{v}\rho & \text{for } 0 \leq \rho\sigma \\ \overline{v}(2\sigma - \rho) & \text{for } \sigma \leq \rho \leq 1 \end{cases} \quad (2.83)$$

Thus, we get for the maximums point $\sigma = \frac{1}{2}$.

The three examples above serve as a first introduction and as exemplary velocity function models. After indicating these examples, we now define the solution of the Cauchy problem (2.80) in our situation. For this purpose the different lines $I_i = [a_i, b_i] \subset \mathbb{R}$, $i = 1, \ldots, N$, $a_i < b_i$ or $b_i = \infty$ are modeled according to the conservation equation (2.80) in conjunction with condition (2.83). The network evolution is due to solution ρ_i defined on $[0, \infty[\times I_i$. The ρ_i are weak entropy solution on the transmission line I_i, i.e.

$$\int_0^\infty \int_{a_i}^{b_i} \left(\rho_i \frac{\partial \varphi}{\partial t} + f(\rho_i) \frac{\partial \varphi}{\partial x} \right) dxdt = 0 \quad (2.84)$$

for all $\varphi : [0, \infty[\times I_i \longrightarrow \mathbb{R}$ smooth, positive with compact support in $]0, \infty[\times [a_i, b_i[$. This reflects the weak solution condition. The usual entropy condition is formulated in the form of the so called Kruzkov entropy.

Definition 2.70. *Let the scalar version of (2.80) with initial condition (2.87) be given. Let $\rho(t,x)$ ($t \in [0,T]$, $x \in \mathbb{R}$) be a weak solution. Then we say that ρ satisfies the* Kruzkov entropy admissibility, *provided*

$$\int_0^T \int_{\mathbb{R}} \left(|\rho - k| \frac{\partial \hat{\varphi}}{\partial t} + \operatorname{sgn}(\rho - k)(f(\rho) - f(k)) \frac{\partial \hat{\varphi}}{\partial x} \right) dxdt \geq 0$$

for every $k \in \mathbb{R}$ and every C^1–function $\varphi \geq 0$ with compact support in $[0, T[\times \mathbb{R}$. One should compare this with (2.85).

Basic is the following theorem to determine a Kruzkov entropy.

Theorem 2.71. *Let $\rho = \rho(t,x)$ be a piecewise C^1-solution of the scalar version of (2.80) with initial data (2.87). Then ρ is Kruzkov entropy admissible, if and only if along any line of jump $x = \eta(t)$ the following condition holds:*

$$f(\alpha\rho^+ + (1-\alpha)\rho^-) \geq \alpha F(\rho^+) + (1-\alpha)f(\rho^-), \text{ for } \rho^- < \rho^+$$
$$f(\alpha\rho^+ + (1-\alpha)\rho^-) \leq \alpha F(\rho^+) + (1-\alpha)f(\rho^-), \text{ for } \rho^- > \rho^+$$

where $\rho^- := \rho(t, \eta(t)-)$ and $\rho^+ := \rho(t, \eta(t)+)$.

The condition in the previous theorem merely states that in the first case the graph of f lies above the line connecting ρ^- and ρ^+, while in the second case the picture is reverse.

The weak entropy condition reads for each node as

$$\int_0^\infty \int_{a_i}^{b_i} \left(|\rho_i - k|\frac{\partial \varphi}{\partial t} + \text{sgn}(\rho_i - k)(f(\rho_i) - f(k))\frac{\partial \varphi}{\partial x} \right) dxdt \geq 0 \qquad (2.85)$$

for $k \in \mathbb{R}$ and $\hat{\varphi} : [0, \infty[\times I_i \longrightarrow \mathbb{R}$ smooth, positive with compact support in $]0, \infty[\times]a_i, b_i[$. Finally, the definition of the whole network requires the sum over all lines – incoming as outgoing – between the different nodes

$$\sum_{j=1}^{n+m} \left(\int_0^\infty \int_{a_j}^{b_j} \left(\rho_j \frac{\partial \varphi_j}{\partial t} + f(\rho_j)\frac{\partial \varphi_j}{\partial x} \right) dxdt \right) = 0 \qquad (2.86)$$

where φ_j, $j = 1, \ldots, n + m$ are smooth function with compact support in $]0, \infty[\times]a_j, b_j]$ if $j = 1, \ldots, n$ (these represents the incoming lines) and for the outgoing ones in $]0, \infty[\times [a_j, b_j[$, if $j = n + 1, \ldots, n + m$. We also stress that the solution should be smooth across the junctions (nodes)

$$\varphi_i(\cdot, b_i) = \varphi_j(\cdot, a_j), \ \frac{\partial \varphi_i}{\partial x}(\cdot, b_i) = \frac{\partial \varphi_j}{\partial x}(\cdot, a_j), \ i = 1, \ldots, n, \ j = n+1, \ldots, n+m$$

Remark 2.72. The equations (2.84) and (2.86) represent the 'weak solution' resp. 'solutions in the distributive sense' for ρ. This is the reason, why we require in the equation the property 'smooth with compact support' for the φ functions.

Remark 2.73. Suppose the vector $\rho = (\rho_1, \ldots, \rho_{n+m})$ is a weak solution (see remark 2.72 above) at the junction, so that every function $\rho_i(t, \cdot)$ has bounded variation (see definition below). Then the function fulfills the so called *Rankine-Hugoniot* condition

$$\sum_{i=1}^n f(\rho_i(t, b_i-)) = \sum_{j=n+1}^{n+m} f(\rho_i(t, a_j+))$$

for almost all $t > 0$.

Before we continue to establish a concept for solutions we introduce certain definition and results concerning general solutions for partial differential equations. We have to mention that, by no means, we could give a complete introduction in this field and refer to the literature, which is certainly huge (see e.g. [80]). Since we are only concerned with the conservation equation (2.80), we will give the definitions in respect of this particular situation. We start with the concept of *weak solution*.

2.9 Network Models

Definition 2.74. Let $u_0 \in L^1_{loc}(\mathbb{R}, \mathbb{R}^n)$ and $T > 0$. A function $\rho : [0, T] \times \mathbb{R} \longrightarrow \mathbb{R}^n$ is a weak solution to the Cauchy problem (2.80) with initial data $\rho(0, x) = u_0(x)$, if ρ is continuous as a function from $[0, T]$ into L^1_{loc} and if for every C^1 function φ with compact support and $\text{supp}(\varphi) \subset \,]-\infty, T[\times \mathbb{R}$ we have

$$\int_0^T \int_{\mathbb{R}} \left(\rho \cdot \frac{\partial \varphi}{\partial t} + f(\rho) \cdot \frac{\partial \varphi}{\partial x} \right) dx dt + \int_{\mathbb{R}} u_0(x) \cdot \varphi(0, x) dx = 0$$

The equality required for a function to be weak solution to (2.80)

$$\rho(0, x) = u_0(x), \text{ for a.a. } x \in \mathbb{R} \qquad (2.87)$$

A first condition imposed on weak solutions is originated in physics, the so called *entropy* (see e.g. [63]).

Definition 2.75. A C^1-function $g : \mathbb{R}^n \longrightarrow \mathbb{R}$ is an entropy for the system (2.80), if it is convex, and if there exists a C^1 function $q : \mathbb{R}^n \longrightarrow \mathbb{R}$, such that

$$Df(u) \cdot \nabla g(u) = \nabla q(u)$$

for all $u \in \mathbb{R}^n$ (where 'D' is the differential operator). The function q is called the entropy flux for g. The pair (g, q) is said to be entropy-entropy flux pair for (2.80).

Definition 2.76. A weak solution ρ to the Cauchy problem (2.80) with initial data (2.87) is said entropy admissible, if for all $\psi \geq 0$ with compact support in $[0, T[\times \mathbb{R}$ and any entropy-entropy flux pair (g, q) we have

$$\int_0^T \int_{\mathbb{R}} \left(g(\rho) \frac{\partial \psi}{\partial t} + q(\rho) \frac{\partial \psi}{\partial x} \right) dx dt \geq 0$$

We shortly illustrate the concept for the scalar case, i.e. $n = 1$.

Example 2.77. Let the scalar version of (2.80) and (2.87) be given. In this case of $f : \mathbb{R} \longrightarrow \mathbb{R}$ the entropy respectively entropy flux assumes the form

$$g'(u) f'(u) = q'(u)$$

Thus, we can express the entropy flux with the help of the entropy g by

$$q(u) = \int_{u_0}^{u} g'(s) f'(s) ds$$

with $u_0 \in \mathbb{R}$.

Fundamental for dealing the solution in the network problem is the formulation of the Cauchy problem in terms of the Riemann problem, which indicates that the initial data is of Heaviside form. For this let $\Omega \subset \mathbb{R}^n$ be an open set and let $f : \Omega \longrightarrow \mathbb{R}^n$ be smooth. We suppose that the problem (2.80) is purely hyperbolic.

Definition 2.78. *A Riemann problem concerning the Cauchy problem (2.80) has the initial data u_0 of the form*

$$u_0(x) = \begin{cases} u^- & \text{for } x < 0 \\ u^+ & \text{for } x > 0 \end{cases}$$

This initial condition just tells us, that the starting density before a node N_k (here expressed by x) is u_-, while after passing the node it turns into u_+. So, we incorporate a certain discontinuity at the nodes, representing edges, servers or routers.

The Riemann problem is the key step for solving the Cauchy problem. In fact, one establishes the so called wave-front tracking method described as follows:

- Approximate the initial data by piecewise constant data.
- At every point solve the respective Riemann problem.
- Approximate the exact solution with the help of Riemann problems with piecewise constant functions and glue them together to a function until two wave fronts interact.
- Repeat this steps inductively starting from the interaction time.
- Prove that the sequence converges to a limit function and prove that this is an admissible entropy weak solution.

The next definition is required to find a space, where solutions to the Riemann problem can be established, and where one can find the space for suitable convergence.

Definition 2.79. *Let $I \subset \mathbb{R}$ be an interval and $g : I \longrightarrow \mathbb{R}$ be a function. The total variation of the function g is defined by (Klammer ergaenzt)*

$$\|g\| = \sup \left(\sum_{i=1}^{N} |g(x_j) - g(x_{j-1})| \right)$$

where the supremum is taken over all choices $x_0 < x_1 < \ldots < x_N$ of arbitrary points in I ($N \in \mathbb{N}$). The function g is said of bounded variation, *if $\|g\| < \infty$. The space of functions of bounded variation on I is called $BV(I)$ and is a Banach space equipped with the norm $\|\cdot\|$, defined above (variational norm).*

Fundamental for the application is Helly's theorem (see e.g. [74]). We state it in the for our purpose required form.

Theorem 2.80. *Let $h_n : I \longrightarrow \mathbb{R}^n$ be s sequence of uniformly bounded function is the space of function of bounded variations, i.e. there are constant $L, M > 0$ such that*

- *$\sup_{n \in \mathbb{N}} \|h_n\| \leq L$ and*
- *$\sup_{x \in I} \sup_{n \in \mathbb{N}} |h_n(x)| \leq M$.*

Then there exists a subsequence (h_{n_k}) and a function of bounded variation $h : I \longrightarrow \mathbb{R}^n$, such that

- $\lim_{k \to \infty} h_{n_k}(x) = h(x)$ for all $x \in I$,
- $\|h\| \leq L$ and
- $\sup_{x \in I} |h(x)| \leq M$.

2.9.6 Riemann Solvers

This concept was originated by Picolli and Garavello (see e.g. [204, 62]). First, we call the Cauchy problem in (2.80) with corresponding initial data, which are constant along the transmission lines, a *Riemann problem*. In this respect we define a *Riemann solver*.

Definition 2.81. *(see e.g.[62]) Let J be a junction. A map*

$$RS : [0,1]^n \times [0,1]^m \longrightarrow [0,1]^n \times [0,1]^m$$

which associates initial data $\rho_0 = (\rho_{1,0}, \ldots, \rho_{n+m,0})$ at the junction J to a vector $\tilde{\rho} = (\tilde{\rho}_1, \ldots, \tilde{\rho}_{n+m})$ in such a way that the solution on an incoming transmission line I_i ($i = 1, \ldots, n$) is given by the wave solution $(\rho_{i,0}, \tilde{\rho}_i)$ and on the outgoing line I_j ($j = n+1, \ldots, n+m$) by the wave $(\tilde{\rho}_j, \rho_{j,0})$. In addition, the consistency condition has to hold

$$RS(RS(\rho_0)) = RS(\rho_0) \qquad (2.88)$$

Remark 2.82. The condition (2.88) is necessary to provide uniqueness.

Remark 2.83. Suppose we assume that $RS(\rho) = \rho'$ and $RS(\rho') = \rho$ for $\rho \neq \rho'$. If we want to solve the Riemann problem with the datum ρ, one should apply the boundary datum ρ' at the junction. If ρ' gets into the lines, we have to go back to ρ and so on. Then a solution would not exist.

If we have obtained the Riemann solver, we can define the admissible solution and J.

Definition 2.84. *Suppose we have assigned a Riemann solver RS. Let $\rho = (\rho_1, \ldots, \rho_{n+m})$, so that $\rho_i(t, \cdot)$ be of bounded variation for all $t \geq 0$. Then ρ is an admissible solution of (2.80) related to RS at the junction J if and only if we have:*

- *ρ is a weak solution at the junction J and*
- *for almost every t we have by setting*

$$\rho_J(t) = (\rho_1(\cdot, b_1-), \ldots, \rho_n(\cdot, b_n-), \rho_{n+1}(\cdot, a_{n+1}+), \ldots, \rho_{n+m}(\cdot, a_{n+m}+))$$

$$RS(\rho_J(t)) = \rho_J(t)$$

We apply the definition of Riemann solver for the two basic scenarios of transmission networks: We assume the following two different routing algorithm:

(RA1) Assume the following two conditions

(A) The traffic from incoming transmission lines being distributed on outgoing transmission lines according to fixed coefficients.
(B) Let (A) be assumed. Then the router chooses to send packets in order to maximize the flux (i.e. the number of packets processed).
(RA2) The number of both incoming and outgoing packets through the junction are maximized.

Riemann Solver for Model (RA1)

We start with an algorithm to solve the problem (RA1). For getting an idea in the technique to solve the Riemann problem, we sketch the method but refer to the literature for more details see [52, 55, 204]). The network in consideration consists of a junction J with n transmission lines with incoming traffic and m lines for outgoing traffic. Assumption (A) implies that we choose a transmission matrix

$$\mathcal{A} = (\alpha_{ji})_{j=1,\ldots,m}^{i=1,\ldots,n} \in \mathbb{R}^{m,n}, \text{ with constraints } 0 < \alpha_{ji} < 1, \sum_{j=n+1}^{m+n} \alpha_{ji} = 1$$

for each $i = 1,\ldots,n$ and $j = n+1,\ldots,m+n$. The α_{ji} gives the percentage of packets from the incoming line i to the outgoing one j. We indicate for $i = 1,\ldots,n$ the density of the incoming line i with

$$(t,x) \ni \mathbb{R}_+ \times I_i \longmapsto \rho_i(t,x) \in [0,1]$$

and for $j = n+1,\ldots,n+m$ as

$$(t,x) \ni \mathbb{R}_+ \times I_j \longmapsto \rho_j(t,x) \in [0,1]$$

for the density of the outgoing line j. We depict this situation in figure 2.40.

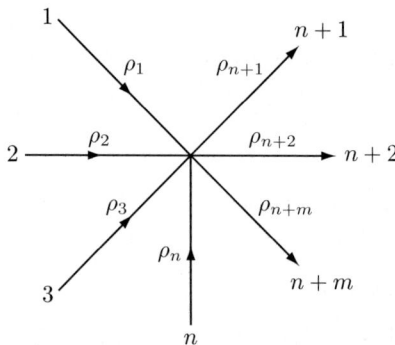

Fig. 2.40. A typical example of a junction (node) for n incoming traffic and m outgoing lines

Definition 2.85. *Define a map* $\tau : [0,1] \longmapsto [0,1]$ *such that*
- $f(\tau(\rho)) = f(\rho)$ *for all* $\rho \in [0,1]$ *and*
- $\tau(\rho) \neq \rho$ *for all* $\rho \in [0,1] \setminus \{\sigma\}$.

We see that such a function is well-defined with the property

$$0 \leq \rho \leq \sigma \Leftrightarrow \sigma \leq \tau(\rho) \leq 1$$
$$\sigma \leq \rho \leq 1 \Leftrightarrow 0 \leq \tau(\rho) \leq \sigma$$

To formulate the main result for this Riemann solver, we have to impose some additional properties for the matrix $\mathcal{A}$ (this is automatically fulfilled in case of $m = n$). Denote by $\{e_1, \ldots, e_n\}$ the canonical basis of $\mathbb{R}^n$ and for any subset $V \subset \mathbb{R}^n$ let $V^\perp$ its orthogonal complement. Let H_i be the hyperplane orthogonal to the basis vector e_i, i.e. $H_i = \{e_i\}^\perp$ and let $\alpha_j = \{\alpha_{j1}, \ldots, \alpha_{jn}\} \in \mathbb{R}^n$. Define $H_j = \{\alpha_j\}^\perp$. Furthermore select tupels of indices $k = (k_1, \ldots, k_l)$ with $1 \leq l \leq n-1$, such that $0 \leq k_1, \ldots, k_l \leq n+m$, define the set of all these tupels by $\mathcal{K}$ and build finally for $k \in \mathcal{K}$ the space $H_k = \bigcap_{r=1}^{l} H_r$. Let $\mathbf{1} = (1, \ldots, 1) \in \mathbb{R}^n$. For the following theorem we need to assume

$$\mathbf{1} \notin H_k^\perp, \text{ for all } k \in \mathcal{K} \tag{2.89}$$

We can interpret the technical assumption (2.89) as follows: Condition (2.89) on the matrix $\mathcal{A}$ cannot hold for a situation with two incoming and one outgoing line.

We are now ready for the main theorem for the Riemann solver in case of algorithm (RA1).

Theorem 2.86. *(see [55]) Consider a junction J of n incoming and m outgoing lines. Assume that the flux* $f : [0,1] \longrightarrow \mathbb{R}$ *satisfies the conditions (2.83) and let (2.89) be fulfilled. For all initial conditions* $\rho_{1,0}, \ldots, \rho_{n+m,0} \in [0,1]$, *there exists a unique admissible centered weak solution* $\rho = (\rho_1, \ldots, \rho_{n+m})$ *of (2.80) at the junction J so that*

$$\rho_1(0, \cdot) = \rho_{1,0}, \ldots, \rho_{n+m}(0, \cdot) = \rho_{n+m,0}$$

In addition, there exists a unique $(n+m)$-tupel $(\tilde{\rho}_1, \ldots, \tilde{\rho}_{n+m}) \in [0,1]^{n+m}$, *so that for* $i = 1, \ldots, n$

$$\tilde{\rho}_i \in \begin{cases} \{\rho_{i,0}\} \cup \,]\tau(\rho_{i,0}), 1] & \text{for } 0 \leq \rho_{i,0} \leq \sigma \\ [\sigma, 1] & \text{for } \sigma \leq \rho_{i,0} \leq 1 \end{cases} \tag{2.90}$$

and for $j = n+1, \ldots, n+m$

$$\tilde{\rho}_j \in \begin{cases} [0, \sigma] & \text{for } 0 \leq \rho_{j,0} \leq \sigma \\ \{\rho_{j,0}\} \cup [0, \tau(\rho_{i,0})[& \text{for } \sigma \leq \rho_{i,0} \leq 1 \end{cases} \tag{2.91}$$

On each incoming line I_j the solution consists of a single wave $(\rho_{i,0}, \tilde{\rho}_i)$, where for each outgoing one I_j the solution exists for the single wave $(\tilde{\rho}_j, \rho_{j,0})$

Riemann Solver for Model (RA2)

This traffic model works with priority and certain traffic distribution such as 'differentiating traffic'. Thus, we have to introduce the following parameters, describing these traffic imposements. To keep the description as easy as possible and to get the decisive idea, we consider a junction J with $n = m = 2$, i.e. two incoming and outgoing lines. So, we have only one priority parameter $q \in\ ]0,1[$ and one traffic distribution parameter $\alpha \in\ ]0,1[$. Denote by $\rho_i(t,x)$, $i = 1, 2$, resp. $\rho_j(t,x)$, $j = 3, 4$, the traffic density of the incoming resp. outgoing lines. Let $(\rho_{1,0}, \rho_{2,0}, \rho_{3,0}, \rho_{4,0})$ be the initial value. We define two maximal incoming $\gamma_i^{\max}$ (for $i = 1, 2$) resp. outgoing $\gamma_j^{\max}$ (for $j = 3, 4$) flux

$$\gamma_i^{\max} = \begin{cases} f(\rho_i, 0) & \text{for } \rho_{i,0} \in [0, \sigma] \\ f(\sigma) & \text{for } \rho_{i,0} \in [\sigma, 1] \end{cases}$$

and

$$\gamma_j^{\max} = \begin{cases} f(\sigma) & \text{for } \rho_{i,0} \in [0, \sigma] \\ f(\rho_i, 0) & \text{for } \rho_{i,0} \in [\sigma, 1] \end{cases}$$

These values describe the maximal flux, which can be obtained by a single wave solution on each transmission line, incoming as outgoing. We denote $\gamma_{\text{in}}^{\max} = \gamma_1^{\max} + \gamma_2^{\max}$, $\gamma_{\text{out}}^{\max} = \gamma_3^{\max} + \gamma_4^{\max}$ and finally

$$\gamma = \min(\gamma_{\text{in}}^{\max}, \gamma_{\text{out}}^{\max})$$

The aim is, to have γ as flux through the junction J. As in theorem 2.86 we need to determine $\tilde{\gamma}_i = f(\tilde{\rho}_i)$ for $i = 1, 2$. For getting simple wave with the suitable velocities, i.e. negative on incoming, and positive on outgoing lines, we obtain the constraints (2.90) and (2.91). To tackle the problem, we distinguish, due to the definition, two cases:

- $\gamma_{\text{in}}^{\max} = \gamma$ and
- $\gamma_{\text{in}}^{\max} > \gamma$.

In the first case, obviously we have $\tilde{\gamma}_i = \gamma_i^{\max}$ for $i = 1, 2$. Thus, we analyze the second case with the help of the priority parameter q. Consider (γ_1, γ_2) and define the following lines

$$\tau_q : \gamma_2 = \frac{1-q}{q}\gamma_1$$
$$\tau_\gamma : \gamma_1 + \gamma_2 = \gamma$$

Let

$$\Omega = \{(\gamma_1, \gamma_2); 0 \leq \gamma_i \leq \gamma_i^{\max},\ i = 1, 2\}$$

we can detect two cases:

- P belongs to Ω and

- P does not lie in Ω.

In the first case $P = (\tilde{\gamma}_1, \tilde{\gamma}_2)$ and in the second case $Q = (\tilde{\gamma}_1, \tilde{\gamma}_2)$ where $Q = \mathrm{Pr}_{\Omega \cap \tau_\gamma}(P)$ is the projection on the convex set $\Omega \cap \tau_\gamma$. The argument can be transferred to n incoming lines. In the case $\mathbb{R}^n$ the line τ_q is given according to $\tau_q = t v_q$ ($t \in \mathbb{R}$), where the velocity fulfills $v_q \in \Delta_{n-1}$ for the set

$$\Delta_{n-1} = \left\{ (\gamma_1, \gamma_n); \gamma \geq 0, \ i = 1, \ldots, n, \ \sum_{i=1}^{n} \gamma_i = 1 \right\}$$

as $(n-1)$-dimensional simplex and

$$H_\gamma = \left\{ (\gamma_1, \ldots, \gamma_n); \sum_{i=1}^{n} \gamma_i = \gamma \right\}$$

as hyperplane with $\gamma = \min(\sum_{\text{in}} \gamma_i^{\max}, \sum_{\text{out}} \gamma_j^{\max})$. Because $v_q \in \Delta_{n-1}$, we have a unique point $P \in \tau_q \cap H_\gamma$. Suppose $P \in \Omega$, then we choose $(\tilde{\gamma}_1, \ldots, \tilde{\gamma}_n) = P$. In case $P \notin \Omega$ we pick $(\tilde{\gamma}_1, \ldots, \tilde{\gamma}_n) = Q$, where similar as before $Q = \mathrm{Pr}_{\Omega \cap H_\gamma}$. As known from functional analysis the projection is unique, since the set $\Omega \cap H_\gamma$ is a closed and convex subset of H_γ.

Remark 2.87. As alternative definition of $(\tilde{\gamma}_1, \ldots, \tilde{\gamma}_n)$ in the case of $P \notin \Omega$ we can pick the vertices of the convex set $\Omega \cap H_\gamma$.

Lemma 2.88. *The condition (2.88) required for the RS in definition 2.81 holds for the RS defined above for the algorithm (RA2).*

Estimates on Densities and Existence of Solutions

Up to now we showed that Riemann solver exists in both cases. But, we need to solve the original Cauchy problem (2.80). Here, we will use the concept of function with bounded variation and a certain uniform boundedness condition to show that a suitable sequence of RS converges. Again we split our consideration into two cases (RA1) and (RA2) and start with the algorithm (RA1).

Lemma 2.89. *There exists an $L > 0$ such that for all times $t \geq 0$*

$$\|f(\rho(t, \cdot))\| \leq e^{Lt} \|f(\rho(0+, \cdot))\| \leq e^{Lt} (\|f(\rho(0, \cdot))\| + 2N f(\sigma))$$

where N is the total number of lines in the network.

A proof can be found in [55].
Condition (2.83) is responsible that the next lemma can be proved.

Lemma 2.90. *The function f satisfies condition (2.83). Then there is an $L > 0$ such that for all $t \geq 0$*

$$\|\rho(t,\cdot)\| \leq \|\rho(0,\cdot)\| + 2N \left(\frac{e^{Lt} f(\sigma)}{\bar{v}} + 1 \right)$$

where N is the total number of transmission lines in the network, $\bar{v}$ the mean velocity.

We are ready for the theorem, which is based on the Lipschitz continuous dependence in L_{loc}, theorem 2.80 and the lemmata 2.88 and 2.90.

Theorem 2.91. *Let a telecommunication network (I, J) be given and assume (2.83) and (2.89). Then under the assumption (RA1) and given $T > 0$ for every initial data there exists an admissible solution to the Cauchy problem (2.80) on the interval $[0, T]$.*

It should be mention that we have to choose the locally convex space L_{loc} instead of L_1, since there does not exist any Lipschitz continuity in general (see [55]).

We turn now to the algorithm (RA2). Condition (2.83) is responsible that the next lemma can be proved.

Lemma 2.92. *The function f satisfies condition (2.83). Then there is an $L > 0$ such that for all $t \geq 0$*

$$\|\rho(t,\cdot)\| \leq \|\rho(0,\cdot)\| + 2N \left(f(\sigma)\bar{v} + 1 \right)$$

where N is the total number of transmission lines in the network, $\bar{v}$ the mean velocity.

We are ready for the theorem, which is based on the Lipschitz continuous dependence in L_{loc}, theorem 2.80 and the lemmata 2.88 and 2.92.

Theorem 2.93. *Let a telecommunication network (I, J) be given and assume (2.83) and (2.89). Then under the assumption (RA2) and given $T > 0$ for every initial data there exists an admissible solution to the Cauchy problem (2.80) on the interval $[0, T]$.*

Remark 2.94. We did not go into details for the proofs, since the interested reader can study them in the original literature (see [52, 55, 204]).

2.9.7 Stochastic Velocities and Density Functions

This section is an outlook and should give a first approach to incorporate stochastic perturbations into the deterministic setting of the previous conservation model. We introduce the model just for getting understanding; there will be no evaluations or mathematical solution, since at the moment we have not developed the technical methods for solving. This will be done at the end of the monograph, because the required technique would surpass the scope of the book at the present state. In addition, there is still data evaluation to be done for verifying the model, which indicates that the concept is still in state of development.

The starting point is the macrodynamic view given by the conservation equation (2.80)

$$\frac{\partial}{\partial t}\rho(t,x) + \frac{\partial}{\partial x}f(\rho(t,x)) = 0$$

We rewrite it in the form

$$d\rho(t,x) = -\frac{\partial}{\partial x}f(\rho(t,x))dt$$

expressing that a small change in time of the density of the traffic depends proportional on the time interval dt with factor $-\frac{\partial}{\partial x}f(\rho(t,x))$, expressing that the change is negative proportional on the change of the flux on the different nodes or line spots x.

Now we suppose that beside the TCP algorithm, expressed by the probability function $p(\rho)$ as done in the examples 2.67 to 2.69, we insert a stochastic perturbation, which is triggered by the underlying links. Since we want to keep the model on a simple level and since we had worked in the small scale influence by the function p, we consider a perturbation by a fractional Brownian motion, whose scale is determined proportional to the flux derivation

$$-a\frac{\partial}{\partial x}f(\rho(t,x)), \ a \in \mathbb{R}_+$$

Thus, we obtain the equation

$$d\rho(t,x) = -\frac{\partial}{\partial x}f(\rho(t,x))dt - a\frac{\partial}{\partial x}f(\rho(t,x))dB_t^{(H)}$$

Since we have not yet introduced the fractional Brownian motion (FBM) at that point, we refer to the later chapter 3 for detailed description. We just collect some fact on the FBM relevant for the IP traffic. As we already mentioned in the introductory chapter 1 and there especially in the section 1.4, the IP traffic reveals a so called long-range dependency. Roughly speaking, this mean that changes now perform influences on changes of later times and this dependency decays slowlier than linear. E.g. the classical Brownian motion or even the Poisson process have independent increments, thus no existing dependency. In addition, the variance of the traffic is not decaying linear as in the Brownian motion case but hyperbolic. All these phenomena can be covered by the perturbation with the FBM up to a certain extend. Hence, though we have already incorporated a stochastic component in the pure deterministic framework of the conservation law with the probability function $p(\rho)$, the long-range dependence is not explained by the above model, presented in section 2.9.4. The part

$$a\frac{\partial}{\partial x}f(\rho(t,x))dB_t^{(H)}$$

means that the perturbation is driven by the FBM and the magnitude depends on a scaling factor a and the change in the flux $\frac{\partial}{\partial x}f(\rho(t,x))$ depending on the density and the space variable x, thus the location.

2.10 Matrix-Analytical Methods

In this last section we give a short introduction to the matrix analytical methods. In contrast to the models based on the statistical evaluation and its key properties, like self-similarity and long-range dependence, which lead to approaches based on stochastic processes and its analysis, in this part we still remain in the classical environment, expressed by the Kendall notation. Neuts and Lucantoni are two of the major representatives in this area. At the end, after the introduction of the toolbox, we demonstrate the application to the IP network. It should be emphazised that the models will mimic the long-range dependence resp. the heavy-tail phenomena of the data transfer, using a more complicated Poisson arrival process. As Lucantoni et al. put it, the models based on perturbation of stochastic processes are asymptotical self-similar, as all continuous model inherit this property, while the models based on the matrix analytic model are claimed to be exact. Our description is based on the monographs of Tran-Gia [254, 255] and the original literature of Neuts [187] and Lucantoni [166, 167, 168].

2.10.1 Phase Distribution

The basic concept consists in describing non-Markovian components, as they appear in the serving time processes of the data transfer (see section 2.7.5), a more complex system based on pure Markovian processes, in particular by the Poisson process. This leads to combine *phases* of the Poisson process, and brings up the notion of the *phase distribution*

Erlang-k Distribution as Motivation

The phase representation of various RV as well as the phase distribution of a general distributed RV is done by exponential distributed phases. To get the idea we start with the Erlang distribution. We summarize the facts given by the figure 2.41. In both cases we find an Erlang distributed RV:

- Phase representation: This is basically a sequence of λ-exponential distributed phases: The RV starts in phase 1 at the end, enters into phase 2 and so forth. After the j-th phase it enters the phase $(j+1)$, again exponential distributed with the same λ.
- Phase distribution: After the phase k the process enters the so called *absorbing state* $k+1$. Then, a new event occurs. Leaving the absorbed state, means a restart of the process in the state 1. Here, the process starts always in 1 $(p_1 = 1, p_i = 0, \ i \neq 1)$.

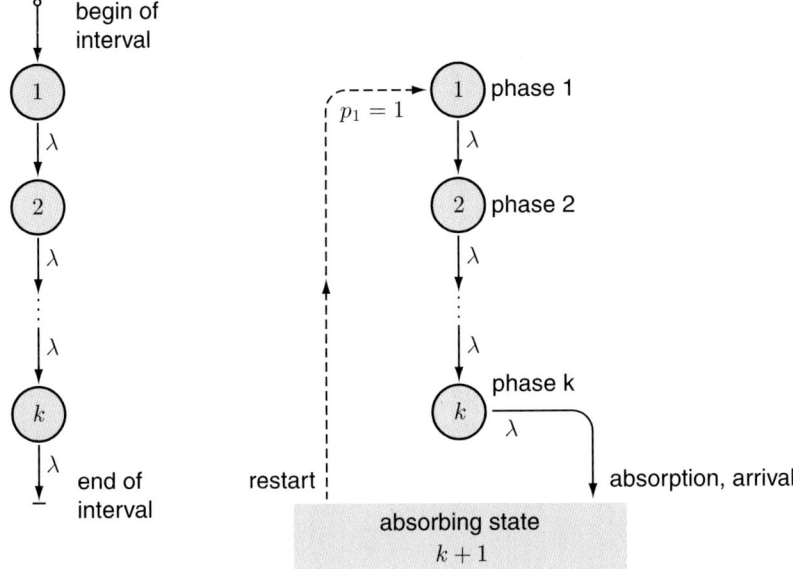

Fig. 2.41. In the left figure we show the scheme of an Erlang-k distribution. The right one is the phase distribution representation of an Erlang-k distribution

Definition of Phase Distributions

We have two basic phases – the *transient state* and the *absorbing state*. The process is based on a finite and irreducible Markov chain in continuous time, called the *guided Markov chain*. The transition between two transient phases i and j is described by transition probability densities q_{ij} ($i, j \in \{1, \ldots, k\}$). The process can move into the absorbing state (only one state) with rate ω_i from the transient state i. We write them into the *absorption vector*

$$\Omega = \begin{pmatrix} \omega_1 \\ \omega_2 \\ \vdots \\ \omega_k \end{pmatrix}$$

The decisive difference to the Erlang-k distribution is the introduction of the absorbing state. The process stops in the absorbing state, but restarts in state i as arrival event at the end of the serving period. The process restarts independent of the phase before the absorption occurs with rate p_i into state i. The time spent in the absorbing state depends on the process – if we want to model serving period, then the restart will coincide with the beginning of the serving period. The duration of the absorption will be the time of the serving process. In contrast, if we model the interarrival time process, then the restart occurs immediately after the absorption and the restart happens timeless.

We define the *new start vector* as

$$\Pi = \{p_1, p_2, \ldots, p_k, p_{k+1}\}$$

The vector includes the probability distribution of the restart with the completeness relation

$$\sum_{j=1}^{k+1} p_k = 1 \quad \text{or equivalently} \quad \Pi e = 1$$

where $e = (1, 1, \ldots, 1) \in \mathbb{R}^{k+1}$. The probability p_{k+1}, describing the restart directly after the absorbing state, is used by compound RV to reflect the transition between the components. Usually we set $p_{k+1} = 0$ and obtain the modified new start vector

$$\Pi = \{p_1, p_2, \ldots, p_k\}, \text{ with } \sum_{j=1}^{k} p_j = 1$$

The transition probability densities fulfill

$$q_{jj} = -\sum_{i=1, i \neq j}^{k} q_{ji} - \omega_j = -\lambda_j$$

where λ^{-1} describes the mean time spent in phase i. Similar to the pure Markov process we define the transition matrix

$$Q = \begin{pmatrix} Q_0 & \Omega \\ \mathcal{O} & 0 \end{pmatrix} \in \mathbb{R}^{(k+1) \times (k+1)} \tag{2.92}$$

Note that, since $Q_0 \in \mathbb{R}^{k \times k}$, $\Omega \in \mathbb{R}^{k \times 1}$ and $\mathcal{O} \in \mathbb{R}^{1 \times k}$ (zero matrix), we get $Q \in \mathbb{R}^{(k+1) \times (k+1)}$. Since the Markov chain is irreducible, we have $q_{ii} < 0$ and the matrix Q_0 is regular. We can determine uniquely the absorption vector Ω

$$Q_0 e + \Omega = 0 \quad \text{resp.} \quad \Omega = -Q_0 e$$

The pair (Π, Q_0) determines uniquely the phase distribution. Hence, we denote by $PH(\Pi, Q_0)$ the given phase distribution (fig. 2.42).

Distribution Function of a Phase Distributed Random Variable

How does a distribution function for a $PH(\Pi, Q_0)$-distributed random variable looks like? The random variable ξ describes the time from the restart until the absorption. If ξ describes the interarrival time of the arrival process, then ξ is nothing else than the time between two different absorption time spots. Let $X(t)$ denote the state of the process, describing the particular

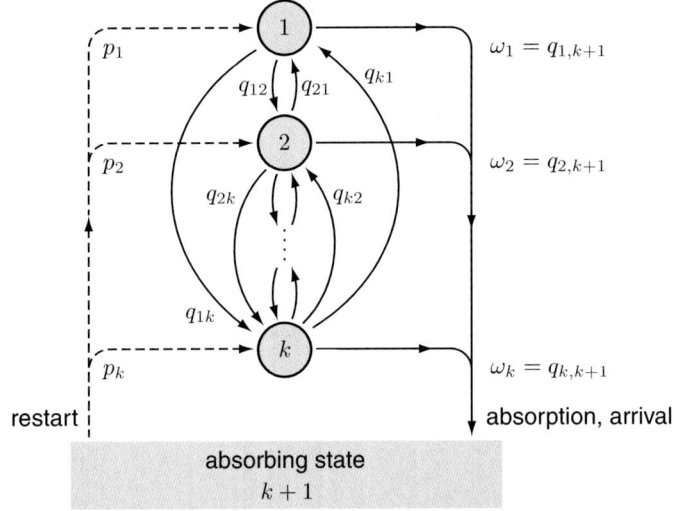

Fig. 2.42. Scheme of a typical phase distribution (PH)

phase, i.e $X(t) = 1, \ldots, k, k + 1$. Suppose at time $t = 0$, we have a restart. Then we can describe the random variable as follows

$$f_\xi(t) = (\xi \leq t) = \mathbb{P}(\text{phase process at time } t \text{ is in the absorbing state})$$
$$= \mathbb{P}(X(t) = k + 1)$$

It is clear that the process can pass through several transient states until it hits the absorption state. To get full information about the distribution function, we have to analyze the transition probabilities from the restart up to the absorption state (fig. 2.43). We assume that the process $(X(t))$ is homogeneous. Again we follow the standard technique for state equations and state probabilities (see e.g. [141, 238]).

Let $\mathcal{P}(t)$ denote the matrix of transition probability densities designed for the interval $]0, t[$, describing two succeeding states of absorption. According to the Kolmogorov backward equations we obtain

$$\frac{d\mathcal{P}(t)}{dt} = Q \cdot \mathcal{P}(t) \tag{2.93}$$

Since we set $\mathcal{P}(0) = \text{id}$, we obtain as solution for the matrix valued differential equation (2.93)

$$\mathcal{P}(t) = e^{Qt} = \text{id} + Qt + Q^2 \frac{t^2}{2!} + Q^3 \frac{t^3}{3!} + \ldots$$

By the block structure of Q according to (2.92) and since $\Omega = -Q_0 e$ we conclude

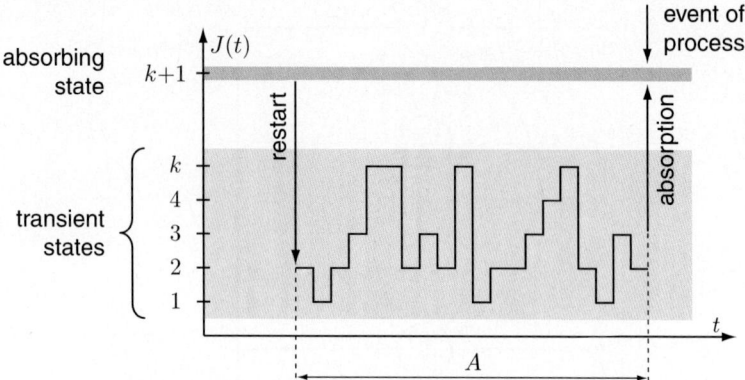

Fig. 2.43. Realization of a phase distributed RV

$$e^{Qt} = \begin{pmatrix} e^{Q_0 t} & e - e^{Q_0 t} e \\ \mathcal{O} & 1 \end{pmatrix}$$

The vector

$$\begin{pmatrix} e - e^{Q_0 t} e \\ 1 \end{pmatrix}$$

as element in $\mathbb{R}^{k+1}$ and its components reflect the conditional probability for the fact that, after the start in state i at time 0, the process is in the absorbing state at time t. Applying left multiplication with the restart vector Π, we obtain the distribution function of the random variable ξ

$$F_\xi(t) = 1 - \Pi e^{Q_0 t} e \quad \text{PH distribution function}$$

We differentiate to obtain the density function, called *PH-density function*

$$f_\xi(t) = \frac{dF_\xi(t)}{dt} = -\Pi e^{Q_0 t} Q_0 e = -\Pi e^{Q_0 t} \Omega$$

We compute the Laplace transform

$$\mathcal{L}_{f_\xi}(s) = \int_0^\infty f_\xi(t) e^{-st} dt \tag{2.94}$$

$$= p_{k+1} + \Pi \left(s \cdot \text{id} - Q_0\right)^{-1} \cdot \Omega \quad \text{Laplace transform of a PH-CDF}$$

Renewal Property

Since the restart vector is independent of the absorption phase, the phase distributed arrival process enjoys at the arrival time spots the renewal property. The interarrival intervals, which indicates the time between two absorption times, are stochastic independent. The sequence of absorptions time spots consists of a classical renewal process, as defined in section 2.6

2.10.2 Examples for Different Phase Distributions

Exponential Distribution

We start with the exponential distribution, which has as phase distribution the representatives
$$\Pi = 1 \quad \text{and} \quad Q_0 = (-\lambda)$$
The corresponding arrival process is a Poisson process. Inserting the parameters into equation (2.94), we obtain the Laplace transform
$$\mathcal{L}_{f_\varepsilon}(s) = \frac{\lambda}{\lambda + s}$$

Erlang-k Distribution

Selecting the Erlang-k distribution, we conclude for its representation as phase distribution
$$\Pi = \{1, 0, \ldots, 0\}$$
and the $k \times k$-matrix
$$Q_0 = \begin{pmatrix} -\lambda & \lambda & 0 & \ldots & 0 & 0 \\ 0 & -\lambda & \lambda & \ldots & 0 & 0 \\ 0 & 0 & -\lambda & \ldots & 0 & 0 \\ \vdots & \vdots & \vdots & \ddots & \vdots & \vdots \\ 0 & 0 & 0 & \ldots & -\lambda & \lambda \\ 0 & 0 & 0 & \ldots & 0 & -\lambda \end{pmatrix}$$

Hyperexponential Distribution

As a well-known fact, the hyperexponential distribution can be considered as k-parallel exponential distributed phases with different rates. The phase j possesses a rate λ_j, which is selected with probability p_j. Hence, we obtain the following representation for the phase distribution
$$\Pi = \{p_1, p_2, \ldots, p_k\}$$
and
$$Q_0 = \begin{pmatrix} -\lambda_1 & 0 & \ldots & 0 \\ 0 & -\lambda_2 & \ldots & 0 \\ \vdots & \vdots & \ddots & \vdots \\ 0 & 0 & \ldots & -\lambda_k \end{pmatrix} = \text{diag}(-\lambda_1, \ldots, -\lambda_k)$$
The matrix reveals that there is no transition between the transient phases.

Interrupted Poisson Processes

This a particular interested example for the simple description of IP traffic. It forms the basic idea, which we will transform in the next chapter for the on-off models in the packet oriented traffic. We select a Poisson process with rate λ_2, modulated with another independent process in an 'on-off' type manner. The 'off' phase V_1 and the 'on' phase of the modulated process consists of exponential distributed interarrival times, given by the parameters γ_1 and γ_2, i.e.

$$F_{\xi_{\text{off}}} = 1 - e^{-\gamma_1 t} \quad \text{and} \quad F_{\xi_{\text{on}}} = 1 - e^{-\gamma_2 t}$$

During the 'On'-phases we generate events with rate λ_2, while during the 'Off'-phases nothing happens (fig. 2.44). This process is called *interrupted*

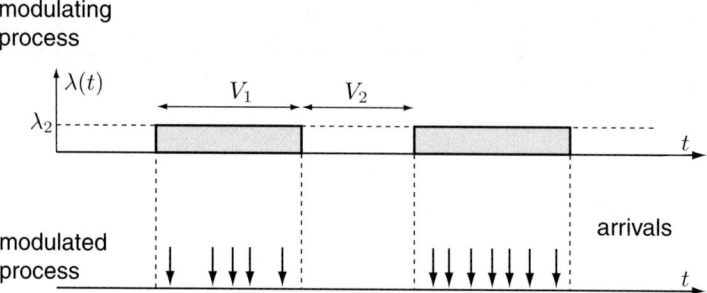

Fig. 2.44. The interrupted Poisson process

Poisson process (IPP). The IPP is a renewal and phase distributed process and can be characterize by the parameters

$$\Pi = \{0, 1\}$$

and

$$Q = \begin{pmatrix} Q_0 & \Omega \\ \mathcal{O} & 0 \end{pmatrix} = \begin{pmatrix} -\gamma_1 & \gamma_1 & 0 \\ \gamma_2 & (-\gamma_2 - \gamma_1) & \lambda_2 \\ 0 & 0 & 0 \end{pmatrix}$$

The transient phase 1 corresponds to the 'off' phase of the modulated process. The probability to leave this state depends only on the exponential distributed phase duration with rate γ_1. The transient phase is the 'on' phase. Here, we meet two possibilities: either we have absorption with rate λ_2, or we encounter the end of the phase with rate γ_2. Since we have an arrival process modeled, we see a restart after the absorption, which runs into phase 2 again (fig. 2.45). Thus, we have a classical modified renewal process.

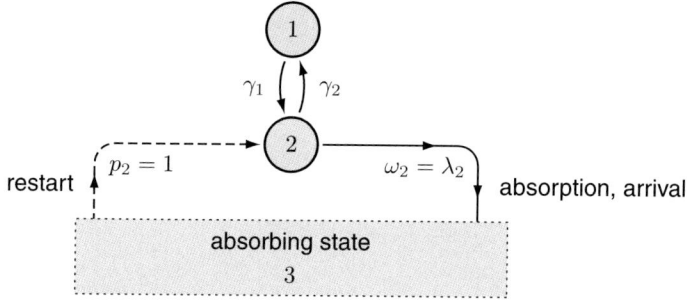

Fig. 2.45. Scheme of the phase distribution of the interrupted Poisson process

Time Discrete Phase Distribution

For describing communication networks and for indicating the key values of performance some times, one has to map IP data traffic with constant duration. Here, the *time discrete phase distribution* (D-PH) is very suitable.
The time axis will be discretized by Δt, where this reflects the duration of transient phase. The decisive difference consists in the fact that all events happens at discrete time spots The duration of a transient phase lasts Δt. The transition behavior is characterized by the transition probability, not by transition rates. The transition matrix reads as

$$Q = \begin{pmatrix} Q_0 & \Omega \\ \mathcal{O} & 1 \end{pmatrix}$$

where Q lies in $\mathbb{R}^{(k+1)\times(k+1)}$ and is a stochastic matrix, since the sum over a row has to match 1. The submatrix Q_0 is substochastic, i.e. the sum over all element of a row is less or equal to 1 and reflects the transitions between the single transient states. The restart vector Π is a row vector and indicates the restart probability in the time continuous case. We have

$$\Omega = e - Q_0 e = \begin{pmatrix} 1 - \sum_{j=1}^{k} Q_0(1 \cdot j) \\ \vdots \\ 1 - \sum_{j=1}^{k} Q_0(k \cdot j) \end{pmatrix}$$

The time from the restart to the absorption state is modeled by the discrete random variable X. Using a discrete phase distribution model the restart occurs immediately after the absorption. The time X coincides with the interarrival time. If we model the serving time with the discrete model, the restart time spot is determined by the serving time. We can compute the distribution of X resp. the probability that the absorption occurs i time intervals after the restart according to

$$x(j) = \mathbb{P}(X = j) = \Pi Q_0^{j-1} \Omega, \; j > 0$$
$$x(0) = p_{k+1}$$

2.10.3 Markovian Arrival Processes

This section will bring a generalization of the phase distribution, called *Markovian Arrival Process (MAP)*, which is based on original work of Neuts and Lucantoni ([187, 188, 166, 167]). They include in addition general correlated arrival processes without renewal property.

Definition

As we did for the phase distribution, we start with an irreducible Markov chain. Beside the guiding chain we will consider an embedded one in discrete time, which will describe the jump behavior. In figure 2.46 we sketch the significance of the embedded Markov chain. Different to the phase distribution, the restart probability depends on the phase directly before the absorption. This indicates the advantage: the MAP can be used for modeling segmented dependent and correlated processes.

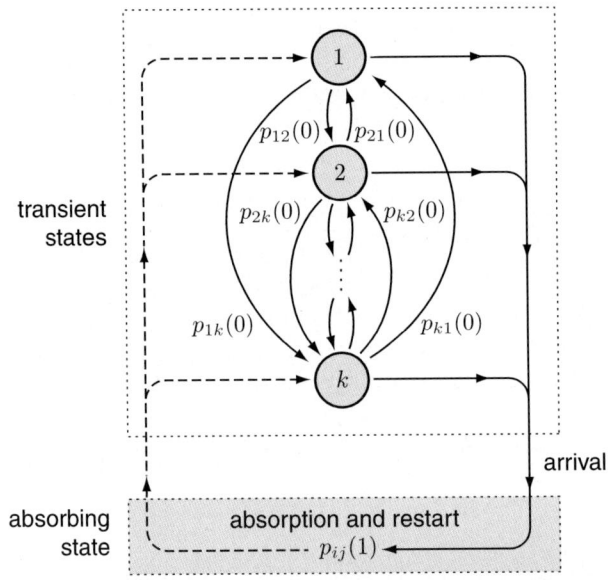

Fig. 2.46. The Markovian arrival process with its state transitions

Transition Properties

The state space of the embedded Markov chain consists of k transient phases and one absorbing one. The duration of each transient state is exponential distributed with rate λ_i:

- Transient transition: This happens with a probability $p_{ij}(0), i \neq j$ between the transient state i and j without generating a new arrival event.
- Transition with absorption and restart: The transition probability $p_{ij}(1)$ describes the transition of the transient state i into the restart in phase j, passing the absorption state and generating an arrival event.

The transition probabilities fulfill the completeness property

$$\sum_{i=1, i\neq j}^{k} p_{ji}(0) + \sum_{i=1}^{k} p_{ji}(1) = 1, \text{ for } 1 \leq j \leq k \qquad (2.95)$$

and built up the $k \times k$ arrival matrices $\mathcal{A}_0$ resp. $\mathcal{A}_1$

$$\mathcal{D}_0 = \begin{pmatrix} -\lambda_1 & \lambda_1 p_{12}(0) & \cdots & \lambda_1 p_{1k}(0) \\ \lambda_2 p_{21}(0) & -\lambda_2 & \cdots & \lambda_2 p_{2k}(0) \\ \vdots & \vdots & \ddots & \vdots \\ \lambda_k p_{k1}(0) & \lambda_k p_{k2}(0) & \cdots & -\lambda_k \end{pmatrix}$$

$$\mathcal{D}_1 = \begin{pmatrix} \lambda_1 p_{11}(1) & \lambda_1 p_{12}(1) & \cdots & \lambda_1 p_{1k}(1) \\ \lambda_2 p_{21}(1) & \lambda_2 p_{22}(1) & \cdots & \lambda_2 p_{2k}(1) \\ \vdots & \vdots & \ddots & \vdots \\ \lambda_k p_{k1}(1) & \lambda_k p_{k2}(1) & \cdots & \lambda_k p_{kk}(1) \end{pmatrix}$$

The transition matrix $\mathcal{D}_0$ belongs to the case that a phase transition happens without simultaneously generating a new arrival event. The matrix $\mathcal{D}_1$ reflects the case that at the time spot of the phase transition and arrival event occurs. We assume the regularity of the matrix $\mathcal{D}_0$. The summation matrix

$$\mathcal{D} = \mathcal{D}_0 + \mathcal{D}_1$$

is the generator of the guiding Markov chain. $(\mathcal{D}_0, \mathcal{D}_1)$ is called *representative of the Markovian arrival process*. We can define similar to the distribution function a generating function for the arrival matrices of the arrival process

$$\mathcal{D}_{EF}(z) = \sum_{j=0}^{\infty} \mathcal{D}_j z^j = \mathcal{D}_0 + z\mathcal{D}_1$$

setting $\mathcal{D}_j = 0$ for $j \geq 2$.

Semi-Markov Property and Markov Renewal Processes

The time intervals between two succeeding phases of absorption are not necessarily identical distributed, since the probability for a restart depends of the phase before absorption. We consider the arrival process as *semi Markov process*. Let (ξ_n, X_n) be the sequence of pairs describing the time between the $(n-1)$-th and n-th absorption, and X_n the phase immediately after the n-th absorption, then we obtain a Markovian renewal process. The past of the whole process is included completely in the embedding time spots of the state (ξ_n, X_n). The sequence (ξ_n, X_n) describes a Markov chain.

Phase Distribution as Example of MAP

We give the phase distribution, indicating its representatives (Π, Q_0). For this, we define the arrival matrices to ensure the Markov property of the arrival process

$$\mathcal{D}_0 = Q_0$$
$$\mathcal{D}_1 = -Q_0 e \Pi = \Omega \Pi$$

Markov Modulated Poisson Processes

The *Markov Modulated Poisson Process* (MPP) is often used for correlated traffic streams in communication networks. It is a double stochastic process, where the arrival rate is described by a Markov chain. The modulated MMPP possesses k transient phases, where the duration of each is exponential distributed with rate γ_i. Suppose the process is in phase i. Then, arrival events will occur according to a Markov process with rate λ_i. The guiding modulated Markov process is based on a time continuous Markov chain with generator Γ. We can characterize it by the matrix Γ and the rate matrix Λ

$$\Lambda = \mathrm{diag}(\lambda_1, \lambda_2, \ldots, \lambda_k)$$

Other than the case of the interrupted Poisson process, the MMPP is not a renewal process. With the above matrices Γ and Λ we can indicate the representatives of the MMPP (see also Fischer and Meier-Hellstein [89] and Lucantoni [167])

$$\mathcal{D}_0 = \Gamma - \Lambda$$
$$\mathcal{D}_1 = \Lambda$$

For the special case of an interrupted Poisson process defined by

$$\Gamma = \begin{pmatrix} -\gamma_1 & \gamma_1 \\ \gamma_2 & -\gamma_2 \end{pmatrix} \quad (2.96)$$

and

$$\Lambda = \begin{pmatrix} 0 & 0 \\ 0 & \lambda_2 \end{pmatrix}$$

we obtain with equation (2.96) the MAP representatives

$$\mathcal{D}_0 = \Gamma - \Lambda = \begin{pmatrix} -\gamma_1 & \gamma_1 \\ \gamma_2 & -\gamma_2 - \lambda_2 \end{pmatrix}, \quad \mathcal{D}_1 = \Lambda = \begin{pmatrix} 0 & 0 \\ 0 & \lambda_2 \end{pmatrix}$$

This representative can also be deduced by the phase distribution of the IPP

$$\Pi = \{0, 1\}, \quad Q_0 = \begin{pmatrix} -\gamma_1 & \gamma_1 \\ \gamma_2 & -\gamma_2 - \lambda_2 \end{pmatrix}$$

and the equivalence of the phase distribution of the MAP, expressed in equation (2.95)

$$\mathcal{D}_0 = Q_0 = \begin{pmatrix} -\gamma_1 & \gamma_1 \\ \gamma_2 & -\gamma_2 - \lambda_2 \end{pmatrix}, \quad \mathcal{D}_1 = -Q_0 e\Pi = \begin{pmatrix} 0 & 0 \\ 0 & \lambda_2 \end{pmatrix}$$

Batch Markovian Arrival Processes

With the help of the matrices $\mathcal{D}_0$ and $\mathcal{D}_1$ we describe transitions without and with arrival events. Analogously we define the matrices $\mathcal{D}_n$ indicating the case of n arrivals. For this purpose we denote by $p_{ij}(n)$, $i \neq j$ the transition probability of phase i, passing through an absorption state to a restart into phase j, where n demands appear simultaneously. The probability fulfills the completeness relation

$$\sum_{i=1, i \neq j}^{k} p_{ji}(0) + \sum_{n=1}^{\infty} \sum_{i=1}^{k} p_{ji}(n) = 1, \text{ for } 1 \leq j \leq k$$

The corresponding arrival matrices $\mathcal{D}_0$ and $\mathcal{D}_n$ are defined by

$$\mathcal{D}_0 = \begin{pmatrix} -\lambda_1 & \lambda_1 p_{12}(0) & \ldots & \lambda_1 p_{1k}(0) \\ \lambda_2 p_{21}(0) & -\lambda_2 & \ldots & \lambda_2 p_{2k}(0) \\ \vdots & \vdots & \ddots & \vdots \\ \lambda_k p_{k1}(0) & \lambda_k p_{k2}(0) & \ldots & -\lambda_k \end{pmatrix}$$

$$\mathcal{D}_n = \begin{pmatrix} \lambda_1 p_{11}(n) & \lambda_1 p_{12}(n) & \ldots & \lambda_1 p_{1k}(n) \\ \lambda_2 p_{21}(n) & \lambda_2 p_{22}(n) & \ldots & \lambda_2 p_{2k}(n) \\ \vdots & \vdots & \ddots & \vdots \\ \lambda_k p_{k1}(n) & \lambda_k p_{k2}(n) & \ldots & \lambda_k p_{kk}(n) \end{pmatrix}$$

The above defined process is called *Batch Markovian Arrival Process (BMAP)*. This process enables to describe the group arrivals, which are suitable to model the buffers of multiplexers in communication networks. A good reference for a more detailed description of BMAP is [166, 167].

Time Discrete Markov Arrival Processes

Basic of the *Time Discrete Markov Arrival Process* (D-MAP) is the consideration of a discretized time axis by the step width Δt, which indicates the duration of each transient phase. Hence, an absorption occurs only at the end of each phase. Again we have, as in the D-PH-case, transition probabilities substituting the rates. Thus, the matrices $\mathcal{D}_0$ and $\mathcal{D}_1$ are stochastic matrices, indicating the respective transition probabilities. The summation matrix

$\mathcal{D} = \mathcal{D}_0 + \mathcal{D}_1$ is again stochastic, and by the element $\mathcal{D}[ij]$ we mean the probability that the phase j was the successor of transient state i. The matrix $\mathcal{D}$ is assumed to be irreducible and aperiodical. This indicates that the matrix

$$(\mathrm{id} - \mathcal{D}_0)^{-1} = \sum_{k=0}^{\infty} \mathcal{D}_0^k$$

exists. Especially $(1 - \mathcal{D}_0)^{-1}\mathcal{D}_1$ describes the probability that a period with arrival finishes sometimes with the arrival. The above properties ensure that $\mathcal{D}$ has eigenvalue 1 and, hence, let π be a left eigen vector of $\mathcal{D}$, i.e. satisfying

$$\pi = \pi\mathcal{D}, \text{ with } \pi e = 1 \quad (2.97)$$

indicating the stationary probability to remain in a transient phase. Let λ be the column vector, representing the phase depending arrival rates. We have

$$\lambda = \mathcal{D}_1 e \quad (2.98)$$

As mean aggregated arrival rates we conclude

$$\lambda = \pi \cdot \lambda = \pi\mathcal{D}_1 e$$

Example 2.95. Cyclic on-off process: This Markovian arrival process is particular suitable for the characterization of general processes with the correlation property. We describe in 2.47 an example of time discrete process with a cycle of 5 slots – 3 full and 2 empty ones.

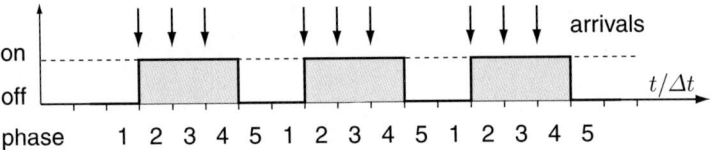

Fig. 2.47. Realization of a phase distributed RV

These 5 slots represent 5 transient phases

$$\mathcal{D}_0 = \begin{pmatrix} 0 & 0 & 0 & 0 & 0 \\ 0 & 0 & 0 & 0 & 0 \\ 0 & 0 & 0 & 0 & 0 \\ 0 & 0 & 0 & 0 & 1 \\ 1 & 0 & 0 & 0 & 0 \end{pmatrix}$$

$$\mathcal{D}_1 = \begin{pmatrix} 0 & 1 & 0 & 0 & 0 \\ 0 & 0 & 1 & 0 & 0 \\ 0 & 0 & 0 & 1 & 0 \\ 0 & 0 & 0 & 0 & 0 \\ 0 & 0 & 0 & 0 & 0 \end{pmatrix}$$

2.10.4 Queueing Systems MAP/G/1

An important approach consists in the generalization of the M/G/1 system. We will basically proceed as in section 2.7.3 and use the embedded Markovian renewal process. More and detailed description can be found in the literature (see [166, 167, 168]). In the figure 2.48 we indicate the following model MAP/G/1-∞, i.e with infinite queueing space:

> We have the two $k \times k$ matrices $\mathcal{D}_0$ and $\mathcal{D}_1$ indicating the transition probability of the transient phases with corresponding mean arrival rate λ. The serving time η is general distributed.

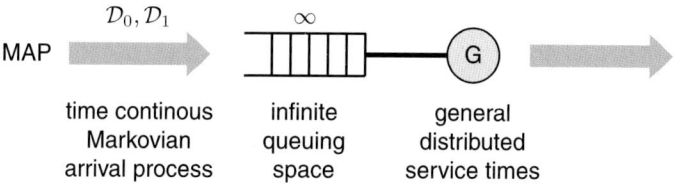

Fig. 2.48. Realization of a phase distributed RV

The analysis is more or less equivalent to the classical M/G/1 system. But it does not suffice to know the state condition resp. the amount of demands at the embedded time spots. In addition, we need information on the relation of the arrival process to the embedded time spots.

We consider the system in the interval $]0, t]$. As usual, we denote by $N(t)$ the number of arrivals, the phase at time t is denoted by $X(t)$. The triple $(N(t), X(t))_{t>0}$ is a time continuous and state discrete process. The process is Markovian and the generator is the matrix

$$Q_A = \begin{pmatrix} \mathcal{D}_0 & \mathcal{D}_1 & 0 & 0 & \ldots \\ 0 & \mathcal{D}_0 & \mathcal{D}_1 & 0 & \ldots \\ 0 & 0 & \mathcal{D}_0 & \mathcal{D}_1 & \ldots \\ 0 & 0 & 0 & \mathcal{D}_0 & \ldots \\ \vdots & \vdots & \vdots & \vdots & \ddots \end{pmatrix} \qquad (2.99)$$

We already divided the process into two levels:

- the level of events resp. demands and
- the level of the phase description of the MAP arrival process.

Block Structure of Matrices

The matrix in (2.99) reveals a typical block structure and reflects the form of the process $(N(t), X(t))_{t>0}$. On the level of demands we recognize the arrival matrices due to the MAP arrival process. Defining a *niveau* as the set of all states, where the demands are not changed, these so called niveaus of the process correspond to the block structures of row and columns. Since the MAP-arrival process enjoys k transients, we have k elementary states. A change in the states on the level of demands corresponds with passing from niveau to another, while a change in a phase is equivalent to passing between two transient states.

The basic idea behind the block structure is the linearization of the two dimensional processes $(N(t), X(t))_{t>0}$ and enables a simplified notation of the transition probabilities, given by the above generator Q_A. We can find a comparison with the formulas of the one dimensional case in the literature e.g. in [167].

Transition Equations

We denote by $\mathcal{R}(t)$ the transition probability for the process $(N(t), X(t))_{t>0}$. It is stochastic and enjoys block structure as well, like the matrix Q_A.

$$\mathcal{R}(t) = \begin{pmatrix} \mathcal{R}_0(t) & \mathcal{R}_1(t) & \mathcal{R}_2(t) & \mathcal{R}_3(t) & \cdots \\ 0 & \mathcal{R}_0(t) & \mathcal{R}_1(t) & \mathcal{R}_2(t) & \cdots \\ 0 & 0 & \mathcal{R}_0(t) & \mathcal{R}_1(t) & \cdots \\ 0 & 0 & 0 & \mathcal{R}_0(t) & \cdots \\ \vdots & \vdots & \vdots & \vdots & \ddots \end{pmatrix}$$

The blocks $\mathcal{R}_n(t)$ are matrices representing the transition probability

$$\mathcal{R}_n(t)[ij] = \mathbb{P}\left(N(t) = n, \ J(t) = j \mid N(0) = 0, \ J(0) = i\right)$$

The matrix $\mathcal{R}_n(t)$ is called *counting function* of the arrival process. Since $\mathcal{R}(t)$ and the generator Q_A satisfy the Kolmogorov forward equation, $\mathcal{R}_n(t)$ fulfills the matrix-valued initial value problem

$$\frac{d}{dt}\mathcal{R}(t) = \mathcal{R}(t)Q_A, \text{ for } t \geq 0 \quad \text{and} \quad \mathcal{R}(0) = \mathrm{id}$$

By inserting the matrices $\mathcal{R}(t)$ and Q_A, we obtain using the block structure

$$\frac{d}{dt}\mathcal{R}_n(t) = \mathcal{R}_n(t) \cdot \mathcal{D}_0 + \mathcal{R}_{n-1} \cdot \mathcal{D}_1, \text{ for } n \geq 1, \ t \geq 0 \quad \text{and} \quad \mathcal{R}_0(0) = \mathrm{id}$$
(2.100)

A formal transformation of the counting function $\mathcal{R}_n(t)$ with the help of the generating function reveals

$$\mathcal{R}_{EF}(z,t) = \sum_{k=0}^{\infty} \mathcal{R}_k(t) z^k$$

With the help of equation (2.100) we conclude

$$\frac{\partial}{\partial t}\mathcal{R}_{EF}(z,t) = \mathcal{R}_{EF}(z,t) \cdot \mathcal{D}_{EF}(z) \quad \text{and} \quad \mathcal{R}_{EF}(z,0) = \text{id}$$

where

$$\mathcal{D}_{EF}(z) = \sum_{k=0}^{\infty} \mathcal{D}_k z^k = \mathcal{D}_0 + z\mathcal{D}_1$$

Thus, it follows

$$\mathcal{R}_{EF}(z,t) = e^{\mathcal{D}_{EF}(z)t}, \quad \text{for } |z| \leq 1, t \geq 0 \tag{2.101}$$

the last equation helps to determine the transition probability matrix, depending on the pair $(\mathcal{D}_0, \mathcal{D}_1)$ of the MAP arrival process.

Example 2.96. Poisson process: Equation (2.101) will be discussed using the Poisson arrival process. For this case we have $\mathcal{D}_0 = (-\lambda)$ and $\mathcal{D}_1 = (\lambda)$ and the transition matrix $\mathcal{R}(t)$ is nothing else than the counting function of the Poisson process, i.e.

$$\mathcal{R}_n(t) = \frac{1}{n!}(\lambda t)^n e^{-\lambda t}$$

Inserted into equation (2.101), gives

$$\mathcal{R}_{EF}(z,t) = e^{(-\lambda + \lambda z)t}$$

This examples demonstrates clearly that the MAP is the matrix generalization of the Poisson process. The important information of the transition behavior in the MAP/G/1 system will be attacked in the following section.

Embedded Markovian Renewal Processes

Let F_η be the distribution function of the serving time. We consider a sequence of time spots $t_0, \ldots, t_\nu$, immediately after a certain end of a serving period. Since the process $(N(t), X(t))_{t>0}$ enjoys the Markov property at these time spots, we can embed a Markov process at these times.

The state of this embedded Markov process is determined by $(N(t_\nu), X(t_\nu))_{\nu \geq 0}$ and attains values in $\{(n,j); n \geq 0, 1 \leq j \leq k\}$. The component $(N(t_\nu))$ reflects a number of changes in the system until time t_ν, while $X(t_\nu)$ indicates the transient state at time t_ν. We assume that the start of the process is set at $t = 0$.

Let us now analyze the transitions. We start with two succeeding embedded time spots t_ν and $t_{\nu+1}$ and let $t = t_{\nu+1} - t_\nu$. The probability for the transition of the states $[N(t_\nu) = n_\nu, X(t_\nu) = x_\nu]$ to $[N(t_{\nu+1}) = n_{\nu+1}, X(t_{\nu+1}) = x_{\nu+1}]$ during the time t is given by the following matrix, enjoying block structure as well

$$\mathcal{P}(t) = \begin{pmatrix} \Omega_0(t) & \Omega_1(t) & \Omega_2(t) & \Omega_3(t) & \ldots \\ \Upsilon_0(t) & \Upsilon_1(t) & \Upsilon_2(t) & \Upsilon_3(t) & \ldots \\ 0 & \Upsilon_0(t) & \Upsilon_1(t) & \Upsilon_2(t) & \ldots \\ 0 & 0 & \Upsilon_0(t) & \Upsilon_1(t) & \ldots \\ \vdots & \vdots & \vdots & \vdots & \ddots \end{pmatrix}$$

Here, $\Upsilon_n(t)$ as well as $\Omega_n(t)$ are again matrices defined as follows:

- $\Upsilon_n(t)[ij]$: Probability for the transition: At time 0 we find the end of a serving period. The system is not empty, and immediately afterwards a new serving period starts. This serving period finishes at least at time t. At time 0 the MAP arrival process is in state i, at time t in phase j. We encounter n new arrivals in the interval $]0, t]$.
- $\Omega_n(t)[ij]$: Probability for the following transition: A serving period ends at time 0 and the system is empty and runs through a vacant phase until the next arrival. Immediately afterwards, the serving period starts and finishes at least at time t. At time 0 the MAP arrival process is in phase i and at t in phase j. During the time interval $]0, t]$ we detect $(n+1)$ demands in the system, thus, at the end of the period n demands remain.

Both matrices fully describe the possibilities, how the transition from one embedded time spot to the other during the time period t occurs. We can compute the matrix Υ_n from the counting function $\mathcal{R}_n$ according to

$$\Upsilon_n(t) = \int_0^t \mathcal{R}_n(s) dF_\eta(s)$$

The matrix Ω_n can be computed from the matrix Υ_n, by initially inserting a vacant period. We can explain the relationship using suitable transformations as follows. For this, we construct the Laplace-Stieltjes transform the transition matrices, where we consider the integrand as matrix-valued

$$\Phi_{\Upsilon_n}(s) = \int_0^\infty e^{-st} d\Upsilon_n(s), \quad \Phi_{\Omega_n}(s) \int_0^\infty e^{-st} d\Omega_n(s)$$

The expression will be transformed to the so called double transform (DT), using matrix generating function according to the arrivals

$$\Upsilon_{DT}(z,s) = \sum_{n=0}^\infty \Phi_{\Upsilon_n}(s) z^n, \quad \Omega_{DT}(z,s) = \sum_{n=0}^\infty \Phi_{\Omega_n}(s) z^n \quad (2.102)$$

We have as usual $|z| \leq 1$ and $\mathrm{Re}(s) \geq 0$.

First, we are interested in the stationary distributions, i.e. the asymptotical behavior for the time $t \to \infty$. We can consider this after the swing in. We obtain

$$\mathcal{P} = \mathcal{P}(\infty) = \lim_{t\to\infty} \mathcal{P}(t) = \begin{pmatrix} \Omega_0 & \Omega_1 & \Omega_2 & \Omega_3 & \dots \\ \Upsilon_0 & \Upsilon_1 & \Upsilon_2 & \Upsilon_3 & \dots \\ 0 & \Upsilon_0 & \Upsilon_1 & \Upsilon_2 & \dots \\ 0 & 0 & \Upsilon_0 & \Upsilon_1 & \dots \\ \vdots & \vdots & \vdots & \vdots & \ddots \end{pmatrix}$$

where we define the stationary transition probabilities

2.10 Matrix-Analytical Methods

$$\Upsilon_n = \Phi_{\Upsilon_n}(0) = \Upsilon_n(\infty) = \lim_{t \to \infty} \Upsilon_n(t) \tag{2.103}$$

$$\Omega_n = \Phi_{\Omega_n}(0) = \Omega_n(\infty) = \lim_{t \to \infty} \Omega_n(t)$$

Note that $\mathcal{P}$ is a stochastic matrix, representing the transition probabilities. Formally for later purposes we introduce the sum of the matrices (Υ_n) resp. (Ω_n)

$$\Upsilon = \sum_{n=0}^{\infty} \Upsilon_n = \lim_{s \to 0, z \to 1} \Upsilon_{DT}(z, s), \quad \Omega = \sum_{n=0}^{\infty} \Omega_n = \lim_{s \to 0, z \to 1} \Omega_{DT}(z, s)$$

Again the matrices Υ and Ω are stochastic. From the definition and the properties of the limit of transformation we can determine the form of Υ

$$\Upsilon = \lim_{s \to 0, z \to 1} \Upsilon_{DT}(z, s) = \lim_{s \to 0, z \to 1} \int_0^\infty e^{-st} e^{\mathcal{D}_{EF}(z) t} dF_\eta(s)$$
$$= \int_0^\infty e^{\mathcal{D} t} dF_\eta(s)$$

From the double transformation

$$\Omega_{DT}(z, s) = \frac{1}{z}(s \operatorname{id} - \mathcal{D}_0)^{-1}(\mathcal{D}_{EF}(z) - \mathcal{D}_0) \Upsilon_{DT}(z, t)$$

we conclude by the limit process $s \to 0$ and $\to 1$

$$\Omega_n = -\mathcal{D}_0^{-1} \mathcal{D}_1 \Upsilon_n$$

and

$$\Omega = \left(\operatorname{id} - \mathcal{D}_0^{-1} \mathcal{D}\right) \Upsilon$$

It is possible to show that the stationary probability π, to remain in the particular transient state defined in equation (2.97), also fulfills the relation

$$\pi = \pi \Upsilon$$

We can formally obtain from the double transform $\Upsilon_n DT(z, s)$ the average of (Υ_n) with respect to n

$$\mathbb{E}\Upsilon = \sum_{n=0}^{\infty} n \Upsilon_n e = \lim_{z \to 1} \left(\frac{\partial}{\partial z_n} \Upsilon_{DT}(z, 0) \right) \cdot e = \rho e + (\operatorname{id} - \Upsilon)(e\pi - \mathcal{D})^{-1} \lambda$$

The row vector $\lambda = \mathcal{D}_1 \cdot e$ represents the phase depending arrival rates of the MAP arrival process and is defined in equation (2.98). The mean number of arrivals during the serving period is identical with the mean traffic load ρ. We have

$$\pi \mathbb{E}\Upsilon = \rho$$

Stationary State Probabilities

For the determination of the state vector $\mathcal{X}$ at the embedded time spots, which means the ends of the serving periods, we consider the following eigen value problem
$$\mathcal{X} = \mathcal{X} \cdot \mathcal{P}, \text{ with } \mathcal{X} \cdot e = 1$$
Note the stability restriction $\rho < 1$. In addition, the row vector $\mathcal{X}$ is according to the matrix $\mathcal{P}(t)$ split into block of size $1 \times k$

$$\mathcal{X} = \{\mathcal{X}_0, \mathcal{X}_1, \ldots\} \tag{2.104}$$

The subvector $\mathcal{X}_\nu$ corresponds to the level ν and indicates the phase depending probability that the system will be on the level ν at the serving end. From the block structure of $\mathcal{P}$ we obtain the subvector

$$\mathcal{X}_\nu = \mathcal{X}_0 \Omega_\nu + \sum_{n=1}^{\nu+1} \mathcal{X}_n \Upsilon_{\nu+1-n}, \text{ for } \nu \geq 0 \tag{2.105}$$

The component $\mathcal{X}_\nu[j]$ indicates the probability that the system contains ν demands after the serving period and the MAP process is in phase j. The above equations reveals the close relationship to the classical M/G/1 system, which we repeat again (see also (2.27))

$$x(\nu) = x(0)\gamma(\nu) + \sum_{j=1}^{\nu+1} x(j)\gamma(\nu + 1 - j)$$

Let us shortly describe the equivalence of (2.104) with (2.105):

- The vector $\mathcal{X}_\nu$ of the MAP/G/1 system correspond with the with state probability $x(\nu)$ in the M/G/1 case.
- The matrices Υ_ν, Ω_ν with $\gamma(\nu)$ as the number of Poisson arrivals during a serving period.

How can we solve the linear system (2.104)? First, using the basic cycle, we determine the subvector $\mathcal{X}_0$. This is start for the iterative algorithm to get successively $\mathcal{X}_1, \mathcal{X}_2 \ldots$. The stationary vector $\mathcal{X}^*$ will be obtained via $\mathcal{X}$ as seen in the next section.

Fundamental Period

For further investigations we introduce some notions (fig. 2.49):

- First passage time: this is the time period for the transition of level ν_1 to level ν_2, i.e. the time interval for the first entering the level ν_1 until the first arrival in level ν_2 We denote it by $[\nu_1, \nu_2]$.

2.10 Matrix-Analytical Methods

- Back step: This is the time period between the arrival in the level $\nu > 0$ and the immediately following enter into level $\nu - 1$ ends. The transition time is also called *fundamental period*.
- The first passage time $[0,0]$ is called *fundamental cycle* and consists of a vacant period and the serving period of serving process.

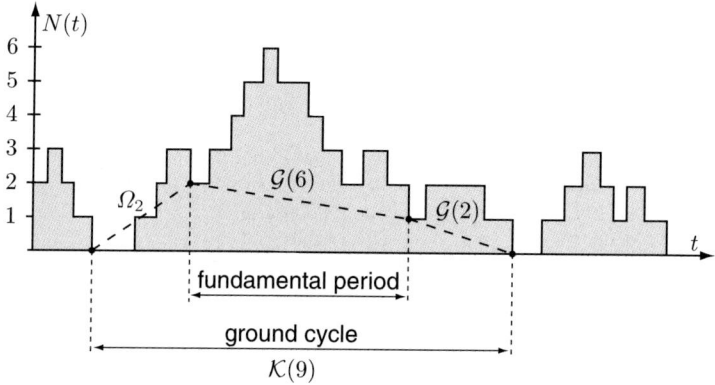

Fig. 2.49. Fundamental cycle and period

With $\mathcal{G}(r) \in \mathbb{R}^k \times \mathbb{R}^k$ we denote the probability that the fundamental period consists of r serving periods. As the definition reveals, this means that during $[\nu, \nu - 1]$ exactly r demands are served. Obviously we have a matrix of probabilities, since we have to encounter the phases of the arrival process at the beginning and at the end of the serving period. In this way, the component $\mathcal{G}(r)[ij]$ indicates the probability that the fundamental period consists of r serving periods and at the beginning resp. at the end of the fundamental period the MAP arrival process is in phase i resp. j. Figure 2.49 depicts some examples of fundamental periods. In figure 2.50 we show in an example the realization of a fundamental period, where from level 4 to 3 we have $r = 8$ serving periods. In the general case the fundamental period of r serving periods consists of:

- A serving period where n demands arrive ($n \geq 0$). This is characterized by matrix Υ_n and
- n fundamental periods consisting of $(r-1)$ serving periods. This is describe by the convolution $\mathcal{G}^{(n)}(r-1)$.

We obtain in matrix notation

$$\mathcal{G}(1) = \Upsilon_0$$

$$\mathcal{G}(r) = \sum_{n=0}^{\infty} \Upsilon_n \mathcal{G}^{(n)}(r-1), \quad r > 1$$

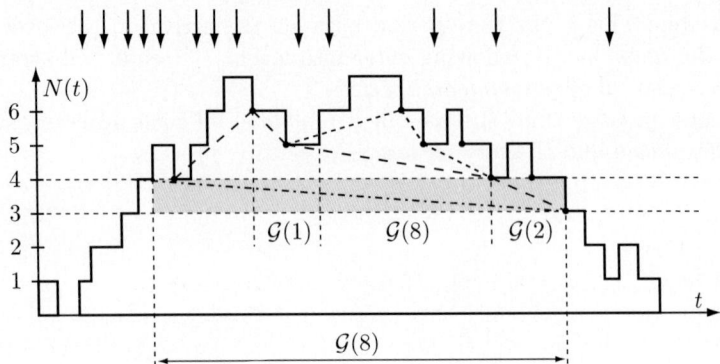

Fig. 2.50. Structure of the fundamental period

Here, we set $\mathcal{G}^{(0)} = \text{id}$ and $\mathcal{G}^{(0)}(r) = \mathcal{O}$ for $r > 0$. After a transformation, using the generating function, we obtain

$$\mathcal{G}_{EF}(z) = \sum_{r=0}^{\infty} \mathcal{G}(r) z^r = z \sum_{n=0}^{\infty} \Upsilon_n \left(\mathcal{G}_{EF}(z) \right)^n \qquad (2.106)$$

The consideration of the limit $z \to 1$, we conclude

$$\mathcal{G} = \mathcal{G}_{EF}(1) = \sum_{r=0}^{\infty} \mathcal{G}(r)$$

and with the help of equation (2.106) we obtain

$$\mathcal{G} = \sum_{n=0}^{\infty} \Upsilon_n \mathcal{G}^n \qquad (2.107)$$

The above equation gives us the tool to compute $\mathcal{G}$ successively. Some comments on the matrix are in order:

- The component $\mathcal{G}[ij]$ reflect the probability that at the beginning resp. at the end of a fundamental period the MAP arrival process is in phase i resp. j.
- The matrix $\mathcal{G}$ is stochastic, irreducible and a uniquely determined fix point vector γ, i.e. the solution of

$$\gamma = \gamma \mathcal{G}, \text{ with } \gamma \cdot \mathbf{e} = 1$$

With the matrix $\mathcal{G}$ we finally found the central player in the matrix analytic method, and the interested reader can consult [187]. We use the vector $\mathbb{E}_{\mathcal{G}}$, the *mean value vector of the fundamental period*, without derivation and compute it according to

$$\mathbb{E}_{\mathcal{G}} = \left(\frac{d}{dz}\mathcal{G}_{EF}(z)\right)\bigg|_{z=1} \cdot e = (\text{id} - \mathcal{G} + e \cdot \gamma)(\text{id} - \Upsilon + (e - \mathbb{E}_{\Upsilon})\gamma)^{-1} e$$

Similar to the matrix $\mathcal{G}$ the component $\mathbb{E}_{\mathcal{G}}[i]$ indicates the mean number of serving time in a fundamental period starting at i.

Fundamental Cycle

We defined a fundamental cycle $\mathcal{K}_0$ by the first passage time $[0,0]$, which is also denoted as the recurrence time in level 0. As we remember, it consist of a vacant and a serving period. The matrix $\mathcal{K}(r) \in \mathbb{R}^k \times \mathbb{R}^k$ reflects the probability that $\mathcal{K}_0$ includes r serving periods. Similar as before the element $\mathcal{K}(r)[ij]$ indicates that the time period $\mathcal{K}_0$ has r serving periods, while at the start resp. at the end of the fundamental cycle the MAP arrival process is in phase i resp. j. In figure 2.49 we have depicted an example with 9 serving times. After each vacant period a serving period starts, while 2 demands arrive. This is described by a matrix Ω_2. The process gets to level 2 afterwards. To get to level 2, the process has to pass through 2 fundamental cycles, including over all 8 serving periods. The path is built up by the matrices $\mathcal{G}(6)$ and $\mathcal{G}(2)$. The interval $\mathcal{K}_0$ in questions can be described by the matrix multiplication $\Omega_2 \cdot \mathcal{G}(6)\mathcal{G}(2)$.

In the general case the fundamental cycle, which includes r serving units and is represented by the matrix $\mathcal{K}(r)$ consists of:

- A vacant period succeeding a first serving period, during which n demands arrive, and which is characterized by a matrix Ω_n and
- n fundamental periods, including $r-1$ serving periods. This is described by the convolution of $\mathcal{G}^{(n)}(r-1)$.

By considering all possible paths which are needed for the realization of a basic cycle, we obtain the so called *matrix distribution of the basic cycle*

$$\mathcal{K}(r) = \sum_{k=0}^{\infty} \Omega_k \mathcal{G}^{(k)}(r-1), \; r \geq 1$$

We obtain by applying a transformation

$$\mathcal{K}_{EF}(z) = \sum_{r=0}^{\infty} \mathcal{G}(r) z^r = z \sum_{n=0}^{\infty} \Omega_n \left(\mathcal{G}_{EF}(z)\right)^n$$

Analogously as before we get by the limit procedure $z \to 1$

$$\mathcal{K} = \mathcal{K}_{EF}(1) = \sum_{r=0}^{\infty} \mathcal{K}(r) \qquad (2.108)$$

Equation (2.108) implies

$$\mathcal{K} = \sum_{n=0}^{\infty} \Omega_n \mathcal{G}^n$$

And guess again, we give some comment on $\mathcal{K}$: the component $\mathcal{K}[ij]$ indicates the probability that at the start resp. at the end of a fundamental cycle $\mathcal{K}_0$ the MAP arrival process is in state i resp. j. We obtain for the matrix $\mathcal{K}$

$$\mathcal{K} = -\mathcal{D}_0^{-1}(\mathcal{D}_{EF}(\mathcal{G}) - \mathcal{D}_0) = \text{id} - \mathcal{D}_0^{-1}\mathcal{D}_{EF}(\mathcal{G})$$

where we defined $\mathcal{D}_{EF}(\mathcal{G}) = \mathcal{D}_0 + \mathcal{D}_1 \mathcal{G}$. Note the fact that the fix point vector of $\mathcal{D}_{GF}(\mathcal{G})$ is identical with that of $\mathcal{G}$

$$\gamma = \gamma \mathcal{D}_{EF}(\mathcal{G})$$

Not surprising, we find the matrix $\mathcal{K}$ stochastic, irreducible and with a unique fix point vector

$$\kappa = \kappa \mathcal{K}, \text{ with } \kappa \cdot e = 1$$

The component $\kappa[j]$ tells us the stationary probability that at the start of a fundamental cycle $\mathcal{K}_0$ the MAP arrival process is in state j.

State Probabilities

For the computation of the vector $\mathcal{X}$ we need the mean value vector $\mathbb{E}_{\mathcal{K}}$

$$\mathbb{E}_{\mathcal{K}} = \left(\frac{d}{dz}\mathcal{K}_{EF}(z)\right)\bigg|_{z=1} \cdot e$$
$$= -\mathcal{D}_0^{-1}(\mathcal{D} - \mathcal{D}_{EF}(\mathcal{G}) + \lambda \cdot \gamma)(\text{id} - \Upsilon + (e - \mathbb{E}_\Upsilon)\gamma)^{-1} e$$

We know the game already – here are some comments on $\mathbb{E}_{\mathcal{K}}$: the component $\mathbb{E}_{\mathcal{K}}[i]$ indicates the mean serving time in a fundamental cycle $\mathcal{K}_0$, starting in phase i. With this we can determine the mean number $\bar{\eta}$ of serving times in the fundamental cycle

$$\bar{\eta} = \kappa \cdot \mathbb{E}_{\mathcal{K}}$$

We can compute the vector $\mathcal{X}_0$ with the help of the mean vector $\mathbb{E}_{\mathcal{K}}$

$$\mathcal{X}_0 = \frac{\kappa}{\bar{\eta}} = \frac{\kappa}{\kappa \cdot \mathbb{E}_{\mathcal{K}}}$$

After some sophisticated arrangements we conclude

$$\mathcal{X}_0 = \frac{1-\rho}{\lambda}\gamma(-\mathcal{D}_0)$$

We can determine the vectors $\mathcal{X}_\nu$ recursively

$$\mathcal{X}_\nu = \left(\mathcal{X}_0 \overline{\Omega}_\nu + \sum_{n=1}^{\nu+1} \mathcal{X}_n \overline{\Upsilon}_{\nu+1-n}\right)(\text{id} - \overline{\Upsilon}_1)^{-1}, \text{ for } \nu \geq 0 \qquad (2.109)$$

2.10 Matrix-Analytical Methods

where in the case of a pure MAP/G/1 queueing system we have

$$\overline{\Upsilon} = \sum_{\nu=n}^{\infty} \Upsilon_\nu \mathcal{G}^{\nu-n} \quad \text{and} \quad \overline{\Omega_n} = \sum_{\nu=n}^{\infty} \Omega_\nu \mathcal{G}^{\nu-n}$$

Define $\mathcal{X}_{GF}(z) = \sum_{\nu=0}^{\infty} \mathcal{X}_\nu z^\nu$, we have

$$\mathcal{X}_{GF}(z)(z\mathrm{id} - \Upsilon_{GF}(z)) = \mathcal{X}_0 (z\Omega_{GF}(z) - \Upsilon_{GF}(z))$$
$$= -\mathcal{X}_0 \mathcal{D}_0^{-1} \mathcal{D}_{GF}(z) \Upsilon_{GF}(z)$$

where according to equation (2.102) we get

$$\Upsilon_{GF}(z) = \sum_{n=0}^{\infty} \Upsilon_n z^n \quad \text{and} \quad \Omega_{GF}(z) = \sum_{n=0}^{\infty} \Omega_n z^n$$

By equation (2.109), we can explicitly determine the complete vector of the stationary state probabilities at the embedded time spots.

State Probabilities at Arbitrary Time Spots

As already done, $\mathcal{X}^*$ denotes the stationary distribution of the state at arbitrary time spots. We write $\mathcal{X}^* = (\mathcal{X}_0^*, \mathcal{X}_1^*, \ldots)$, where each component is a $1 \times k$-matrix. Here is the relationship between $\mathcal{X}$ and $\mathcal{X}^*$

$$\mathcal{X}_0^* = \lambda \mathcal{X}_0 \mathcal{D}_0^{-1} = (1-\rho)\gamma$$

$$\mathcal{X}_{\nu+1}^* = \left(\sum_{n=0}^{\nu} \mathcal{X}_n^* \mathcal{D}_{\nu+1-n} - \lambda(\mathcal{X}_\nu - \mathcal{X}_{\nu+1}) \right)(-\mathcal{D}_0)^{-1} \quad (2.110)$$

The corresponding generating functions reads as

$$\mathcal{X}_{GF}^*(z) \mathcal{D}_{GF}(z) = -\lambda(1-z)\mathcal{X}_{GF}(z)$$

Virtual Queueing Time Distribution Function

With the notion *virtual queueing time* we denote the time a test demand would have to wait. The MAP arrival process does not allow an arrival at all time, thus, we do not have PASTA. We obtain different queueing time distribution for the different arrivals.

Consequently, let $W_j(\cdot)$ denote the distribution of the virtual queueing time, if the MAP arrival process is in phase j. We build the vector

$$\mathbf{W}(t) = \{W_1(t), \ldots, W_k(t)\}$$

The phase independent distribution function of the virtual queueing time is given by

$$W(t) = \mathbf{W}(t) \cdot \mathbf{e}$$

The Laplace-Stieltjes transform reads as

$$\Phi_\mathbf{W}(s) = \int_0^\infty e^{-st} d\mathbf{W}(t), \text{ i.e. } \Phi_\mathbf{W}(s) = \Phi_W(s) \cdot \mathbf{e}$$

Finally, here is the generalization of the known Pollaczek-Khintchine formula for the queueing time, which we met in the M/G/1 system. Without proof we present

$$\phi_\mathbf{W}(s) = s\mathcal{X}_0^* \left(\text{sid} + \mathcal{D}_{GF}(\Phi_\eta(s))\right)^{-1}$$

$$\Phi_\mathbf{W}(0) = \pi$$

With this we obtain as generalization for the virtual queueing time distribution function Φ_W in the MAP/G/1 system

$$\Phi_W(s) = \Phi_\mathbf{W}(s) \cdot \mathbf{e} \qquad (2.111)$$
$$= s\mathcal{X}_0^* \left(\text{sid} + \mathcal{D}_{GF}(\Phi_\eta(s))\right)^{-1} \cdot \mathbf{e}$$

Summary of the Algorithm

We start with the MAP/G/1 queueing system. As usual we denote by $\mathcal{D}_0$ and $\mathcal{D}_1$ the distribution matrices of the MAP arrival process. Let η be the serving time. We present the steps for analyzing the virtual queueing time:

a) Calculation of the matrices (Υ_n) according to equation (2.103)
b) Computation of the matrix $\mathcal{G}$. According to Neuts [187] we use an expression equivalent to (2.107), which enables iteration to determine $\mathcal{G}$

$$\mathcal{G}_{\nu+1} = (\text{id} - \Upsilon)^{-1} \sum_{n=0, n\neq 1}^\infty \Upsilon_n \mathcal{G}_\nu^n$$

where ν denotes the iteration index. The start matrix $\mathcal{G}_0$ uses the matrix Υ. Thus, the iterative determined matrices $\mathcal{G}_1, \mathcal{G}_2, \ldots$ are stochastic. We can stop the iteration provided

$$\max_{i,j} \left| \mathcal{G}_\nu[ij] - \left(\sum_{n=0}^\infty \Upsilon_n \mathcal{G}_\nu^n \right)[ij] \right| < \epsilon$$

The fix point vector γ can be computed, using an iteration method as well.
c) With the help of equation (2.109) we obtain $\mathcal{X}_0$ and thus, the other vectors $\mathcal{X}_\nu, \nu > 0$.
d) Computation of the vectors $\mathcal{X}_\nu^*$ using (2.110).
e) Knowing $\mathcal{X}_0^*$, we can compute the virtual queueing time distribution by (2.111).

2.10.5 Application to IP Traffic

In this subsection we want to apply the above theoretical concepts to the IP traffic modeling. We refer mainly to a paper of Heyman and Lucantoni [108]. The approach is a MMPP model to describe the IP traffic on the packet resp. byte level. Instead of using self-similar processes, as we will introduce in the next chapter, one of the characteristics of IP traffic, obtained by statistical evaluations, this traffic phenomena is mimicked by the MMPP model. As seen in the next chapter the IP-modeling, using self-similar processes, describes the traffic asymptotically. The model presented here claims to be exact, proved by fitting the traces and statistical resp. numerical evaluation. In fact, by using MMPP the relevant part of the trace showing heavy-tail behavior is quite good fitted by the MMPP, where the underlying example is an ATM trace. In several papers [12, 225] the MMPP example is fitted to certain IP traces to mimic the LRD behavior over several magnitudes. But for this model as for the other ones we have to stress that it depends on the traces and the view point – long-range behavior, i.e. averaging over large scales, self-similarity or the behavior on small scales – there is no model fitting for all kind of traces. With this and the others again we open windows, looking at some aspects of the traces.
But enough! Let's start with the description.

Introduction

We can detect two fundamental applications of queueing models: sizing the line transport and fitting the buffer space in routers. Actually we will deal with this in more detail in the last chapter, but since the application of the matrix analytic methods to the IP traffic should be given, we present the discussion here. As already mentioned, the content is an extraction of the paper [108]. The basic situation can be described by several traffic steams, each is modeled by a MMPP, combined together. This can happen, when at the provider's several streams at the router are compounded to use afterwards a single high speed link. As we demonstrated in the first chapter, IP traffic can not be correctly interpreted by the Poisson or even purely Markovian queueing model. Before we start with the more mathematical advanced models in the next chapter, we still stick to the Poisson approach, but now as a matrix model MMPP, as e.g originally demonstrated in papers like [108, 189]. In papers like [12, 225] the behavior is captured on several scales.

Fitting IP Traces and D-MMPP Model

Before giving the basic idea we state some alternatives to the MMPP model.

Hidden Markov Model

A Hidden Markov model, which is a discrete-time process, where the parameters, controlling the distributions (as interarrival and serving time) are governed by a (hidden) Markov chain, is an alternative way to the MMPP. An MMPP is a point process, where the key random variables are the interarrival times. The phases of the MMPP are continuous-time Markov chain. Now we slightly change and suppose that the phases are described by a discrete time Markov chain, where the rates behave as in the case of the MMPP. Thus, the phase changes happen as an integer multiple of a fundamental time interval, which is called *phase transition epochs*. Though Markov chain is involved, it is not a Markovian process, since it is not memoryless, it is regenerative at any transition time spot. Otherwise the technique can be copied from the standard MMPP. We call this process D-MMPP, which is a special case of the BMAP, the *batch Markovian arrival process* (see [168]). Another approach, using matrix methods, consists in the

Interrupted Poisson Processes

In the paper [15] with the help of an maximum likelihood estimator (see section 4.1.4) they use superstition of IPP to reproduce a MMPP. This requires that the arrival times are given in addition to the counts, which should be avoided, since it requires to use too much information. In the paper [157] the time between the phases are modeled with a non exponential distribution. The traces, the paper [108] uses, does not fit into this approach.

D-MMPP Model

For the D-MMPP model we define a $n \times n$ transition probability matrix $\mathcal{P}$. After each discrete step of Δt we have a Poisson arrival with rate λ_i, where i is the step before the transition (Markov chain). We define the arrival vector $\boldsymbol{\lambda} = (\lambda_1, \ldots, \lambda_n)^T$, where we assume $\lambda_1 \geq \lambda_2 \geq \ldots \lambda_n$. Note that n is the number of states in the Markov chain. We have a D-MMPP model, which coincides with a discrete BMAP (resp discrete BMAP) arrival process with representation matrices D_k, where

$$D_k = \text{diag}\left(e^{-\lambda_1}\frac{\lambda_1^k}{k!}, \ldots, e^{-\lambda_n}\frac{\lambda_n^k}{k!}\right) \cdot \mathcal{P}$$

Let's define the irreducible, stochastic matrix $\mathbb{D} = \sum_{k=0}^{\infty} D_k$. We have a fix point or stationary probability vector $\boldsymbol{\pi}$ and mean arrival rate vector $\boldsymbol{\lambda}$ computed as

$$\sum_{k=0}^{\infty} k D_k \mathbf{e} = (\lambda_1, \lambda_2, \ldots, \lambda_n) = \boldsymbol{\lambda}$$

How can we fit now a data trace to the above model? As always stressed before, the data trace has to be suitably described by an MMPP or D-MMPP. A two state MMPP cannot describe a highly bursty data trace, since we have a big gap in between. This can easily be seen by the simple Gaussian approximation of the Poisson distribution. As well known fact, 95% of the probabilty is in a range of deviation 1.95. The mean and variance of the Poisson distribution let say is λ. For large λ Poisson and Gaussian distribution function are close, where F_P is within $\lambda \pm a\sqrt{\lambda}$, with $a = 2$. Assume we have a large peak to mean ratio, then the two phases λ_1 and λ_2 differ heavily and $\lambda_2 + 2\sqrt{\lambda_2}$ is much less than $\lambda_1 - 2\sqrt{\lambda_1}$. And hence, the traffic in between is not matched by the model.

1. Choosing the arrival rates: We use rates $\lambda_1 \geq \lambda_2 \geq \ldots \geq \lambda_N$, where N has to be specified. λ_1 should describe the large observation.

$$\lambda_1 + 2\sqrt{\lambda_1} = \text{peak of data trace}$$

   which gives the possibility that a sample path may cover data larger than the observed ones. We get for a solution

$$\lambda_1 = (\sqrt{1 + \text{peak}} - 1)^2$$

   covering the lower bound of the peak data by $\lambda_1 - 2\sqrt{\lambda_1}$, which will be the upper bound for the data covered by λ_2.

$$\lambda_2 + 2\sqrt{\lambda_2} = \lambda_1 - 2\sqrt{\lambda_1}$$

   The solution is

$$\lambda_2 = (\sqrt{\lambda_1} - 2)^2$$

   Now we know the game and repeat the procedure for the other λ_i, where we stop, when meeting the smallest data trace.

   We have to give some notes on some key parameter of the algorithm. First why choosing $a = 2$? A smaller number yields more states and a bigger number gives a larger overlap algorithm based on this first iteration as given in [107].

2. Fitting the Markov chain: By the algorithm the smallest data trace is covered by the algorithm stops, and we have found N. Now we construct the Markov chain transition matrix for the phases. For this assume that $\{x_i, i = 1, 2, \ldots, T\}$ are the observations. The phase φ_i is associated with the observation x_i in the following way

$$\lambda_j - a\sqrt{\lambda_j} < x_i \leq \lambda_j + a\sqrt{\lambda_j} \Rightarrow \varphi_i = j$$

   The set $\{\varphi_i; i = 1, \ldots, N\}$ can be interpreted as observations on the phase process. Define the matrix $\mathcal{P} = (p_{ij})$ where

$$p_{ij} = \frac{\text{number of transitions from } i \text{ to } j}{\text{number of transitions out of } i}$$

Fitting of N-State MMPP

Though the D-MMPP/D/1-queueing system falls into the paradigm of M/G/1, its computational treatment and storage is elusive and hence, it is not as suitable for highly structured traffic, since the transition matrices turn relatively quickly to be large. So one can avoid this by applying the continuous MMPP queueing system resp. MAP (see [168]).

As we know from the generator or rate matrix for the phase transition, $\mathcal{Q} = (q_{ij})$ is the basic ingredient. There is a simple way to attach the transition matrix of the D-MMPP system to the matrix Q by

$$q_{ij} = p_{ij}, \ i \neq j, \ q_{ii} = p_{ii} - 1$$

This implies that both system enjoy the same sojourn time in phases i, for the probability of given states and the steady-state distribution. Thus, we get

$$D_1 = \text{diag}(\boldsymbol{\lambda}) \quad \text{and} \quad D_0 = Q - D_1$$

using the standard notation for MAP. The alert reader may ask, why first doing the computation of the D-MMPP and not directly of the MMPP. Well, in real traffic with high speed links it is easier to obtain traffic measurements over counts than over the amount of individual packets.

Mean Queue Length at Arrivals

We denote by W_V the virtual and by W_A the waiting time seen at an arrival in the MAP/D/1 system with mean service time set to 1. The results in [167, 168] reveal

$$\mathbb{E}(W_V) = \frac{3\rho - 2\mathbf{b}D_1\mathbf{e}}{2(1-\rho)}$$

$$\mathbb{E}(W_A) = \frac{1 - (\mathbf{b}D_1\mathbf{e})}{\rho + \mathbb{E}(W_v)} + \mathbb{E}(W_v)$$

with $\mathbf{b} = ((1-\rho)\mathbf{g} + \boldsymbol{\pi}D_1)(\mathbf{e}\boldsymbol{\pi} + D_0 + D_1)^{-1}$ being the stationary probability vector of the irreducible stochastic matrix $\mathcal{G}$, being the solution of the fundamental equation

$$G = e^{D_0 + D_1 G} \tag{2.112}$$

and $\boldsymbol{\pi}$ is the stationary vector (or fix point vector) of the generator $D_0 + D_1$. To compute $\mathcal{G}$ from (2.112), we start the iteration with $\mathcal{G}_0 = \mathbf{e}\boldsymbol{\pi}$ and use about $100,000$ steps to achieve 10 decimal accuracy. The Little formula implies finally the mean queue length

$$\mathbb{E}(Q_V) = \rho\mathbb{E}(W_V) \quad \text{and} \quad \mathbb{E}(Q_A) = \rho\mathbb{E}(W_A)$$

Example 2.97. In [107] the following trace is investigated to fit the D-MMPP. The depicted figure show a quite good description of the examined traffic to the D-MMPP model. The trace was observed over a period of three hours on a IP backbone link. The data measures packets per 500 ms and mean packet length of 1000 byte. The mean data rate is 360 packets per 500 ms or 5,76 Mb/s. The authors used the algorithm LAMBDA to fit the D-MMPP model. In the example 21 states were needed for the description.

The autocorrelation existing in the data trace is not used directly for fitting the model. The autocorrelation for the rough data is decaying more rapidly than the smoothed one. It does not affect the queueing behavior, since the link utilization is higher than the simulated one. For testing the ability of the D-MMPP model, the author estimated the mean packet delay by simulating an infinite buffer queue with deterministic (constant) service time. The model as well as the data trace fit quite good in the range for a traffic utilization of $\rho \in\ ]0, 0.6[$. The fact that the accuracy declines heavily, is also due to the TCP effects and leads to an application of other methods like the multifractals in the next chapter.

A second test consist in the investigation of the packet loss. Here, up to the traffic load $\rho = 0.72$, both the applied MMPP models and the data trace fit quite accurately. More on the detailed data traces and results the reader can find in [107].

Limiting Access Line Speeds

As we showed in the previous chapter 3 and will again deal with in the next chapter 3, the Poisson process is inadequate to cover the bursty character of the data traffic.

If we assume that the original traffic is well modeled by the D-MMPP, it is not difficult to represent the same process transmitted over the access line with limit rates by another D-MMPP model, but now, with all states of the original D-MMPP model with higher rates than the peak rate are included into one particular state (see the main result in theorem 2.98).

So, let a D-MMPP system with transition probability matrix $\mathcal{P}$ and arrival rate vector $\boldsymbol{\lambda} = (\lambda_1, \ldots, \lambda_n)^T$ be given, where $\lambda_1 \geq \lambda_2 \geq \ldots \geq \lambda_n$. The D-MMPP approximates a bursty traffic entering an output port, while the peak rate of the flow is limited by the rate of the link. We provide now a D-MMPP model to reflect the effect of limiting the peak.

Choose the stationary (resp. fixpoint vector) $\boldsymbol{\pi}$ of $\mathcal{P}$. This indicates

$$\boldsymbol{\pi} = \boldsymbol{\pi}\mathcal{P}, \ \boldsymbol{\pi}\mathbf{e} = 1$$

Then, we have $\lambda^* = \langle \boldsymbol{\lambda}, \boldsymbol{\pi} \rangle$ for the arrival rate in the D-MMPP system. Suppose the peak rate is $\overline{\lambda}$, with $\lambda_i < \overline{\lambda} < \lambda_1$ for $i > 1$. With T_1 we denote the generic sojourn time in state 1 of the Markov chain. If the system is in state 1, this means by definition that the arrival rate is higher than the line can process. The expected number of arrivals in state 1 is

$$\lambda_1 \mathbb{E}(T_1) = \frac{\lambda_1}{1 - P_{11}}$$

Now we assume an infinite buffer, implying that the data is stored without loss (one could incorporate a loss factor, as e.g. by setting the output rate to the carried load on the link).

As defined above, the actual rate that the packets are processed is $\overline{\lambda}$. Hence, to conserve the expected number of packets during a sojourn time in a state is to replace λ_1 in the original D-MMPP model by $\overline{\lambda}$ and $\mathcal{P}'$ instead of $\mathcal{P}$, provided the mean arrival rate is conserved. The same holds for the behavior during a sojourn time. Because $\overline{\lambda} < \lambda_1$ and the others are fixed, we have $P'_{11} > P_{11}$. Thus, the average sojourn time in state 1 increases. In the sequel we derive the necessary parameters to approximate the traffic by the D-MMPP model. We pick an index m, such that $\lambda_m \geq \overline{\lambda}$ and $\overline{\lambda} > \lambda_{m+1}$ Let $k = n - m$ and split $\mathcal{P}$ and $\boldsymbol{\lambda}$ according to

$$\mathcal{P} = \begin{pmatrix} S & U \\ V & T \end{pmatrix}, \quad \boldsymbol{\lambda} = \begin{pmatrix} \boldsymbol{\lambda}_S \\ \boldsymbol{\lambda}_T \end{pmatrix}$$

where S is a $m \times m$ matrix, T a $k \times k$ matrix and $\boldsymbol{\lambda}_S$ resp. $\boldsymbol{\lambda}_T$ are $m \times 1$ resp. $k \times 1$ vectors.

The idea is to include all states with rates larger or equal to the peak into one particular state with peak rate as the corresponding rate and keep the rest behavior untouched. The sojourn time in the state of the peak rate should be enlarged to provide an unchanged overall mean rate.

We split the vector $\boldsymbol{\pi} = (\boldsymbol{\pi}'_S, \boldsymbol{\pi}'_T)$ and normalise the first component by $\boldsymbol{\pi}_S = \frac{\boldsymbol{\pi}'_S}{\langle \boldsymbol{\pi}'_S, \mathbf{e} \rangle}$. The main theorem reads as follows.

Theorem 2.98. *The D-MMPP with the required property has the transition probability matrix*

$$\mathcal{P}' = \begin{pmatrix} 1-p & p\mathbf{u} \\ \mathbf{v} & T \end{pmatrix} \tag{2.113}$$

where the mean arrival rate vector is $\boldsymbol{\lambda}' = (\overline{\lambda}, \boldsymbol{\lambda}_T^T)^T$ and where

$$\mathbf{u} = \frac{\boldsymbol{\pi}_S (id - S)^{-1} U}{\boldsymbol{\pi}_S (id - S)^{-1} U \mathbf{e}}$$

$$\mathbf{v} = V\mathbf{e}, \quad p = \frac{\lambda^* - \overline{\lambda}}{\beta \lambda^* - \gamma}$$

$$\beta = -\mathbf{u}(id - T)^{-1}\mathbf{e} \quad \text{and} \quad \gamma = -\mathbf{u}(id - T)^{-1}\boldsymbol{\lambda}_T$$

Proof. To get some insight we sketch the proof. Evidently the traffic described by the D-MMPP model according to (2.113) has peak rate $\overline{\lambda}$, otherwise no behavior is changed from the original one. The vector $\mathbf{v}$ is constructed, such that the rate into the peak state from less-peak-state i is the same as the rate into states with rates larger or equal to the peak state in the original model.

The probability for getting into a less-than-peak-rate-state for the first time in state j under the premise that the process starts in a larger-than-peak-rate-state i, can be detected by the (i,j) element of the matrix $(\text{id} - S)^{-1}U$ (see [139, 97]).

We can calculate directly the stationary (or fixed point) probability vector of the matrix $\mathcal{P}'$ and obtain

$$\boldsymbol{\alpha} = (1 + p\mathbf{u}(\text{id} - T)^{-1}\mathbf{e})^{-1}(1, p\mathbf{u}(\text{id} - T)^{-1})$$

with the average arrival rate $\boldsymbol{\alpha}\boldsymbol{\lambda}'$. If we redefine this as λ^* and solve for p, we get the final expression of the theorem. □

Example 2.99. The above trace is used for the investigation of limited access lines as well. Remember the mean traffic of 5.76 MBit/s. On a 45 MBit/s line only little effect is seen, while by decreasing the capacity to 34.6 MBit/s and 23 MBit/s limiting the peak rate a certain level of smoothing is achieved. To achieve the average of the original trace, as the peak gets more limited, the D-MMPP has to stay at the peak rate for longer time.

Further Literature

There are a lot of monographs dealing with classical traffic theory. We used the monographs of [254, 255], which we think, give a good introduction to the different topics of the area. Classical queueing theory for the traffic theory are presented in [20, 87, 142]. For the general description of stochastic analysis we recommend for further reading [38, 75, 208, 237] and for analysis [222].

3
Mathematical Modeling of IP-based Traffic

> *As far as theorems in mathematics concern the reality, they are uncertain and as far as they are exact, they do not reflect the reality.*
>
> Albert Einstein (20th century)

3.1 Scalefree Traffic Observation

3.1.1 Motivation and Concept

As we already pointed out in section 1.4, the Bellcore experiment of the Leland group raised doubts on applying the classical traffic theory of telecommunication. We briefly summarize the facts and the major ingredients, which lead to the application of new mathematical models.

Observations

Among several observations of the Leland group [160], one of the basic ideas of them was that statistical key values like variance and correlation of the traffic amount could not be explained by the Poisson based models, being used in traffic modeling since the days of Erlang. In addition, IP traffic issues like data amount or interarrival time perform no significant change on different scales. As we already realized in the previous chapter, the IP traffic is highly not deterministic and thus, the Leland group looked for mathematical concepts to describe the traffic. One way of finding the appropriate model was to average the traffic amount:

> They started with the certain time spots i and evaluated the traffic amount at that particular time. Since they were not interested in the special chosen time i, they averaged over a time period of length m, i.e. they built disjoint blocks of length m
>
> $$X^{(m)}(i) = \frac{1}{m}\left(X_{(i-1)m+1} + X_{(i-1)m+2} + \ldots + X_{im}\right)$$
>
> now depending on the reference time spots i and looking back the period of m subdivided times.

They did it with increasing length m and realized that the picture of the evaluated traffic looked similar. This lead to a concept of *self-similarity*, already used by Mandelbrot and Van Ness in the 1960ties to describe asset pricing in finance. Roughly speaking – it will be done mathematical rigorous below in section 3.2 – this means that the traffic behavior does not depend on the scale. It is *scale free* or *fractal*. But, as later investigation will indicate, the traffic is not exactly self-similar, i.e. does not match the mathematical model exactly. In fact, the mentioned picture of successive averaging can be misleading as the example of the geometric Brownian motion shows. It is the asymptotic self-similarity, i.e. the compounded data process after successive averaging over large intervals, which coincides in the limit with the well-treated concept of drifted self-similarity. Figure 1.9 resp. 1.10 in chapter 1 indicates the process of averaging over several time scales and reveal the fact that the real data traffic does not change its shape for different scales.

Example: Heavy-Tail Distribution of File Sizes

As Crovella and Bestavros argued in a later paper [59], one reason may result in the high data rates, which they investigated exemplarily for the World Wide Web. It was proven in [250], based on a simple physical on-off model, that the traffic leads asymptotically to the above mentioned self-similarity, provided the on-periods are modeled with a distribution without existing variance.
So, one possibility in explaining the phenomena of self-similarity would be to detect a heavy-tail distribution for the on-periods (see section 2.7.4).
In section 4.1 we give an example of measured data traffic, which demonstrate that the size is distributed according to a heavy-tail distribution.

Example: Data Traffic

But still there might be some hope for the use of the classical Poisson process. Could it be possible that the Poisson process perform asymptotical self-similarity, too? But, as we already showed in the figure 1.9 in section 1.4, the asympotic of the Poisson process, i.e. the averaging of the blocks $X^{(m)}(i)$ does not lead to the respective observed figure of the IP traffic. In the Poisson case the bursts are leveled out.
That the real data traffic is not matched by the Poisson process but instead of the so called self-similar processes is reflected in the figures 1.9 and 1.10 in chapter 1.

Example: Protocol Influence

Later, when the models of Norros or Willinger et al. were well established and the World Wide Web 'got on the victory line', the protocol seemed to have more influence on the traffic as one has thought of, and it is not only the statistically gained asymptotic self-similarity, which performs the IP-based traffic picture. This lead and continues with the incorporation of protocol influences into the framework of self-similarity. The pure self-similarity aspect more or less leads to a leveling of the traffic and gives way for applying the central limit theory, which is a good initiator for the Gaussian distribution (see section 3.2.3). But, the more spiky traffic seems not to match the picture of the Gaussian processes. Here, we encounter the so called multifractal traffic models, which will be the issue in section 3.8.

3.1.2 Self-Similarity

In section 3.1.1 we briefly described the IP traffic observation, e.g. obtained by the Leland group. In the following section we will introduce and sketch the main results on the mathematical concept of self-similarity. But first, we summarize two methods of examination and formulate suitable notions for the IP traffic:

- Instead of looking at the traffic X_t on a link during and interval $[0,t]$ one considers the increment $X_t - X_s$ for $s < t$. Since this is certainly depending of the interval, one has to divide by its length $t - s$. As we saw the behavior for different lengths of the interval of the measured traffic does not change. So we have a similarity for the shape of the traffic. This leads to the following definition.

  Definition 3.1. *Let $(X_k)_{k\in\mathbb{N}}$ be a discrete stationary process of second order with expectation 0. We call $X^{(m)}$ the averaged process with $i = 1, 2, \ldots,\ m = 1, 2, \ldots$ and define it according*

  $$X^{(m)}(i) = \frac{1}{m}\left(X_{(i-1)m+1} + X_{(i-1)m+2} + \ldots + X_{im}\right) \qquad (3.1)$$

- A useful basic number is the development of the variance for the averaged process $X^{(m)}(i)$. In fact, with growing block lengths the process should approach an ideal self-similar process, which we introduced above. But in addition it should confirm the observed measurements. Hence, We define the block process $(X^{(m)}(i))_{i\in\mathbb{N}}$ to be *asymptotic self-similar* if there is a constant $H \in\ ]0,1[$ (the *Hurst exponent*) provided

  $$\lim_{m\to\infty} \frac{\mathrm{Var}(X^{(m)}(k))}{\sigma^2 m^{2H-2}} = 1,\ 0 < \sigma^2 < \infty \qquad (3.2)$$

  or equivalent via the covariance structure

  $$\begin{aligned}\lim_{m\to\infty} r^{(m)}(k) &= \lim_{m\to\infty} \mathbb{C}\mathrm{or}\left(X^{(m)}(k), X^{(m)}(k+1)\right) \qquad (3.3)\\ &= \left((k+1)^{2H} - 2k^{2H} + (k-1)^{2H}\right),\ k\in\mathbb{N}\end{aligned}$$

Interpretation: The idea underlying this concept for the IP traffic is the following: The equation (3.2) and (3.3) hold for stochastic processes, which exhibit a self-similar structure (see theorems 3.3 and 3.11). IP traffic does not reveal an exact self-similar structure but by averaging over all scales we can detect in the limit such a self-similarity. Hence, a good mathematical approach is an approximation of the traffic by a suitable self-similar process.

3.2 Self-Similar Processes

3.2.1 Definition and Basic Properties

We start with the mathematical concept of the *self-similarity*. The crucial point interesting for the IP-based traffic is the behavior for large times, i.e. the time depending development of a process. We meet the notion of self-similarity, which we will introduce and which offers the best possibility to describe the IP traffic (as e.g. data traffic in the Internet) for asymptotic large times. Thus, we have to examine some of the basic properties more closely and start with the treatment of the important class of self-similar processes. From the motivation of traffic observation we can extract the definition of self-similarity, which was originally introduced by Mandelbrot and Van Ness ([174]) calling it *self affine*.

Definition 3.2. *A process* $(X_t)_{t \in I}$ $I \subset \mathbb{R}$ *is called* exact self-similar, *if one finds for all choices of factors* $a > 0$ *a* $b(a) > 0$ *such that, for the random vectors*

$$(X_{at_0}, \ldots, X_{at_n}) \quad \text{and} \quad b(a)(X_{t_0}, \ldots, X_{t_n}) \tag{3.4}$$

coincide in distribution for all times $t_0, \ldots, t_n$. *Here, I could be discrete or a continuum like* $I = \mathbb{R}, [0, \infty[$.

The equation (3.4) raises the question, whether the factor $b(a)$ could be written in the form a^H, with a fixed $0 < H$. In other words, does for a self-similar process (X_t) exists a $H > 0$, such that for all $a > 0$ and all times $t_0, \ldots, t_n$ the finite marginal distributions of

$$(X_{at_0}, \ldots, X_{at_n}) \quad \text{and} \quad a^H(X_{t_0}, \ldots, X_{t_n}) \tag{3.5}$$

coincide? Indeed Lamperti [154] solved this question under certain additional assumption.

Theorem 3.3. *(Lamperti) Let* $(X_t)_{t \in [0,\infty[}$ *be a self-similar non trivial process (i.e., not almost surely constant), whose paths are stochastic continuous at* $t = 0$ *(i.e,* $\mathbb{P}(|X_h - X_0| \geq \epsilon) \to 0$ *for all* $\epsilon > 0$ *and* $h \to 0$*), then there is exactly one* $H \geq 0$ *with* $b(a) = a^H$ *for all choices of* $a > 0$.

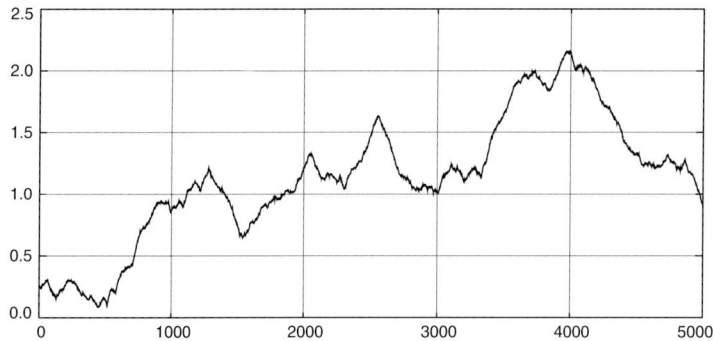

Fig. 3.1. Path of a fractional Brownian motion with $H = 0.8$

Remark 3.4. In the sequel we list several basic properties of self-similar processes:

a) The value H is called *Hurst exponent*. Theorem 3.3 shows, that the Hurst exponent H is unique. Thus, we will call a process $(X_t)_{t \in I}$ an H-ss process, provided (3.4) resp. (3.5) holds ('ss' for 'self-similar').
b) The relation (3.4) (i.e. equality in distribution) expresses the invariance of scales and not any change of the corresponding paths. It should be emphasized that representing self-similar processes cannot be done by zooming in and out of the paths. The scaling represents an equality in distribution, since the paths are a realization due to the randomness, though it can be used for a first presentation.
c) Usually one considers t as time variable and X_t as the space variable. Hence, the self-similarity means that zooming in the time dimension with a factor a implies an enlarging of the space vector by a factor a^H, always under consideration of a realization $\omega \in \Omega$ and the equality in distribution.
d) Self-similarity of a process is useful for the path simulation of X_t, since knowing the path on $[0, 1]$, enables to stretch it by a factor a^H on the scale at.
e) Choosing $t_0 = 0$ and $H > 0$ the relation $X_{a0} = a^H X_0$ reveals that $X_0 = 0$ almost surely. Under the assumption of theorem 3.3 we can deduce an even stronger property
$$H = 0 \Leftrightarrow X_t = X_0 \text{ a.s.}$$
Due to the last two properties we will consider in the sequel H-ss processes with $H > 0$, whose paths are stochastic continuous at 0. There exist processes, which are self-similar for $H = 0$, but which do not satisfy $X_0 \neq 0$ almost surely (see [77, S. 3]).
f) H-ss processes with stationary increments are briefly called H-sssi processes.
g) An H-sssi process with existing first moment satisfies for all $t \geq 0$

$$\mathbb{E}(X_{2t}) = 2^H \mathbb{E}(X_t) \quad \text{according to self-similarity}$$
$$\mathbb{E}(X_{2t}) = \mathbb{E}(X_{2t} - X_t) + \mathbb{E}(X_t)$$
$$= \mathbb{E}(X_t) + \mathbb{E}(X_t) = 2\mathbb{E}(X_t) \quad \text{because of stat. increments}$$

Thus we have $\mathbb{E}(X_t) = 0$ for all $t \in \mathbb{R}$ using $2^H \mathbb{E}(X_t) = 2\mathbb{E}(X_t)$, provided $H \neq 1$.

h) Furthermore we get for an H-sssi process (X_t) and $t \in \mathbb{R}$

$$X_{-t} \stackrel{e)}{=} X_{-t} - X_0 \stackrel{g),d}{=} X_0 - X_t = -X_t$$

i) Suppose an H-sssi process (X_t) has a finite second moment. Then with h) we have $\sigma^2 = \mathrm{Var}(X_1) = \mathbb{E}(X_1^2)$ provided $H < 1$ (remember that for $H < 1$ it is $\mathbb{E}(X_t) = 0$)

$$\mathbb{E}(X_t^2) \stackrel{h)}{=} \mathbb{E}(X_{|t|\mathrm{sign}(t)}^2) = |t|^{2H} \mathbb{E}(X_{1\cdot\mathrm{sign}(t)}^2) \stackrel{h)}{=} t^{2H} \mathbb{E}(X_1^2) = t^{2H} \sigma^2$$

If $\sigma^2 = \mathbb{E}(X_1^2) = 1$, then (X_t) is called a *standard process*.

j) Provided $\mathbb{E}(|X_1|) < \infty$, we deduce according to the stationary increments

$$\mathbb{E}(|X_2|) = \mathbb{E}(|X_2 - X_1 + X_1|) \leq \mathbb{E}(|X_2 - X_1|) + \mathbb{E}(|X_1|) = 2\mathbb{E}(|X_1|)$$

Self-similarity implies

$$\mathbb{E}(|X_2|) = 2^H \mathbb{E}(|X_1|)$$

Finally it follows $2^H \leq 2$ which is equivalent to $H \leq 1$.

Example 3.5. Brownian motion: For this we introduce a Gaussian process, which is a process with finite dimensional normal distribution. The Brownian motion is determined as a process with continuous paths, $X_0 = 0$, $m(t) = \mathbb{E}(X_t) = 0$ and a special covariance function $\gamma(s,t) = \mathbb{C}\mathrm{ov}(X_s, X_t) = t$. If (X_t) is a H-ss process (not necessarily H-sssi!), then

$$m(t) = t^H \mathbb{E}(X_1) \quad \text{and} \quad \gamma(s,t) = s^{2H} \mathbb{C}\mathrm{ov}(X_{\frac{t}{s}}, X_1), \; s \neq 0$$

This equation can be deduced easily as exercise. For the Brownian motion we get $H = \frac{1}{2}$ as Hurst parameter. In other words: Choosing an H-ss Gaussian process with continuous paths and $H_0 = 0$, then we have exactly a Brownian motion provided a Hurst parameter $H = \frac{1}{2}$. Gaussian processes with $H \neq 0$ will be called *fractional Brownian motion* (FBM). We will intensively investigate those processes and their importance for modeling IP traffic in section 3.2.2.

Example 3.6. According to a result of Lamperti [154] each self-similar process is the weak limit of certain normalized sum processes. The Brownian motion is such a limit: It is the weak limit according to the central limit theorem or more precisely the invariance principle of Donsker [76]. In a separate section 3.2.3 we will transfer this concept of weak limits to other limit distribution than the normal distribution. This will lead to limit processes called α-stable processes ($\alpha < 2$) which are performing an important feature for IP traffic: They are self-similar with heavy-tail distribution, which do not have a finite variance. Hence, those processes show in addition a so called *long-range dependence* (LRD).

Example 3.7. An example of a not self-similar process, revealing a similar time structure and thus is important for modeling heavy-tail phenomena is the *geometric fractional Brownian motion* and consider them in the special section 3.2.2. We will in addition introduce shortly the concept of *stochastic differential equation*.

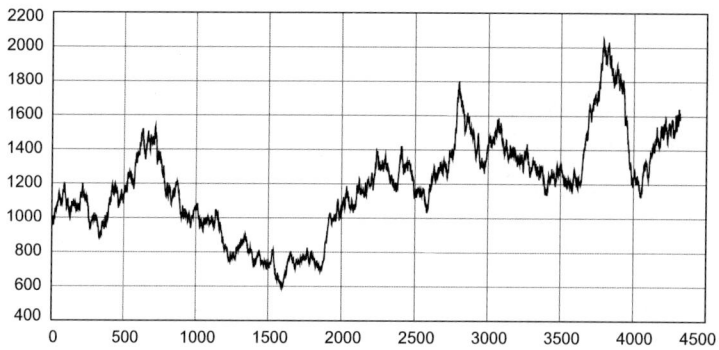

Fig. 3.2. Path of a geometric Brownian motion

Example 3.8. α-stable motions: These processes start at 0, have piecewise continuous paths, independent and stationary increments and α-stable finite dimensional distributions. For $\alpha < 2$ we have pure jump processes, for $\alpha = 2$ we get the Brownian motion with $H = \frac{1}{2}$. These processes will be investigated in detail in section 3.2.3. As the Brownian motion the α-stable motions (or in general the Lévy processes) have nowhere differentiable paths almost surely – a result due to Mandelbrot and Van Ness [174] which holds for all important self-similar processes. In case of $\alpha < 2$ (in particular for all non Brownian motions) these processes have no continuous paths: All paths have countable many jumps. The figures 3.3 demonstrate two examples of self-similar processes, the so called linear fractional α-stable process. Particularly in the right figure the jumps show the discontinuity (see section 3.2.3).

Example 3.9. Connection to stationary processes: Suppose (X_t) is an H-ss process. Then the process $Y_t = e^{-Ht} X_{e^t}$ is a stationary process. We remind that stationary processes play a decisive rôle in the consideration of serving and interarrival times. Vice versa every stationary process can be transformed into an H-ss process. The result is again due to Lamperti [154]. For more details consult [77, Th. 1.5.1].

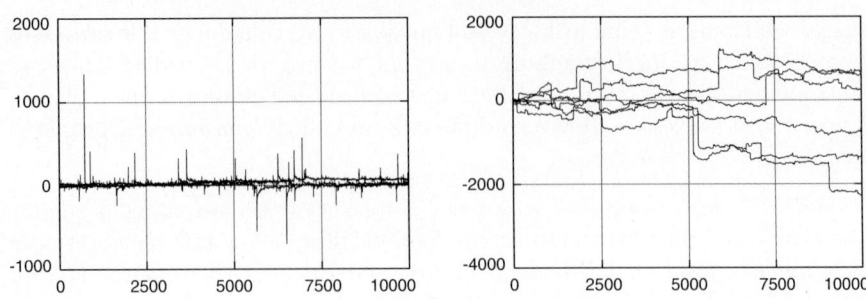

Fig. 3.3. Paths of α-stable processes with $\alpha = 1.5$ and $H = 0.2$ (left) and $H = 0.7$ (right)

Example 3.10. As we know from the classical traffic theory, processes with stationary increments play an important rôle. Thus, we will stick in particular to H-sssi processes. For these H-sssi processes hold the following equation, describing the stationary increments:

$$(X_{t_0+h} - X_h, ..., X_{t_n+h} - X_h) = (X_{t_0}, ..., X_{t_n}) \text{ in distribution}$$

In the case of Gaußprocesses this characterizes exactly the fractional Brownian motion.

A decisive connection will be discovered later with heavy-tail distribution. Since the notion of self-similarity is basic for the description of IP-based traffic, we will cite some important facts and follow the line of the monograph of Embrechts and Maejima [77]. A fundamental result concerns the form of the covariance function $\gamma(s,t) = \mathbb{C}\text{ov}(X_s, X_t)$ of an H-ss process. It is due to Taqqu [240]. We provide a proof for exercise.

Theorem 3.11. *Let $(X_t)_{t \in [0,\infty[}$ an H-ss process with stationary increments and $H < 1$. Suppose $\mathbb{E}(X_1^2) < \infty$. Then*

$$\gamma(s,t) = \mathbb{E}(X_s X_t) = \frac{1}{2}\left(t^{2H} + s^{2H} - |t-s|^{2H}\right) \mathbb{E}(X_1^2)$$

Proof. We compute (here we apply the identity $ab = \frac{1}{2}(a^2 + b^2 - (a-b)^2)$):

$$\mathbb{E}(X_s X_t) = \frac{1}{2}\left(\mathbb{E}(X_t^2) + \mathbb{E}(X_s^2) - \mathbb{E}((X_t - X_s)^2)\right) \quad (3.6)$$

$$= \frac{1}{2}\left(\mathbb{E}(X_t^2) + \mathbb{E}(X_s^2) - \mathbb{E}(X_{|t-s|}^2)\right) \quad \text{because of stat. increments}$$

$$= \frac{1}{2}\left(t^{2H} + s^{2H} - |t-s|^{2H}\right) \mathbb{E}(X_1^2)$$

Here we used the fact:

$$\mathbb{E}(X_t^2) = \mathbb{E}((t^H)^2 X_1^2) = t^{2H} \mathbb{E}(X_1^2)$$

Again because of the stationary increments and since $X_0 = 0$, we have $\mathbb{E}(X_t) = 0$ according to the assumption $H < 1$ with remark 3.4 g). This implies
$$\gamma(s,t) = \mathbb{E}(X_s X_t)$$
□

From (3.6) we deduce in the case $H = 1$ (in (3.6) we did not use the restriction $H < 1$)
$$\mathbb{E}(X_s X_t) = st\mathbb{E}(X_1^2)$$
and hence
$$\mathbb{E}((X_t - tX_1)^2) = \mathbb{E}(X_t^2) - 2t\mathbb{E}(X_t X_1) + t^2\mathbb{E}(X_1^2) = (t^2 - 2tt + t^2)\mathbb{E}(X_1^2) = 0$$
Thus we can deduce for $H = 1$ immediately under the condition $\mathbb{E}(X_1^2) < \infty$
$$X_t = tX_1 \text{ holds for all } t \in \mathbb{R} \text{ a.s.}$$

It seems obvious that for an Hurst exponent $H \in \,]0,1[$ one defines a fractional Brownian motion $(B_t^{(H)})$ as real Gaussian process with continuous paths, $\mathbb{E}(B_t^{(H)}) = 0$ and covariance function
$$\mathbb{E}\left(B_t^{(H)} B_s^{(H)}\right) = \frac{1}{2}\left(t^{2H} + s^{2H} - |t-s|^{2H}\right) \mathbb{E}\left((B_1^{(H)})^2\right)$$

We will provide an integral representation of the fractional Brownian motion in section 3.2.3 for $H \in \,]0,1[$. A proof that by the above definition the FBM is self-similar to H, is given in [77, p. 6].
We summarize a list of results on self-similar processes ([77, p. 19]).

Proposition 3.12. *Let* $(X_t)_{t \in [0,\infty[}$ *be an* H-*sssi process with* $H > 0$ *and not for* $t > 0$, *i.e., is not* $\mathbb{P}$-*f.s. constant. Then:*

a) *If* $\mathbb{E}(|X_1|^\delta) < \infty$ *for a* $0 < \delta < 1$, *then* $H < \frac{1}{\delta}$ *(see [169])*.
b) *From* $\mathbb{E}(X_1) < \infty$ *follows* $H \leq 1$.
c) *If* $\mathbb{E}(|X_1|) < \infty$ *and* $0 < H < 1$, *then* $\mathbb{E}(X_t) = 0$ *for all* $t \geq 0$ *(see remark 3.4 g))*.
d) *With* $\mathbb{E}(|X_1|) < \infty$ *and* $H = 1$ *follows* $X_t = tX_1$ *almost surely*.

This implies immediately the following result.

Corollary 3.13. *Let* $(X_t)_{t \in [0,\infty[}$ H-*sssi with* $H > 1$. *Then* $\mathbb{E}(|X_1|^{\frac{1}{H}}) = \infty$.

Proof. Suppose $\mathbb{E}(|X_1|^{\frac{1}{H}}) < \infty$. Since $\delta = \frac{1}{H} < 1$ proposition 3.12 shows $H < \frac{1}{\delta} = H$, a contradiction. □

3.2.2 Fractional Brownian Motion

The Brownian motion can be considered as the 'mother of all processes'. This follows in particular by the Gaussian marginal distribution, which are one of best investigated one. A central result says, that a Gaussian process (i.e. a process with Gaussian marginal distributions) is determined by its covariance structure, i.e. by the function $\gamma(s,t)$. In particular for a given symmetric function, one finds exactly one particular Gaussian process. Since we are interested in self-similar processes, more precisely in H-sssi processes, by theorem 3.11 we know, how a Gaussian H-sssi has to look like and hence is uniquely determined. Thus, we give the following definition.

Definition 3.14. *We call a continuous Gaussian H-sssi process $B_t^{(H)}$ with $0 < H < 1$ fractional Brownian motion. It is standard, if $\mathbb{E}(B_1^{(H)}) = 1$.*

Remark 3.15. We briefly comment on the exposed values of H, namely $H = 1$ and $H = \frac{1}{2}$:

- We have excluded $H = 1$, since then every H-sssi process has the form tX_1.
- For $H = \frac{1}{2}$ we have

$$\gamma(s,t) = \mathbb{E}(B_s^{(H)}, B_t^{(H)}) = \begin{cases} \min(|s|,|t|) & \text{for } s \cdot t \geq 0 \\ 0 & \text{else} \end{cases}$$

We are now able to give a first characterizing theorem.

Theorem 3.16. *Let $(X_t)_{t \in \mathbb{R}}$ be a process with stationary increments. Furthermore we have:*

- *$(X_t)_{t \in \mathbb{R}}$ is Gaussian distributed with mean 0 and $X_0 = 0$.*
- *$\mathbb{E}(X_t^2) = \sigma^2 |t|^{2H}$, for a $\sigma > 0$ and $0 < H < 1$.*

Then $(X_t)_{t \in \mathbb{R}}$ is a FBM to the exponent H.

The discrete process $(\Delta(n))_{n \in \mathbb{Z}}$ from theorem 3.24 is called in case of a Gaussian process (X_t) also *fractional Gaussian noise*.

In the representation resp. simulation of processes properties of corresponding paths play an important rôle. A result of Kolmogorov [98, S. 281, corollary 7.2.51] states, that a process $(X_t)_{t \in \mathbb{R}}$ has continuous paths, provided that there are $\delta \geq 1$, $\nu > 1$ and $c > 0$, with

$$\mathbb{E}\left(|X_{t_1} - X_{t_2}|^\delta\right) \leq c|t_1 - t_2|^\nu, \text{ for all } t_1, t_2 \in \mathbb{R} \tag{3.7}$$

For a FBM $(B_t^{(H)})$ we choose a $\delta \geq 1$ with $\delta > \frac{1}{H}$. Since the FBM is an H-sssi process, we have

$$\mathbb{E}\left(|B_{t_2}^{(H)} - B_{t_1}^{(H)}|^\delta\right) = \mathbb{E}\left(|X_1|^\delta\right) |t_2 - t_1|^{H\delta} \tag{3.8}$$

and thus the continuity. But there is more

Theorem 3.17. *(Theorem of Adler) Let $0 < H$ and $\delta > 0$. Then there are constants $h > 0$ and $A > 0$, such that*

$$|B_{t_2}^{(H)} - B_{t_1}^{(H)}| \leq A|t_2 - t_1|^{H-\delta} \text{ a.s.}$$

for all $0 \leq t_1, t_2 \leq 1, |t_1 - t_2| < h$. On the other hand one finds for all $\delta > 0$ a sufficient small $h > 0$ with

$$\sup\{|B_t^{(H)} - B_s^{(H)}|; |t - s| \leq h\} \geq Kh^{H+\delta}$$

for all $s, t \in [a, b]$ and all $K < \infty$ a.s.

What can be said about the differentiability of the paths? For this we have to consider the differential quotient

$$\frac{B_{t_1}^{(H)} - B_{t_2}^{(H)}}{t_1 - t_2}$$

for $t_1 \to t_2$. The relation

$$\mathbb{E}\left(\left|\frac{B_{t_2}^{(H)} - B_{t_1}^{(H)}}{t_2 - t_1}\right|^2\right) = \sigma^2 |t_2 - t_1|^{2H-2} \to \infty, \text{ for } t_2 \to t_1 \quad (3.9)$$

implies that the path $(B_\cdot^{(H)}(\omega))$ is nowhere differentiable for almost all ω in the case of $H < 1$.

But it can be stated even more. For this we consider the notion of *Hölder continuity*: A function or path $Y(\cdot)$ is Hölder continuous with respect to $\lambda > 0$ at t, if all s in the neighbourhood of t satisfy

$$|Y(t) - Y(s)| \leq C|t - s|^\lambda$$

The bigger the λ, the more smooth the function will be. Then we have the

Proposition 3.18. *The paths of a fractional Brownian motion $(B_\cdot^{(H)}(\omega))$ are for all $\lambda > H$ nowhere (i.e. for no $t \in \mathbb{R}$) Hölder continuous.*

By (3.9) we see tat with small H the right hand side in (3.9) tends to ∞ faster. Thus, for $0 < H < \frac{1}{2}$ we detect a stronger 'zigzag' as for the Brownian motion ($H = \frac{1}{2}$). Simultaneously for $\frac{1}{2} < H < 1$ the path is smoother. Thus, we divide the FBM into

- *antipersistent* for $0 < H < \frac{1}{2}$,
- *chaotic* for $H = \frac{1}{2}$ and
- *persistent* for $\frac{1}{2} < H < 1$.

The stronger 'zigzag' stems from the negative value of the autocorrelation function (see remark to (3.17)) in contrast to the case of the positive correlation in the case $\frac{1}{2} < H < 1$.

In fact, there are non-Gaussian processes with finite variance and H-sssi processes without existing variance. In the section 3.2.3 we briefly consider this phenomena. Detailed description the reader can find in [226].

Representation of the FBM

In section 3.5.1 we need a representation of the fractional Brownian motion using the Fourier transform. Thus, we want to indicate an alternative formula for the FBM $(B_t^{(H)})$. The following representation is cited from the upcoming section 3.2.3, where it is introduced in a broader context

$$B_t^{(H)} = \frac{1}{C(H)} \int_{-\infty}^{\infty} \left((t-x)_+^{H-\frac{1}{2}} - (-x)_+^{H-\frac{1}{2}} \right) dB(x) \qquad (3.10)$$

where B is the Brownian motion and

$$C(H) = \left(\int_0^{\infty} (1+x)^{H-\frac{1}{2}} - x^{H-\frac{1}{2}})^2 dx + \frac{1}{2H} \right)^{\frac{1}{2}}$$

As described in 3.2.3, we have understand (3.10) as Itō integral (see e.g. [134]). We now consider the so called *harmonic representation* of the FBM. For this we will use the form

$$I(\hat{f}_t) = \int_{-\infty}^{\infty} \hat{f}_t(x) d\hat{B}(x) \qquad (3.11)$$

and derive the function $\hat{f}_t(x)$, $(t, x \in \mathbb{R})$. Here, $\hat{f}_t$ is a complex valued function with $\hat{f}_t = f_t^{(1)} + i f_t^{(2)}$, where $f_t^{(1)}, f_t^{(2)}$ is real valued functions, which are $f_t^{(1)}(x) = f_t^{(1)}(-x)$ (axially symmetric) and $f_t^{(2)} = -f_t^{(2)}(-x)$, $x \in \mathbb{R}$ (symmetric w.r.t. the origin). By $\hat{B} = B^{(1)} + i B^{(2)}$ we denote the complex valued version of the Brownian motion. $B^{(1)}$ and $B^{(2)}$ are independent versions of the Brownian motion on $\mathbb{R}_+$ with (considered as measures) $B^{(1)}(A) = B^{(1)}(-A)$ and $B^{(2)}(A) = -B^{(2)}(-A)$, $(A \subset \mathbb{R}_+$ measurable). Then the integral (3.11) will turn to

$$I(\hat{f}_t) = \int_{-\infty}^{\infty} f_t^{(1)}(x) dB^{(1)}(x) - \int_{-\infty}^{\infty} f_t^{(2)}(x) dB^{(2)}(x)$$

We give now the marmonic representation and consult for a rigorous proof to [226, S. 328].

Theorem 3.19. *Let $0 < H < 1$. Then the fractional Brownian motion $(B_t^{(H)})_{t \in \mathbb{R}}$ can be represented in the integral from*

$$B_t^{(H)} = \frac{1}{C_1(H)} \int_{-\infty}^{\infty} \frac{e^{ixt} - 1}{ix} |x|^{-(H-\frac{1}{2})} d\hat{B}(x), \ t \in \mathbb{R} \qquad (3.12)$$

with

$$C_1(H) = \left(\frac{\pi}{H \Gamma(2H) \sin(H\pi)} \right)^{\frac{1}{2}}$$

3.2 Self-Similar Processes

We will give a heuristic derivation of this fact, not being a rigorous proof, since the appearing integrals diverges partly. But with this we show some of the used techniques. We follow the argumentation given in [226] and assume for simplification $H > \frac{1}{2}$ and $t > 0$. If we rewrite (3.10), we obtain

$$B_t^{(H)} = \frac{1}{C(H)} \left(H - \frac{1}{2}\right) \int_{-\infty}^{\infty} \left(\int_0^t (s-x)_+^{H-\frac{3}{2}} ds\right) dB(x)$$

With this the integrand in (3.10) was simply differentiated and then integrated (thus, we have assumed $t > 0$). Furthermore from the above integral we derive:

$$B_t^{(H)} = \frac{1}{C(H)} \left(H - \frac{1}{2}\right) \int_{-\infty}^{\infty} \left(\int_{-\infty}^{\infty} \mathbf{1}_{[0,t]}(s)(s-x)_+^{H-\frac{3}{2}} ds\right) dB(x) \quad (3.13)$$

Considering the integrand, we see that it consists in a convolution of the functions $s \longmapsto \mathbf{1}_{[0,t]}(s)$ and $s \longmapsto s_+^{H-\frac{3}{2}}$. The convolution of the Fourier transform leads as well known to a product. Now the Fourier transform of $\mathbf{1}_{[0,t]}$ reads as

$$\frac{1}{\sqrt{2\pi}} \int_0^t e^{ixs} ds = \frac{1}{\sqrt{2\pi}} \frac{e^{ixt} - 1}{ix}$$

The Fourier transform of $s \longmapsto s_+^{H-\frac{3}{2}}$ does not exist, since $s \longmapsto s_+^{H-\frac{3}{2}}$ is neither square nor simple integrable. Nevertheless we will develop the approach further. A correct but rather technical derivation lies in the suitable cutting of the integrand and an approximation of $s \longmapsto s_+^{H-\frac{3}{2}}$ using functions with non existing Fourier transforms (for the rigorous proof see e.g. [244]):

$$\frac{1}{\sqrt{2\pi}} \int_{-\infty}^{\infty} e^{ixs} s_+^{H-\frac{3}{2}} ds = |x|^{-\left(H-\frac{1}{2}\right)} \sqrt{2\pi} \int_0^{\infty} e^{iu} u^{H-\frac{3}{2}} du \quad (3.14)$$

where we have set $u = xs$. The integral on the right hand side in (3.14) can be considered as the Cauchy value of the integral. It as the value

$$\int_0^{\infty} e^{iu} u^{H-\frac{3}{2}} du = e^{i\frac{H-\frac{1}{2}}{2}\pi} \Gamma\left(H - \frac{1}{2}\right)$$

If f is an integrable function and

$$\hat{f}(s) = \frac{1}{\sqrt{2\pi}} \int_{-\infty}^{\infty} e^{isu} f_t(u) du$$

is the Fourier transform, then we have in distribution

$$\left\{I(f_t) = \int_{-\infty}^{\infty} f_t(x) dB(x),\ t \in \mathbb{R}\right\} \stackrel{d}{=} \left\{I(\hat{f}_t) = \int_{-\infty}^{\infty} \hat{f}_t d\hat{B}(x),\ t \in \mathbb{R}\right\} \quad (3.15)$$

Using the rules for the Fourier transforms $\widehat{(f \cdot g)} = \sqrt{2\pi} \hat{f} \cdot \hat{g}$, we apply them to the integrands in (3.13) and deduce with result (3.15) the fact in theorem 3.19.

3.2.3 α-stable Processes

We start with the definition of an α-stable distributed random variable.

Definition 3.20. *A stable distribution G (and a G-distributed random variable ξ) has the characteristic function*

$$\Phi_G(t) = \mathbb{E}\left(\exp(i\xi t)\right) = \exp\left(i\gamma t - c|t|^\alpha \left(1 - i\beta \mathrm{sign}(t) z(t,\alpha)\right)\right), \ t \in \mathbb{R}$$

Here, γ is a real number, $c < 0$, $\alpha \in\ ]0,2]$, $\beta \in [-1,1]$ and

$$z(t,\alpha) = \begin{cases} \tan(\frac{\pi\alpha}{2}) & \text{for } \alpha \neq 1 \\ -\frac{2}{\pi} \log|t| & \text{for } \alpha = 1 \end{cases}$$

In more detail we will consider and investigate the rôle of α-stable distributed random variables for the description of the IP-based traffic. There, too, the importance of α-stable processes will be revealed. In this section we purely describe the α-stable processes within the framework of self-similar and long-range dependent processes:

- For our purposes we will consider only distributions with $\gamma = 0$.
- An important parameter is α. It determines decisive details of the distribution, as the existing moments, the shape of the complementary distribution (so called tails) and the asymptotic behavior of iid sums.
- The number α is called the *characteristic moment* and the corresponding distribution *α-stable distribution*.
- The case $\alpha = 2$ gives the well-known normal or Gaussian distribution with the characteristic function

$$\Phi_G(t) = \exp\left(-ct^2\right)$$

  Here the random variable has the average 0 and variance $2c$. We see that the normal distribution depends, as special case of α-stable distributions, only on two parameters, mean and variance, while the general is determined by four values. The reason lies in the symmetry of the normal distribution. The general α-stable distribution can be asymmetric for $\alpha < 2$ or is only defined on one half axis (as e.g. for $\alpha < 1$).
- Another important case is $\alpha = 1$, which leads to the so called *Cauchy laws* resp. to the *Cauchy distribution*. The general characteristic function reads in this cases:

$$\Phi_G(t) = \exp\left(-c|t|\left(1 + i\beta\frac{2}{\pi}\mathrm{sign}(t)\log(t)\right)\right)$$

- For fixed α the values c and β determine details of the distribution. The constant c is a scaling number and corresponds in case of the Gaussian distribution (i.e. for finite variance) to the value $c = \frac{\sigma^2}{2}$. If there is no existing second moment (which can be observed in many cases of modeling

IP traffic, i.e. if there is no Gaussian distribution), then it is possible to interpret this in an equivalent way. β determines the range of values of the random variable. The characteristic function is exactly real valued if and only if $\beta = 0$. In this case the random variable is symmetric. Such random variables are called $s\alpha s$ (symmetric α-stable). If $\beta = 1$ and $\alpha < 1$, then the corresponding random variable ξ is positive.

Definition 3.21. *(α-stable motion) A stochastic process $(\xi_t)_{t \in [0,1]}$ with càdlàg paths (i.e. continuous from right with existing limits from left) is called α-stable motion, if:*

- *$\xi_0 = 0$ $\mathbb{P}$-almost surely (the process starts in 0).*
- *The process has independent and stationary increments.*
- *For all $t \in [0, 1]$ the random variable ξ_t has an α-stable distribution with fixed and to t independent parameters $\beta \in [-1, 1]$ and $\gamma = 0$ in the spectral representation.*

Such a process is also called an α-stable Lévy process. If nothing is determined about the marginal distributions and only a) and b) hold, then one simple calls it Lévy process (see for more details [77, p. 8–9]).

We come to a short introduction of the so called *stable processes*. These are rarely investigated in the applied literature, but especially for the investigation of heavy-tail distribution and the long-range dependence those processes are of significant importance. The described phenomena can be summarized under the notion of *extremal events*. Another point to mentioned is the lacking interest of practitioners to use this processes for their modeling, though their properties are widely investigated and known.

Consider a process $(X_t)_{t \in I}$. Then the marginal distribution determine the process. This can be observed e.g. in the case of the Brownian motion, where for each process with normal marginal distributions there is a version with continuous paths. Hence, we have to know the details of the distribution of any finite dimensional vector $(X_{t_1}, \ldots, X_{t_d})$, $t_1, \ldots, t_d$, $d \in \mathbb{N}$, which in turn characterizes the process. We will restrict ourselves to the symmetric α-stable processes (abbreviated $s\alpha s$), since nothing changes by multiplying the process with -1. It only simplifies the representation. We recall that the α-stable distribution of a random variable ξ has a characteristic function of the form $\mathbb{E}(e^{it\xi}) = e^{-c|t|^\alpha}$ ($c > 0$, $0 < \alpha \leq 2$, the case $\alpha = 2$ implies Gaussian distribution).

To introduce the definition of an α-stable distribution of a random vector, we require the scalar product on $\mathbb{R}^d$:

For $\mathbf{x} = (x_1, \ldots, x_d)$, $\mathbf{y} = (y_1, \ldots, y_d) \in \mathbb{R}^d$, define the scalar product

$$\langle \mathbf{x}, \mathbf{y} \rangle = x_1 y_1 + x_2 y_2 + \ldots + x_d y_d$$

Similar to the introduction of the α-stable distribution we define the distribution of a random vector $\mathbb{X} \in \mathbb{R}^d$ by the characteristic function of the distribution F of $\mathbb{X}$. For $0 < \alpha < 2$ we call $\mathbb{X} \in \mathbb{R}^d$ a $s\alpha s$-random vector, if the characteristic function has the form

$$\mathbb{E}(e^{i\langle \mathbf{t}, \mathbb{X}\rangle}) = \exp\left(-\int_{S^{d-1}} |\langle \mathbf{t}, \mathbf{y}\rangle|^\alpha dm_s(\mathbf{y})\right), \ \mathbf{t} \in \mathbb{R}^d$$

where $m_s(\cdot)$ is a symmetric probability measure on 'the unit sphere' $S^{d-1} = \{y = (y_1, \ldots, y_d); \|y\|_2^2 = y_1^2 + \ldots + y_d^2 = 1\}$, i.e. $m_s(A) = m_s(-A)$, for $A \subset \mathbb{R}^n$ and $\|y\|_2^2 = 1$ for all $y \in A$. The measure m_s is called *spectral measure* of the random vector $\mathbb{X}$.

If $(X_1, \ldots, X_d)$ are independent $s\alpha s$-random variables with characteristic functions $\mathbb{E}(e^{itX_j}) = \exp(-c_j|t|^\alpha)$, $c_j > 0$, $j = 1, \ldots, d$ and if we define $\mathbb{X} = (X_1, \ldots, X_d)$ then we have as spectral representation of the random vector

$$\mathbb{E}\left(e^{i\langle \mathbf{t}, \mathbb{X}\rangle}\right) = \prod_{j=1}^d \mathbb{E}\left(e^{it_j X_j}\right) = \exp\left(-\sum_{j=1}^d c_j |t_j|^\alpha\right), \ \mathbf{t} = (t_1, \ldots, t_d)$$

After this preliminaries we start with the introduction of $s\alpha s$ processes.

Definition 3.22. *A stochastic process* (X_t) *with $s\alpha s$-stable marginal distribution is called an $s\alpha s$ process.*

From the definition we see that linear combinations of $s\alpha s$ processes are again $s\alpha s$. A particular example of an $s\alpha s$ process is already known to us: the $s\alpha s$ motion, whose most prominent member is the Brownian motion. A $s\alpha s$ process (X_t) is an $s\alpha s$ motion, if and only if:

- $X_0 = 0$ almost surely and (X_t) is stochastic continuous at $t = 0$ and
- (X_t) has independent and stationary increments and
- (X_t) has càdlàg paths almost surely.

A $s\alpha s$ motion is also called a $s\alpha s$-stable Lévy process. In particular any $s\alpha s$ motion prevails a self-similar structure, i.e., we have for all $c > 0$ and $t_1, \ldots, t_d$

$$(X_{ct_1}, \ldots, X_{ct_d}) \text{ is distributed as } c^{\frac{1}{\alpha}}(X_{ct_1}, \ldots, X_{t_d})$$

The exponent $\frac{1}{\alpha}$ is noting else than the Hurst parameter. We abbreviate an H-sssi $s\alpha s$ process with $(L_{H,\alpha}(t))$. It holds

$$L_{H,\alpha}(t) - L_{H,\alpha}(s) \text{ is distributed like a } s\alpha s \text{ random variable}$$

with variance $|t-s|^H \text{Var}(L_{H,\alpha}(1))$. The increments of $L_{H,\alpha}$ are exactly independent if we have $H = \frac{1}{\alpha}$ (and thus the case of a $s\alpha s$ motion). This generalizes the notion of the Brownian motion for $H = \frac{1}{2}$. Simply for a $s\alpha s$ motion we write $L_H = L_{\frac{1}{\alpha}}$.

Since for the description of the packet switched traffic $s\alpha s$ processes (and in particular the fractional Brownian motion) play a decisive rôle, we will sketch a constructive approach.

In case of $\alpha = 2$ we receive the Brownian motion. Why? Hence, the process has independent increments for $\alpha = 2$. But for $\alpha < 2$ the process $(X_t^{(\log)})$ do not have independent increments and thus, it is no longer an α-stable Lévy process.

- Another interesting process without independent increments can be found in [140]. It is the limit of stochastic path processes, where the limit is represented by a stochastic integrands (see also [77, p. 40–41]).
- All processes under considerations exhibited self-similarity with stationary increments. If in addition, these increments are independent, we called them Lévy process (or in particular $s\alpha s$- motion). The marginal distribution are determined uniquely. Dropping the 'independence' and 'stationarity' this is no longer the case and a sufficient classification is not at hand. But for self-similar processes with independent increments, shortly H-ssii, we can formulate again some facts (see also [77, p. 57–62]). Since the correlation structure reads as $\mathrm{Cor}(X_t - X_s, X_v - X_u) = 0$ for $0 \le s < t < u < v < \infty$, these H-ssii processes are inadequate for describing long-range dependence and thus, IP-based traffic.
- We cite in this context an interesting example of an H-ssii process in [196]: Let $(B_t^{(H)})_{t \ge 0}$ be a fractional Brownian motion with $\frac{1}{2} < H < 1$. Define:

$$X_t = \int_0^t x^{\frac{1}{2}}(t-x)^{\frac{1}{2}-H} dB^{(H)}(x), \ t \ge 0$$

Norros et al. showed in [196] that $(X_t)_{t \ge 0}$
- is Gaussian,
- is $(1-H)$ self-similar,
- has independent increments, which are not stationary,
- is a martingale.

Closing this section we classify the up to now considered self-similar processes. For this we divide the particular processes by the property of their paths. If two processes $(X_t)_{t \in I}$ and $(Y_t)_{t \in I}$, $(I \subset \mathbb{R})$ are equal in distribution (i.e. have the same marginal distribution), we say each of them is a version of the other. It is possible to investigate certain properties of the paths:

A   There is a version with continuous paths.
B   Property A is not true, but there exists a version with right continuous paths having limits from the left. As we know they are called càdlàg.
C   Each version of the process is nowhere bounded, i.e. unbounded on each interval.

A classification of the up to now considered processes can be done in the following way:

- In class A we find the Brownian motion, generally the FBM and the linear $s\alpha s$ motion with $\frac{1}{\alpha} < H < 1$ (see (3.8) and the definition of a $s\alpha s$ motion).
- In B there are all non Gaussian Lévy processes.

- In C are the log-fractional stable motions and the linear $s\alpha s$ motion for $0 < H < \frac{1}{\alpha}$.

Property A holds for linear fractional stable motions and can be proved with (3.7) analogously to the FBM case. The property B follows from the definition. How can one prove property C for a particular process? For this we pick a $s\alpha s$ process in the integral representation

$$X_t = \int_{-\infty}^{\infty} f(t,x) dM_\alpha(x)$$

where M_α is a $s\alpha s$-random measure and $f(t,\cdot) \in L^\alpha(\mathbb{R}, \mathcal{B}, m)$ with a suitable control measure m. Fortunately we can state the following theorem for classification.

Theorem 3.23. *Let $0 < \alpha < 2$. Suppose there is a countable dense subset $T^* \subset \mathbb{R}$ such that for all pairs $a < b$ the integral has value*

$$\int_a^b \sup_{t \in T^*} |f(t,x)|^\alpha dm(x) = \infty$$

Then the process (X_t) has property C.

For the case of the log-fractional and linear fractional stable motion one chooses $T^* = \mathbb{Q}$.

3.3 Long-Range Dependence

One of the major differences of the IP traffic, observed e.g. by the Leland group, compared to the Poissonian approach consists in the so called *long-range dependence*. As we mentioned already in section 2.1.3, the memoryless behavior of the classical telecommunication traffic results in the use of the exponential distribution and hence, the Poisson process. Based on the Erlang model it turns out that the resulting Poisson process prevails independent increments. This means that the change at the instant moment does not influence later periods. The statistical evaluation of the IP traces, as in the LAN, WAN or for specific applications such as World Wide Web, reveals two major issues for the transmitted data amount A_t, different to the classical modeling:

a) The variance $\mathbb{V}(A_t) \sim t^p$ with $p > 1$, showing a hyperbolic shape.
b) For $r(n) = \mathbb{E}((A_{n+1} - A_n)A_1))$ the series $\sum_{n \in \mathbb{N}} r(n) = \infty$ does not converge.

Figure 1.11 reflects the fact of the long-range dependence of real data traffic best, in contrast to the short-range dependent Poisson process.
Though the Brownian motion is a fractal (i.e. self-similar) process, the increments are independent (hence item b) fails) and the variance $\mathbb{V}(B_t) = t$

behaves consequently linear. Again Mandelbrot and Van Ness considered the feature interesting enough for their economic descriptions, and called the phenomena in b) 'Joseph-effect'. There is a quite interesting equivalent way to look at b): If ones considers the spectral density of the underlying process (A_t), one will in fact detect, as we will point later in section 3.3.1, that the spectral density has a singularity at 0. This means that with low frequency the cycles may be arbitrary long, which in turn can be interpreted for the IP traffic in such a way that the data amount may be transmitted in larger time intervals and that this may happen in arbitrary long cycles. That is again a confirmation for the fact that a large queueing time may trigger larger ones in the future. Considering subexponential distribution this is nothing new, as we have already pointed out in section 2.7.4. Short-range depending processes, especially those with independent increments like Poisson process, Brownian motion, or the ARMA time series, show the different behavior close to the frequency 0 by cycles of bounded length.

3.3.1 Definition and Concepts

After the preceeding section, where we detect the long-range dependence in the IP traffic, we proceed in this subsection with the presentation of the mathematical concept and indicate the connection to the self-similarity. We return to the concept of fractional Gaussian noise and define for an H-sssi process $(X_t)_{t \in I}$

$$\Delta(n) = X_{n+1} - X_n, \ n \in \mathbb{Z} \quad \text{Increments}$$
$$r(n) = \mathbb{E}(\Delta(0)\Delta(n)), \ n \in \mathbb{Z}$$

The (discrete) process describes the change within one time unit at the discrete times $n \in \mathbb{N}$. The next theorem describes that this changes depend heavily on the Hurst exponent H.

Theorem 3.24. *The sequence* $(\Delta(n))_{n \in \mathbb{N}}$ *is stationary and we have*

$$\begin{cases} r(n) \sim H(2H-1)n^{2H-2}\mathbb{E}(X_1^2), \text{ for } n \to \infty \text{ and } H \neq \frac{1}{2} \\ r(n) = 0, \text{ for } n \geq 1 \text{ and } H = \frac{1}{2} \end{cases} \quad (3.17)$$

We give a proof for this fact to get some insight in the structure of H-sssi processes.

Proof. We prove 3.17. From remark 3.4 e) follows $X_0 = 0$ almost surely. Since $\mathbb{E}(|X_1|) < \infty$, the fact that the second moment exists and by assumption $0 < H < 1$, follows from proposition 3.12 c), that $\mathbb{E}(X_t) = 0$ for all $t \geq 0$. This implies $r(\Delta(n)) = r(n) = \gamma(0, n)$ and we can use theorem 3.11

$$\begin{aligned} r(n) &= \mathbb{E}\left(\Delta(0)\Delta(n)\right) = \mathbb{E}\left(X_1(X_{n+1} - X_n)\right) \quad (3.18) \\ &= \mathbb{E}(X_1 X_{n+1}) - \mathbb{E}(X_1 X_n) \\ &= \frac{1}{2}\left((n+1)^{2H} - 2n^{2H} + (n-1)^{2H}\right)\mathbb{E}(X_1^2) \end{aligned}$$

where we applied theorem 3.11 suitably on the last equation. For $H = \frac{1}{2}$ immediately we deduce $r(n) = 0$. The assertion for $H \neq \frac{1}{2}$ is a simple calculation and left to the reader. □

Remark 3.25. According to the above asymptotic we can summarize for $0 < H < 1$:

a) If $0 < H < \frac{1}{2}$, then $2H - 2 < -1$ and thus, the series $\sum_{n=0}^{\infty} |r(n)| < \infty$ converges. In addition, we have $r(n) < 0$ (negative correlation). This can be deduced from the last equation in the above proof: the function $x \longmapsto x^{2H}$ is concave for $0 < H < \frac{1}{2}$.
b) For $H = \frac{1}{2}$ the discrete process $(\Delta(n))_{n \in \mathbb{Z}}$ is uncorrelated.
c) If $\frac{1}{2} < H < 1$, then $2H - 2 > -1$ and hence, the series

$$\sum_{n=0}^{\infty} r(n) = \infty \tag{3.19}$$

is no longer convergent. In this case the function $x \longmapsto x^{2H}$ is convex, which implies that the right side of the equation in the proof is a positive number (we can call it positive correlation).

The phenomena in item c) is usually called *long-range dependence*. It should be emphasized that this definition is only valid for processes with existing second moment, especially for Gaussian processes). Thus, we will introduce some different notions for the long-range dependence. This can be compared with the different definitions of stationary processes, depending whether the underlying process has a second moment or not. We deduce the following definitions of long-range dependence according to the above explanations.

Definition 3.26. *A (discrete) process (or time series) (X_k) is called long-range dependent, if there exists a $H \in]\frac{1}{2}, 1[$, such that*

$$r(k) = \mathcal{O}(k^{2H-2}) \tag{3.20}$$

If one considers the 'block process' according to (3.1), then this means

$$r^{(m)}(k) = \mathcal{O}(m^{2H-2})$$

In this case we have the asymptotic for $k, m \to \infty$. A slightly more general concept for long-range dependence is given in the next definition.

Definition 3.27. *A process (or time series) (X_k) is called* long-range dependent in the general sense, *if*

$$\sum_{k \in \mathbb{N}} |r(k)| = \infty \quad \text{if the process is discrete} \tag{3.21}$$

or

$$\int_0^\infty |r(t)| dt = \infty \quad \text{if the process is time continuous}$$

(see the above remark 3.25 and (3.19)).

It should be mentioned that the definition 3.26 implies the one of 3.27. It always depends on the situation, which approach is chosen. In this situation, too, one can consider the compounded process and hence, we have to replace $r(k)$ by $r^{(m)}(k)$.

Remark 3.28. If the second moment does not exists, a different definition has to be used. As we discussed before in more details, most considered processes are embedded into the general context of the α-stable processes, where $0 < \alpha \leq 2$. The variance exists at most in the case of $\alpha = 2$. In the general situation the notion of long-range dependence can be defined by fixing

$$H > \frac{1}{\alpha}$$

In the case of $\alpha = 2$ the definition coincides with the usual given definition above. But for $\alpha \leq 1$ we cannot proceed that way, since there is no H-sssi process for $H > \max(1, \frac{1}{\alpha})$, hence in particular for $\alpha \leq 1$, as we know by [77, Theorem 3.6.1].

An alternative approach for defining the long-range dependence for processes without existing variance was introduced by Heyde und Yang [106].

Definition 3.29. *If $(X_n)_{n \in \mathbb{N}}$ is a stationary sequence with mean 0, then this sequence is called* long-range dependent in the sense of the Allan-variance (LRD-SAV), *provided the quotient*

$$\frac{(\sum_{k=1}^{m} X_k)^2}{\sum_{k=1}^{m} X_k^2} \to \infty$$

in probability for $m \to \infty$.

In this context we find the following result interesting.

Proposition 3.30. *Suppose $(X_t)_{t \in [0,\infty[}$ is a H-sssi process and we have the relation $\mathbb{E}(|X_1|^\delta) < \infty$ for a $\delta \in \,]0, 2[$, then (X_t) is LRD-SAV, provided $H > \frac{1}{\delta}$.*

This is again a reason for defining the long-range dependence of a general α-stable process via the condition $H > \frac{1}{\alpha}$, $\alpha \in \,]1, 2[$.
There are a number of alternative not necessarily equivalent definitions of the long-range dependence. For special processes these notions are equivalent. Thus, we will discuss in the sequel mainly processes with existing variance. With the above notation we look at a H-sssi process (X_t) with $\mathbb{E}(X_1^2) < \infty$. As mentioned above we define $r(n) = \mathbb{E}(X_{n+1}X_1) - \mathbb{E}(X_n X_1)$, $(X_0 = 0!)$. Hence, the long-range dependence is exactly determined by the autocorrelation function $\gamma(n, m) = \gamma(|n - m|)$, $n, m \in \mathbb{N}$ (see the definition of autocorellation function and of stationary process, e.g. [38, 208]). This is the reason why we treat $r(n)$ in the following instead of $\gamma(n, m)$.

We know for a H-sssi process that $\gamma(k), (k = |n - m|)$ behaves asymptotically up to constant as k^{2H-2}. Let $f : [-\pi, \pi] \longrightarrow \mathbb{C}$ be the spectral density of the autocovariance $(\gamma(k))$, which is expressed according to

$$\gamma(k) = \int_{-\pi}^{\pi} e^{ikt} f(t) dt$$

(this is nothing else than the discrete Fourier transform). Then we have

$$f(t) = \frac{1}{2\pi} \sum_{k=-\infty}^{\infty} e^{-ikt} \gamma(k), \ t \in [-\pi, \pi] \quad (3.22)$$

The long-range dependence expressed in 'time' k will be transferred to the spectral density f close to 0, since low frequencies, i.e. values of t close to 0 imply big jumps in time. According to the result of (see [226]) we have indeed

$$f(t) \sim \text{const} \cdot t^{1-2H}, \ \text{for } t \to 0 \quad (3.23)$$

This means for $\frac{1}{2} < H < 1$, that the spectral density has a *singularity* at $t = 0$ (remark that in this case $1 - 2H < 0$). Figure 3.5 shows a fractional white noise $Z_t = X_{t+1} - X_t$ for a FBM (X_t).

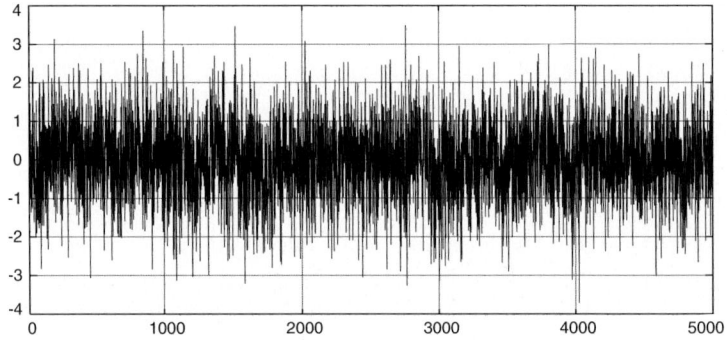

Fig. 3.5. Fractional Brownian motion with $H = 0.6$

According to the remark of the result (3.17) the divergence of the series $\sum_{n \in \mathbb{N}} r(n)$ determines the long-range dependence, though $r(n)$ converges to 0 with the order of $2H - 2$. We will use the following definitions and concepts to describe the long-range dependence

$$\gamma(n) \sim n^{-\alpha} L_1(n), \ n \to \infty \text{ for a } 0 < \alpha < 1$$
$$f(t) \sim t^{-\beta} L_2(t), \ t \to 0 \text{ for a } 0 < \beta < 1 \quad (3.24)$$
$$\sum_{k=0}^{n} \gamma(k) \sim n^{\vartheta} L_3(n), \ n \to \infty \text{ for a } 0 < \vartheta < 1$$

$L_i \in \mathcal{R}_0$, $i = 1, 2, 3$ are slowly varying functions. The parameters α, β and ϑ measure the degree of long-range dependence. The larger or higher the values of the parameters the more dependence we observe.

As seen in section 2.7.4, the function $U(x) = x^p L(x)$, $x > 0$, $L \in \mathcal{R}_0$ play a decisive rôle. Since we have for the quotient $\lim_{x \to \infty} \frac{U(tx)}{U(x)} = t^p$ for all $t > 0$, the slowly varying function L does not exhibit any influence for large values of x. Is it possible to specify the relationship of the parameters? The following result gives an answer [241, Prop. 4.1].

Proposition 3.31. *Suppose the autocovariance $(\gamma(n))$ is finally monotone for $n \to \infty$, then the conditions in (3.24) are equivalent and we have*

$$\vartheta = 1 - \alpha, \quad L_1(x) = \frac{1}{2}(1 - \alpha) L_3(x)$$

and

$$\beta = 1 - \alpha$$

$$L_2(x) = \frac{1}{2\pi} \Gamma(\vartheta + 1) \sin \frac{\pi(1-\varphi)}{2} L_3\left(\frac{1}{x}\right) = \frac{1}{\pi} \Gamma(1-\alpha) \sin \frac{\pi \alpha}{2} L_1\left(\frac{1}{x}\right)$$

It seems natural to define the short-range dependence. If a process is not long-range dependence then it is called *short-range dependent*. Defining $r(k) = \mathbb{E}(X_{k+1} X_1) - \mathbb{E}(X_k X_1)$, then it follows

$$\sum_{k \in \mathbb{N}} |r(k)| < \infty \text{ or in the case of continuous time } \int_0^\infty |r(t)| dt < \infty$$

This happens e.g. if the autocorrelation decays exponentially, or if

$$r(k) = \mathcal{O}(\rho^{-k}) \text{ with a suitable constant } 0 < \rho < 1$$

By the property $r(k) = \mathcal{O}(\rho^{-k})$ we have for $\sum_{k \in \mathbb{N}} |r(k)|$ a geometric series, and thus, $\sum_{k \in \mathbb{N}} |r(k)| < \infty$. The difference of long and short-range dependence (shortly LRD resp. SRD) can be described by the convergence of the series $\sum_{k \in \mathbb{N}} |r(k)|$ or the existence of the integral $\int_0^\infty r(t) dt$. The next sections considers the special case of a discrete time series, which can handle both phenomena long and short-range dependence: the fractional ARMA time series.

3.3.2 Fractional Brownian Motion and Fractional Brownian Noise

We start with a first derivation of the above result 3.31 and apply it to the FBM. Proposition 3.31 can be specified for the FBM. As we already considered the discrete process of the increments (in this case the fractional white noise) $\Delta(n) = X_{k+1} - X_k$ of a H-sssi process stated in theorem 3.24, we investigated the different concept of long-range dependence. For the FBM we can state the following fact concerning long-range dependence.

Theorem 3.32. *Let $(B_t^{(H)})$ be a FBM with $\frac{1}{2} < H < 1$. Then the three concepts in (3.24) are equivalent and we have (with the notation from (3.24))*

$$\vartheta = 2H - 1, \quad L_3(x) = \sigma^2 2H$$
$$\alpha = 2 - 2H, \quad L_1(x) = \sigma^2 H(2H - 1)$$
$$\beta = 2H - 1, \quad L_2(x) = \sigma^2 \pi^{-1} H \Gamma(2H) \sin \pi H$$

First, we look in detail at the fractional Brownian noise. We start with a particular FBM $(B_t^{(H)})_{t \in \mathbb{R}}$ and give the following definition.

Definition 3.33. *A sequence $Z_j = B_{j+1}^{(H)} - B_j^{(H)}$ with $j \in \mathbb{Z}$ is called* fractional white noise *or* fractional Brownian noise *(FGN). If $\sigma_0^2 = \mathbb{V}ar(Z_j) = 1$ then we call it* standard fractional Brownian noise. *Often one uses the terminology* fractional Gaussian noise.

Indeed it is possible to define the fractional Brownian noise in continuous time, i.e.

$$Z_t = B_{t+1}^{(H)} - B_t^{(H)}, \quad t \in \mathbb{R}$$

Because most applications in IP traffic are modeled in discrete time and most phenomena can be described there, we will stick to our original definition in discrete time.

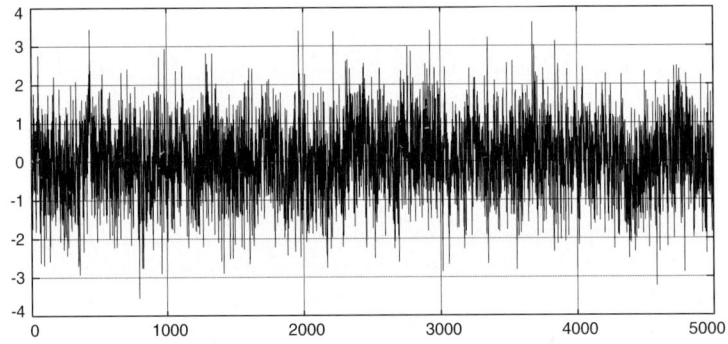

Fig. 3.6. Fractional Brownian noise with $H = 0.7$.

Since the FBM is a process with stationary increments, the FGN is a stationary process.

In our short introduction to H-sssi processes we cited general results for the process of increments. In the case of the FGN we briefly note them in this particular situation. For this we define by $\gamma(k) = r(k) = \mathbb{E}(Z_0 Z_j)$ as the covariance function done in theorem 3.24.

Theorem 3.34. *The FGN has a covariance function*

3.3 Long-Range Dependence

$$r(k) = \frac{\sigma_0^2}{2}\left(|k+1|^{2H} - |k|^{2H} + |k-1|^{2H}\right), \quad k \in \mathbb{Z} \qquad (3.25)$$

and a spectral density f

$$f(\lambda) = \frac{\sigma_0^2}{C_1(H)^2}|\exp(i\lambda) - 1|^2 \sum_{k=-\infty}^{\infty} \frac{1}{|\lambda + 2\pi k|^{2H+1}}$$

$$= \frac{\sigma_0^2 \int_0^\infty \cos(x\lambda)\left(\sin^2\left(\frac{x}{2}\right)\right) x^{-2H-1} dx}{\int_0^\infty \left(\sin^2\left(\frac{x}{2}\right)\right) x^{-2H-1} dx}, \quad -\pi \leq \lambda \leq \pi$$

As partially introduced before, it is possible to elaborate some results:

- $r(k)$ behaves for $k \to \infty$ as $f(\lambda)$ for $\lambda \to 0$, i.e. as a power of k resp. as λ.
- In the case of $\frac{1}{2} < H < 1$ we have long-range dependence, i.e. the series $\sum_{k=-\infty}^{\infty} |r(k)| = \infty$ diverges. Equivalently the spectral density reveals a singularity at 0.
- The case $0 < H < \frac{1}{2}$ indicates short-range dependence, which is for most considerations in IP traffic not relevant. Nevertheless it is interesting in some sense. We have $\sum_{k=-\infty}^{\infty} |r(k)| < \infty$ and because of the telescope sum of the sequence $r(k)$ it follows $\sum_{k=-\infty}^{\infty} r(k) = 0$. This is also expressed by the property of the spectral density f because $f(\lambda) \to 0$ if $|\lambda| \to 0$. Since the constant $H(2H-1)$ is negative, we can conclude with 3.24, that for large k the correlation $r(k)$ is negative. Indeed we have $r(k) < 0$ for all $k \neq 0$ (see [35]). Hence

$$0 = \sum_{k=-\infty}^{\infty} r(k) = r(0) + 2\sum_{k=1}^{\infty} r(k) = \sigma_0^2 + 2\sum_{k=1}^{\infty} r(k)$$

and thus

$$\sum_{k=1}^{\infty} r(k) = -\frac{\sigma_0^2}{2}$$

- The FGN serves as classical counterexample to the central limit theorem in the case of $H \neq \frac{1}{2}$. For a suitable sequence of coefficients a_n the sequence $\frac{1}{a_n}\sum_{k=1}^{n} Z_k$ converges for $n \to \infty$ to a non trivial random variable. Nevertheless it is not possible to choose the same coefficients $a_n = \sqrt{n}$ as in the classical central limit theorem. It is necessary to pick $a_n = n^H$ for large n. We have $\frac{1}{n^H}\sum_{k=1}^{n} Z_j = \frac{1}{n^H}B_n^{(H)} \stackrel{d}{=} B_1^{(H)}$. In general the process $\frac{1}{n^H}\sum_{k=1}^{[nt]} Z_j \to B_t^{(H)}$ converges for all t, where $[x]$ is the entire function, i.e. the largest integer less than x.
- Because of the long-range dependence the normalized sum of polynomials in (Z_k) does not converge to a Gaussian process.

If we consider the covariance (3.25) of the FGN, then we observe the well-known asymptotic behavior $r(k) \sim Ck^{2H-2}$ for $k \to \infty$, which expresses the long-range dependence in the case $\frac{1}{2} < H < 1$. In a lot of applications one is interested in this asymptotic and wants to use it, since equality (3.25) is not suitable for modeling. A possible approach can be seen in using gliding averages, i.e. use the approach of FARIMA time series, a concept, which we will discuss in detail later. These Gaussian sequences reveal the form

$$X_k = \sum_{j=-\infty}^{k} a_{k-j}\epsilon_j$$

where the sequence (ϵ_j) is an iid $\mathbb{N}(0,1)$ sequence of random variables. Then we have (as cited in theorem 3.37 in detail)

$$r(k) = \mathbb{E}(X_0 X_k) = \sum_{j=-\infty}^{k} a_{-j} a_{k-j} \sim C k^{2H-2}$$

Finally, we give a detailed description of long-range dependence of a Gaussian series (see also theorem 3.24 resp. 3.32) and show that, as in the case of FGN, the central limit theorem does not hold necessarily.

Theorem 3.35. *Let $0 < H < 1$ and (X_j) be a stationary Gaussian time series with autocovariance $r(k) = \mathbb{E}(X_0 X_k)$, which fulfills one of the following cases:*

- *If $\frac{1}{2} < H < 1$, then*

$$r(k) \sim c k^{2H-2}, \ k \to \infty, \ c > 0$$

- *If $H = \frac{1}{2}$, then*

$$\sum_{k=1}^{\infty} |r(k)| < \infty \quad \text{and} \quad \sum_{k=-\infty}^{\infty} r(k) = c > 0$$

- *If $0 < H < \frac{1}{2}$, then*

$$r(k) \sim c k^{2H-2}, \ k \to \infty, \ \text{with } c < 0 \text{ and } \sum_{k=-\infty}^{\infty} r(k) = 0$$

Then the marginal distributions of the process $\{\frac{1}{n}\sum_{k=1}^{[nt]} X_k; 0 \le t \le 1\}$ converge to those of $\{\sigma_0 B_t^{(H)}; 0 \le t \le 1\}$, where $(B_t^{(H)})$ is a standard fractional Brownian motion. For the constant σ_0 we have

$$\sigma_0^2 = \begin{cases} H^{-1}(2H-1)^{-1}c & \text{for } \frac{1}{2} < H < 1 \\ c & \text{for } H = \frac{1}{2} \\ -H^{-1}(2H-1)^{-1}c & \text{for } 0 < H < \frac{1}{2} \end{cases}$$

More on mathematical treatments for the fractional Brownian motion can be found in [33, 34, 114, 113, 35] and the forthcoming book [35].

3.3.3 Farima Time Series

The Gaussian or fractional white noise $(\Delta(n))_{n\in\mathbb{N}}$ (FGN) (see theorem 3.24) characterizes the long-range dependence. It is suitable for describing the behavior of the IP-based traffic over large time intervals. Changing to small time fractions, then the observed or estimated autocovariance differs from the Gaussian or the FGN. At the end of section 3.8.6 we will consider physically based models of the IP traffic, which encounter this two different phenomena: a long-range one and a multifractal one on short scale. A first step to overcome this drawback of FGN-models well be done in this section by the introduction of special time series, the so called FARIMA models. Here, it will already be possible to model the long-range dependence as well as the calibration of parameters for the small time scales. The FARIMA models (*fractional autoregressive moving average*) enlarge the concept of the ARMA time series. For a deeper understanding of the FARIMA time series we will deduce its definition in detail. Thus, we need some preliminaries.

Definition 3.36. *A sequence* $(X_t)_{t\in\mathbb{Z}}$ *is called* linear Gaussian sequence, *if there is a sequence* $(a_j)_{j\in\mathbb{Z}}$, *such that*

$$X_t = \sum_{j=-\infty}^{\infty} a_{t-j}\epsilon_j = \sum_{j=-\infty}^{\infty} a_j \epsilon_{t-j}, \ t \in \mathbb{Z} \qquad (3.26)$$

where $\sum_{j=-\infty}^{\infty} |a_j| < \infty$, *and* $(\epsilon_j)_{j\in\mathbb{Z}}$ *is a sequence of iid Gaussian distributed random variables, which are called* generator *or* increments.

We summarize some properties:

- The sequence $(X_t)_{t\in\mathbb{Z}}$ is called *causal*, provided $a_j = 0$ for all $j < 0$.
- A linear Gaussian sequence (X_t) is stationary.
- If the $(\epsilon_j)_{j\in\mathbb{Z}}$ are distributed according to $\mathcal{N}(\mu,\sigma^2)$ with $\mu \neq 0$, then we require $\sum_{j=-\infty}^{\infty} |a_j| < \infty$
- If in addition the sequence is $\mathcal{N}(\mu,\sigma^2)$ with $\mu = 0$, then stressing $\sum_{j=-\infty}^{\infty} a_j^2 < \infty$ is sufficient.

It is not necessary to choose the increments (ϵ_j) in (3.26) to be Gaussian distributed. It is possible to use α-stable increments, which is of importance for modeling IP-based traffic, where the incoming traffic need not to have finite variance. In the sections 3.3.3 and 4.2 we will discuss in more detail special α-stable increments and judge the different possibilities to calibrate the different parameters for the IP-based traffic. If the mean of ϵ_1 exists, one needs the absolute convergence of $\sum_{j=-\infty}^{\infty} |a_j| < \infty$. Suppose the variance of ϵ_1 is finite, then it suffices that $\sum_{j=-\infty}^{\infty} a_j^2 < \infty$. If the (ϵ_j) are Gaussian distributed, so is the time series (X_t). But if the (ϵ_j) are e.g. exponential distributed, it is no longer possible to specify the distribution (X_t), since linear combinations of exponential distributed random variables are not necessarily exponential distributed. The behavior of mean and variance is the same as in the Gaussian case.

The FARIMA series will be introduced as a special case of the linear Gaussian models. But the concept can be transferred to other distributions. These time series will bear the same name. We will use the common notation FARIMA$[p, d, q]$, with $p, q \in \mathbb{N}_0$ and $d \in \mathbb{R}$ and consider the case $p = q = 0$ first, in other words FARIMA$[0, d, 0]$. Demonstrating how to construct a FARIMA$[0, d, 0]$-series for a given $d \in \mathbb{R}$, we will have to make some preparations. Since the time series play a decisive rôle in describing IP-based traffic, we will deduce the construction in more detail. First we choose $d = 0, 1, \ldots$, thus, a non-negative integer. Let $(X_t)_{t \in \mathbb{Z}}$ be a discrete process. We define for (X_t)

$$\Delta^0 X_t = X_t, \ \Delta^1 X_t = X_t - X_{t-1}, \ \Delta^2 X_t = \Delta(\Delta X_t) = \Delta(X_t - X_{t-1}), \ldots$$

It is possible to write the operator Δ^j, $j = 0, 1, 2, \ldots$ in the form

$$\Delta^j = (Id - A)^j, \ j = 1, 2 \ldots$$

where $A^j X_t = X_{t-j}$ represents a shift operator with $Id = A^0$. We have e.g.

$$\Delta^2 X_t = \Delta(X_t - X_{t-1}) = X_t - 2X_{t-1} + X_{t-2}$$
$$= (Id - 2A + A^2)X_t = (Id - A)^2 X_t$$

We call a linear Gaussian sequence $(X_t)_{t \in \mathbb{Z}}$ a FARIMA$[0, d, 0]$ time series, if

$$\Delta^d X_t = \epsilon_t, \text{ for all } t \in \mathbb{Z} \tag{3.27}$$

How can we define the operator Δ^d for $d \in \mathbb{R}$? For this we rewrite (3.27), multiply

$$\Delta^d X_t = (Id - A)^d X_t = \epsilon_t, \text{ for all } t \in \mathbb{Z}$$

with $(Id - A)^{-d}$ and deduce $X_t = (Id - A)^{-d} \epsilon_t$ if $(Id - A)^{-1}$ exists (e.g. if $\|A\| < 1$). In this case the series is representable as $(Id - A)^{-d} = \sum_{j=0}^{\infty} b_j A^j$, where the series

$$X_t = (Id - A)^{-d} \epsilon_t = \sum_{j=0}^{\infty} b_j A^j e_j = \sum_{j=0}^{\infty} b_j e_{t-j}$$

converges. This condition is too strong; thus, we we will deduce – also for further purposes – a weaker convergence. For a complex number $z \in \mathbb{C}$ with $|z| < 1$ we can represent $(1 - z)^{-d}$ in a power series

$$(1 - z)^{-d} = \sum_{j=0}^{\infty} b_j z^j$$

Remark that the function $z \longmapsto (1-z)^{-d}$ is analytic for $|z| < 1$, i.e., representable in a power (see e.g. [92, S. 174 f.]). In this representation we have $b_0 = 1$, and for the other coefficients $j \in \mathbb{N}$ it holds

3.3 Long-Range Dependence

$$b_j = \prod_{k=1}^{j} \frac{k-1+d}{k} = \frac{\Gamma(j+d)}{\Gamma(j+1)\Gamma(d)} \qquad (3.28)$$

where $\Gamma(x) = \int_0^\infty t^{x-1} e^{-t} dt$, $x > 0$ the Gamma function. The Gamma function satisfies the functional equation

$$\Gamma(x+1) = x\Gamma(x) \qquad (3.29)$$

and the well known *Stirling formula*

$$\Gamma(x) \sim \sqrt{2\pi} e^{-x+1} (x-1)^{x-\frac{1}{2}}, \quad x \to \infty$$

Applying (3.29) to the representation (3.28), then one has for large $j \to \infty$

$$b_j \sim \Gamma(d)^{-1} j^{d-1}$$

Hence $\sum_{j=1}^\infty b_j^2 < \infty$ converges (the convergence is determined by large j), if $\sum_{j=1}^\infty j^{2(d-1)} < \infty$ holds. And thus, by a criterion for convergent series $2(d-1) < -1 \Leftrightarrow -\infty < d < \frac{1}{2}$. In this case $\sum_{j=0}^\infty b_j \epsilon_j$ converges in the space $L_2(\Omega)$. Remark that in this space (ϵ_{t-j}) consists of an orthogonal system and, as shown above, we have $\sum_{j=1}^\infty b_j^2 < \infty$. Hence, we can define for $-\frac{1}{2} < d < \frac{1}{2}$ the representation

$$X_t = (Id - A)^{-d} \epsilon_t = \sum_{j=0}^\infty b_j \epsilon_{t-j}$$

and we have $(Id - A)^d X_t = \epsilon_t$. This enables us easily to enlarge the definition of a FARIMA[0,d,0] time series for all $d \in \mathbb{R}$, since a general $d \in \mathbb{R}$ can be written in the form $d = d' + d''$, where $d' \in \mathbb{Z}$ and $-\frac{1}{2} \le d'' \le \frac{1}{2}$ (e.g. $d = 1.7 = 2 + (-0.3)$). In general we write

$$d = \left[d + \frac{1}{2}\right] + \left(d - \left[d + \frac{1}{2}\right]\right)$$

where $[x]$ = the smallest integer less or equal to x. For this we remark that $-\frac{1}{2} \le d - [d + \frac{1}{2}] < \frac{1}{2}$. Suppose $d \ge \frac{1}{2}$, then define the FARIMA[0, d, 0] process (X_t) by

$$(Id - A)^{d - [d + \frac{1}{2}]} \left((Id - A)^{[d + \frac{1}{2}]} X_t \right) = \epsilon_t, \ t \in \mathbb{Z}$$

This means that the process $Z_t = (Id - A)^{[d + \frac{1}{2}]} X_t$ is a FARIMA[0, $d - [d + \frac{1}{2}]$, 0] time series. But Z_t can be defined for integer exponents (see (3.27)). For $d < -\frac{1}{2}$ the process can explicitly written in the form

$$X_t = (Id - A)^{-[d + \frac{1}{2}]} \left((Id - A)^{-d + [d + \frac{1}{2}]} \epsilon_t \right), \ t \in \mathbb{Z}$$

The introduction of the FARIMA time series represents a time continuous process (X_t) using a series of Gaussian distributed random variables, thus, discretized. In the context of self-similarity and long-range dependence we are especially interested how the covariance function $\gamma(k) = \mathbb{E}(X_0 X_k)$ looks like for large numbers $k \to \infty$. For this we give the following fundamental result without detailed proof.

Theorem 3.37. *Let $(X_t)_{t \in \mathbb{Z}}$ be a FARIMA$[0, d, 0]$ time series with $-\frac{1}{2} < d < \frac{1}{2}$, $d \neq 0$. Then we have*

$$\gamma(0) = \sigma^2 \frac{\Gamma(1-2d)}{\Gamma(1-d)^2}$$

$$\gamma(k) = \sigma^2 \frac{(-1)^k \Gamma(1-2d)}{\Gamma(k-d+1)\Gamma(1-k-d)} \quad (3.30)$$

$$= \sigma^2 \frac{\Gamma(k+d)\Gamma(1-2d)}{\Gamma(k-d+1)\Gamma(d)\Gamma(1-k-d)} \sim c|k|^{2d-1}$$

for $k \to \infty$, where $c = \gamma(0) \frac{\Gamma(1-2d)}{\Gamma(d)\Gamma(1-d)}$.

Proof. We do not present the full proof, but present the argument for the asymptotic in (3.30) as exercise

$$\sigma^2 \frac{\Gamma(k+d)\Gamma(1-2d)}{\Gamma(k-d+1)\Gamma(d)\Gamma(1-k-d)} = \sigma^2 \frac{\Gamma(1-2d)}{\Gamma(d)\Gamma(1-d)} \frac{\Gamma(k+d)}{\Gamma(k-d+1)}$$

$$= c \frac{\Gamma(k+d)}{\Gamma(k-d+1)} \sim c e^{-2d+1}(k-1+d)^{k+d-\frac{1}{2}}(k-d)^{-k+d-\frac{1}{2}}$$

$$\sim c e^{-2d+1}(k^2-d^2)^{-\frac{1}{2}}(k^2-d^2)^d \left(\frac{k-1+d}{k-d}\right)^k \sim c|k|^{2d-1}$$

since $\left(\frac{k-1+d}{k-d}\right)^k \to e^{2d-1}$, for $k \to \infty$. □

We know that a stationary process $(X_t)_{t \in \mathbb{Z}}$ is long-range dependent, if $\sum_{k=0}^{\infty} |\gamma(k)| = \infty$ for $\gamma(k) = \mathbb{E}(X_t X_{t+k})$. But we have

$$\sum_{k=1}^{\infty} k^{2d-1} = \infty \Leftrightarrow 2d - 1 \geq -1 \Leftrightarrow d \geq 0$$

Thus, we see by theorem 3.37 that this holds exactly for $0 < d < \frac{1}{2}$. How do we detect the relationship with the Hurst exponent? From equation (3.17) we know for a H-sssi process $(X_t)_{t \in \mathbb{Z}}$

$$\gamma(k) = \mathbb{E}(X_t X_{t+k}) \sim c k^{2H-2}, \text{ for } k \to \infty \quad (3.31)$$

Comparing the behavior for large k in (3.30) with (3.31), this implies

$$2H - 2 = 2d - 1 \Leftrightarrow H = d + \frac{1}{2}$$

Thus, we see that long-range dependence of a FARIMA$[0, d, 0]$ time series determined by $0 < d < \frac{1}{2}$ which is equivalent to the fact that the Hurst exponent satisfies $\frac{1}{2} < H < 1$. There is another common feature for FARIMA time series and long-range dependent processes: As shown in [226] (see also 3.35), we have for a Gaussian FARIMA$[0, d, 0]$ time series $(X_s)_{s \in \mathbb{Z}}$ ($0 < d < \frac{1}{2}$) the weak convergence

$$\frac{1}{n^H} \sum_{s=1}^{[nt]} X_s \longrightarrow B_t^{(H)}, \text{ for } n \to \infty$$

hence, in the marginal distribution. Here $H = d + \frac{1}{2}$ and $(B_t^{(H)})_{t \geq 0}$ is a fractional Brownian motion.

We start now with the general case of a Gaussian FARIMA$[p, d, q]$ time series. For this we split the general case again. First let $d = 0$. This coincides with the well known case of an ARMA(p, q) time series (*autoregressive moving average*).

Definition 3.38. *A time series $(X_t)_{t \in \mathbb{Z}}$ is called an ARMA(p, q) process, provided there are numbers $\phi_1, \ldots, \phi_p$ and $\psi_1, \ldots, \psi_q$ with*

$$X_t - \phi_1 X_{t-1} - \ldots - \phi_p X_{t-p} = \epsilon_t - \psi_1 \epsilon_{t-1} - \ldots - \psi_q \epsilon_{t-q}, \ t \in \mathbb{Z} \quad (3.32)$$

where (ϵ_t) is an independent distributed sequence of $\mathcal{N}(0, \sigma^2)$ random variables.

The idea is to incorporate the 'past' of a time series $(X_t)_{t \in \mathbb{Z}}$ by representing it with the help of a suitable sequence of independent centralized Gaussian variables. If $p = 0$, then we have a MA(q) process (moving average). In the case of $q = 0$, it is called an AR(p) process (autoregressive). As in the derivation of the FARIMA[0,d,0] time series we define for certain operators representing the left resp. the right side of (3.32) $\Phi_p(A) = Id - \phi_1 A - \ldots - \phi_p A^p$ resp. $\Psi_q(A) = Id - \psi_1 A - \ldots - \psi_p A^q$. Then we rewrite (3.32) into

$$\Phi_p(A) X_t = \Psi_q(A) \epsilon_t \quad (3.33)$$

We want a representation of (X_t) in dependence of a Gaussian sequence, as in the case of the FARIMA$[0, d, 0]$ time series, i.e., we want to solve (3.33) to X_t in the form

$$X_t = \Phi_p(A)^{-1} \Psi_q(A) \epsilon_t \quad (3.34)$$

by getting an expression like

$$X_t = \sum_{j=0}^{\infty} b_j \epsilon_{t-j}$$

For this we have to consider complex valued polynomials and define (we use the notation of the operators) the polynomial $\Phi_p(z) = 1 - \phi_1 z - \phi_2 z^2 -$

... $- \phi_p z^p$ of degree p, which possesses in the complex plain $\mathbb{C}$ the complex roots (points of value 0 for the polynomial) $r_1, \ldots, r_p$. Then we can write the polynomial in the form

$$\Phi_p(z) = \left(1 - \frac{z}{r_1}\right) \cdots \left(1 - \frac{z}{r_p}\right) \tag{3.35}$$

This implies

$$\Phi_p(z)^{-1} = \left(1 - \frac{z}{r_1}\right)^{-1} \cdots \left(1 - \frac{z}{r_p}\right)^{-1}$$

Since the operator A has the norm 1, we want to know in which cases

$$\left(1 - \frac{z}{r_i}\right)^{-1} = \sum_{j=1}^{\infty} \left(\frac{1}{r_i}\right)^j z^j, \ i = 1, \ldots, p$$

converges for $|z| = 1$. This happens exactly, if $|r_i| > 1$ ($i = 1, \ldots, p$, geometric series!). With this we stress that the roots of the polynomial in (3.35) lie outside the unit circle. Thus it is possible to define

$$\left(Id - \frac{A}{r_i}\right)^{-1} = \sum_{j=1}^{\infty} \left(\frac{1}{r_i}\right)^j A^j$$

and apply it to a $L_2(\Omega)$-integrable time series. The equation (3.34) is well-defined. Similar we define the polynome

$$\Psi_q(z) = 1 - \psi_1 z - \psi_2 z^2 - \ldots - \psi_q z^q$$

and demand, that $\Phi_p(z)$ and $\Psi_q(z)$ do not have common roots and in addition, all roots of $\Psi_q(z)$ lie outside the unit circle. The last property is important for estimation, since one considers ϵ_t as realization of the X_t.

Example 3.39. Let us start with an AR(2) time series (X_t), which has according to statistical evaluations (see section 4.2) the following form

$$X_t - \phi_1 X_{t-1} - \phi_2 X_{t-2} = \epsilon_t, \ t \in \mathbb{R}$$

Hence, we have to consider polynomial $p(z) = 1 - \phi_1 z - \phi_2 z^2$, $\phi_2 \neq 0$. The roots are

$$r_{1/2} = -\frac{\phi_1}{2\phi_2} \pm \sqrt{\left(\frac{\phi_1}{2\phi_2}\right)^2 + \frac{1}{\phi_2}}$$

Then, we get $|r_i| > 1$ for $i = 1, 2$, provided $\frac{1}{\phi_2} > 1$. The form of the time series turns into

$$X_t = \left(1 - \frac{A}{r_1}\right)^{-1} \left(1 - \frac{A}{r_2}\right)^{-1} \epsilon_t$$

$$= \left(\sum_{i=0}^{\infty} \left(\frac{1}{r_1}\right)^i A^i\right) \left(\sum_{j=1}^{\infty} \frac{1}{r_2} A^j\right) \epsilon_t$$

$$= \sum_{n=0}^{\infty} \sum_{k=0}^{n} \left(1 - \frac{1}{r_1}\right)^{n-k} \left(1 - \frac{1}{r_2}\right)^k \epsilon_{t-n}$$

An important observation is the decay of the covariance function. It is easy to demonstrate and left to the reader as exercise that there is an $s \in {]0,1[}$ with

$$|r(k)| \leq \mathrm{const} \cdot s^{|k|} = \mathrm{const} \cdot e^{-|k|\log\frac{1}{s}}$$

(see also [42]). As seen above we detect an exponential decay, which in turn describes the SRD phenomena of the ARMA(p,q) time series. The advantage of the general FARIMA$[p,d,q]$ time series lies in the combination of the LRD property of the FARIMA$[0,d,0]$ and the SRD phenomena in the ARMA(p,q) part. Thus, we have to consider the general case finally.

Definition 3.40. *A time series $(X_t)_{t \in \mathbb{Z}}$ is a FARIMA$[p,d,q]$ time series, if the representation*

$$\Phi_p(A)\Delta^d X_t = \Psi_q(A)\epsilon_t, \; t \in \mathbb{Z}$$

holds.

Thus we combine both above cases and receive the following representation under the made presumptions

$$X_t = \Phi_p(A)^{-1}\Psi_q(A)\Delta^{-d}\epsilon_t \qquad (3.36)$$

If $d < \frac{1}{2}$ and the polynomial $\phi_p(z)$ does not have a root within the unit circle, then the time series in (3.36) is well-defined, causal and stationary. If in addition $\Psi_q(z)$ has roots only outside the unit circle, the representation can be reversed. The covariance structure depends only on d and the roots of $\Phi_p(z)$ and $\Psi_q(z)$.

The increments in the serving part of the IP-based traffic, i.e. the connection times, data rates and the amount of TCP-connections, are mostly heavy-tail distributed. These distributions include the Pareto- resp. all general α-stable distributions, as will be treated in the section 3.4.1. But we cannot transfer the results, especially (3.30), literally. In this respect the long-range dependence is only valid for $d \in [0, 1-\frac{1}{\alpha}[$. The connection between α and the Hurst exponent can be stated as

$$H = d + \frac{1}{\alpha} \qquad (3.37)$$

(see [226, S. 382]). For further results of FARIMA-series with α-stable increments the reader should consult the monograph [226, section 7.13].

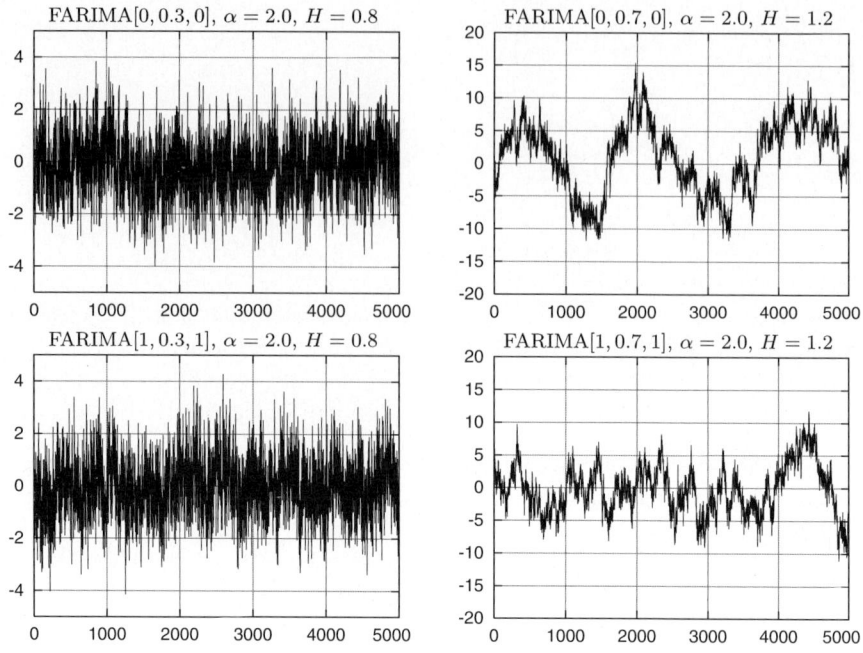

Fig. 3.7. Realizations of Gaussian FARIMA time series ($\alpha = 2$) with $H = 0.8$ resp. $H = 1.2$ and different p, q

3.3.4 Fractional Brownian Motion and IP Traffic – the Norros Approach

As we will derive in section 3.4.4, averaging of a simple on-off model over time will lead in the limit with certain assumptions on the serving time to the fractional Brownian motion resp. fractional white noise. This is the decisive reason, why the FBM is an important tool for modeling the traffic load $A(t)$. Before describing the classical Norros approach [190], we shall give notes on the basic concept of the Gaussian model. Though the approach is still very popular in the literature, some critics has to be mentioned, since the traffic does not behave Gaussian in principle. This is based on reasons, which will be discussed in the section on multifractal approach later (see section 3.8). First of all the Gaussian model implies that with the same probability negative values could occur (locally the FBM has the expected value 0!). Then the traffic flows described by the FBM model are assumed to be independent of the transport protocol, which is for most application very unrealistic. A Gaussian traffic flow is mainly balanced or averaged and hence, reveals only a few bursts, which is especially not true for TCP-based traffic. Thus, there are only a few queueing models with pure Gaussian input.
Nevertheless the attractivity of Gaussian based models will still prevail that for once a many-user-traffic is balanced by averaging according to the central

3.3 Long-Range Dependence

limit theorem. On the other hand an individual treatment of each single flow of IP packets would be too complex to gain sufficient exactness in the description, as e.g. seen in the video applications. The simple structure of mean and covariance of the FBM in junction with the stationary increments simplifies the modeling of aggregated traffic.

This section is devoted to an introduction an partially a summary of some important observations, as the fractional Brownian motion (FBM) could be applied in modeling the data traffic, especially in a LAN, and as the FBM describes the scenarios appropriately. A lot of these results go back to the original literature due to Norros, which we cite and comment on. It was Mandelbrot and Van Ness [174], who discovered the great value of the FBM in explaining the so called *long-range dependence*. At that time, they surely used the FBM in modeling the stock prices and not data traffic in IP networks.

Remark 3.41. We first give some initial remarks concerning the situation, which will be investigated:

- A lot of data is sent in small independent burst in the considered networks. The bursts consist of IP packets, which in turn contain UDP-datagrams or TCP-segments. We already describe this in the section 1.1 intensively.
- In packet switched traffic no constant data rate could be detected, in contrast to the circuit switched traffic. This property can be especially observed for the TCP-based traffic (see section 1.1), who prevents by the mechanism for congestion control an overflow at the routers mainly by dynamical adjustment of the sent data rates.
- Beside transmission times in packet switched traffic there are silent moments. Thus, the on-off model is the adequate approach as initial model, where the level of aggregated traffic is asymptotically reached, and hence, the FBM can be used.
- The time structure depends on several factors: E.g. from the implemented protocols, the day time (different peaks in different user scenarios) and the application (see table 1.1) and sections 1.1 and 1.3).
- In the packet switched traffic a unified modeling is practically impossible, since the traffic depends on a lot of components.
- The Leland group at Bellcore discovered 1992 that the data traffic in a LAN can be described very well by self-similarity (or more appropriate by the asymptotical self-similarity).
- We use the so called Gaussian model, i.e. processes with Gaussian mariginal distribution. We give a survey on models with non Gaussian marginal in subsequent sections (see e.g. section 3.8) with special focus on multi-fractal models.

We proceed to the mathematical description of the approach and split the approach in several steps, motivated by the described observations of the Bellcore experiment:

- With A_t we denote as usual the traffic amount (e.g. bit rates) within the half open interval $[0,t[$, $t \in]0,\infty[$ resp. $]t,0]$, $t \in]-\infty,0[$. Let $A_{s,t} = A_t - A_s$ be the traffic within the half open interval $[s,t[$, $s < t$.
- We assume that the process (A_t) has stationary increments. In the case of the Gaussian distribution we have $\mathrm{Var}(A_t) < \infty$, i.e., $\mathbb{E}(A_t^2) < \infty$ and $\mathbb{E}(A_t) = mt$ because of the stationarity. We have
$$\mathrm{Cov}(A_s, A_t) = \frac{1}{2}\left(v(t) + v(s) - v(t-s)\right)$$
where we set $v(t) = \mathrm{Var}(A_t)$.
- We call a process *short-range dependent*, provided the correlation fulfills
$$\mathrm{Cor}(A_{cs,ct}, A_{cu,cv}) \longrightarrow 0, \text{ for } c \to \infty \text{ and all } s < t \le u < v$$
Otherwise we call the process *long-range dependent*. This definition coincides with the one in section 3.3, as one easily can verify.
- If (A_t) is short-range dependent, then $\mathrm{Var}(A_t)$ is asymptotically linear, i.e. $\mathrm{Var}(A_t) \sim ct$ for $t \to \infty$.
- If the process (A_t) has independent increments, as e.g. the Poisson process or the Brownian motion, then the variance $\mathrm{Var}(A_t)$ is asymptotically linear – these processes are indeed short-range dependent due to the Markov-property.
- The measurements of Bellcore provided for the function $v(t)$ an asymptotic of degree t^p with $1 < p < 2$, as we e.g. already found in our theoretical considerations in section 1.1 resp. 1.4
- On the other hand the self-similarity of the process (A_t) provided as we realized in section 3.3) that
$$v(A_{ct}) = (ct)^{2H} = c^{2H} v(A_t)$$
for a suitable $H \in]\frac{1}{2}, 1[$. Thus, A_{ct} and $c^H A_t$ enjoy the same correlation structure, and these self-similar processes are long-range dependence, except they have independent increments. This property is already known to us.
- In section 3.2.2 on fractional Brownian motion we remarked that the FBM can be introduced via the correlation structure, something which is impossible in the general case of α-stable processes. Thus, the Gaussian processes are the best and only choice and this is one major reason for selecting the FBM.
- The FBM (Z_t) was introduced in section 3.1.2 via the autocorrelation function.
- Considering the covariance of the fractional white noise (see theorem 3.24), then one discovers for non overlapping intervals, i.e. for all $s < t \le u < v$ a similar structure
$$\mathrm{Cor}(Z_t - Z_s, Z_v - Z_u)$$
$$= \frac{1}{2}\left((v-s)^{2H} - (u-s)^{2H} + (u-t)^{2H} - (v-t)^{2H}\right)$$

3.3 Long-Range Dependence

- As already remarked the FBM with Hurst exponent $H \neq \frac{1}{2}$ lacks some major properties of the usual processes used in telecommunication, like the Poisson process, which include e.g.: The FBM does not enjoy the Markov property and is not a semi-martingale, which is important for the classical form of the stochastic integration (for the definition of semi-martingale see e.g. [208]). As the Brownian motion the FBM has continuous but nowhere differentiable paths. In addition the FBM looks smoother, if simulated, (see section 3.2.2 resp. (3.9) with proposition 3.18).
- At last we remark that the FBM is ergodic though the strong correlation of the increments, i.e the correlation of the fractional white noise $Z_{t+1} - Z_t$.

After the mathematical description of the Bellcore observation we will now introduce the traffic model, which is basic for the traffic model of Norros ad which will be dealt of in the next parts of this section:

- The process (A_t) describing the data amount will be defined as follows

$$A_t = mt + \sqrt{am} Z_t, \ t \in \,]-\infty, \infty[\tag{3.38}$$

where (Z_t) is the standard FBM. The three parameters allow the following interpretation: $m > 0$ is the average input rate and $a > 0$ represents the variance of the traffic. The FBM Z_t has no physical unit – one has to norm the parameter t, to indicate a time unit. Keeping things simple, we skip this fact.
- The factor $\sqrt{m}$ can be explained by *superposition*: Suppose $A_t = \sum_{i=1}^n A_t^{(i)}$ shows the aggregated sum of n traffic streams with the same parameters a and the fractional Brownian motion with the same Hurst exponent H, but with individual rates m_i, then m can be written as the sum $m = \sum_{i=1}^n m_i$, where $A_t = mt + \sqrt{ma} Z_t$. The figures 3.8 until 3.11 show on the one hand the measured traffic and below the simulated paths according to the Norros model. Very clearly one recognized that the Norros model represents very well the structure of the traffic for large scales (at most 100 ms).
- The parameters a and H describe the 'quality' and m the 'quantity' of the traffic.

Finally we give some comments on the selection of the process A_t. In classical telecommunication one uses the Poisson process for modeling the arrival and serving processes. As known one can represent the Poisson process (N_t) on the form

$$N_t = mt + M_t$$

where $M_t = N_t - mt$ is a martingale. Let us consider the martingale more closely. For $c \to \infty$ we have the convergence

$$\frac{N_{ct} - cmt}{\sqrt{cm}} \to B_t$$

$((B_t)_{t \geq 0}$ Brownian motion), as the theorem of Donsker or the functional central limit theorem from section 3.2.3 tell us. Thus, we can in general write the

Poisson process in the form

$$N_t = mt + \sqrt{m}B_t$$

In the case of the self-similar traffic the data process (A_t) is modeled analogously. We exchange the Brownian motion by the FBM and introduce an additional parameter a for better modeling.

With the help of a flow model in traffic theory we will try to give some practical meaning to the parameters a and H. In addition, we will reveal the connection between 'self-similarity' and 'heavy-tail' distributed serving time. Especially, we will point out, using the classical traffc model M/G/1, how the heavy-tail distributed serving time influences the self-similarity resp. long-range dependence.

Description of the Model

We start with describing the situation, useful for the derivation of the Norros model, i.e. the perturbation with the fractional Brownian motion. For this purpose we describe the traffic with the help of the classical traffic model M/G/1:

- As arrival process we choose a Poisson process with parameter λ.
- The arrival rate for each 'Burst' B_n is given by $r > 0$.
- We assume a joint distribution of the amount of burst,i.e $F_B(x) = \mathbb{P}(B \leq x)$.
- The length of each burst (i.e the serving time) $T_n = \frac{B_n}{r}$. This gives a joint distribution of the serving time T: $F_T(t) = \mathbb{P}(T \leq t) = F_B(rt)$.

Let us denote by N_t the number of bursts, which are served up to time t. We use the traffic model M/G/∞, which is a model with infinite serving space. As shown in section 2.4.2, the demand process will be stationary, if the serving rate $\mu = \mathbb{E}(B)$ is finite. In this case we get

$$\mathbb{E}(N) = \lambda \mathbb{E}(T)$$

by the theorem of Little. We get for the covariance (see e.g. [57])

$$\mathrm{Cov}(N_t, N_{t+h}) = \lambda \int_h^\infty F_B^c(rs)ds$$

The observation tells us that the traffic is long-range dependent. Hence to have the system M/G/∞ 'long-range dependent' (see the definition in section 3.3), we need $r(t) = \mathrm{Cov}(N_0, N_t)$

$$\mathbb{E}(T^2) = \int_0^\infty r(t)dt$$
$$= \int_0^\infty \mathrm{Cov}(N_0, N_t)dt$$
$$= \lambda \int_0^\infty \int_t^\infty F_B^c(rs)ds = \infty$$

3.3 Long-Range Dependence

which coincides with the above definition, since in our case the covariance is not integrable. This implies, that B has *no* finite variance. Furthermore we denote with
$$R_t = rN_t$$
the arrival rate of the fow process at time t. The R has a mean arrival rate of
$$m = \mathbb{E}(R) = r\mathbb{E}(N) = r\lambda\mathbb{E}(T) = \lambda\mathbb{E}(B) = rb$$
The use of the letter m is intended to show the connection with the approach of the traffic amount process A_t. The aggregated process A_t has the form
$$A_t = \int_0^t R_s ds$$
As variance of A_t we deduce
$$\operatorname{Var}(A_t) = \operatorname{Cov}\left(\int_0^t R_s ds, \int_0^t R_u du\right)$$
$$= 2\lambda rb \int_0^t \int_0^s \mathbb{P}(U > ru) ds du \qquad (3.39)$$
$$= \lambda rb\left(2t\mathbb{E}\left(\frac{U}{r}\wedge t\right) - \mathbb{E}\left(\frac{U}{r}\wedge t\right)^2\right)$$

Here U is a random variable, which is distributed according to the complementary distribution of B, i.e.
$$\mathbb{P}(U \leq u) = \frac{1}{b}\int_0^u F_B^c(s) ds$$
$$= \frac{1}{\mathbb{E}(B)}\int_0^u F_B^c(s) ds$$
$$= F_{I,B}(u)$$

The infinite variance of B implies an infinite value of expectation of U, because
$$\int_0^\infty \mathbb{P}(U > u) du = \int_0^\infty F_{I,B}^c(u) du \qquad (3.40)$$

and the right integral is finite, if and only if the variance of B is finite. From the equation (3.40) in junction with (3.39) the following can be deduced for the variance of the traffic load $\operatorname{Var}(A_t)$. We assume that complementary serving time B is heavy-tailed with $F_B^c(t) \sim Ct^{-\alpha}$ and $\alpha < 2$. Then the variance does not exists and by a result of Karamata 2.49 we have $F_{I,B}^c(u) \sim C'u^{-\alpha+1}$. The integration in (3.40) provides for $x \to \infty$ the asymptotic
$$\int_0^x F_{I,B}^c(u) du \sim C'' x^{-\alpha+2}$$

again by the result of Karamata (*see* theorem 2.49). With (3.39) we get

$$\mathbb{V}\mathrm{ar}(A_t) \sim \mathrm{const} \cdot t^{-\alpha+3} \tag{3.41}$$

Suppose now that $\alpha \in \,]1, 2[$, so is $-\alpha + 3 \in \,]1, 2[$ which is again a hint for the long-range dependence, as we have pointed out at the beginning of this section. We compare this with the result (3.54) from section 3.4.4. There we had

$$v(t) = \mathbb{V}\mathrm{ar}(A_t) \sim \sigma^2 t^{2H}$$

where we deduce according to (3.55) $H = \frac{3-\min(\alpha_{\mathrm{on}}, \alpha_{\mathrm{off}})}{2}$. If $\alpha = \min(\alpha_{\mathrm{on}}, \alpha_{\mathrm{off}})$, then by (3.54) together with (3.55) it follows

$$v(t) \sim \sigma^2 t^{-\alpha+3}$$

what is exactly (3.41). This is again a justification for the approach of Norros: As with the help of the on-off model and the deduced asymptotic according to (3.56), as with the classical approach using the traffic theory model $M/G/\infty$, we receive the same values of α and H.

As we already stated in the previous section, the figures 3.8 to 3.11 illustrate evidently that the Norros model provides a suitable description of measured traffic – here volume of data per time unit – over several timescales.

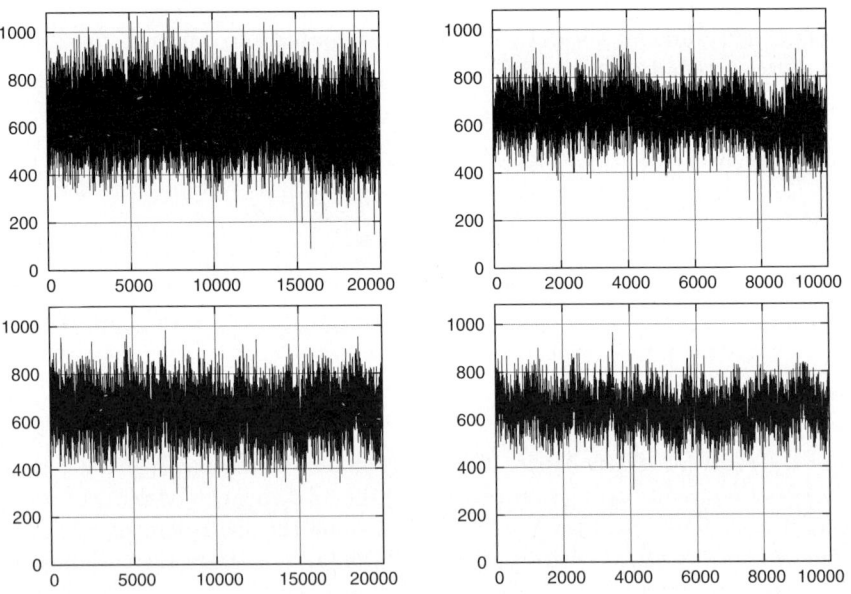

Fig. 3.8. Measured traffic (above) compared to simulated traffic according to the Norros model (below), scaling $t = 100$ ms (left) and $t = 200$ ms (right)

3.3 Long-Range Dependence   225

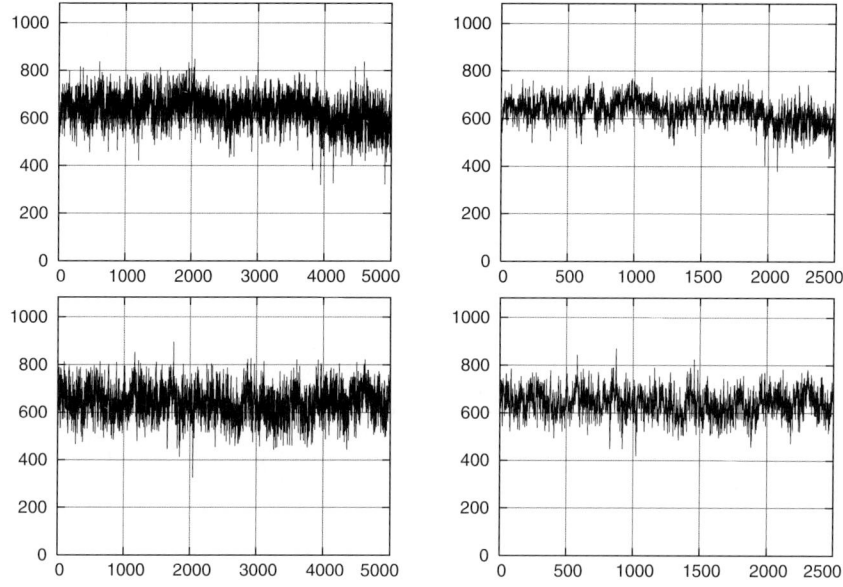

Fig. 3.9. Measured traffic (above) compared to simulated traffic according to the Norros model (below), scaling $t = 400$ ms (left) and $t = 800$ ms (right)

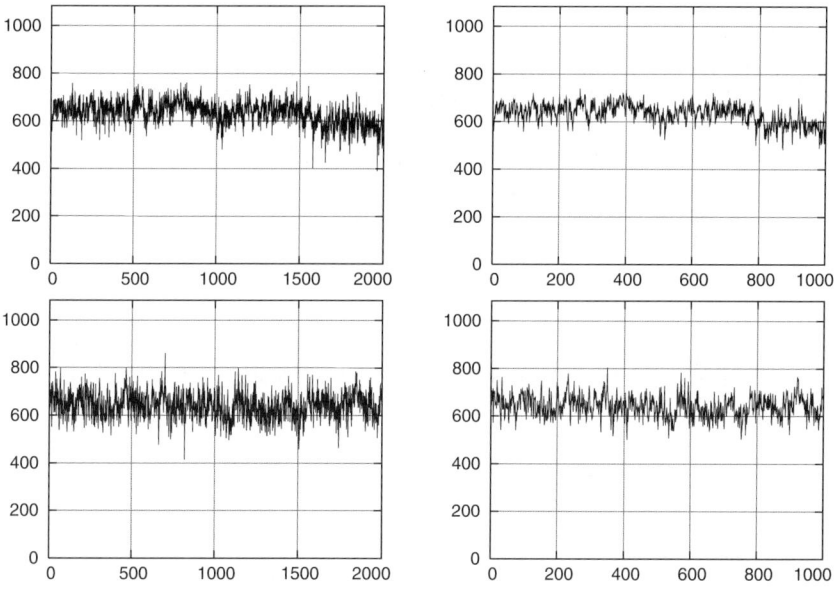

Fig. 3.10. Measured traffic (above) compared to simulated traffic according to the Norros model (below), scaling $t = 1000$ ms (left) and $t = 2000$ ms (right)

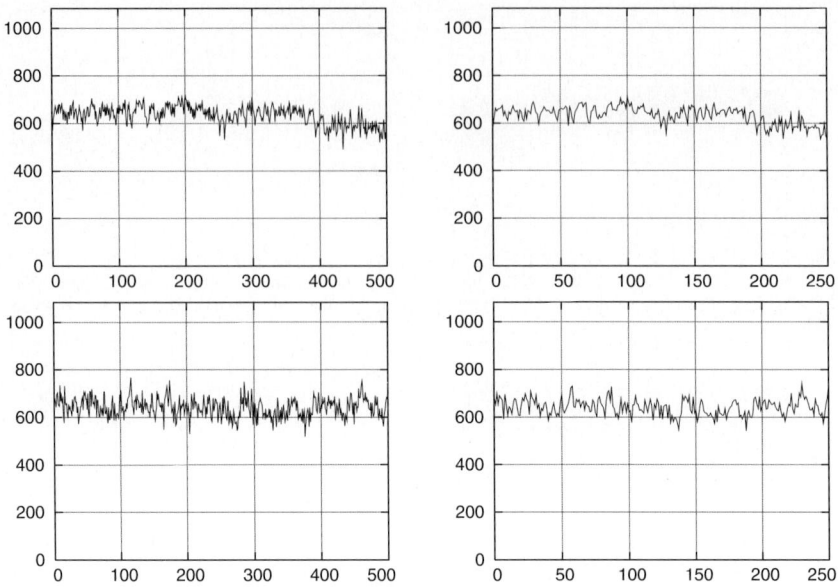

Fig. 3.11. Measured traffic (above) compared to simulated traffic according to the Norros model (below), scaling $t = 4000$ ms (left) and $t = 8000$ ms (right)

3.4 Influence of Heavy-Tail Distributions on Long-Range Dependence

3.4.1 General Central Limit Theorem

We already introduced the α-stable distributions and α-stable processes. In this section we give a derivation and motivation for the α-stable distribution. We shortly return and remind to the central limit theorem:

> For a sequence of iid random variables (ξ_k) and its corresponding partial sums $(S_n) = (\sum_{k=1}^{n} \xi_i)$ hold $\sigma^2 = \mathrm{Var}(\xi_1) < \infty$
>
> $$\frac{1}{\sigma \sqrt{n}} (S_n - n\mathbb{E}(\xi_1)) \longrightarrow \mathcal{N}(0, 1) \text{ in distribution}$$

As seen in the above examples we cannot assume a finite variance $\mathrm{Var}(\xi_1) < \infty$ for the amount of transmitted data in general. How do we have to change the central limit theorem? Closely related to this fact is the characterization of those distributions G, for which hold:

> For all random variable ξ, ξ_1, ξ_2, distributed according to G, and for all number $c_1, c_2 \geq 0$, there are numbers $b(c_1, c_2) > 0$ and $a(c_1, c_2)$, such that

3.4 Influence of Heavy-Tail Distributions on Long-Range Dependence

$$c_1\xi_1 + c_2\xi_2 \stackrel{d}{=} b(c_1,c_2)\xi + a(c_1,c_2) \qquad (3.42)$$

in distribution.

The left side is the convolution of two with c_i multiplied random variables. E.g. the convolution of two Poisson distributed random variables are again Poisson distributed. Nevertheless (3.42) is not true for the Poisson distribution.

Theorem 3.42. *A distribution G is stable, if and only if for independent and G-distributed random variables ξ, ξ_1 and ξ_2 the relation (3.42) holds.*

Generalizing (3.42) on n identical and independent random variables and thus, the corresponding identity (3.42) for

$$S_n = \xi_1 + \ldots + \xi_n \stackrel{d}{=} b_n \xi + a_n$$

we get solving

$$b_n^{-1}(S_n - a_n) \stackrel{d}{=} \xi$$

in distribution. This is valid for all $n \in \mathbb{N}$, i.e. convergence in distribution, since the right side stays constant. The following question comes up: Does the stable distribution exactly characterize this convergence? The answer will be a *yes*.

Theorem 3.43. *The class of stable distributions coincides with the class of those distributions for which after appropriate norming and centralizing the sequence $(b_n^{-1}(S_n - a_n))$ converges within this class in distribution.*

We formulate the major question: What for condition do the iid ξ_k for a given α-stable distribution G_α have to fulfill and which values for b_n and a_n do we have to choose, to get

$$b_n^{-1}(S_n - a_n) \stackrel{d}{\to} G_\alpha \qquad (3.43)$$

thus convergence in distribution? Suppose the random variables ξ_k are distributed according to F, then we call F in the domain of attraction of G_α, if (3.43) hold (written as $F \in DA(G_\alpha)$ or if we do not stress on the distribution G_α simply $F \in DA(\alpha)$).

We have to distinguish two cases: $\alpha = 2$ and $\alpha < 2$. For the case $\alpha = 2$ (i.e. for example if G_2 is Gaussian) follows that $F \in DA(G_2)$ if and only if for a random variable ξ, distributed according to F, one of the following cases holds:

- $\mathbb{E}(\xi^2) < \infty$ or
- $\mathbb{E}(\xi^2) = \infty$ and $\mathbb{P}(|\xi| > x) = o(x^{-2} \int_{|y| \leq x} y^2 dF(y))$ (o is the small Landau-symbol).

The case $\alpha < 2$ has to be treated differently. For this we need the notion of *slowly varying function* from classical analysis, which we treated already in section 2.7.4. In the literature (e.g [76]) one finds more concerning this topic. We give more references below.

Definition 3.44. *A function L on $[0, \infty[$ is called* slowly varying *or belonging to the class $\mathcal{R}_0$, if we have for all $t > 0$*

$$\lim_{x \to \infty} \frac{L(tx)}{L(x)} = 1$$

Suppose $F \in DA(G_\alpha)$, then $\mathbb{P}(|\xi| > x) = x^{-\alpha} L(x), (x > 0)$, with L slowly varying, and

$$\frac{x^2 \mathbb{P}(|\xi| > x)}{\int_{|y| \leq x} y^2 dF(y)} \to \frac{2 - \alpha}{\alpha}, \text{ for } x \to \infty \qquad (3.44)$$

We can formulate the following cases. Let $F \in DA(G_\alpha)$ (and ξ distributed according to F). Then

$\mathbb{E}(|\xi|^\delta) < \infty$, for $\delta < \alpha$
$\mathbb{E}(|\xi|^\delta) = \infty$, for $\delta > \alpha$ and $\alpha < 2$, in particular $\text{Var}(\xi) = \infty$
$\mathbb{E}(|\xi|) < \infty$, for $\alpha > 1$
$\mathbb{E}(|\xi|) = \infty$, for $\alpha < 1$

We are now prepared to answer the major question.

Theorem 3.45. *General central limit theorem: Let $F \in DA(G_\alpha)$ and $\alpha \in]0, 2]$:*

- *Suppose $\mathbb{E}(\xi^2) < \infty$ (i.e. the variance σ^2 exists). Then it follows with Φ the Gaussian distribution*

$$\left(\sigma n^{\frac{1}{2}}\right)^{-1} (S_n - n\mu) \xrightarrow{d} \Phi$$

- *If $\mathbb{E}(\xi^2) = \infty$ and $\alpha = 2$ hold or if $\alpha < 2$, then we get*

$$\left(n^{\frac{1}{\alpha}} L(n)\right)^{-1} (S_n - a_n) \xrightarrow{d} G_\alpha \qquad (3.45)$$

where G_α is an α-stable distribution, L a suitable slowly varying function and

$$a_n = \begin{cases} \mathbb{E}(\xi) = \mu & \text{for } \alpha \in]1, 2] \\ 0 & \text{for } \alpha \in]0, 1[\\ 0 & \text{for } \alpha = 1 \text{ and } F \text{ is symmetrical} \end{cases}$$

The norming factor according to (3.45) is give by $b_n^{-1} = (n^{\frac{1}{\alpha}} L(n))^{-1}$. Of course it is difficult to determine the function L. Thus, one is interested to find possibilities to choose $b_n = cn^{\frac{1}{\alpha}}$ instead of $b_n = (n^{\frac{1}{\alpha}} L(n))$. This is the case e.g. if $\mathbb{E}(\xi^2) < \infty$ resp. if ξ are α-stable distributed.

We introduce a notion for the general case. The random variables (ξ_k) with corresponding distribution function F lies *in the domain of normal attraction* of an α-stable distribution G_α (in sign $F \in DNA(\alpha)$), if in (3.45) $b_n = cn^{\frac{1}{\alpha}}$ can be chosen instead of $(n^{\frac{1}{\alpha}} L(n))$ ($c > 0$ suitable). The next result characterizes the distributions $F \in DNA(G_\alpha)$.

3.4.3 Heavy-Tail Distributions in On-Off Models

Simple On-Off Models

For a better understanding of the aggregated traffic and for parallel investigation of the impact of subexponential distribution we consider a simple on-off model of data transmission. We want to illustrate the above processes – FBM and α-stable processes. The model goes back to Taqqu et al. (see [264]). For modeling it starts on the application level (see figure 1.5). We have a switching of active and silent intervals at the user (sender) and receiver. During the active phase packets of fixed defined length are sent, while during the silent phase nothing happens. The structure of transmission does not enter the model – we solely assume a fixed number of users – sender and receiver. This is an alternating renewal process, which we fix in a mathematical model. The sequence of times $(T_n)_{n \in \mathbb{N}}$ is given, and forms a discrete stationary renewal process. With F we denote the distribution of the interarrival times (after the first arrival). This leads to a distribution of the variable T_1 (and hence because of the stationarity also for the other times), which has the form (remember the theorem 2.22 and theorem 2.13)

$$F_{T_1}^c(t) = \mathbb{P}(T_1 > t) = F_I^c(t),\ t \geq 0$$

where $F_I(t)$ is the integrated complementary distribution function of F, i.e.

$$F_I(t) = \frac{1}{\mu} \int_0^t F^c(x) dx$$

Assuming that there exists a first moment of the interarrival times, let $F, F_I \in \mathcal{S}$.

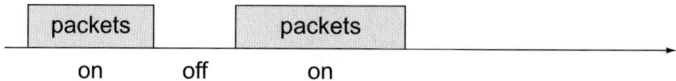

Fig. 3.12. Schema of a simple on-off model

Let $(W(t))_{t \geq 0}$ be the stochastic process defined in the following way

$$W(t) = \begin{cases} 1 & \text{if } t \text{ falls in an } \textit{on} \text{ interval} \\ 0 & \text{if } t \text{ falls in an } \textit{off} \text{ phase} \end{cases}$$

The phases alternate, $W(t)$ is a renewal process during the on-phase with $W(t) = 1$ for these t. In a similar way we have consequently a reward 0 and so on. After an 'on' interval follows an 'off' interval and vice versa. We show, that the process $(W(t))$ is long-range dependent in the sense of the definition introduced in section 3.1.2, i.e. that the auto correlation function $\mathbb{C}\mathrm{or}(W(0), W(t))$ decay more slowly than exponential.

Theorem 3.53. *Let arrival times $(T_n)_{n\in\mathbb{N}}$ be a stationary renewal process, whose interarrival times are distributed according to $F \in \mathcal{S}$ and $F_I \in \mathcal{S}$. Let $(W(t))_{t\geq 0}$ be the above defined stationary renewal process with transmission probability p_{ij}, $i,j = 0,1$ (i.e. $p_{ij} = \mathbb{P}(W(T_n) = j \,|\, W(T_{n-1}) = i)$, $n = 1,2,\ldots$) and stationary remain probability π_i, for $i = 0,1$. Then*

$$\mathbb{C}ov(W(0), W(t)) \sim \mathbb{V}ar(W(0)) F_I^c(t), \ t \to \infty$$

Note that $F^c(t) = o(F_I^c(t))$, which means that the integrated distribution shows a stronger heavy-tail behavior than the original distribution of the interarrival times. For large times t the correlation $\mathbb{C}or(W(0), W(t))$ behaves roughly speaking according to theorem 3.53 like the probability, that the on-phase, which is active at time 0, will hold on at time t. Since we assumed the heavy-tail property for F_I^c, then the autocorrelation decays consequently subexponential. We can formulate some special cases, which are especially useful for the applications, as we will discuss in the next subsection.

Corollary 3.54. *Suppose the complementary distribution function has the form $F^c(x) = x^{-\alpha} L(x)$, $x > 0$ with a slowly varying function L and $\alpha > 1$. Then we have for $t \to \infty$:*

$$\mathbb{C}ov(W(0), W(t)) \sim \mathbb{V}ar(W(0)) \frac{1}{(\alpha-1)\mu} t^{-(\alpha-1)} L(t)$$

Obviously for $1 < \alpha < 2$ the auto correlation function is no longer integrable. Hence we have long-range dependence in the stronger sense. Since by the assumption the 'on' as the 'off' phases possesses the same distribution in this model, one has to consider the approach as not consistent with the observation in telecommunication. But it is suitable for illustrating the influence of heavy-tail distributions on the long-range behavior. The model will be extended in the next subsection.

On-Off Models with Different Distribution for On and Off Phases

The non-negative iid random variables $(\xi_{\text{on}}, \xi_n)_{n\in\mathbb{N}}$ describe the on-periods, the non-negative iid $(\eta_{\text{off}}, \eta_n)_{n\in\mathbb{N}}$ the 'off' intervals. By F_{on} resp. F_{off} we denote the distribution function of the 'on' resp. 'off' periods. We assume the existence of the first moments $\mu_{on} = \mathbb{E}(\xi_{\text{on}})$ and $\mu_{\text{off}} = \mathbb{E}(\eta_{\text{off}})$ with $\mu = \mu_{\text{on}} + \mu_{\text{off}}$. The above renewal process $(W(t))$ has interarrival times, distributed as $\xi_{\text{on}} + \eta_{\text{off}}$. Thus, each renewal time is a starting point for a new 'on' phase and the interarrival times consists exactly of an 'on' and an 'off' interval. If we see an 'off' period, then the renewal time T_1 appears immediately after an 'off' phase. Otherwise if we see an 'on' interval, then T_1 appears after the next 'off' period. To describe the interval $[0, T_1[$, we introduce the random variables ξ_I, η_I and B, which are stochastic independent to the triple $(\eta_{\text{off}}, \xi_n, \eta_n)$. Here ξ_I has the integrated complementary distribution function

3.4 Influence of Heavy-Tail Distributions on Long-Range Dependence

$$F_{\text{on,I}}(x) = \frac{1}{\mu_{\text{on}}} \int_0^x F_{\text{on}}^c(t)dt$$

η_I the corresponding distribution function

$$F_{\text{off,I}}(x) = \frac{1}{\mu_{\text{off}}} \int_0^x F_{\text{off}}^c(t)dt$$

and B is a Bernoulli distributed random variable with $\mathbb{P}(B=1) = \frac{\mu_{\text{on}}}{\mu}$. Then we can describe the time T_1 in the form

$$T_1 = B(\xi_I + \eta_{\text{off}}) + (1-B)\eta_I$$

Our process $(W(t))$ can be represented as

$$W(t) = B\mathbf{1}_{[0,\xi_I[}(t) + \sum_{n=1}^{\infty} \mathbf{1}_{[T_n, T_n + \xi_{n+1}[}(t), \quad t > 0$$

where (T_n) is the sequence of the renewal times.

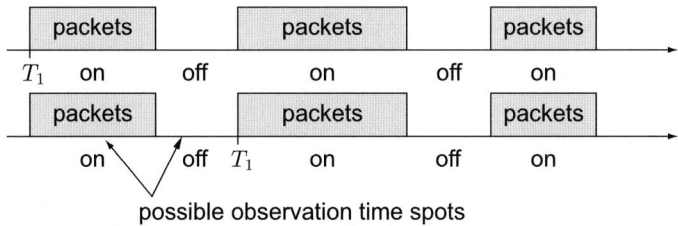

Fig. 3.13. Scheme of a simple on-off model with observation time spot

Suppose $t \geq T_1$, then we have

$$W(t) = \begin{cases} 1 & \text{for } T_n \leq t < T_n + \xi_{n+1} \\ 0 & \text{for } T_n + \xi_{n+1} \leq t < T_{n+1} \end{cases}$$

For $t \in [0, T_1[$ we get

$$W(t) = \begin{cases} 1 & \text{provided } B = 1 \text{ and } 0 \leq t < \xi_I \\ 0 & \text{else} \end{cases}$$

With this construction $(W(t))$ is stationary in the stronger sense, since $(W(t))$ inherits this property from the renewal times (T_n) (see e.g. the result for renewal processes from theorem 2.22). In addition we deduce

$$\mathbb{P}(W(t) = 1) = \mathbb{E}(W(t)) \tag{3.47}$$

$$= \mathbb{P}(B=1)\mathbb{P}(\xi_I > t) + \sum_{n=1}^{\infty} \mathbb{P}(T_n \leq t < T_n + \xi_{n+1})$$

By the elementary renewal theorem we deduce for the renewal function H_1 (note that $W(t)$ is stationary and see theorem 2.22)

$$H_1(t) = \sum_{n=1}^{\infty} \mathbb{P}(T_n \leq t) = \frac{t}{\mu}, \quad t > 0 \qquad (3.48)$$

We want to quantify the representation of $\mathbb{P}(W(t))$. The infinite sum in (3.47) can be rearranged for $t > 0$ in the form

$$\sum_{n=1}^{\infty} \int_0^t F_{on}^c(t-u) \mathbf{1}_{\{T_n \leq u\}} d\mathbb{P}(u) \stackrel{(3.48)}{=} \int_0^t F_{on}^c(t-u) dH_1(u)$$

$$= \frac{1}{\mu} \int_0^t F_{on}^c(t-u) du$$

$$\stackrel{(3.48)}{=} \frac{\mu_{on}}{\mu} F_{on,I}(t)$$

With the definition of the random variable B we deduce

$$\mathbb{E}(W(t)) = \mathbb{P}(B=1)\mathbb{P}(\xi_I > t) + \frac{\mu_{on}}{\mu} F_{on}^c(t)$$

$$= \frac{\mu_{on}}{\mu} \left(\mathbb{P}(\xi_I > t) + \mathbb{P}(\xi_I \leq t)\right) = \frac{\mu_{on}}{\mu}$$

We are ready to state the first major result of the model.

Theorem 3.55. Let $F_{on}^c(t) = t^{-\alpha} L(t)$, $t > 0$, where $L \in R_0$ and $\alpha \in\ ]1, 2[$. Furthermore let $F_{off}^c(t) = o(F_{on}^c(t))$ for $t \to \infty$ and $\xi_{on} + \eta_{off}$ not trivial (i.e. not $\mathbb{P}$-a.s. 0). Then we have

$$\mathbb{C}ov(W(0), W(t)) \sim \frac{\mu_{off}^2}{(\alpha-1)\mu^3} t^{-(\alpha-1)} L(t), \ for\ t \to \infty$$

The proof is based on deep results from renewal theory for heavy-tail distributions.

We define the aggregated input process up to time t

$$A(t) = \int_0^t W(u) du \qquad (3.49)$$

During the time interval $[0, 1]$ the traffic load $A(t)$ is streaming into a buffer or storage. Since $\mathbb{E}(W(t)) = \frac{\mu_{on}}{\mu}$, by the lay of great numbers we have

$$\frac{A(t)}{t} \longrightarrow \frac{\mu_{on}}{\mu} \text{ a.s., for } t \to \infty$$

Let r be the serving rate, where we write $r(x) = r$, if the buffer is not empty (i.e $x > 0$). This means that there is transmission data in the network. Furthermore let $r(x) = 0$, in case if the buffer is empty. For the stability of the

3.4 Influence of Heavy-Tail Distributions on Long-Range Dependence

system we need that $\frac{\mu_{\text{on}}}{\mu} < r < 1$ (otherwise the growth gets unstable). Next let $V(t)$ denote the storage process in the buffer, which can be regarded as its load at time t. It will be described be the stochastic differential equation

$$dV_t = dA(t) - r(V(t))\,dt \qquad (3.50)$$

with a random variable V_0 as initial value, choosing for $A(t)$ a general stochastic process, one can not give the equation (3.50) a rigorous definition. But we have to understand the integral in (3.49) pathwise (i.e. for each $\omega \in \Omega$) and the equation (3.50) can be considered as pathwise differential equation. The equation (3.50) means, that the change in the storage process is on the one hand determined by the change in the load process minus the serving rate, where the serving rate depends on the instantaneous amount of storage. During the 'on' period the incoming data rate of $(1-r)$ will be put into the buffer, while during the silent period data with a rate of r is released. The sequence (T_n) consists of the renewal times of the storage process $(V(t))_{t\geq 0}$, stationary ergodic. The difference of $V(T_n)$ and $V(T_{n+1})$ between the renewal times T_n and T_{n+1} can be written in the form

$$V_{T_{n+1}} = \left(V_{T_n} + (1-r)\xi_{n+1} - r\eta_{n+1}\right)^+, \quad n \in \mathbb{N}$$

The increments have the expected value of

$$\mathbb{E}\left((1-r)\xi_{n+1} - r\eta_{n+1}\right) = (1-r)\mu_{\text{on}} - r\mu_{\text{off}} = \mu_{\text{on}} - r\mu < 0$$

Thus, V_{T_n} satisfies the recursive equation

$$V_0 = 0$$
$$V_{T_{n+1}} = \left(V_{T_n} + (1-r)\xi_{n+1} - r(T_{n+1} - T_n)\right)^+$$
$$= \left(V_{T_n} + (1-r)\xi_{n+1} - r\eta_{n+1}\right)^+$$

For the asymptotic we get with the help of [11, Lemma 11.1.1]

$$V_{T_n} \xrightarrow{d} V_{T_\infty} = \sup_{n \geq 1} \sum_{i=1}^{n} \left((1-r)\xi_{i+1} - r\eta_{i+1}\right)^+$$

The stationary waiting time distribution will be denoted by π and the load rate as usual by $\rho = $ arrival rate/serving rate $= \frac{(1-r)\mu_{\text{on}}}{r\mu_{\text{off}}}$. Then we get for the waiting time distribution the following theorem.

Theorem 3.56. *Denoting by π the distribution function of V_{T_∞} and setting $\rho = \frac{(1-r)\mu_{\text{on}}}{r\mu_{\text{off}}}$ we have*

$$\pi \in \mathcal{S} \Leftrightarrow F_{\text{on,I}} \in \mathcal{S} \Rightarrow \pi^c(x) \sim \frac{\rho}{1-\rho} F^c_{\text{on,I}}\left(\frac{x}{1-r}\right), \text{ for } x \to \infty$$

Suppose we choose F^c in the form $x^{-\alpha}L(x)$, then by the theorem of Karamata (see theorem 2.49) it follows the next corollary.

Corollary 3.57. *If $F^c(x) = x^{-\alpha}L(x)$ for $\alpha > 1$ and $L \in \mathcal{R}_0$ slowly varying ($x > 0$), then we have*

$$\pi^c(x) \sim \frac{\rho}{1 - \rho\,\mu_{\text{on}}(\alpha - 1)} (1-r)^{\alpha-1} x^{-(\alpha-1)} L(x) =: b x^{-(\alpha-1)} L(x), \text{ for } x \to \infty$$

The cyclic behavior of the buffer content reaches its maximum not at the times T_n, but at $T_n + \xi_{n+1}$. The following result is hence not surprisingly. It reveals that $V(t)$ shows a larger tail as V_{T_n}.

Theorem 3.58. *Let $F^c_{\text{on}}(x) = x^{-\alpha}L(x)$, $x > 0$ for $\alpha > 1$ and L a slowly varying function. Let b as in 3.57. Then it follows $V_t \xrightarrow{d} V_\infty$. For V_∞ we get the distribution*

$$\mathbb{P}(V_\infty > x) \sim \left(b + \frac{(1-r)^{\alpha-1}}{\mu(\alpha - 1)}\right) x^{-(\alpha-1)} L(x), \ x \to \infty$$

Figure 3.14 shows an example of an (empirical) heavy-tail distribution of transmitted video files in the WWW traffic in a logarithmic scale. It reveals for an $\alpha = 1.3$, which justifies the application of the upper result in the WWW traffic.

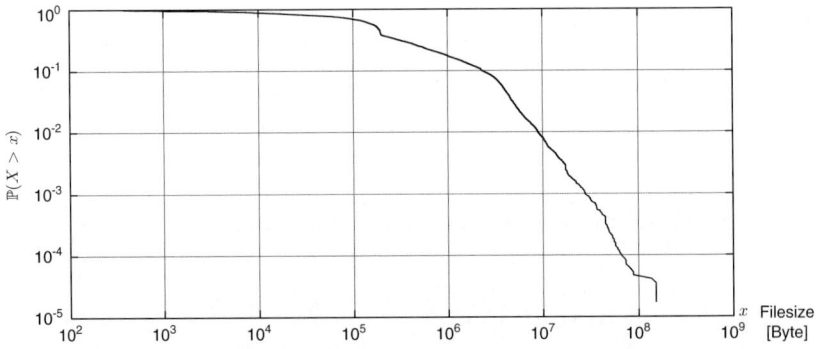

Fig. 3.14. CCDF of measured video file sizes in WWW traffic

3.4.4 Aggregated Traffic

Willinger et al. [263, 264] were the first to derive from a simple on-off model self-similarity and long-range dependence of the data traffic. We will follow their arguments and hence, consider a traffic load process $A(t)$ in more detail. The self-similar process can be regarded as superstition of many simple

3.4 Influence of Heavy-Tail Distributions on Long-Range Dependence

renewal processes with renewal values 0 or 1, where one can describe the construction in the following way. The on-off periods alternate without overlapping. Thus, we choose the following approach for the distribution of the random variables ξ_{on} and η_{off}

$$F^c_{\text{on}}(x) = x^{-\alpha_{\text{on}}} L_{\text{on}}(x), \ 1 < \alpha_{\text{on}} \leq 2 \tag{3.51}$$
$$F^c_{\text{off}}(x) = x^{-\alpha_{\text{off}}} L_{\text{off}}(x), \ 1 < \alpha_{\text{off}} \leq 2 \tag{3.52}$$

where L_{on} and L_{off} are slowly varying function. For each time the load process $A(t)$ represents the amount of the 'on' phase. Such a model can represent a network with several server, where every server transmit data with constant rate or not.

We keep in mind the results of the previous sections, where we derived a waiting queue estimation for serving resp. connection time on the basis of the classical traffic theory. But we also know that there are subexponential distribution (like the lognormal distribution), which can not be represented in the form (3.51). These distributions did not indicate long-range dependence in the M/G/∞ model. First we compute the variance the load process $A(t)$

$$v(t) = \text{Var}\,(A(t)) = \text{Var}\left(\int_0^t W(u)du\right) \tag{3.53}$$
$$= 2 \int_0^t \left(\int_0^u \gamma(v)dv\right) du$$

where
$$\gamma(v) = \text{Cov}\,(W(0), W(v)) = \mathbb{E}\,(W(v)W(0)) - \mathbb{E}\,(W(0))^2$$

describe the covariance function of $W(t)$. The last equation in (3.53) can be seen in the following way: Remembering the result from theorem 3.55, it tells us roughly speaking:

Suppose $F^c_{\text{off}}(x) = o(F^c_{\text{on}}(x))$, then for the covariance function holds

$$\gamma(v) \sim \text{const} \cdot v^{-\alpha_{\text{on}}-1} L(v), \ v \to \infty$$

with a slowly varying function L.

Extending and generalizing the result from theorem 3.55 to $A(t)$, we have

$$v(t) \sim \sigma^2 t^{2H} \tag{3.54}$$

with a constant $\sigma > 0$ and H as Hurst exponent given by

$$H = \frac{3 - \min(\alpha_{\text{on}}, \alpha_{\text{off}})}{2} \tag{3.55}$$

The proof to (3.54) resp. (3.55) is slightly complicated and for the details the reader is referred to the original literature (see [250]).

The asymptotic in (3.54) is crucial for the subsequent results. Easily seen the case $\alpha_{\text{on}} = \alpha_{\text{off}} = 2$ implies a Hurst parameter $\frac{1}{2}$, what immediate leads to the Brownian motion in the case of Gaussian processes. These results are valid for just one user. So we ask, what will happen in the case of arbitrary many? We investigate several asymptotics, i.e. the behavior for large time resp. averaging over large time intervals and over large number of users. The results reflect an asymptotic behavior, which means, we get an *asymptotic self-similarity*. For this let M be the number of users and respectively $W^{(m)}(t)$ the load at time t for the m-th user. If we assume an identical distribution for the single user (homogeneous sources), then we receive for the variance $\text{Var}(W^{(m)}(t)) = \text{Var}(W(t))$ of the m-th user. First we increase the number of users resp. applications. Since we do not observe any influence of time, we can apply the central limit theorem. Then we get for the convergence in distribution according to the central limit theorem (note that it holds $\text{Var}(W(t)) < \infty$)

$$\lim_{M \to \infty} \frac{1}{M^{\frac{1}{2}}} \sum_{m=1}^{M} \left(W^{(m)}(t) - \mathbb{E}(W^{(m)}(t)) \right) = G(t), \ t \geq 0$$

Here, $G(t)$ is a Gaussian random variable with mean 0, i.e. $(G(t))_{t \geq 0}$ is a centralized Gaussian process. It is stationary, since $W^{(m)}(t)$ is stationary for all $m \in \mathbb{N}$, and for the covariance function we deduce $\gamma(t)$ (identical distribution!). We continue with the time scaling. For this we stretch the time by the factor $n > 0$, as shown in the figure in section 3.1.2. For the derivation we choose a value $1 < \alpha \leq 2$ and compute the normalized integral

$$n^{-\frac{1}{\alpha}} \int_0^{nt} G(u) du$$

What value does α actual have? For the answer we transform

$$\int_0^{nt} G(u) du = \int_0^{nt} \lim_{M \to \infty} \frac{1}{M^{\frac{1}{2}}} \sum_{m=1}^{M} \left(W^{(m)}(u) - \mathbb{E}\left(W^{(m)}(u) \right) \right) du$$

$$= \lim_{M \to \infty} \frac{1}{M^{\frac{1}{2}}} \int_0^{nt} \sum_{m=1}^{M} \left(W^{(m)}(u) - \mathbb{E}\left(W^{(m)}(u) \right) \right) du$$

$$= \lim_{M \to \infty} \frac{1}{M^{\frac{1}{2}}} \sum_{m=1}^{M} \left(\int_0^{nt} W^{(m)}(u) - \mathbb{E}\left(W^{(m)}(u) du \right) \right)$$

$$= \lim_{M \to \infty} \frac{1}{M^{\frac{1}{2}}} \sum_{m=1}^{M} \left(A(nt) - \mathbb{E}\left(A(nt) \right) \right)$$

The process $(A(t))$ has a covariance function of the form σt^{2H}. To get convergence in the distributive sense and that the limit process reveals the corresponding covariance structure the value has to be $\alpha = \frac{1}{H}$. Hence, it follows

3.4 Influence of Heavy-Tail Distributions on Long-Range Dependence

$$\lim_{n\to\infty} \frac{1}{n^H} \int_0^{nt} G(u)du = \sigma B_H(t), \ t \geq 0 \tag{3.56}$$

Here the process $\sigma B_H(t)$ possesses Gaussian marginal distributions, stationary increments (since this is true for the integral) and the variance can be written in the form σt^{2H}. But this meets exactly the definition of the fractional Brownian motion with Hurst exponent H (see in addition theorem 3.16).

The result tells us that by appropriate normalization, assuming the number of users tending to infinity and finally averaging over large time intervals, then the load process converges towards a FBM with Hurst parameter H. What happens, if one switches the order of convergence, i.e. first averaging over the time intervals (hence for each single user) and then tending the number of users towards infinity? Then the results looks different. Let's consider it more closely. For this we choose $\alpha = \min(\alpha_{on}, \alpha_{off})$ and deduce for the convergence in distribution

$$\lim_{M\to\infty} \lim_{n\to\infty} \frac{1}{M^{\frac{1}{\alpha}}} \frac{1}{n^{\frac{1}{\alpha}}} \int_0^{Tt} \sum_{m=1}^M \left(W^{(m)}(u) - \mathbb{E}\left(W^{(m)}(u)\right)\right) du = \mathrm{const} \cdot S_\alpha, \ t \geq 0$$

where $(S_\alpha(t))_{t\geq 0}$ is an α-stable Lévy process (or an α-stable motion) with in general infinite variance.

Expressing the result concerning the load for all users, thus, $A^{(M)}(t) = \int_0^t \sum_{m=1}^M W^{(m)}(u)du$, so we get the different convergences in distribution

$$\lim_{n\to\infty} \lim_{M\to\infty} \frac{A^{(M)}(nt) - \mathbb{E}\left(A^{(M)}(nt)\right)}{n^H \sqrt{M}} = \sigma B_H(t), \ t \geq 0$$

and

$$\lim_{M\to\infty} \lim_{n\to\infty} \frac{A^{(M)}(nt) - \mathbb{E}\left(A^{(M)}(nt)\right)}{n^H \sqrt{M}} = \mathrm{const} \cdot S_\alpha(t), \ t \geq 0$$

A more detailed derivation can be found in [264].

What happens, if one performs the limits simultaneously? For this we choose a function $M(n)$ with values in the natural numbers, not decreasing and tending towards ∞ for $n \to \infty$. There are two cases to distinguish:

- If $\lim_{n\to\infty} \frac{M(n)}{n^{\alpha-1}} = \infty$ (e.g. for $M(n) = n^\kappa$, $\kappa > \alpha - 1$), then the number of users grows faster than the width of the time windows. We get

$$\lim_{n\to\infty} \frac{A^{(M(n))}(nt) - \mathbb{E}\left(A^{(M(n))}(nt)\right)}{n^H \sqrt{M(n)}} = \sigma B_H(t)$$

- Considering the other extreme, i.e. $\lim_{n\to\infty} \frac{M(n)}{n^{\alpha-1}} = 0$ (e.g. for $M(n) = n^\kappa$, $\kappa < \alpha - 1$), then we stretch the time interval faster than the number of users increases, we conclude

$$\lim_{n\to\infty} \frac{A^{(M(n))}(nt) - \mathbb{E}\left(A^{(M(n))}(nt)\right)}{n^H \sqrt{M(n)}} = \mathrm{const} \cdot S_\alpha(t)$$

Of course we ask, in which way this model reflects the reality resp. in which situations this model can be applied best.

On-Off Models with Poisson Arrivals – Embedding into M/G/∞ Approach

The model based on Cox assumes a infinite source. With an intensity λ the sources send data according to Poisson arrival. The distribution F_{on} describes the transmission time and is heavy-tailed according to

$$F_{on}^c(x) \sim cx^{-\alpha}, \quad \alpha \in \,]1,2[$$

Thus, we have an average transmission time μ_{on}, without finite variance (since $\alpha < 2$). The single transmissions of the sources are iid. After the transmission each source is silent. This model can be used to describe simple applications. For this we attach to each application its transmission time, which enters the network according to a Poisson arrival process and which is send through the network according to a heavy-tail distributed connection time. This model is in a certain sense equivalent to the one from section 3.4.4, because the rate λ corresponds to the limit $M \to \infty$. Remembering the derivation of the Poisson distribution, the relationship is evident. In correspondence to section 3.4.4 let $W(u)$ be the amount of active sources at time $u > 0$. Then we define

$$A(t) = \int_0^t W(u)du \qquad (3.57)$$

as aggregated load at time t. It is clear that $A(t)$ depends on the arrival process and thus on λ. If one generally denotes with $\lambda(T)$ the aggregated arrival rate up to time T, then one has to require first

$$\lim_{T \to \infty} \frac{\lambda(T)}{T^{\alpha-1}} = \infty$$

and one has the convergence in distribution

$$\lim_{n \to \infty} \frac{A(nt) - \mathbb{E}(A(nt))}{\sqrt{\lambda T^{\frac{3-\alpha}{2}}}} = C_1 B_t^{(H)}, \quad t \geq 0 \qquad (3.58)$$

where $(B_t^{(H)})$ is a fractional Brownian motion with Hurst exponent $H = \frac{3-\alpha}{2}$. The arrival rate at time T is $\lambda(T)$ (hence, not stationary), and we can consider this value as expected value of the intensity of the counting process $N(t)$, i.e., $\lambda(T) = \frac{\mathbb{E}(N(T))}{T}$. Thus, comparing with the renewal theorem 2.15 the condition (3.57) is a weighting growing condition for the interarrival times.
But if we have

$$\lim_{T \to \infty} \frac{\lambda(T)}{T^{\alpha-1}} = 0$$

then we can deduce again the convergence in distribution

$$\lim_{n\to\infty} \frac{A(nt) - \mathbb{E}\left(A(nt)\right)}{\sqrt{\lambda T}^{\frac{3-\alpha}{2}}} = C_2 S_\alpha(t), \ t \geq 0$$

with an α-stable motion. According to the theorem of Little in Proposition 2.5 we get $\mathbb{E}(A(t)) = \mu_{on}\lambda t$. This convergence (3.58) was first deduced by Kurtz [152], while the treatment of the cases is due to [181].

The representations in the section 3.4.4 serve showing that using the classical approaches in traffic theory and the on-off model we get in the limit self-similar processes exhibiting long-range dependence. Thus, for modeling the IP traffic one immediately uses these processes as starting point. But one needs for consideration of the serving times and the physical structure of the traffic the approach of the classical traffic or the on-off models. Without them a satisfaction a use of these processes would be unthinkable.

3.5 Models for Time Sensitive Traffic

3.5.1 Multiscale Fractional Brownian Motion

With this section we start transferring the Norros model with the Brownian motion as driving process to a multiscale view. Later we go on to the multifractal picture.

Trying to model the time sensitive traffic, one faces clear presumption to guaranty a necessary minimal quality. Though at this stage we do not care for performance, this model of multiscale Brownian motion can give some good insight, how a model can be applied for maintenance resp. improvement of the QoS of the transmitted data and services. For this a necessary lower bound D_{req} will be introduce, which describes the maximal allowed delay. Simultaneously we indicate by $d^{i,j}$ the delay between the POPs i and j. This represents a random variable, computed from the model. Afterwards we define a *bound* or *threshold* $\epsilon > 0$, resulting from the necessity and QoS requirements and stress

$$\mathbb{P}\left(d^{i,j} > D_{\text{req}}\right) < \epsilon \qquad (3.59)$$

The condition (3.59) describes the required quality.

In section 1.1 we already mentioned the modeling and dimensioning in a IP network and remarked the two fundamental time scalings. Here, we present a approach due to Diot, Fraleigh and Tobagi [95], for going a first step towards modeling on *several* scales. The mentioned model, as a example, is based on the approach due to Norros, which we dealt with before. Thus, we use the notations from section 5.1. The queueing length at time $t = 0$ is given by

$$Q_0 = \sup_{t \geq 0}(A_t - Ct)$$

where C is the constant capacity in the network. The probability that the queue surpasses a length x, can be computed according to

$$\mathbb{P}(Q_0 > x) = \mathbb{P}\left(\sup_{t \geq 0}(A_t - Ct) > x\right)$$

which turned out to be difficult. Hence, we have derived in section 5.1.1 an lower estimation

$$\mathbb{P}(Q_0 > x) = \mathbb{P}\left(\sup_{t \geq 0}(A_t - Ct) > x\right) \qquad (3.60)$$
$$\geq \sup_{t \geq 0} \mathbb{P}\left((A_t > x + Ct)\right)$$

This estimation is kind of rough, but as seen in 5.1.1, it turns out well in logarithmic scale for large values of x. Computing the value b (e.g. in byte) the waiting time for an IP packet, we get the amounts:

- $\frac{x}{C}$ as aggregated time of the waiting queue and
- $\frac{b}{C}$ as serving time in the network.

The distribution of the stationary waiting time $\mathbb{W}_\infty$ can be determined immediately from the waiting queue distribution

$$\mathbb{P}(\mathbb{W}_\infty > d) = \sup_{t \geq 0} \mathbb{P}(A_t > C(t+d))$$

Since we are not interested in the serving time and since it is significant smaller than the waiting time, we do not care about it any longer. In figure 3.15 the empirical distribution function of the averaged traffic was computed (from time intervals 100 ms until $10{,}000{,}000$ ms). The x-axis represents the corresponding data rate in Mbit/s. We have an OC-3 connection with a data rate of 155 Mbit/s. Each IP packet has a maximal volume of $1{,}500$ byte. Thus, the pure transmission time for a single IP packet with maximum length over an OC-3 connection takes only 77 μs. We realize that the empirical distribution function resembles a Gaussian distribution (in particular the figures for $1{,}000$ and $10{,}000$ ms).

The correlation of the FBM $\gamma(t)$ is, as known, decaying with scale t^{2H-2} (see theorem 3.32). As we discussed before, we can model the data traffic with the help of the one dimensional FBM very well on large scales. The multifractal wavelet model respectively the cascade model from sections 3.8.2 and 3.8.5 try to incorporate both scales, the large and the small scales. Since these models are introduced for the analysis on small scales, and for large networks this distinction is not necessary needed, we will proceed with an simplified multiscale model.

In figure 3.16 the variance of the averaged traffic over different scales t is shown. As already seen in figure 3.15, there is an edge between the time scale of 100 ms and the other, which represents a qualitative difference. Thus, it is understandable in figure 3.16 that the edge in the curve appears in the estimated traffic at that point where the influence of the protocol enters. This motivates the approach to use not only a single Hurst parameter H, but to distinguish at least two sales. We will present the analytic part in the sequel.

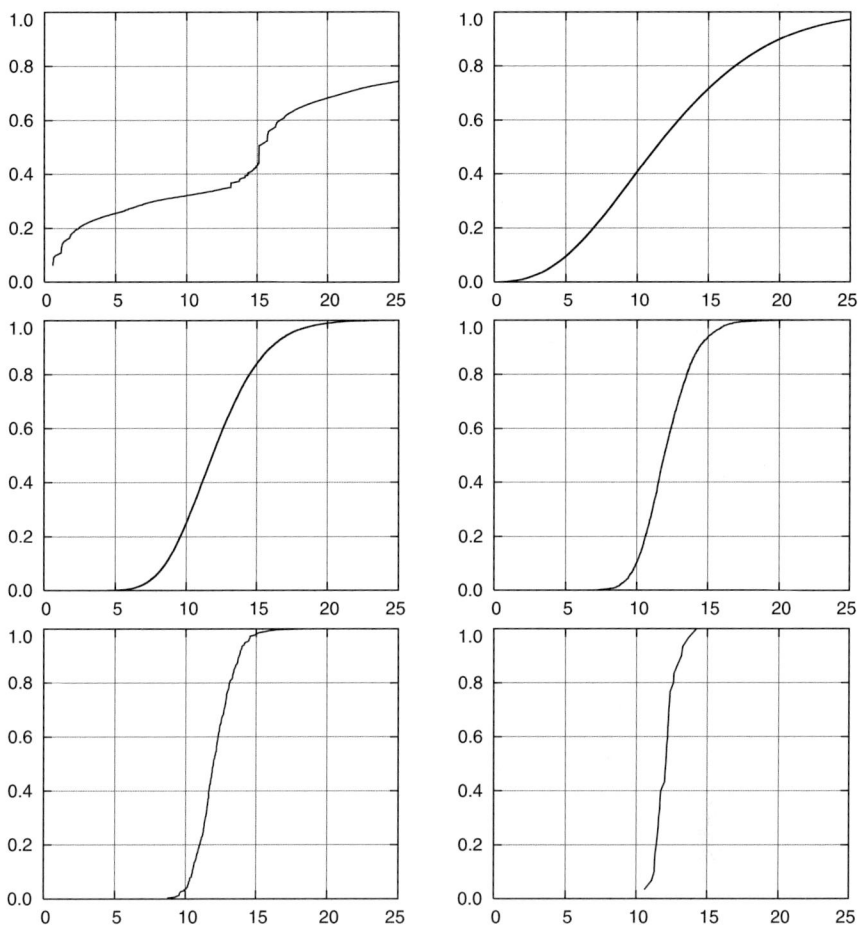

Fig. 3.15. Empirical distribution of data amounts in the time range from 16:00 to 16:05 for the different scalings $t = 10^2$, 10^3, 10^4, 10^5, 10^6, and 10^7 ms (from above left to bottom right)

(M_K)-Fractional Brownian Motion

The multiscale-FBM is an extension of the standard FBM, where one does not have a fixed Hurst parameter H, but where one introduces several time dependent exponents into the model. For this let H_0 be the Hurst parameter for large times and H_1 for small time scales. It is possible to achieve a similar description using different FBM approaches; we will come back to this issue, when dealing with multifractal models (see also the remark at the end of the section). This model, described here, has the decisive advantage of stationary increments, while this property is not fulfilled by other approaches, as we will see later. There, only locally the FBM is reflected, with its properties of

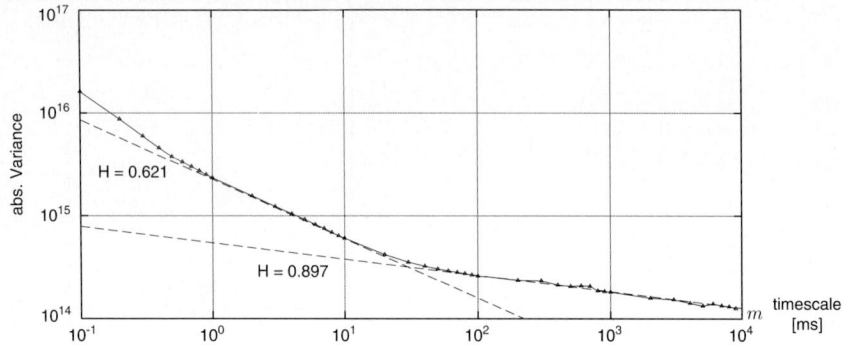

Fig. 3.16. Variance of the aggregated traffic over different scales according to the traffic from figure 3.15

self-similarity and stationary increments. This suffices to recapture the basic results for H-sssi processes from section 3.1.2 as seen later in section 3.8.1.

Description of the Model

We use the representation of an FBM by its Fourier transformation:

$$B_H(t) = \int_{-\infty}^{\infty} \frac{e^{i\omega t} - 1}{C(H)|\omega|^{H+\frac{1}{2}}} d\hat{B}(\omega)$$

where $dB(\omega)$ is the Wiener measure, $d\hat{B}(\omega)$ its Fourier transform (see 3.19) and $C(H) = (\frac{\pi}{H\Gamma(2H)\sin(H\pi)})^{\frac{1}{2}}$ (see the representation (3.12) from section 3.2.2). The (M_K)-Fractional Brownian Motion $((M_K)$-FBM) is the generalization of this representation, where H is a function of time resp. the inverse of its frequency. We have to stress that other than in section 3.8.1 we used the Fourier transform, where directly the definition of the FBM enters. Thus we define the (M_K)-FBM process (X_η) by

$$X_\eta(t) = \int_{-\infty}^{\infty} \frac{e^{i\omega t} - 1}{\eta(\omega)} d\hat{B}(\omega), \quad -\infty < t < \infty$$

We have:

- For $i, \ldots, K$ ($K \in \mathbb{N}$ represents the number of the Hurst parameter) let $\omega_i, a_i, H_i \in \mathbb{R}_+ \times \mathbb{R}_+ \times]\frac{1}{2}, 1[$, such that

$$\eta(\omega) = \frac{C(H_i)|\omega|^{H_i+\frac{1}{2}}}{\sqrt{a_i}}$$

with $\omega_i \leq \omega < \omega_{i+1}$ and the property $0 = \omega_0 < \omega_1 < \ldots < \omega_K < \omega_{K+1} = \infty$.

- In addition, $\eta(-\omega) = \eta(\omega)$ so that the integral can be defined for $\omega < 0$.

Using the Fourier transform the small times correspond to large frequencies, and vice versa one maps higher time scales to small frequencies. This defined process has Gaussian marginal distributions and stationary increments (see. [226]). We get for the variance at time difference $\delta > 0$

$$\operatorname{Var}(\delta) = \mathbb{E}\left((X_\eta(t+\delta) - X_\eta(t))^2\right)$$
$$= 4 \sum_{j=0}^{K} \delta^{2H_j} \frac{a_j}{C(H_j)^2} \int_{\delta\omega_j}^{\delta\omega_{j+1}} \frac{1 - \cos u}{u^{2H_j+1}} du$$

3.5.2 Norros Models for Differentiating Traffic

The priority traffic is only considered here as compounded traffic and separated in individual flows, though in routers the single flows are separately determined and classified according to the priority classes. The priority traffic is built up in different classes, which are adjoined a special QoS. In addition, this can be implemented if there is an admissible description.

We use as stochastic driving process the FBM, i.e. the Gauß, since we may assume that the traffic is approximative Gaussian distribution according to the central limit theorem. Within the elastic traffic the queueing length analysis is not important, since TCP imposes a strong influence on the end-to-to connections. Differently we detect the situation for the time sensitive traffic. There, the queueing is decisive and for maintaining the QoS its understanding is crucial. In the priority traffic the FIFO queueing are of no meaning. Thus, it is important, how the queueing behave, provided the incoming traffic is Gaussian.

We start with the description of the fundamental model according to Norros:

a) We consider n different priority classes, which are classified according to its importance: Class 1 has highest priority, class n the lowest.
b) We consider a link and denote by $(A_t^{(j)})_{t\in\mathbb{R}}$ the aggregated traffic on this link, belonging to calls j. We simple set $A_0^{(j)} = 0$.
c) The increment, i.e. the additional incoming traffic in the interval $]s,t]$, is computed according to $A^{(j)}(s,t) = A_t^{(j)} - A_s^{(j)}$, $-\infty < s < t < \infty$.
d) Mixing different classes i, j, we obtain accordingly

$$A_t^{(i,j)} = A_t^{(i)} + A_t^{(j)}$$

e) We assume that the particular processes $(A_t^{(j)})_{t\in\mathbb{R}}$ are stochastic independent. Every process $A_t^{(j)}$ is modeled as continuous Gaußprocess with stationary increments. This lead to the expression

$$A_t^{(j)} = m_j t + X_t^{(j)}, \quad \operatorname{Var}(X_t^j) = \gamma_j(t)$$

where $X_t^{(j)}$ are centralized Gaussian process with covariance function (thus an FBM)

$$\text{Cov}\left(X_s^{(j)}, X_t^{(j)}\right) = \gamma(s,t) = \frac{1}{2}\left(\gamma_j(s) + \gamma_j(t) - \gamma_j(s-t)\right)$$

To exclude certain pathological cases, we assume

$$\text{There is an } \alpha \in \,]0,2[: \lim_{x \to \infty} \frac{\gamma_j(x)}{x^\alpha} = 0, \text{ for all } i = 1, \ldots, n$$

Here, m_j is the mean data rate. We may not assume $m_j = 0$, since each class enjoys a different load rate. For the Gauß-traffic the values m_j and γ_j, are easy to estimate.

f) By C we denote the link capacity.

Inductively we define the queueing process and start with class 1

$$Q_t^{(1)} = \sup_{s \le t}\left(A^{(1)}(s,t) - C(t-s)\right)$$

Combining class 1 and class 2, we obtain

$$Q_t^{(1,2)} = \sup_{s \le t}\left(A^{(1,2)}(s,t) - C(t-s)\right)$$

Thus, the waiting queue for class 2 can be written according to

$$Q_t^{(2)} = Q_t^{(1,2)} - Q_t^{(1)}$$

Generally we may say

$$Q_t^{(j+1)} = Q_t^{(1,\ldots,j+1)} - Q_t^{(1,\ldots,j)}$$

We sum up all classes to

$$Q_t = \sup_{s \le t}\left(\sum_{j=1}^n A^{(i)}(s,t) - C(t-s)\right)$$

Note that modeling using Gaussian processes the waiting queue can attain negative values – an undesired effect, which do not have a decisive influence on the further analysis. Nevertheless we can avoid this effect by using the discrete time approach for describing the traffic, to keep the individual waiting queue from becoming negative.

The virtual queueing time of class j at time t will be denoted by $V_t^{(j)}$. This is the time τ, necessary for a unit entering the system at time t leaving the system at time $t+\tau$. We can describe $V_t^{(j)}$ differently. It is the smallest τ, such that the whole buffered traffic of higher or equal priority as j and the new incoming traffic in interval $[t, t+\tau]$ will be transmitted with higher priority than j before $t+\tau$. In addition

$$V_t^{(1)} = \frac{Q_t^{(1)}}{C}, \quad V_t^{(j+1)} = \inf\left\{s \geq 0; Q_t^{(1,\ldots,j+1)} + A^{(1,\ldots,j)}(t, t+s) \leq Cs\right\}$$

For technical reasons we need the space of paths

$$C_w(\mathbb{R}) = \left\{f: \mathbb{R} \longrightarrow \mathbb{R}; f \text{ is steady}, f(0) = 0, \lim_{x \to \pm\infty} \frac{f(x)}{1+|x|}\right\}$$

On $C_w(\mathbb{R})$ we define a suitable norm

$$\|f\|_w = \sup\left\{\frac{f(x)}{1+|x|}; x \in \mathbb{R}\right\}$$

With this $(C_w(\mathbb{R}), \|\cdot\|_w)$ will be a separable Banach space (see e.g.[74]). Simultaneously we fix the probability space

$$\Omega = C_w^n(\mathbb{R}) \quad \text{the } n\text{-times cartesian product}$$

and

$$\mathcal{F} = \mathbb{B}(\Omega) \quad \text{the Borel } \sigma\text{-algebra on } C_w^n(\mathbb{R})$$

(see [74]). Furthermore we choose the probability measure $\mathbb{P}$ as unique measure on Ω in such a way that the random variable $X_t^{(j)}(f_1, \ldots, f_n) = f_j(t)$ $j = 1, \ldots, n$, define independent Gaussian processes with covariance function $\gamma_j(\cdot, \cdot)$. With other words we choose a projection on the j-th component of the Gaussian margibnal distribution with given covariance matrix. This is as well known uniquely determined (see e.g. [38, 98]). The necessary probability reads as

$$\lim_{t \to \infty} \frac{X_t^{(j)}}{t} = 0 \text{ a.s.}$$

(due to the definition of the space $C_w(\mathbb{R})$).

If we want to obtain informations for the queueing time $\{Q_t > x\}$, i.e., on the probability that the buffer containment lies above a treshold x, we transform the set to

$$\{Q_t > x\} = \left\{\sup_{s \leq t} \frac{X_t - X_s}{x + (C-m)(t-s)} > 1\right\} \tag{3.61}$$

The process

$$Y_s^{(x,t)} = \frac{X_t - X_s}{x - (C-m)(t-s)}$$

is a centralized Gaussian process. Hence, (3.61) is equivalently with determining the marginal distribution of the supremum $\sup_{s \leq t} Y_s^{(x,t)}$ of a centralized Gaussian process. Numerical results are complicated, as we will treat and see in the section 5.1.1. But as there we can find a lower bound

$$\mathbb{P}(Q_t > x) \geq \sum_{s \leq t} \mathbb{P}\left(Y_s^{(x,t)} > 1\right) = \Psi^c\left(\frac{x + (C-m)t^*}{\sqrt{\gamma(t^*)}}\right)$$

where Ψ^c is the complemented distribution function of the standard Gaussian distribution and $t^* > 0$ is chosen such that the expression

$$\frac{(x + (C_m)t)^2}{\gamma(t)}, \quad t > 0$$

will be minimized ($t^* = t_x^*$ depends on x). The value t^* characterizes the time in question that a waiting queue of length x can be observed. Building the logarithm we have the approximation

$$\mathbb{P}(Q_t > x) \approx \exp\left(-\frac{((x - C - m)t^*)^2}{2\gamma(t^*)}\right) \tag{3.62}$$

In simulations it can verified that (3.62) is also an upper bound, without finding a rigorous mathematical proof. Another possibility consists in approximating a discrete waiting queue $(Q_t)_{t \in \mathbb{N}}$ by appropriate multiplication with a scalar p, in such a way that

$$p \cdot \lim_{x \to 0+} \exp\left(-\frac{((x - C - m)t^*)^2}{2\gamma(t_x^*)}\right) \approx \mathbb{P}(Q_t > x)$$

There are good approximations for a discrete waiting queue not being empty

$$\mathbb{P}(Q_t > 0) \approx 2\mathbb{P}(A_\delta > C\delta)$$

($\delta > 0$ is the length of a time interval). We have to keep an eye on the geometry of the set $\{Q_t > x\}$. It is the union of the sets of the form $\{A(s,t) - C(t-s) > x\}$ over s. This in turn are half spaces and thus, the complement of a convex sets, containing the origin.

First we want to look at the case, how the waiting queue of size x builds up in our model. The waiting queue principle does not influence this at this stage since we have a queueing system and the over all traffic is not divided in classes. First we realize how we detect the most probable paths.

Proposition 3.59. *The most probable path vector f_x^* in $\{Q_0^{(1,\ldots,k)} \geq x\}$ has the form*

$$-\frac{x + (C-m)(-t_x)}{\sum_{j=1}^k \gamma_j(t_x)} (\gamma_1(t_x, \cdot), \ldots, \gamma_k(t_x, \cdot))$$

where t_x minimizes the expression

$$w(t) = \frac{(x + (C-m)(-t))^2}{\sum_{j=1}^k \gamma_j(t)}$$

3.5 Models for Time Sensitive Traffic

Let us briefly discuss the notation of the most probable path vector. Basis is the principle of largest deviation for Gaussian distribution in Banach spaces (see e.g. [44] and [191]). As e.g described in [74], we can find for the autocovariance functions $\gamma_i(\cdot, \cdot)$ of the particular Gaussian processes $(X_t^{(i)})$ Hilbert spaces $\mathcal{H}_i \subset C_w(\mathbb{R})$. The cartesian product $\mathcal{H} = \mathcal{H}_1 \times \cdots \times \mathcal{H}_n$ will be equipped with the scalar product

$$\langle (g_1, \ldots, g_n), (h_1, \ldots, h_n) \rangle_{\mathcal{H}} = \sum_{i=1}^{n} \langle g_i, h_i \rangle_{\mathcal{H}_i}$$

to define again a Hilbert space $\mathcal{H} \subset C_w^n(\mathcal{R})$. In our situation we define for $\omega \in \Omega$, i.e., $\omega = (g_1, \ldots, g_n) \in C_w^n(\mathbb{R})$:

$$I(\omega) = \begin{cases} \frac{1}{2} \|\omega\|_{\mathcal{H}}^2 & \text{for } \omega \in \mathcal{H} \\ \infty & \text{else} \end{cases}$$

The significance lies in the statement

$$\text{if } F \subset \Omega \text{ is closed:} \quad \limsup_{k \to \infty} \log \mathbb{P}\left(\frac{X}{\sqrt{n}} \in F\right) \leq - \inf_{\omega \in F} I(\omega)$$

$$\text{if } G \subset \Omega \text{ is open:} \quad \liminf_{k \to \infty} \log \mathbb{P}\left(\frac{X}{\sqrt{n}} \in G\right) \geq - \inf_{\omega \in G} I(\omega)$$

To provide a possible most sharp estimation we have to find a path ω, minimizing I (because of inf). This path will be called the *most probable path*. We can consider $\exp(-I(\omega))$ as probability density (similar to the exponential distribution) with respect to $\mathbb{P}$. Hence, the most probable path means maximizing this density. The interested reader is referred to the literature for more information [9, 176, 266].

The most probable path vector, minimizing $I(\omega)$ in the set

$$\left\{ A^{(1,\ldots,k)}(t,0) \geq y \right\}, \ t < 0 \text{ and } y > -mt$$

can be expressed as conditional expectation

$$\mathbb{E}\left((X_s^{(1)}, \ldots, X_s^{(k)}) | A^{(1,\ldots,k)}(t,0) = y \right)$$

The reason for this is based on the fact that the conditional distribution of a Gaussian random vector w.r.t. a linear condition is gain Gaussian and the expectation equals the value, where the density attains its maximum. Figure 3.17 illustrates the most proabale vector f^* graphically.

Example 3.60. We choose $C = 1$, $m_1 = 0.2$, $m_2 = 0.4$, $\gamma_1(t) = |t|^{1.7}$, and $\gamma_2(t) = [t]_1[1 - t]_1$, where $[t]_1 = t \mod 1$. Thus, the first class is driven by a FBM, while the second is a periodical Brownian motion. In this case the minimizing function $w(t)$ is not convex and thus, t_x is not unique.

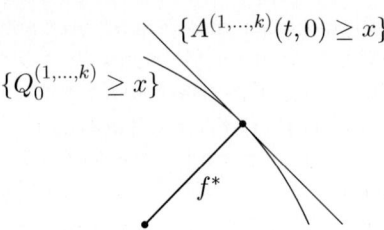

Fig. 3.17. Scheme of the most probable path vector as the perpendicular on the tangent at the circle

Now we turn to the case of an empty buffer and want to determine the waiting queue. In different articles it was stated that the waiting queue in a two class system can be approximated by the one of the lower class 2. This can be very applied for thso called processor sharing models. If the most probable path vector f_x^* is determined according to proposition 3.59, then we set $Q_0^{(1)}(f_x^*) = 0$ and get

$$\mathbb{P}\left(Q^{(2)} \geq x\right) \approx \mathbb{P}\left(Q^{(1,2)} \geq x\right) \approx \exp\left(-\frac{1}{2}\|f_x^*\|_R^2\right) = \exp\left(-\frac{1}{2}h(t_x)\right)$$

Tough in the analytical approach it is often difficult to prove whether the vector corresponding to x is the most probable, we can determine approximately for two classes. We call this approach the *coarse empty link approximation(ELA)*. For this we set

$$f_{x,1}^*(t_x, 0) + m_1|t_x| \leq C\mu_1|t_x| \tag{3.63}$$

i.e., that the most probable incoming rate of class 1 is smaller than the guaranteed part in the interval $[t_x, 0]$. For this $\mu_1 = 1$ is defined for the priority class 1.

We can deduce the following interesting facts. As usual we assume a system with two priority classes. Without loss of generality let $m_1 = 0$. The condition (3.63) reads as

$$\frac{x}{t_x} - m_2 \leq \frac{\gamma_2(t_x)}{\gamma_1(t_x)}C$$

In particular we have $m_2 \geq \frac{x}{t_x}$, where on the other hand t_x depends on the presumptions. If the variance of the particular load processes has simple structure, e.g. $\gamma_i(t) = a_i\gamma(t)$, $i = 1, 2$ with a fixed function γ, then t_x is independent of the constants a_i. If m_2 surpasses a certain threshold we encounter roughly speaking an ELA case independent of the variational coefficients a_i. This can be e.g. observed, if both driving fractional Brownian motions enjoy the same Hurst exponent H. Then we deduce $t_x = -\frac{Hx}{(1-H)(C-m_2)}$ and hence, we get the threshold $m_2 \geq \frac{1-H}{C}$. The higher the H the more bursty the traffic and the lower the threshold, which will lead to a larger waiting queue for the class 2.

We turn now to the *full link approximation (FLA)*. Again we examine only two traffic classes. It is possible to transfer the results to a processor sharing model, but the expression will turn significant more complex and the estimations are no longer reliable. We will give a heuristic approach. In a processor sharing system we have two weights g_1 and g_2, where we assume $g_1 = 1$ in in a priority system. Let us determine $\mathbb{P}(Q_0^{(2)} \geq x)$. First, we compute the most probable path f_x^* for the set $\{Q_0^{(1,2)} \geq x\}$. Suppose we have $Q_0^{(1)}(f_x^*) = 0$, we can return to the determination of the empty buffer. Hence, we assume $Q_0^{(1)}(f_x^*) > 0$. This is extraordinary in a priority system. In processor sharing systems this is valid at least for one class.

The idea reads as follows. Every additional superfluous built up waiting queue diminishes the probability for the path vector f_x^*. Since we want to stress that $Q_0^{(2)}$ is large, we have to impose that $Q_0^{(1)}$ close to 0 for the optimal path $f_x^* = \omega \in \Omega$. Hence, a waiting queue in class 2 is at least of size x, provided the class1 uses the guaranted traffic amount $g_1 C$ without waiting queue and in class 2 we ave a garanteed capacity *plus* the surplus x. We simplify a little bit more and set for $t < 0$ the condition

$$\begin{cases} A^{(1)}(t,0) = g_1 C|t| \\ A^{(2)}(t,0) = g_2 C|t| + x \end{cases} \tag{3.64}$$

Now we determine the most probable path, satisfying (3.64), and minimizing the norm w.r.t. t. We summarize the above considerations in the following theorem.

Theorem 3.61. *Full link approximation (FLA): For the most probable path vector f_x^{vla}, fulfilling (3.64) holds*

$$f_x^{vla}(\cdot) = \left(\frac{(g_1 C - m_1) t_x^*}{\gamma_1(t_x^*)} \gamma_1(t_x^*, \cdot), \frac{-x + (g_2 C - m_2) t_x^*}{\gamma_2(t_x^*)} \gamma_2(t_x^*, \cdot) \right)$$

where t_x^* is the value minimizing the function

$$V(t) = \frac{(g_1 C - m_1)^2 t^2}{\gamma_1(t)} + \frac{(x - (g_2 C - m_2) t)^2}{\gamma_2(t)}$$

Remark 3.62. In the case that both classes are driven by a Brownian motion the (FLA) reflects the exact most probable path in the not-(ELA). In general the path in (FLA), subordinated to the class 1, does not use the whole guaranteed capacity on the interval $]t_x^*, 0[$. Hence, a part of the calls-2 traffic is wasted and there is a small waiting queue in class 1, while the one in class 2 turns out smaller.

As already mentioned simulations show that

$$\mathbb{P}(Q_0^{(2)} \geq x) \approx \exp\left(-\frac{1}{2} \|f_x^{vla}\|_R^2\right) = \exp\left(-\frac{1}{2} v(t_x^*)\right)$$

serves as an upper bound. For a further approximation, but this time for the lower bound, we obtain

$$\mathbb{P}(Q_0^{(2)} \geq x) = \mathbb{P}(E) = \Psi^c \left(\sqrt{\frac{((g_1 C - m_1)t_x^*)^2}{\gamma_1(t_x^*)} + \frac{(x + (g_2 C - m_2)t_x^*)^2}{\gamma_2(t_x^*)}} \right)$$

We want to list the steps, how the waiting queue in a priority resp. processor sharing system can be determined. For this purpose we approximate the set $\{Q_t > x\}$ in five steps:

a) If $j = 1$, i.e., we consider $\{Q_t^{(1)} > x\}$, and the server has priority, then we apply the approximation (3.62) and substitute m by m_1 resp. γ by γ_1.
b) We determine the most probable path vector f_x^* corresponding to the results $\{Q_0^{(1,\ldots,n)} > x\}$. This will be handled in the sequel.
c) If $Q_0^{(1,\ldots,n)\setminus\{i\}}(f_x^*) = 0$, then proceed to step d), otherwise to step e).
d) If the buffer is empty, thus $Q_0^{(1,\ldots,k)\setminus\{i\}}(f_x^*) = 0$, then f_x^* is the most probable path vector in $\{Q_t^{(i)} > x\}$ and we use $\exp(-\frac{\|f_x^*\|_R^2}{2})$ as approximation of the probability. Afterwards stop the algorithm (approximation of the empty buffer). This approximation uses the most probable path vector and it is as exactly as the approximation (3.62).
e) Full link approximation: We determine a f_{vla}, so that the only positive waiting queue is $Q^{(i)}$. Afterwards we use the approximation $\exp(-\frac{\|f_{vla}\|_R^2}{2})$. In contrast to step d) f_{vla} is only the heuristic approximation of the most probable path vector.

Example 3.63. We want to give two examples of queueing systems: a homogeneous and a heterogeneous one. The results of the analytical models are compared with the help of measurements from simulations. The random middlepoint method is used (see [25]).

Beside the used resp. observed parameters the calibration of the driving fractional Brownian motion is important. We use in both classea the same Hurst parameter, thus, $\gamma_i(t) = \sigma_i^2 t^{2H}$, with $H \in\,]0,1[$. Now we fix a x and obtain for $\{Q_0^{(1,2)} \geq x\}$

$$t_x = \frac{-Hx}{(1-H)(C-m)}$$

The ELA-criterion (3.63) provides for the waiting queues of classes 2 (Klammern pruefen, 2 eingefuegt):

$$(\mu_1 C - m_1)|t_x| \geq -f_{x,1}^*(t_x) = \frac{(x + (C-m))\,|t_x|\gamma_1(t_x)}{\gamma(t_x) + \gamma_2(t_x)}$$

$$= \frac{(x + (C-m))\,|t_x|\sigma_1^2}{\sigma_1^2 + \sigma_2^2}$$

Inserting t_x, we get

3.5 Models for Time Sensitive Traffic

$$\frac{(\mu_1 C - m_1)H}{C - m} \geq \frac{\sigma_1^2}{\sigma_1^2 + \sigma_2^2} \tag{3.65}$$

The expression separates the serving rates and the mean on the left side from the variances on the right one. If (3.65) is satisfied, it follows

$$\mathbb{P}\left(Q^{(2)} \geq x\right) \approx \mathbb{P}\left(Q^{(1,2)} \geq x\right)$$

$$\approx 2\Phi^c\left(\frac{C - m}{\sqrt{\sigma_1^2 + \sigma_2^2}}\right) \exp\left(\frac{-(C - m)^{2H} x^{2-2H}}{2\phi(H)^2(\sigma_1^2 + \sigma_2^2)}\right)$$

where $\varphi(H) = H^H(1-H)^{1-H}$ and Φ^c are the complementary standard Gaussian distribution. $2\Phi^c$ is necessary to obtain in simulations with a resolution of $\Delta = 1$ approximately a correct probability that the buffer is not empty. If (3.65) does not hold, then we turn to the coarse ELA. We have to consider the set $\{-A_t^{(1)} \geq \mu_1 C |t|, -A_t^{(2)} \geq \mu_2 C |t|\}$. Then the square R-norm of most probable path is f_x^*

$$\|f_x^*\|_r^2 = \frac{(\mu_1 C - m_1)^2}{\sigma_1^2}|t|^{2-2H} + \frac{(x - (\mu_2 C - m_2)t)^2}{\sigma_2^2|t|^{2H}}$$

For the wanted minimum we get $\hat{t}_x = -\vartheta x$, where ϑ is the solution of a quadratic equation

$$\vartheta = \frac{a_2 + \sqrt{a_2^2 + 4aH}}{2a_1}$$

with

$$a_1 = \left(\frac{(\mu_1 C - m_1)^2}{\sigma_1^2} + \frac{(\mu_2 C - m_2)^2}{\sigma_2^2}\right)(1 - H), \quad a_2 = \frac{(\mu_2 C - m_2)(2H - 1)}{\sigma_2^2}$$

Then we can determine the queue probability

$$\mathbb{P}(Q_0^{(2)} \geq x)$$

$$\approx -\frac{1}{2}\left(\frac{(\mu_1 C - m_1)^2}{\sigma_1^2} + \frac{(\mu_2 C - m_2)^2}{\sigma_2^2}\vartheta^{2-2H} + \frac{((\mu_2 C - m_2)\vartheta + 1)^2}{\sigma_2^2\vartheta^{2H}}x^{2-2H}\right)$$

The right side tends towards 1 for $x \to 0$. Keeping to the resolution for the simulation at $\Delta = 1$, we have to choose the suitable prefactor

$$p_2 = \min\left(2\Phi^c\left(\frac{\mu_2 C - m_2}{\sigma_2}\right), 2\Phi\left(\frac{C - m}{\sqrt{\sigma_1^2 + \sigma_2^2}}\right)\right)$$

The prefactor expresses the worst case. Independent of the classes 1 and 2 at least $\mu_2 C$ capacity has to be reserved. With this the non-idle probability cannot exceed than the one in the single classes with guaranty $\mu_2 C$. If on the other hand the guaranty is close to the mean rate or even smaller, then one should not use the non-idle probability of the whole waiting queue. The waiting queue of class 1 can of course be treated simply by exchanging the rôles. As concrete example we use $m_1 = 1$, $m_2 = 2$, $\sigma_1^2 = 1$, $\sigma_2^2 = 2$, $H = 0.75$, and $C = 4$ to study the influence of the parameter μ_i.

We turn to the second example, the heterogeneous traffic. Here, we can find explicit formulas. But, using numerical methods we can compute approximations. The following numerical examples are based on the length of waiting queues. We consider a general processor sharing knot, which is serving two independent Gaussian flows. Then, we assume that class 1 is LRD, while class 2 is SRD. Our parameters reads as $m_1 = 2$, $m_2 = 1.5$, $C = 5$ and

$$\gamma_1(t) = 6(t - 1 + e^{-1}), \quad v_2(t) = \begin{cases} \frac{9}{8}t^2 - \frac{1}{8}t^3 & \text{for } t \in [0, 1] \\ \frac{1}{8} - \frac{9}{8}t + 2t^{\frac{3}{2}} & \text{for } t > 1 \end{cases}$$

Here the class-2 traffic is Poisson with Pareto bursts. For the exemplary values for the priority we choose

$$\mu = (\mu_1, \mu_2) = (1, 0) \quad \text{resp.} \quad \mu = (0.5, 0.5) \quad \text{resp.} \quad \mu = (0.3, 0.7)$$

The first case describes a pure priority. Here, we can use a simple queueing formula, by imposing the FIFO principle with input $A_t^{(1)}$. But with the coarse FLA for the classes 1 and the ELA for class 2 we can obtain good approximations.

The second case is the most interesting one. The most probable path vector for the class 1 only exists on the interval $[0, 5]$. For $x \in\ ]5, 67]$ both classes are sent into the queueing and in the case $x > 67$ we have only one queue for the class 2. Thus, we get the classification given in table 3.1.

Table 3.1. Example for the classification of priority traffic

Class	$[0, 5]$	$[5, 67]$	$]67, \infty[$
1	ELA	coarse FLA	coarse VLA
2	coarse FLA	coarse FLA	ELA

The second case demonstrates the opposite to the first case. The guaranteed load of class 1 is smaller than its the mean rate and thus, we can approximate f_x^* using ELA. Similarly class 2 will be approximated by FLA. For the non-idle case we can determine the following probabilities

$$\tilde{p}(\Delta) = 2\Phi^c\left(\frac{(C-m)\Delta}{\sqrt{\gamma_1(\Delta) + \gamma_1(\Gamma)}}\right)$$

$$\tilde{p}_i(\Delta, \mu) = \min\left(2\Phi^c\left(\frac{(\mu_i C - m)\Delta}{\sqrt{\gamma_i(\Delta)}}\right), \tilde{p}(\Delta)\right), \quad i = 1, 2$$

As above Δ is the step width of the simulation. Because both used Gaussian processes are absolutely continuous, no probability converges towards 1, if $x \to 0$. Thus, we have to weight the non-idle probability suitably

$$\frac{(x+(C-m)t_x)^2}{\gamma(t_x)} = \frac{4\gamma(t_x)}{\gamma'(t_x)^2}$$

where $\gamma(t) = \gamma_1(t) + \gamma_2(t)$ is differentiable at t_x. Since from $x \to 0$ we conclude the converges $t_x \to 0$, we obtain

$$\lim_{x \to 0} \frac{(x+(C-m)t_x)^2}{\gamma(t_x)} = \frac{18}{33}$$

For the coarse FLA we cannot proceed similarly, but we have to apply numerical methods. Thus, first the limits of

$$L_1(\mu) = \lim_{x \to 0} \frac{(x - (\mu_1 C - m_1)\tilde{t}_x)^2}{\gamma_1(\tilde{t}_x)} + \frac{(\mu_2 C - m_2)^2 \tilde{t}_x^2}{\gamma_2(\tilde{t}_x^2)}$$

$$L_2(\mu) = \lim_{x \to 0} \frac{(\mu_1 C - m_1)^2 \tilde{t}_x^2}{\gamma_1(\tilde{t}_x)} + \frac{(x - (\mu_2 C - m_2)\tilde{t}_x)^2}{\gamma_2(\tilde{t}_x^2)}$$

have to be determined. If this is done, we can compute the prefactor

$$p(\Delta) = e^{\frac{33}{36}} \tilde{p}(\Delta)$$
$$p_1(\mu, \Delta) = e^{\Omega_1(\mu)} \tilde{p}_1(\mu, \Delta)$$
$$p_2(\mu, \Delta) = e^{\Omega_2(\mu)} \tilde{p}(\mu, \Delta)$$

The prefactor p is used for the ELA estimation and for p_1 resp. p_2 in the corresponding coarse FLA.

3.6 Fractional Lévy Motion in IP-based Network Traffic

3.6.1 Description of the Model

Starting point is the consideration of the α-stable processes. We shortly review the definition.

Definition 3.64. *Let $(X_t)_{t \in I}$ be stochastic process. Then*

- *$(X_t)_{t \in I}$ is a linear fractional stable motion, if for given $a, b \in \mathbb{R}$, $ab \neq 0$ it possesses a representation*

$$X_t(H, a, b, \alpha) = \int_{-\infty}^{\infty} \left(a \left((t-x)_+^{H-\frac{1}{2}} - (-x)_+^{H-\frac{1}{2}} \right) \right.$$
$$\left. + b \left((t-x)_-^{H-\frac{1}{\alpha}} - (-x)_-^{H-\frac{1}{\alpha}} \right) \right) dZ_\alpha(x)$$

where $(Z_\alpha(t))$ is a sαs-Lévy process, i.e. a process with α-stable finite marginal distributions, H-sssi with independent increments. A linear fractional stable motion is a H-sssi sαs process.

- $(X_t)_{t \in I}$ is a fractional Lévy process, if it has a representation

$$X_t(H, \alpha) = \int_0^\infty \frac{1}{\Gamma\left(H + \frac{1}{2} - \frac{1}{\alpha}\right)} (t-x)^{H - \frac{1}{\alpha}} dZ_\alpha(x)$$

where $(Z_\alpha(t))$ is a sαs-Lévy motion, i.e. a process with α-stable finite marginal distributions, H-sssi with independent increments. A fractional Lévy process (FLM) has stationary increments and is H-sssi.

The FLM is a generalization of the FBM. As we know the path are no longer continuous – they exhibit infinite countable many jumps. Since the fractional Lévy process is easier to handle we stick for the IP-based traffic modeling to the FLM. We start with the distribution density of an α-stable Lévy motion $(Z_\alpha(t))$. For given $c > 0$ and $0 < \alpha \leq 2$ we have

$$p_\alpha(x,t) = \frac{1}{2\pi} \int_{-\infty}^\infty \exp\left(-c|u|^\alpha t\right) \exp(-iux) du$$

We cite some properties.

Theorem 3.65. *The FLM is a H-sssi with Hurst exponent $H \in [\frac{1}{\alpha}, 1[$.*

Proposition 3.66. *We have for a FLM (X_t) with subordinated LM $(Z_\alpha(t))$ and a Z_α-integrable function f*

$$\mathbb{E}\left(\exp\left(-ix \int_0^t f(u) dZ_\alpha(u)\right)\right) = \exp\left(-c|x|^\alpha \int_0^t f^\alpha(u) du\right), \text{ for all } x \in \mathbb{R}$$

We get for the density function of a FLM the following theorem.

Theorem 3.67. *The density function $p_{\alpha,H}(x,t)$ of a FLM is*

$$p_{\alpha,H}(x,t) = \frac{1}{2\pi} \int_{-\infty}^\infty \exp\left(-c'|u|^\alpha t^{\alpha(H-\frac{1}{2})+1}\right) \exp(iux) du$$

where $c' = \frac{c}{\Gamma^\alpha(H + \frac{1}{2} - \frac{1}{\alpha})\alpha H}$.

3.6.2 Calibration of a Fractional Lévy Motion Model

We shortly outline the concept to calibrate the FLM-model according to the observed traffic and divide the process into several steps:

1. First we use the standard methods to estimate the mean $\overline{m}$ and the variance $\overline{c}$ of the traffic traces. For more detailed description the reader should consult the section 4.1.

2. Next we use (5.13) to compute the value $X_t(\alpha, H)$ according to the approach
$$X_t(\alpha, H) = (\bar{c}m)^{-\frac{1}{\alpha}} (A(t) - mt)$$
Here, $A(t)$ represents the measured accumulated traffic.
3. The crucial part consists in the estimation of the heavy-tail parameter α for the observed traffic traces (e.g. the interarrival time or the amount of data). In addition we need the Hurst exponent H', computed according to the standard measures as done in section 4.2. Thus, we can compute by theorem (3.65) the Hurst parameter of the Lévy motion Z_α. According to proposition 3.66 we can estimate the parameter α. Another possibility is to use the increments and estimate via the Crovella and Taqqu tool (see 4.1.3) the slope of traffic difference.
4. Finally after obtaining the calibrated model we simulate the FLM-model using the Taqqu tool, described in [234] for matching the observed traffic with the model.

3.7 Fractional Ornstein-Uhlenbeck Processes and Telecom Processes

In the last section we considered the generalization of the Norros approach to model bursty LRD traffic using a process with sαs marginal distributions instead of a Gaussian one. This process is called fractional Lévy process or motion (FLM). In this section we stick to the same framework of LRD and present another view for bursty traffic, namely the fractional Ornstein-Uhlenbeck processes, introduced for network modeling e.g. by Wolpert and Taqqu [269]. Let us first sketch the abstract network terms, which are going to be subject of the investigation:

- The instantaneous work-load or rate $X'(t)$ at time t.
- The cumulated work-load $X(t) = \int_0^t X'(s)ds$. Both describe it in a network with a random incident flux of work with varying intensities κ and durations τ which arrive randomly at times σ.
- Finally the weighted workload $X(\theta) = \int_\mathbb{R} \theta(s)X'(s)ds$ for a suitable space Θ of weighting function θ. We will see the technical reason for this later below.

One of the key items is the distribution for the intensity κ, which will be a so called infinitely-divisible distribution (abbr. by ID). The different distributions will lead to the different fractional Ornstein-Uhlenbeck processes.

3.7.1 Description of the Model

We start with a random measure $M(d\sigma, d\tau)$ on $\mathbb{R} \times \mathbb{R}_+$, which marginal distributions are infinitely-divisible. Later examples concern of Gaussian, sαs, gamma and in general Lèvy random measures. As shortly described in the above introduction we consider the following three situations

$$X'(t) = \int_{-\infty}^{\infty}\int_{0}^{\infty} \mathbf{1}_{]\sigma,\sigma+\tau]}(t) dM(\tau) dM(\sigma) \tag{3.66}$$

$$X(\theta) = \int_{-\infty}^{\infty}\int_{0}^{\infty}\int_{\sigma}^{\sigma+\tau} \theta(s) ds dM(\tau) dM(\sigma) \tag{3.67}$$

$$X(t) = \int_{-\infty}^{\infty}\int_{0}^{\infty} \text{vol}\left(]\sigma,\sigma+\tau]\cap]0,t]\right) dM(\tau) dM(\sigma) \tag{3.68}$$

whereas 'vol' means the volume of a set. We briefly interpret these three equations:

- Equation (3.66) describes the instantaneous workload $X'(t)$ at time t, which is the cumulated sum of all measurements of $M(\sigma,\tau)$ built up over all times starting at $\sigma < t$ until $\sigma + \tau \geq 0$.
- The equation (3.68) shows the accumulated workload from time 0 until t, represented as integral over $X'(s)$.
- The equation (3.67) finally describes the weighted aggregated work by a function $\theta(s)$. If we choose $\theta = \mathbf{1}_{]0,t]}$ (and if the indicator function is an element of Θ), then we return to the case of $X(t)$. Otherwise, if we use as $\theta = \delta_t$ the Dirac function then (again if $\delta_t \in \Theta$) we have the instantaneous rate X'.

3.7.2 Fractional Ornstein-Uhlenbeck Gaussian Processes

In the sequel let $(B_t)_{t\in\mathbb{R}}$ be a standard Brownian motion on a probability space $(\Omega, \mathcal{F}, \mathbb{P})$. For fixed $\lambda, \sigma > 0$ and $t \in \mathbb{R}$ define the Itō integral

$$Z^1(t) = \sigma\sqrt{2\lambda} \int_{-\infty}^{t} \exp(-\lambda(t-s)) dB(s) \tag{3.69}$$

The process $(Z^1(t))_{t\in\mathbb{R}}$ defines the well known Ornstein-Uhlenbeck process with mean 0, stationary and a covariance structure given according to

$$\rho^1(t) = \mathbb{E}\left(Z^1(0) Z^1(t)\right)$$
$$= \sigma^2 2\lambda \int_{-\infty}^{0\wedge t} \exp(-\lambda(t-s))\exp(-\lambda(0-s)) ds = \sigma^2 e^{-\lambda|t|}$$

We define inductively a series of processes derived from (3.69) and indexed by $\nu \geq 2$

$$Z^\nu(t) = \int_{-\infty}^{t} \lambda \exp(-\lambda(t-s)) Z^{\nu-1} ds \tag{3.70}$$

Consequently resubstituting one finds a representation as Itō integral

$$Z^\nu(t) = \sigma\sqrt{2\lambda} \int_{-\infty}^{t} \frac{\lambda^{\nu-1}(t-s)^{\nu-1}}{\Gamma(\nu)} \exp\left(-\lambda(t-s)\right) dB(s) \tag{3.71}$$

Though one can not define $(Z^\nu(t))$ for a $\nu \notin \mathbb{N}$ using the original definition (3.70) with the help of the representation (3.71) one can define it for $\nu > \frac{1}{2}$. In this case – for $\nu > \frac{1}{2}$ – we call the process fractional Ornstein-Uhlenbeck Gaussian process. It is a stationary Gaussian process with mean 0. The covariance structure is given by

$$\begin{aligned}
\rho^{(\nu)} &= \mathbb{E}\left(Z^\nu(0)Z^\nu(t)\right) \\
&= \sigma^2 e^{-\lambda|t|} \frac{2\lambda^{2\nu-1}}{\Gamma(\nu)^2} \int_{-\infty}^{0} (|t|-s)^{\nu-1}(-s)^{\nu-1} \exp(2\lambda s) ds \\
&= \frac{2\sigma^2 e^{-\lambda|t|}}{\Gamma(\nu)^2} \int_{0}^{\infty} (\lambda|t|+x)^{\nu-1} x^{\nu-1} \exp(-2x) dx \\
&= \frac{2\sigma^2}{\Gamma(H-\frac{1}{2})\sqrt{\pi}} \left(\frac{\lambda|t|}{2}\right)^{H-1} \mathcal{K}_{H-1}(\lambda|t|)
\end{aligned}$$

where $H = \nu + \frac{1}{2}$ and $\mathcal{K}_\gamma(u)$ is the modified Bessel function of second kind (see [1]).

3.7.3 Telecom Processes

The heavy-tail renewal process was outlined in section 2.7.4. It was incorporated into the so called on-off models. We saw that in the limit we get after rescaling a FBM as driving process (see (3.56)). The Hurst exponent expressing the self-similarity can be computed according to

$$H = \frac{3-\alpha}{2} \tag{3.72}$$

where $1 < \alpha < 2$ is the heavy-tail exponent of the inter-renewal times. In the first approach we considered constant renewals. We outlined that by assuming a heavy-tailed renewal of exponent $\beta \in]\alpha, 2[$ (being in the domain of attraction of the sαs that the limiting process after rescaling and centering is a process $(Z_t)_{t\geq 0}$), which Levy and Taqqu called *telecom process*, being a sαs process with Hurst exponent

$$H = \frac{\beta - \alpha + 1}{\beta} \tag{3.73}$$

3.7.4 Representations of Telecom Processes

We refer to the abstract approach in section (3.7.3). In [206] Pipiras and Taqqu demonstrated that the telecom process can be represented in the form

$$Z_t = \int_{-\infty}^{\infty} \int_{-\infty}^{\infty} \left(\int_0^t \mathbf{1}_{\{u \leq s \leq v\}} ds \right) dM(u) dM(v) \tag{3.74}$$

where M is a $s\alpha s$ measure with control measure m given by

$$dm(u)dm(v) = \alpha(v-u)_+^{-\alpha-1} dudv \tag{3.75}$$

For more details on controlled $s\alpha s$ random measures on $\mathbb{R}^2$ please refer to section (3.2.3). An interesting observation tells us that for $\beta = 2$ in equation (3.73) gives the well known formula (3.72) for the Hurst exponent and the FBM. In fact, if $\beta \uparrow 2$, then the telecom process in (3.74) converges to the FBM weakly. Towards an interpretation of the telecom process for the describing of network flows we have the following representation of the telecom process in (3.74) using a compensated Poisson measure $\tilde{N}$

$$Z_t \stackrel{d}{=} \iiint_{\mathbb{R}^3} \left(\int_0^t \mathbf{1}_{\{u \leq s \leq v\}} ds \right) w d\tilde{N}(w) d\tilde{N}(u) d\tilde{N}(v) \tag{3.76}$$

where

$$d\tilde{N}(w)d\tilde{N}(u)d\tilde{N}(v) = dN(w)dN(u)dN(v) - d\mu(w)d\mu(u)\mu(v)$$

with

$$dN(w)dN(u)dN(v) \sim \mathcal{P}o(d\mu(w)d\mu(u)d\mu(v))$$

and

$$d\mu(w)d\mu(u)d\mu(v) = |w|^{-\beta-1} dw dm(u) dm(v)$$

The equation $dm(u)dm(v) = \alpha(v-u)_+^{-\alpha-1} dudv$ is the telecom measure according to (3.75). The drawback is that the representation (3.74) and (3.76) are less suitable for an interpretation to IP-based networks. Thus, we need an equivalent way using the so called upstairs representation.

Theorem 3.68. *Let $g : [0, \infty[\longrightarrow \mathbb{R}$ be a non-negative strictly monotone decreasing differentiable function on $\mathbb{R}_+$. Extend g on whole $\mathbb{R}$ by $g(s) = 0$ for $s < 0$. The process*

$$Y_t^{(1)} = \iint_{\mathbb{R}^2} \left(\int_0^t \mathbf{1}_{\{u \leq s \leq v\}} ds \right) dM_\beta^{(1)}(u) dM_\beta^{(1)}(v) \tag{3.77}$$

where $M_\beta^{(1)}$ is a $s\beta s$ measure on $\mathbb{R}^2$ with control measure

$$dm^{(1)}(u)dm^{(1)}(v) = |g'(v-u)| dudv$$

has the same marginal distributions as the process

$$Y_t^{(2)} = \int_{\mathbb{R}^2} \int_0^\infty \left(\int_0^t \mathbf{1}_{\{0 < u \leq g(r-v)\}} dr \right) dM_\beta^{(2)}(u) dM_\beta^{(2)}(v) \tag{3.78}$$

3.7 Fractional Ornstein-Uhlenbeck Processes and Telecom Processes

where $M_\beta^{(2)}$ is a $s\beta s$ measure on $\mathbb{R}^2$ with control measure

$$dm^{(2)}(u)dm^{(2)}(v) = dudv$$

if both processes are well defined. Suppose the process $(Y_t^{(2)})$ is differentiable, i.e. if

$$Y_t' = \int_0^\infty \int_{\mathbb{R}} \mathbf{1}_{\{0<u\leq g(t-v)\}} dM_\beta^{(2)}(u) dM_\beta^{(2)}(v)$$

$$= \int_{-\infty}^t \int_0^{g(t-v)} dM_\beta^{(2)}(u) dM_\beta^{(2)}(v)$$

is well defined. Then the processes $(Y_t^{(1)})$ and $(Y_t^{(2)})$ have the same marginal distributions as

$$Y_t^{(3)} = \int_0^t Y_s' ds = \int_0^t \int_{\mathbb{R}} \int_0^{g(s-v)} dM_\beta^{(2)}(u) dM_\beta^{(2)}(v) ds \qquad (3.79)$$

$$\stackrel{d}{=} \int_0^t \int_{\mathbb{R}} \int_0^{g(s-v)} \int_{\mathbb{R}} w d\tilde{N}(w) d\tilde{N}(u) d\tilde{N}(v) ds$$

where

$$d\tilde{N}(w)d\tilde{N}(u)d\tilde{N}(v) = dN(w)dN(u)dN(v) - d\mu(w)d\mu(u)\mu(v)$$

is the compensated version of a Poisson measure

$$dN(w)dN(u)dN(v) \sim \mathcal{P}o(d\mu(w)d\mu(u)d\mu(v))$$

with control measure

$$d\mu(w)d\mu(u)d\mu(v) = |w|^{-\beta-1} dw du dv \qquad (3.80)$$

3.7.5 Application of Telecom Processes

We return to the equation (3.80) for an interpretation the compensated Poisson measure N generating points (w, u, v). Which interpretation can be given to a particular point (w, u, v)? A point (w, u, v) indicates an arrival at a network at time v for job with rate w, which has a duration τ. The duration is surely randomly with a complementary distribution function

$$\mathbb{P}(\tau > t) = \frac{g(t)}{g(0)}$$

where $g : [0, \infty[\longrightarrow \mathbb{R}$ is a differentiable and strictly monotone decreasing function. We can regard $u = g(\tau)$ as a realization of a uniform random variable on the interval $]0, g(0)]$. This indicates:

The event $\{0 < u \leq g(r-v)\}$ describes that the job generated at time v has not terminated at time τ.

With a given start time v this event occurs with the conditional probability

$$\mathbb{P}\left(0 < u \leq g(r-v)\,|\,v\right) = \frac{g(r-v)}{g(0)}$$

vanishing for $r < v$ and decreasing from 1 at $r = v$ to 0 for $r \to \infty$. This gives a physical explanation for the range of the Poisson measure.

Example 3.69. We want to apply theorem 3.68 to the function $g(s) = (s_+)^{-\alpha}$ if $0 < \beta < \alpha < 2$. The function satisfies the assumption of theorem 3.68, since it is surely strictly monotone decreasing with $g'(s) = -\alpha s^{-\alpha-1}$ on $]0, \infty[$. It can be shown (see [269, p.27]) that equation (3.77) and (3.78) are valid but not the last two in (3.79), since $\int_0^\infty g(s)ds = \infty$, as indication for not differentiability of the process $(Y_t^{(2)})$. To overcome this difficulty we approximate the process by setting

$$g_\epsilon(s) = s^{-\alpha} \wedge \epsilon^{-\alpha} \mathbf{1}_{\{s \geq 0\}}$$

The function g_ϵ is integrable, since $\int_0^\infty g_\epsilon(s)ds = \frac{\epsilon^{1-\alpha}}{\alpha-1} < \infty$ (remember $\alpha > 1$!) So, we have last two equations in (3.79) with g_ϵ instead of g and a differentiable process $(Y_{t,\epsilon}^{(3)})$, which converges in probability to $(Y_t^{(2)})$, if $\epsilon \to 0$. Thus, we can formulate for the telecom process the following corollary

Corollary 3.70. *The telecom process (Y_t) defined by equation (3.74) has the representation*

$$Y_t = \int_{\mathbb{R}} \int_{\mathbb{R}_+} \left(\int_0^t \mathbf{1}_{\{0 < u \leq (r-v)_+^{-\alpha}\}} dr \right) dM_\beta(u) dM_\beta^{(2)}(v)$$

$$\stackrel{d}{=} \lim_{\epsilon \to 0} \int_0^t \left(\int_{\mathbb{R}} \int_0^{g_\epsilon(s-v)} dM_\beta(u) dM_\beta(v) \right) ds$$

$$\stackrel{d}{=} \lim_{\epsilon \to 0} \int_0^t \left(\int_{\mathbb{R}} \int_0^{g_\epsilon(s-v)} \int_{\mathbb{R}} w d\tilde{N}(w) d\tilde{N}(u) d\tilde{N}(v) \right) ds$$

where $g_\epsilon(s) = s^{-\alpha} \wedge \epsilon^{-\alpha} \mathbf{1}_{\{s \geq 0\}}$, M_β is a $s\beta s$ measure with control measure $dudv$ and

$$d\tilde{N}(w) d\tilde{N}(u) d\tilde{N}(v) = dN(w) dN(u) dN(v) - d\mu(w) d\mu(u) \mu(v)$$

is the compensated version of a Poisson measure

$$dN(w) dN(u) dN(v) \sim \mathcal{P}o(d\mu(w) d\mu(u) d\mu(v))$$

with control measure

$$d\mu(w) d\mu(u) d\mu(v) = |w|^{-\beta-1} dw du dv$$

3.8 Multifractal Models and the Influence of Small Scales

Up to know most models are built upon the general central limit theorem (see section 3.4.1) and the result of Willinger et al. as presented in section 3.4.4. Applying the central limit theorem will result in a traffic model, which is self-similar and long-range dependent, but the local structure of the path of the process is still relatively regular. In other words the irregularity of the traffic is identical for all scales. Later we will describe this with the help of the so called deterministic partition function.

As different authors observed the so called monofractality, induced by FBM and FLM, is not quite the true story of the IP traffic. In fact, different scales behave rather differently. If the transport protocol does not exercise a decisive influence on the transmitted traffic, we may stick to the above developed models by perturbations of the FBM and FLM in the suitable way (see sections 3.3.4 and 3.6). But especially the appearance of the TCP changes the IP traffic shape. As described in section 1.1.4, TCP adjusts the transmitted data rate according to a possible congestion in the network. Thus, the already transmitted data as well as the data to be sent, influence the TCP mechanism. This in whole is highly irregular and causes a traffic shape which is quite different to the expected one, using the standard approach due to Norros for example. Since the small scales as well exercise a decisive impact on the traffic behavior as well as its performance (see e.g. section 5.2 for a comparison of the multifractal models to the monofractal with respect to the queueing probability), this behavior has to be taken into account as well, and it is one of the challenging tasks in traffic modeling of today to fit both the long-range behavior, indicated by the large scales and the small scale behavior into one satisfying model. This section will give a short insight of some of the standard approaches and theories, developed in recent years, which are far of being complete, since this area is one of the fastest growing one in the IP traffic theory.

3.8.1 Multifractal Brownian Motion

The first approach for an analysis on different scales and incorporating different Hurst exponents is the following approach using the multifractal Brownian motion. Instead of using a constant H the original model of Norros is modified literately by a selected function $H(t)$.

Definition and Concept

As we know the sample path of a stochastic process (X_t) fulfils a certain Hölder estimation, which we discuss in more detail in section 3.8 and which reads as (see e.g. (3.107)):

There is a function $H(\cdot)$ and a $c > 0$: $|X_{t+h} - X_t| \leq c|h|^{H(t)}$, for $h \to 0$ \hfill (3.81)

The inequality is tight, meaning that the value $H(t)$ is the largest. The FBM fulfils the inequality (3.81) with a constant function H, the Hurst exponent. In general $H : \mathbb{R} \longrightarrow]0,1[$ a is called a *Hölder function*. We start with the definition of the *multifractal Brownian motion*. The concept is derived from the representation of the usual FBM, using an integral representation as we used in section 3.2.3

$$B_t^{(H)} = \frac{1}{\Gamma\left(H + \frac{1}{2}\right)} \int_{\mathbb{R}} \left(\left((t-x)^+\right)^{H-\frac{1}{2}} - \left((-x)^+\right)^{H-\frac{1}{2}}\right) dB(x)$$

$$= \frac{1}{\Gamma\left(H + \frac{1}{2}\right)} \left(\int_{-\infty}^{0} \left((t-x)^{H-\frac{1}{2}} - (-x)^{H-\frac{1}{2}}\right) dB(x)\right.$$

$$\left. + \int_{0}^{t} (t-x)^{H-\frac{1}{2}} dB(x)\right)$$

We turn to the mBm which is a continuous Gaussian process with in general non-stationary increments, in contrast to the FBM

$$W_t^{(H(t))} = \frac{1}{\Gamma\left(H(t) + \frac{1}{2}\right)} \int_{\mathbb{R}} \left(\left((t-x)^+\right)^{H(t)-\frac{1}{2}} - \left((-x)^+\right)^{H(t)-\frac{1}{2}}\right) dB(x)$$

$$= \frac{1}{\Gamma\left(H(t) + \frac{1}{2}\right)} \left(\int_{-\infty}^{0} \left((t-x)^{H(t)-\frac{1}{2}} - (-x)^{H(t)-\frac{1}{2}}\right) dB(x)\right.$$

$$\left. + \int_{0}^{t} (t-x)^{H(t)-\frac{1}{2}} dB(x)\right) \quad (3.82)$$

Though, in general, the mBm does not reaveal even asymptotically stationary increments, we observe a local stationarity, detected as self-similarity as follows

$$\lim_{\gamma \to 0+} \left(\frac{W_{t+\gamma s} - W_t}{\gamma^{H(t)}}\right)_{s>0} = \left(B_{H(t)}(s)\right)_{s>0} \quad (3.83)$$

where (W_t) is a multifractal Brownian motion with Hölder function $H(\cdot)$ and $(B_{H(t)}(s))$ is a FBM with Hurst exponent $H(t)$.

For special properties of the multifractal Brownian motion the reader may consult the literature (see [91, 179]).

Envelope Process

As pointed out in the previous section the multifractal Brownian motion is in general a stochastic process with non-stationary increments. From the general theory in section 3.2.1 we know that H-sssi processes can be easily characterized and worked with. There are a number of models introduced as we already mentioned (see e.g sections 3.8) to incorporate multifractal analysis of the traffic traces. To overcome some difficulties arising from multifractal

Brownian motion, we introduce a stochastic process, easier to handle and easier to understand. Since we are interested in an estimation of necessary bound of the traffic load, this lead to the definition of an envelope process.

Definition 3.71. Let $(X_t)_{t \in I}$ be a stochastic process. We call another process $(Y_t)_{t \in I}$:

- Deterministic envelope, if $X_t \leq Y_t$ for all $t \in I$ almost surely.
- Stochastic envelope, if, for a predefined probability threshold $\epsilon > 0$, $\mathbb{P}(Y_t < X_t) \leq \epsilon$ for all $t \in I$.

Example 3.72. Let $Z_t^H = B_{t+\delta}^H - B_t^H$ (with $\delta > 0$) the increment of a FBM (B_t^H), i.e. the FGN. According to [91], there is a $\kappa > 0 : Z_t^H \leq \kappa H t^{H-1}$ for a suitable small $\delta > 0$. Since we can approximate a mBm locally by a FBM (see (3.83)), we can find an upper estimation for the increments $(Z_t^{H(\cdot)})$ of a mBm $(W_t^{H(\cdot)})$ at time t according to

$$Z_t^{H(t)} \leq \kappa H(t) t^{H(t)-1} \tag{3.84}$$

Consider the Norros model for the accumulated traffic

$$A_t = mt + \sqrt{am} B_t^H \tag{3.85}$$

With the help of (3.85) we deduce by a standard integration a (deterministic) envelope process of the form

$$\hat{A}_t = mt + \kappa \sqrt{am} t^H$$

If we now proceed to the model of using a mBm instead of a FBM in (3.85), then we get again by integration and (3.84)

$$\hat{A}_t = \int_0^t m + \kappa \sigma H(x) x^{H(x)-1} dx \tag{3.86}$$

where $H(\cdot)$ is the Hölder function corresponding to the mBm and $\sigma = \sqrt{am}$. It is natural to ask what form the Hölder function has. In general it possesses a complicated form, if one considers multifractal analysis as done in section 3.8. Often especially for practical reasons one uses polynomials. One should keep in mind that the function has to be integrable. Examples are

$$H(x) = a_2 x^2 + a_1 x + a_0, \ x \in \,]0,1[$$
$$H(x) = b_1 x + b_0, \ x \in \,]0,1[$$

with $a_2, a_1, a_0, b_1, b_0 \in \mathbb{R}$.

Calibration: Examples of IP Traffic Envelopes

If one measures resp. observes IP traffic, the first step for a selection of the mBm is the detection of the multifractal traffic. The decisive tool is the deterministic partition function $T(q)$, given in section 3.8 on multifractal models. Monofractal traffic is linear, as e.g. the classical Norros model shows in this case $T(q) = qH - 1$. If the deterministic function reveals a strictly concave or convex form, this is a clear evidence for multifactal behavior. Before we exhibit the examples, we will present a short summary of the necessary steps to go for calibration and simulation of the observed traffic:

- First we have to detect the 'fractality' of the observed traffic and decide, whether it is mono- or multifractal. This is done by analyzing the shape of the deterministic partition function $T(q)$.
- To fit the suitable multifractal Brownian motion we have to find the right Hölder function $H(x)$. This can be done with the help of the tools mentioned in chapter 4 on statistical methods or as in [47]. In most cases the form of the Hölder function is highly irregular, exhibited by the figures in [179]. With the help of interpolating the original Hölder function using e.g. L_2 approximation by polynomials, we can simplify the Hölder function. Usually one wants to stick to a low degree of the polynomials, keeping the computation as simple as possible. But this has the drawback consisting in the request that a too low degree would not match the multifractal behavior. Usually polynomials of degree between 7 and 15 fit quite good, though for simplicity one selects quadratic polynomials.
- Next we have to compute the aggregated traffic, which will be used for the update to the calibrated model.
- According to the formula (3.86) above, we have to compute $\hat{A}(t)$ and compare this with the (aggregated) real data of the traffic. If the envelop process $\hat{A}(t)$ does not fit with the aggregated traffic, the above steps have to repeated.
- The last step consists in following the line of the section 5.1.4 in the last chapter to gain the key values of queueing and bandwidth and validate this with the real traffic observations. Again this is compared with the observed data.

3.8.2 Wavelet-Based Multifractal Models

Short Introduction to Wavelets

We just give shortly an introduction to the wavelet analysis and refer the reader to the literature for a detailed study. For this purpose we remind to the discrete wavelet transform, which we discussed in the previous section. Let $x : \mathbb{R} \longrightarrow \mathbb{R}$ be a given signal and let $\varphi(\cdot)$ be the Haar mother wavelet. Then we have the wavelet representation:

3.8 Multifractal Models and the Influence of Small Scales

$$x(t) = \sum_{k \in \mathbb{Z}} \left(u_k 2^{-\frac{J_0}{2}} \phi(2^{-J_0}t - k) + \sum_{j=-\infty}^{J_0} w_{j,k} 2^{-\frac{j}{2}} \varphi(2^{-j}t - k) \right) \qquad (3.87)$$

where J_0 is the coarsest scaling, ϕ the scaling function (father-wavelet) corresponding to the Haar wavelet, let u_k be the scaling coefficients and $w_{j,k}$ the wavelet coefficients. The scaling coefficients are a coarse approximation of the signal, while the wavelet coefficients describe the fine structure (parts of the signal with high frequency).

In figure 3.18 the transmitted data amount is splitted into two disjoint interval at each level.

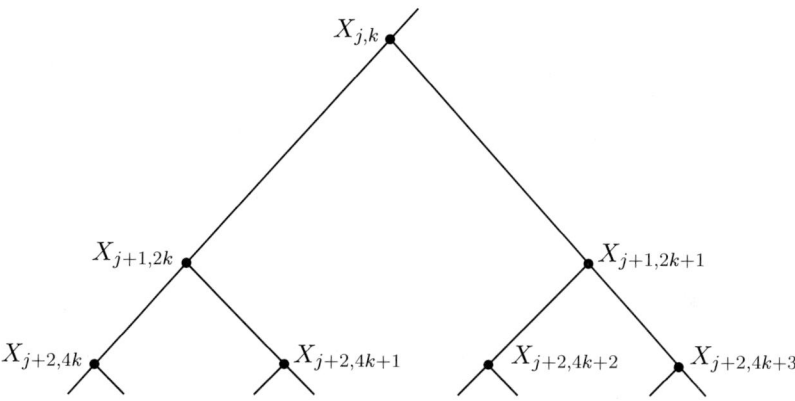

Fig. 3.18. Splitting of transmitted data amount into an dyadic tree. With each level the transmission time is divided into two equal intervals

Describing a LRD signal via a wavelet transform, i.e. by coefficients is the approach, forming a highly correlated signal into an almost uncorrelated one (see section 4.2.7). While the autocorrelation of an FBM is only approximative of order t^{2H-2}, there, the wavelet coefficient is an exact representation of the variance resp. correlation (please note since the FBM is a random variable for each t, so are the corresponding wavelet coefficients because they are computed for each path). For this purpose we use capital letters ($W_{j,k}$ RV for $w_{j,k}$)

$$\mathrm{Var}(W_{j,k}) = \sigma^2 2^{(2H-1)(j-1)} \left(2 - 2^{(2H-1)} \right) \qquad (3.88)$$

As known from the theory of wavelets, it is possible to gain back the original signal from the wavelet coefficients. With this we can get a Gaussian process with in general negative values as well, which is a kind of unrealistic for data traffic. In addition a Gaussian process is not suitable for small scales.

Multifractal Models using Wavelets

We start with describing the basic idea:

- To construct a signal with negative values, we will have to stress certain conditions to the coefficients.
- To obtain the LRD condition we will use the property (3.88).

Wavelet Coefficients for Non-Negative Data

Which conditions do the coefficients $u_j, w_{j,k}$ have to fulfill, to receive the signal x in (3.87) with non negative values? In general one cannot sufficiently answer this for all wavelets in closed form. But for Haar wavelets we obtain

$$u_{j,2k} = 2^{-\frac{1}{2}}(u_{j+1,k} + w_{j+1,k}) \quad (3.89)$$

$$u_{j,2k+1} = 2^{-\frac{1}{2}}(u_{j+1,k} - w_{j+1,k}) \quad (3.90)$$

where we receive the new scaling coefficients recursively and use the old one for the level $k = 0$. If the signal is positive then it follows $u_{j,k} \geq 0$ and

$$|w_{j,k}| \leq u_{j,k} \quad (3.91)$$

Figure 3.19 sketches the fundamental idea of the WIG model.

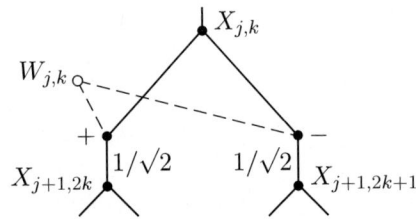

Fig. 3.19. Construction of a WIG-model

Multiplicative Models

The equation (3.91) gives conditions for constructing positive processes. This enables us to build simple multiplicative models. Thus, let $A_{j,k}$ be RV with values in the interval $[-1, 1]$. We define the wavelet coefficients of the process

$$W_{j,k} = A_{j,k} U_{j,k}$$

By (3.89) we obtain

$$U_{j,2k} = 2^{-\frac{1}{2}}(1 + A_{j,k})U_{j+1,k}$$

$$U_{j,2k+1} = 2^{-\frac{1}{2}}(1 - A_{j,k})U_{j+1,k}$$

The $A_{j,k}$ are mostly chosen beta distributed.

In contrast to the WIG model we do not use weighted sums in the MWM, but multiply between each scaling level with certain random variables (here the $A_{j,k}$). Figure 3.20 shows this schematic.

3.8 Multifractal Models and the Influence of Small Scales

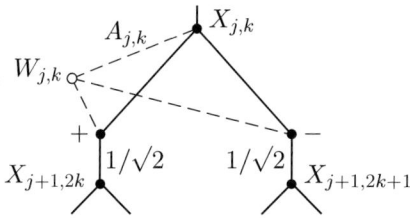

Fig. 3.20. Construction of a multiplicative wavelet model

Alternative Approach to Wavelet Analysis

Beside the analysis using the wavelet transform we will sketch an approach with the help of so called *multiscaled trees*. We remind again to the derivation of the aggregated traffic done in section 3.2.3. Let (X_k) be a stationary LRD process of second order in discrete time. This means that the autocorellation decays according to $r_X(k) = \mathbb{E}\left((X_{t+k} - X_t)^2\right) \sim k^{2H-2}$ for large k and a suitable Hurst parameter $\frac{1}{2} < H < 1$, such that the series $\sum_{k \in \mathbb{N}} r_X(k)$ is not summable. Equivalent to this fact is that for the process summed up in blocks

$$X^{(m)}(k) = X_{(k-1)m+1} + \cdots + X_{km}$$

the variance satisfies $\mathrm{Var}(X(k)^{(m)}) = \mathcal{O}(m^{2H})$ for large $m \in \mathbb{N}$ (the same H and independent of k, since we have a stationary process!). A well known example is the fractional white noise, which we dealt with in the previous sections (3.1.2 and 3.3.2). The increments correspond to the FBM: If $(B_t^{(H)})$ is a fractional Brownian motion with Hurst parameter $H \in \,]\frac{1}{2}, 1[$, then

$$Z_t^{(H)} = B_{t+1}^{(H)} - B_t^{(H)}, \ t \in \mathbb{R}$$

is the corresponding fractional white noise (FGN). We get for the autocorellation (evaluated at discrete times, see (3.18))

$$r_Z(k) = \frac{\sigma^2}{2}\left(|k+1|^{2H} - 2|k|^2 + |k-1|^{2H}\right) \sim k^{2H-2}$$

and the variance of the summed process is (see theorem 3.16b))

$$\mathrm{Var}\left(Z^{(m)}\right) = \sigma^2 m^{2H}$$

The argumentation for the FGN will be transmitted to the observed signal (resp. time series). The blocks $X^{(m)}$ can be ordered in a dyadic tree, which is suitable for synthesizing a process from wavelet coefficients. Let

$$V_{j,k} = X_k^{(2^{m-j})} = V_{j+1,2k} + V_{j+1,2k+1}, \ k = 0,\ldots,2^j - 1 \quad (3.92)$$

Here, the m-th level correspond to the process $X_k^{(2^{m-m})} = X_k^{(1)} = X_k$. The tree is build from the bottom up and the upper level to level $j = 0$ corresponds to the summed process $X_k^{(2^m)}$. The aggregated process forms the bases. It can reflect e.g. the amount of bytes per μs up to the scale k. The whole time period of length 2^n will be splitted in single scales (time spots) $k = 0, 1, \ldots, 2^n - 1$. One can see from the definition of the tree $(V_{j,k})$ in (3.92) that the predecessor is the sum of both successor. This motivated in the following way. On starts with the whole data load, which is described by the coarsest random variable, in our case $V_{0,0}$. Now we proceed with finer scales and the aggregated traffic, which is summed up on one scale is splitted. This can be justified for e.g. by application protocols, like HTTP, and the transport protocols like TCP. They divide the whole traffic in single segments. But this procedure is not deterministic. To generate the observed traffic synthetically, one has to introduce further random variables into the model.

We can construct now the tree from up to down 'synthetic', i.e. one can build up by given wavelet coefficients starting from level $j = 0$, thus $V_{0,0}$, iteratively and generate the corresponding fractional white noise. We consider as above two fundamental different approaches: the WIG and MWM.

Gaussian WIG Model

For this we start with a RV $V_{0,0}$, which is $\mathcal{N}(m2^m, \sigma^2 2^{2mH})$-distributed. Up to now no fine structure is incorporated and only the long-range dependence is exhibited. Furthermore define a certain auxiliary random variables $Z_{j,k}$, which are $\mathcal{N}(0, \sigma^2(2^{2-2H} - 1)2^{2(m-j)H})$-distributed, i.e. for $k = 0, \ldots, 2^j - 1$ the $Z_{j,k}$ are iid. Why did we chose these distributions? First, one uses the discrete recursive definition

$$V_{j+1,2k} = \frac{V_{j,k} + Z_{j,k}}{2}, \quad V_{j+1,2k+1} = \frac{V_{j,k} - Z_{j,k}}{2} \tag{3.93}$$

The figure 3.21 shows on the one hand the splitted traffic into a binomial tree with progressively finer scales and below both approaches for computation.

It can be clearly seen that from a knot in the j-th level the two new knots below are formed with the help of the perturbation $Z_{j,k}$, where one gets the even knots (or left ones) by the arithmetic mean and the odd one (or right one) by the weighted mean. The $Z_{j,k}$ represent the normalized Haar wavelet coefficients of the available data material using wavelet analysis. To gain asymptotically a Gaussian model (i.e. averaging of large times), we have to impose certain conditions to the random variables. For this we choose $V_{0,0}$ and $Z_{j,k}$ Gaussian. Then the $Z_{j,k}$ have to be within the same scaling j identical distributed. In addition the expectations $\mathbb{E}(V_{j,k}) = \mathbb{E}(V_j)$ show up independently of k. It follows

$$\mathbb{V}\mathrm{ar}(V_{j+1}) = \frac{\mathbb{V}\mathrm{ar}(V_j) + \mathbb{V}\mathrm{ar}(Z_j)}{4} \tag{3.94}$$

From these presumptions we get for the RV $V_{0,0}$ and $Z_{j,k}$:

- $V_{0,0}$ is $\mathcal{N}(m2^m, \sigma^2 2^{2mH})$-distributed.

3.8 Multifractal Models and the Influence of Small Scales   275

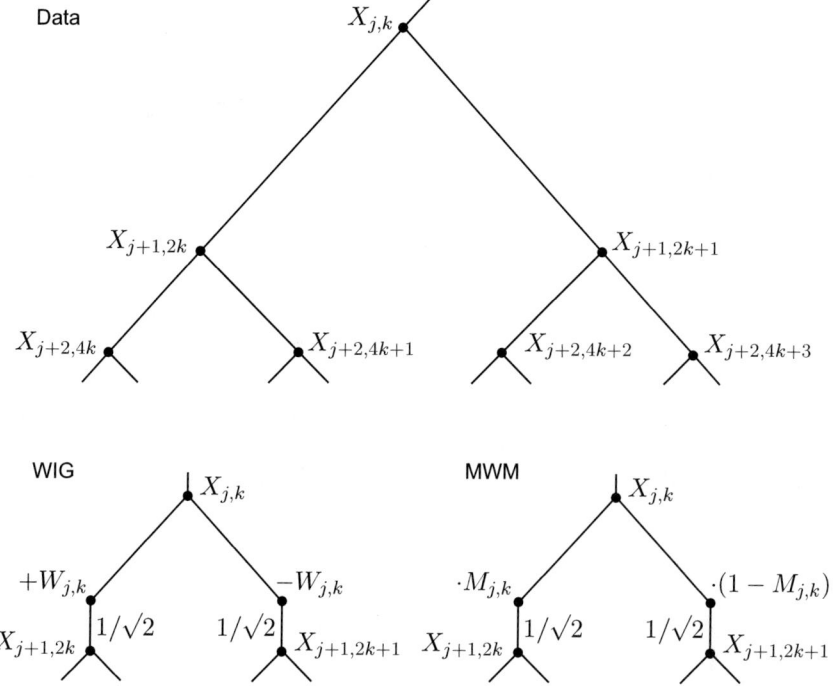

Fig. 3.21. Comparison of data amounts in different scalings, WIG and multiplicative wavelet model (MWM)

- $Z_{j,k}$ are $\mathcal{N}(0, \sigma^2(2^{2-2H} - 1)2^{2(m-j)H})$-distributed, i.e. for $k = 0, \ldots, 2^j - 1$ the $Z_{j,k}$ are iid.

Then, with the help of $(V_{j,k})$, we have constructed a Gaussian process with

$$\mathbb{V}\mathrm{ar}(V_j) = \sigma^2 2^{2(n-j)H} \tag{3.95}$$

Hence, we obtain the same variance as in the FBN case with Hurst exponent H, mean m and Variance σ^2. If one considers this triangle of the $Z_{j,k}$ asymptotical for $j \to \infty$, then the tree converges, i.e. the $V_{j,k}$ converge towards the FGN according to the functional central limit theorem or generalized Donsker principle. We call this model *wavelet-domain independent Gaussian* or simply *WIG* model.

Multifractal Wavelet Model

The disadvantage of the preceding approach lies in the use of the Gaussian marginal distributions, which do not take the bursty character into account, revealing the interarrival times or the load in the network. In addition, the

IP traffic will not be Gaussian on small scales. The basic approach from the WIG model will prevail, but in contrast, we use the iteration

$$V_{j+1,2k} = V_{j,k}M_{j,k}, \quad V_{j+1,2k+1} = V_{j,k}(1 - M_{j,k}) \tag{3.96}$$

Here, the $M_{j,k}$ are symmetric around $\frac{1}{2}$ with values in $]0,1[$ and iid RV for each level j. This assures the stationarity of the process in the level j. Simultaneously let $V_{0,0}$ be positive. We can use beta distributions or simply discrete distributions, which fit in practice. The limit process (for $j \to \infty$) is lognormal for small time scales and bursty, as seen later. This model will be called *multifractal wavelet model (MWM)* and coincides with the general approach of binomial cascades from section 3.8.5.

How do we have to choose the distribution of $V_{0,0}$ and $M_{j,k}$? If we want to obtain again a suitable asymptotic, then we have to use the approach of Gaussian distributions. Since we choose a multiplicative structure we will select a lognormal distribution for the RV $M_{j,k}$, because the product of lognormal distributed RV is again a lognormal distributed RV. A disadvantage of lognormal distributed RV lies in its unboundedness. This is surely not suitable for the IP traffic. Hence we select a *symmetric beta distribution*, i.e.

$$M_{j,k} \text{ is distributed as } \beta(p_j, p_j) \tag{3.97}$$

The variance of a $\beta(p_j, p_j)$ RV is

$$\mathrm{Var}\left(\beta(p_j, p_j)\right) = \frac{1}{4 + 8p_j} \tag{3.98}$$

(see eg. [38]). The parameter p_j has to be chosen, so that it matches the data of the second order statistic (see the definition of the second order statistic 4.1.1 and there theorem 4.9). If we introduce a further parameter for an approach of different distributions, one can match the statistic of higher order. For the initial variable $V_{0,0}$ we can use any distribution with support in the positive real numbers. We choose e.g.

$$V_{0,0} \text{ is distributed as } aM_{-1}$$

where M_{-1} is in turn $\beta(p_{-1}, q_{-1})$-distributed. If the traffic on the coarsest scaling will not be Gaussian, then we pick $p_{-1} \neq q_{-1}$, and M_{-1} is no longer symmetric beta distributed. With sufficient aggregation of the traffic we can assume that the distribution is Gaussian. Then we have $p_{-1} = q_{-1}$.

How can we determine the parameter in (3.98)? We have to compute a and p_j, $j = -1, 0, 1, \ldots, n-1$. It holds

$$a = \mathbb{E}(V_{0,0}) \quad \text{and} \quad p_{-1} = \frac{1}{2}\left(\frac{\mathbb{E}(V_{0,0})^2}{\mathrm{Var}(V_{0,0})} - 1\right)$$

From (3.96) we get for $j = -1, 0, 1, \ldots, n-1$ because of the independence of the $V_{j,k}$ and $M_{j,k}$

$$\mathbb{E}(V_{j+1}^2) = \mathbb{E}(V_{j,k}^2)\mathbb{E}(M_j^2) = \sigma^2 \prod_{i=-1}^{j} \mathbb{E}(M_i^2) \qquad (3.99)$$

With (3.98) and (3.99) it follows

$$p_j = \frac{\mathbb{E}(V_j^2) - 2\mathbb{E}(V_{j+1}^2)}{4\mathbb{E}(V_{j+1}^2) - \mathbb{E}(V_j^2)}, \quad 0 \le j \le n-1$$

Hence, we can obtain the p_j from the estimation of $\mathbb{E}(V_{j+1}^2)$. It is $\mathbb{E}(V_j^2) = 2^{-j}\mathbb{E}(V_{0,0}^2)$, and thus, the determination of the second moment is nothing else than computing the variance. Selecting for a Hurst exponent H the p_j in dependence of H, to incorporate the LRD property, then we get

$$p_j = 2^{2H-2}(p_{j-1}+1) - \frac{1}{2}, \quad j = 0, \dots, n-1$$

Thus, we only need the parameter p_{-1}, H and a.

Traffic Measurements on Small Scales

Up to now we determined the distribution of the multipliers $M_{j,k}$ in connection with the covariance structure of the observed traffic. Now we have to incorporate the small scales in the modeling. We first assume a general distribution and will later fit the data after estimations.

The long-range dependence is described by the Hurst exponent of the second order statistic and by the heavy-tail distribution of the on-off models. With this and the known methods (see e.g. the sections 3.4.3, 4.2 and 5.1.1) we can compute the Hurst exponent. But these information do not suffice to judge correctly the QoS resp. the network performance. Here, we have to consider the small scales.

In general the small time intervals will be given in as RTT (Round Trip Time). But according to the aggregated traffic the situation can change much faster at server, router or on the links. The pure consideration on large scales resp. the asymptotic as in section 3.4.4 is too coarse.

We now use the multifractals. Hence, theoretical results from sections 3.8.2 and 3.8 are applied. We will give references to the results at the corresponding spots.

Considering a path of a general process or an observed time series of an IP-based traffic $(Y(t))_{t \in [0,1]}$, we can describe the change on the small scales using the expression (see the coarse exponents of the increment (3.112))

$$|Y(t+\delta) - Y(t)| \approx \delta^{\alpha(t)}$$

As described in section 3.8.3, a small $\alpha(t)$ indicates that the traffic exhibits high burstiness. We can detect that the difference depends on time t and can assume different values. In the following we try to verify and quantify this by observations, and in addition, to derive a method how to calibrate MW models with the respective information. The decisive basis is formed by the multifractal analysis or the multifractal formalism. There, not only the second order statistic is used, but also the ones of higher order (see again section 3.8.3). This formalism shows the possibility to use non Gaussian models, and it demonstrates, how they can describe the traffic on smaller scales more properly.

Structure on Small Scales

We use the notation from section 3.8.3. For $j \in \mathbb{N}$ let $N^{(n)}(\alpha, 2^{-j})$ denote the number of bursts of magnitude α on the scale 2^{-j} (see (3.113)). We can insert this into a relation of the q-th moment of the increments

$$\mathbb{E}\left(\sum_{k_j=0}^{2^j-1} |Y((k_j+1)2^{-j}) - Y(k_j 2^{-j})|^q\right) \approx 2^{-jT(q)} \qquad (3.100)$$

(see (3.119)). Here, $T(q)$ means the deterministic partition function (3.117). With (3.119) and considering the relation (3.118) resp. the proposition 3.93 we get for the grain-based spectral $N^{(n)}(\alpha, 2^{-j})$ the following relationship

$$N^{(n)}(\alpha, 2^{-j}) \approx 2^{T^*(\alpha)} \qquad (3.101)$$

where $T^*(\alpha) = \inf_q \{q\alpha - T(q)\}$ is the Legendre transform (see (3.119)). We describe in the sections 3.8.3 and 3.8.4 the value $T^*(\alpha)$ in more detail, present further properties and formulate a relationships to other spectrals and functions. We can represent $T^*(\alpha)$ as the smallest distance between the deterministic partition function $T(q)$ and the line $q\alpha$. If $T^*(\alpha)$ is negative, which implies that there is no t with $\alpha(t) = \alpha$. We can denote T^* as the multifractal spectral.

Since in practice it is impossible to consider arbitrary small scales there exists a minimum in the resolution and hence, only a finite number of dyadic time intervals. To determine $T(q)$ from data, we use again a log-log plot (see sections 4.1.2 and 4.1.3). For this we indicate on the x-axis the values of α and map them to the $T^*(\alpha)$ values on the y-axis, where the $T^*(q)$ are computed according to (3.100), (3.101) and theorem 3.95. We can obtain the $T(q)$ for the MWM (see (3.116)). The formula is

$$T(q) = -\frac{1}{j} \log_2 \mathbb{E}\left(\sum_{k=0}^{2^j-1} |V_{j,k}|^q\right) \qquad (3.102)$$

How can we derive the necessary parameter with the help of formula (3.102) to calibrate the MWM? For this we assume that the multipliers $M_{j,k}$ converge for $j \to \infty$ in distribution towards a RV $M \sim \beta(p,p)$. Then we have according to (3.135)

3.8 Multifractal Models and the Influence of Small Scales

$$T_{MWM}(q) = -1 - \log_2 \mathbb{E}(M^q)$$

Hence, it follows by (3.137)

$$T_{MWM}(q) = \begin{cases} -1 - \log_2 \frac{\Gamma(p+q)\Gamma(2p)}{\Gamma(2p+q)\Gamma(p)} & \text{for } q > -p \\ -\infty & \text{for } q \leq -p \end{cases} \quad (3.103)$$

Using instead of the MWM the WIG model with a distribution $V_j \sim \mathcal{N}(0, 2^{-2jH})$, we obtain (see 3.157)

$$T_{WIG}(q) = \begin{cases} qH - 1 & \text{for } q > -1 \\ -\infty & \text{for } q \leq -1 \end{cases}$$

Comparing the form of function $T^*(\alpha)$ with the theoretical computed one $T(q)$, then we realize e.g. in the WIG model

$$T^*(\alpha) = \inf_q (q\alpha - qH + 1) = \inf_{q > -1} (q(\alpha - H) + 1) \quad (3.104)$$

Suppose we have determined a fixed H from the known statistical methods (see section 4.1), then we realize from (3.104) that T^* takes only one value or is linear, since T is linear in α (see e.g. proposition 3.126). The FBM is omnipresent. But the empirical determined curve are strictly concave. This shows clearly that the WIG models do not fit very well to describe the bursty character of the WAN traffic.

How do both models reflect the bursty character resp. the LRD property of the traffic? We will show at selected examples in different scenarios, how the parameter are fitted. The figure 3.22 shows two curves of the measured deterministic grain-based spectral. Clearly we see that there is no linear structure and that thus, the classical approaches with Gaussian models do not match on small scales. The MWM fits decisively well for the values $\alpha \leq 1$.

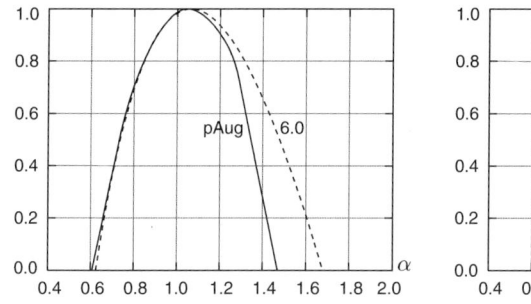
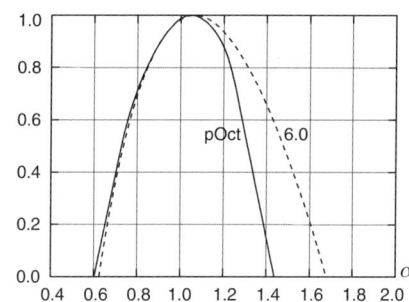

Fig. 3.22. Multifractal spectrum and the Hölder exponent T_α of interarrival times in Bellcore traffic pAug (left) and pOct (right) and fitted MWM model with $p = 6.0$

Remark 3.73. For investigating the irregularity of the processes we will have to employ the multifractal formalism as described above. It relates the amount and degree of irregularity of the paths to the behavior of the moments. Denoting by $\mathbb{P}_j$ the counting measure on $\{1, 2, \ldots, 2^j - 1\}$, then we have for all $q > 0$:

$$\mathbb{P}_j(V_{j,k} \geq 2^{-ja}) = \mathbb{P}_j(V_{j,k}^q \geq 2^{-jqa}) \leq \frac{\mathbb{E}_j(V_{j,k}^q)}{2^{-jqa}} \quad (3.105)$$

We can detect by the principle of large deviation, for which q the estimation is sharp, provided $j \to \infty$. One can consider the right side as the possible, to observe an interarrival time during a realization, which is relatively large compared to the time scale. The most important expression on the right side of (3.105) are the moments. Assuming iid multiplier $M_{j,k}$, we obtain

$$\mathbb{E}_j(V_{j,k}) = (\mathbb{E}_j(M^q))^j$$

Thus, we at a exponential rate of decay for (3.105) in the expression

$$\inf_q (qa + \log_2 \mathbb{E}(M^q))$$

hence, a convex function in a. This implies for the WIG method

$$\mathbb{E}_j(V_{j,k}^q) = c_q \left(2^{-qH}\right)^j$$

where the FGN is applied, or with other words that the $V_{j,k}$ are $\mathcal{N}(0, \sigma^2 2^{-2jH})$-distributed.

The exponential rate of decay in (3.105) will turn to a constant, i.e. we do not observe any decay! According to the classical result of Adler (Satz 3.16) the auto correlation $X^{(m)}$ in realizations reveals mainly a behavior of the order m^H for all t in case of the FGN. But, this cannot be confirmed by the observation of the real data traffic, which hence, looks more similar to the MWM.

3.8.3 Characteristics of Multifractal Models

Singular Measures

The singular measures are fundamental for describing multifractal traffic. Thus, we start with their description in more detail.

Hölder Exponent

We start with the classical concept of continuity and differentiability. The aim is to determine the behavior of a function or process (Y_t) in a neighborhood of a particular point $t \in [0, 1]$. This means, we want to derive an estimation of the form

$$|Y(s) - Y(t)| \leq C|s - t|^h, \quad s \in\,]t - \epsilon, t + \epsilon[, \quad s, t \in [0, 1]$$

(for a fixed $\epsilon > 0$). We have to emphasize that we consider a path of $(Y(t))_{t \in [0,1]}$, hence, the value of the right side is mainly a random variable, thus, depending on $\omega \in \Omega$. The constant C is connected to the point t and is not of particular interest for us. The degree h of the continuity is more important. If $h = 1$, then we call it *Lipschitz continuity*. Since we are in particular interested in the behavior close to t, a large h implies that the curve resp. path looks smoother. The smaller h the more chaotic the behavior will turn out. We will investigate this more closely. In the sequel we will not explicitly emphasize that we consider a path or realization of $(Y(t))$. If $Y(\cdot)$ is continuous at t, then we have $Y(s) - Y(t) = o(1)$, and if $Y(\cdot)$ is even differential at t, then it follows $Y(s) = Y(t) + Y'(t)(t-s) + o(s-t)$. The last fact leads to the well known *Taylor expansion*. One realizes by the last fact that we try to approximate $Y(s)$ by a polynomial. Thus, the following approach is understandable.

Definition 3.74. *Let $Y(\cdot)$ be a function or a path of a stochastic process and let $t \in [0,1]$. Then $Y(\cdot)$ belongs to the class C_t^h, (denoted by $Y \in C_t^h$), if there is a polynomial $P_t(\cdot)$, such that*

$$|Y(s) - P_t(s)| \leq c|s-t|^h \tag{3.106}$$

for s sufficiently close to t. The polynomial depends on the particular point t. We define the local degree of the Hölder continuity *of Y according to*

$$H(t) = \sup\{h > 0; Y \in C_t^h\} \tag{3.107}$$

According to the above considerations it is clear that we try to find a h as large as possible. Suppose Y has a Taylor polynomial of degree n at t, then we have necessarily $[H(t)] = n$, where $[x]$ is the greatest integer smaller or equal to x and $P_t(\cdot)$ is the Taylor polynomial. But this need not to be the case as the example $Y(x) = 1 + x + x^2 + x^{3.9} \sin(\frac{1}{x})$ reveals.

Certainly of particular interest is the case, if the polynomial $P_t(\cdot)$ is constant and thus, necessarily coincides with the value $Y(t)$. Then, we define the Hölder exponent in the following definition.

Definition 3.75. *The function or path $Y(\cdot)$ is* Hölder continuous of degree $h(t)$ at t, *where*

$$h(t) = \liminf_{\epsilon \to 0} \frac{1}{\log_2(2\epsilon)} \log_2 \sup_{|s-t|<\epsilon} |Y(s) - Y(t)|$$

One measures by $h(t)$ the maximal perturbation of $Y(s)$ from the reference point $Y(t)$ in form of a power of two (thus, the logarithm $\log_2$ for the basis 2), and one considers it in dependence of ϵ, thus, the length of the interval (hence, dividing by the logarithm of length of the interval). Considerable simple is the derivation of the fact that if $h < h(t)$ then it holds $|Y(s) - Y(t)| \leq c|s-t|^h$ and $h(t) \leq H(t)$. If we can choose in (3.106) only a constant, then we have of course $h(t) = H(t)$. It follows a useful proposition.

Proposition 3.76. *If $h(t) \notin \mathbb{N}$, then we have P_t in (3.106) as a constant and $h(t) = H(t)$.*

For the applications we have to avoid any 'continuous' selection of the width of the interval. We have to construct a *sequence* of decreasing intervals around t. For this we choose for t and $n \in \mathbb{Z}$ the coefficient

$$k_n(t) = \left[t 2^{-n}\right] \qquad (3.108)$$

Then we have for all $n \in \mathbb{N}$

$$t \in I_{k_n}^{(n)} = [k_n(t) 2^{-n}, (k_n(t)+1) 2^{-n}[$$

With decreasing n the sequence $(I_{k_n}^{(n)})$ forms a sequence of embedded intervals. Using this sequence of intervals, we can find a good approximation of $h(t)$. For this we define the *coarse Hölder exponent of* $(\cdot)$ *at the point* t

$$h_{k_n}^{(n)} = -\frac{1}{n} \log_2 \sup\{|Y(s) - Y(t)|; s \in [(k_n - 1) 2^{-n}, (k_n + 1) 2^{-n}[\}$$

For an illustration note

$$[(k_n - 1) 2^{-n}, (k_n + 1) 2^{-n}[\subset [t - \epsilon, t + \epsilon[$$
$$\subset [(k_{n+2} - 1) 2^{-n+2}, (k_{n+2} + 1) 2^{-n+2}[$$

if one has $2^{-n+1} \leq \epsilon < 2^{-n+2}$. Thus, it follows the next proposition.

Proposition 3.77. *Let $Y(\cdot)$ be a function or path. Then we have for $t \in [0,1]$*

$$h(t) = \liminf_{n \to -\infty} h_{k_n}^{(n)}$$

The Hölder exponent reflects very intrinsically intrinsic how a function behaves close to the reference point. But for the statistical evaluations and thus, the modeling of the IP traffic another exponent is of importance and is more and more used – the wavelet exponent.

Wavelet Exponent

We already exposed that wavelets build a suitable environment to describe signals and data series. Thus, we start with an arbitrary mother wavelet φ and the corresponding scaling function ψ. We fix the wavelet representation and define as already done in section 3.8.2 for $j, k \in \mathbb{Z}$

$$\varphi_{j,k}(t) = 2^{\frac{j}{2}} \varphi(2^j t - k) \quad \text{and} \quad \psi_{j,k}(t) = 2^{\frac{j}{2}} \psi(2^j t - k)$$

We construct for a function or path $Y(\cdot)$ the wavelet coefficient

$$\mathcal{D}_{j,k} = \int_{-\infty}^{\infty} Y(s) \varphi_{j,k}(s) ds \quad \text{resp.} \quad \mathcal{C}_{j,k} = \int_{-\infty}^{\infty} Y(s) \psi_{j,k}(s) ds$$

3.8 Multifractal Models and the Influence of Small Scales   283

Then we have (see section 3.8.2)

$$Y(t) = \sum_{j=J_0}^{\infty} \sum_{k \in \mathbb{Z}} \mathcal{D}_{j,k} \varphi_{j,k}(t) + \sum_{k \in \mathbb{Z}} \mathcal{C}_{J_0,k} \psi_{J_0,k}(t)$$

How can we interpret the particular coefficients? Selecting a mother wavelet, centralized with the frequency f_0, the coefficient $\mathcal{D}_{j,k}$ measures the path around $2^{-j} f_0$ and at the point $2^j k$. The scaling coefficients $\mathcal{C}_{j,k}$ reflect the averaging of the path at the point $2^j t$. With j we describe as well known the scaling: The smaller the j, the finer the scaling will be. The value J_0 indicates the coarsest scaling, which is available and which offers still informations.

Remark 3.78. We detect the difference to the LRD phenomena: While we average traffic load for the LRD consideration over large time intervals and let the scaling width tend to infinity, for the wavelet coefficient we are interested in the *local* behavior. This can already be seen using the Hölder exponent and will be detected later for the wavelet exponent.

For the closer investigation those wavelet with compact support are of interest, and here, we choose the 'most simple' wavelet, the *Haar wavelet*. We can represent the Haar wavelet in the form

$$\varphi(t) = \psi(2t) - \psi(2t-1)$$

where the scaling function ψ is the indicator function of the unit interval. Thus, $J_0 = 0$ is fixed, since we restrict our function resp. the path $Y(\cdot)$ to the interval $[0,1]$ and extend it by 0 elsewhere. The sequence of wavelets can be easily representable: The finer ones 'are halving' the preceding ones. We can describe this fact very easily in a tree. Formally one orders the coefficients in a recursive scheme

$$\mathcal{D}_{j,k} = \frac{1}{\sqrt{2}} (\mathcal{D}_{j+1,2k} + \mathcal{D}_{j+1,2k+1})$$

$$\mathcal{C}_{j,k} = \frac{1}{\sqrt{2}} (\mathcal{C}_{j+1,2k} + \mathcal{C}_{j+1,2k+1})$$

The local behavior of $Y(\cdot)$ can be described very well.

Proposition 3.79. *For fixed $t \in [0,1]$ and a sequence $(k_n = k_n(t))_{n \in \mathbb{Z}}$ as in (3.108) holds: If*

$$|Y(t) - Y(s)| = \mathcal{O}\left(|s-t|^h\right), \text{ for } s \to t \quad (3.109)$$

and φ has a compact support $\int_{-\infty}^{\infty} |\varphi(s)| ds < \infty$, $\int_{-\infty}^{\infty} \varphi(s) ds = 0$, thus, it follows

$$2^{\frac{n}{2}} |\mathcal{D}_{n,k_n}| = 2^n \left| \int_{-\infty}^{\infty} Y(s) \varphi(2^n s - k) ds \right| = \mathcal{O}(2^{-nh}), \text{ for } n \to \infty \quad (3.110)$$

We will give some remarks on this.

Remark 3.80. The equation (3.109) tells us that in a neighbourhood of t the functions change by the order of $|s-t|^h$; hence, a large h implies only a small change and the path is smooth, and vice versa a very small h indicates an abrupt change.

Remark 3.81. The assertion (3.110) shows us an idea, computing the exponent h with the help of the wavelet coefficient this time. Suppressing the Landau symbol $\mathcal{O}$, then we get a equation and solve it for k

$$h = -\frac{1}{n}\log_2\left|2^{\frac{n}{2}}\mathcal{D}_{n,k_n}\right| \tag{3.111}$$

Remark 3.82. The proof of the assertion is still valid, if the wavelet does not have a compact support, but has in turn a sufficient fast decay behavior. For our purposes we skip the details.

Remark 3.83. If we change the wavelet representation, then the finer scaling of the wavelet coefficient will be indicated by decreasing j. Then we have to revert the direction of limiting, i.e. by exchanging $n \to -\infty$. Since this is not practicable, we stay with the usual definition of finer scaling with increasing j.

From remark 3.81 resp. (3.111) we can deduce the definition of the *coarse wavelet coefficient*.

Definition 3.84. *The* coarse wavelet exponent *is defined according to*

$$w_{k_n}^{(n)} = -\frac{1}{n}\log_2\left|2^{\frac{n}{2}}\mathcal{D}_{n,k_n}\right|$$

We call

$$w(t) = \liminf_{n\to\infty} w_{k_n}^{(n)}$$

the local wavelet exponent *at* t.

Summarizing that the Hölder exponent indicates the irregularity of $Y(\cdot)$, but only under the condition that the approximating polynomial is constant. The advantage of the wavelet exponent lies in the independence of polynomial trends, which is caused by the vanishing moments $\int_{-\infty}^{\infty} t^m \varphi(t) dt$. But we can estimate the Hölder exponent only with complicated methods (see [185]).

Considering the Hölder exponent more coarser, and this is important for the so called *increasing* or *monotone stochastic processes*, then we can define a further exponent.

Definition 3.85. *The coarse exponent of the increase for a function or path* $Y(\cdot)$ *is given by*

$$\alpha_{k_n}^{(n)} = \frac{1}{n}\log_2\left|Y\left((k_n+1)2^{-n}\right) - Y\left(k_n 2^{-n}\right)\right| \tag{3.112}$$

3.8 Multifractal Models and the Influence of Small Scales

In the survey section 3.8.2 we selected this particular exponent for the treatment of the IP-based traffic. It can be handled easily because of the simple approach.

Finally analogously to the two preceding exponents we can define the *local exponent of the increments*:

$$\alpha(t) = \liminf_{n \to \infty} \alpha_{k_n}^{(n)}, \ t \in [0,1]$$

Further exponents, as the *Choquet capacity* can be found in [164]. But for our purposes they are not of interest and are not significant for the description of the IP traffic.

Multifractal Analysis – Partition Function

As seen we can quantify the irregularity of a function or path of a stochastic process in different ways. Since we do not differ explicitly between the Hölder-, wavelet- or exponent of the increments, we will denote by

$$s_k^{(n)} \in \left\{ h_k^{(n)}, w_k^{(n)}, \alpha_k^{(n)} \right\}, \ k = 0, \ldots, 2^n, \ n \in \mathbb{Z}$$

a particular coarse exponent for the interval $I_k^{(n)}$. As already mentioned $s_k^{(n)}$ is a random variable in the case of a stochastic process.

It seems clear that for different t we can expect different values of $h(t), w(t)$ and $\alpha(t)$. But how do they behave? How are the values distributed over the interval in question, here $[0, 1]$? The answer is the main aim of the multifractal analysis. It is called for a geometric resp. statistical representation of the function $t \longmapsto s_k(t)$. Roughly speaking we want to determine how often the particular exponent a appears in time, i.e. we have to deal with the question how the set

$$K_\alpha = \{t \in [0,1]; s_k(t) = a\}$$

can be described. We want to give a precise derivation for this heuristic fact.

The Spectral of the Exponent

We define for a number $a \in \mathbb{R}$ two sets

$$E^a = \left\{ t \in [0,1]; \liminf_{n \to \infty} s_{k_n}^{(n)} = a \right\}$$
$$K^a = \left\{ t \in [0,1]; \lim_{n \to \infty} s_{k_n}^{(n)} = a \right\}$$

In general these sets look very 'fractal'. That means, their geometric structure is relatively complex. For a description we introduce the so called *Hausdorff spectrum*.

286   3 Mathematical Modeling of IP-based Traffic

Definition 3.86. *The function*

$$a \longmapsto dim(E^a)$$

is called Haussdorff spectrum, *where $dim(E^a)$ represents the Hausdorff dimension of the set E (see [258]).*

Often in literature the set K^a is considered. But this assumes the convergence of the sequence $s_k^{(n)}$, while the definition of E^a only needs the boundedness from below. Nevertheless the determination of the Hausdorff dimension is difficult in practice and not very practicable. Thus, we introduce a further concept, which 'approximates' the Hausdorff dimension. The definition of the Haussdorff dimension is relatively complex and not of importance for our purposes. Thus, we will not explicitly consider it in our further study and refer the reader to the literature.

Grain-Based Spectrum

We assume the notations from above. Starting point is the determination and computation of

$$N^{(n)}(a,\epsilon) = \mathrm{card}\left(k = 0,\ldots,2^n-1; a-\epsilon \leq s_k^{(n)} \leq a+\epsilon\right) \qquad (3.113)$$

for $\epsilon > 0$. If the random variable satisfies $s_k^{(n)} = \infty$, then we do not consider it in our further countings. 'card' means the number of elements in a set. It is bounded by 2^n for $N^{(n)}(a,\epsilon)$. With the help of the set $N^{(n)}(a,\epsilon)$ we can define the grain-based spectrum.

Definition 3.87. *The* grain-based spectrum *is the function*

$$f(a) = \lim_{\epsilon \downarrow 0} \limsup_{n \to \infty} \frac{1}{n} \log_2 N^{(n)}(a,\epsilon)$$

Useful is the following alternative definition (exchange the supremum by infimum)

$$\underline{f}(a) = \lim_{\epsilon \downarrow 0} \liminf_{n \to \infty} \frac{1}{n} \log_2 N^{(n)}(a,\epsilon)$$

The function offers beside the possibility of the computation of the Hausdorff dimension also another meaning. We want to go into more details in this section. It is for certain importance for modeling the TCP-based IP traffic. We have the following estimations.

Proposition 3.88.

$$dim(E^a) \leq f(a)$$
$$dim(K^a) \leq \underline{f}(a)$$

3.8 Multifractal Models and the Influence of Small Scales

An decisive interpretation will be revealed by the grain-based spectrum, if we examine the principle of large deviation. For this purpose we consider the quotient

$$\frac{N^{(n)}(a, \epsilon)}{2^n}$$

as the probability finding a number $k_n = 0, \ldots, 2^n - 1$, so that for a given function or realization $Y(\cdot)$ the value $s_{k_n}^{(n)}$ lies in the interval $[a - \epsilon, a + \epsilon]$. We want to have a close look at the grain-based spectrum $f(\hat{a})$ for a special point $\hat{a}$. If we assume for simplicity for a small $\epsilon > 0$ $f(\hat{a}) = \frac{1}{n} \log_2 N(\hat{a}, \epsilon)$ and presume that for all $k = 0, \ldots, 2^n - 1$ the value $s_k^{(n)}$ lies in the interval $[\hat{a} - \epsilon, \hat{a} + \epsilon]$, then $f(\hat{a})$ attains the value 1. Then we have in particular

$$\lim_{n \to \infty} s_{k_n}^{(n)} = \hat{a}$$

Of course the reverse direction holds as well. But what happens, if we choose $a \neq \hat{a}$? Then for a particular $\epsilon > 0$ the interval $[a - \epsilon, a + \epsilon]$ does not contain $\hat{a}$ and we have a sequence of coarse exponents $s_{k_n}^{(n)}$, which do not fall into the interval $[a - \epsilon, a + \epsilon]$. Thus, the number $N^{(n)}(a, \epsilon)$ decreases relatively fast with a smaller ϵ and the value of $f(a)$ decreases also fast for an $a \neq \hat{a}$.

We did not give any comments whether we can write instead 'lim sup' in the definition of $f(a)$ simply the limit 'lim'. This we want to investigate now. For this we have to introduce some new notions, which also appear in the construction of IP traffic models (see e.g. the WIG and MWM models).

First we define a random variable $\xi_n = -ns_K^{(n)}$. We remind that we consider a realization or a path of a stochastic process $(Y(t))_{t \in [0,1]}$. Thus, the coarse exponent $s_k^{(n)}$ is a random variable. But for ξ_n we choose K randomly out of the set $\{0, 1, \ldots, 2^n - 1\}$, where we consider for this set the uniform distribution U_n. We replace $N^{(n)}(a, \epsilon)$ by a partition in the following definition.

Definition 3.89. *The partition function τ of the path $Y(\cdot)$ is defined for $q \in \mathbb{R}$ according to*

$$\tau(q) = \liminf_{n \to \infty} -\frac{1}{n} \log_2 S^{(n)}(q)$$

Here, we have

$$S^{(n)}(q) = \sum_{k=0}^{2^n - 1} \exp\left(-nqs_k^{(n)} \log(2)\right) = \sum_{k=0}^{2^n - 1} 2^{-nqs_k^{(n)}} \quad (3.114)$$

$$= 2^n \mathbb{E}_n \left(2^{-nqs_K^{(n)}}\right) = \mathbb{E}_n \left(2^{-q\xi_n}\right)$$

Note that the expectation $\mathbb{E}_n$ is constructed according to the uniform distribution U_n on the set $\{0, 1, \ldots, 2^n - 1\}$ and log is the natural logarithm. We mention the notation $2^{-q\infty} = 0$. How can we represent the grain-based spectrum with the help of the partition function? The next theorem, which has an generalization, gives an answer.

Theorem 3.90. *Assume for all q the limit*

$$\tau(q) = \lim_{n\to\infty} -\frac{1}{n}\log_2 S^{(n)}(q) \tag{3.115}$$

exists and assume that $\tau : \mathbb{R} \longrightarrow \mathbb{R}$ is differentiable. Then there exists also the double limit

$$f(a) = \lim_{\epsilon\downarrow 0}\lim_{n\to\infty}\frac{1}{n}\log_2 N^{(n)}(a,\epsilon)$$

and we have for all $a \in \mathbb{R}$

$$\overline{f}(a) = \underline{f}(a)$$
$$f(a) = \tau^*(a) = \inf_{q\in\mathbb{R}}(qa - \tau(q))$$

Unfortunately the assumption is not fulfilled in most applications and thus, we have to mention the following estimation.

Proposition 3.91. *We have for all $a \in \mathbb{R}$*

$$\overline{f}(a) \le \tau^*(a) \tag{3.116}$$

Often one wants to obtain a good estimation of the partition function and hence, one considers not necessarily a random variable. Thus, we try to investigate the values global, though they depend normally on the path. For this we introduce the deterministic envelope.

Deterministic Envelope

We consider $s_K^{(n)}$ as random variable on the product space $(\Omega \times \{0,1,\ldots,2^n-1\})$. Here, K is uniformly distributed on $\{0,1,\ldots,2^n-1\}$ and independent of $\omega \in \Omega$.

Definition 3.92. *We define the deterministic envelope function of the process $(Y(t))_{t\in[0,1]}$ for $q \in \mathbb{R}$ by*

$$T(q) = \liminf_{n\to\infty} -\frac{1}{n}\log_2 \mathbb{E}_\Omega\left(S^{(n)}(q)\right)$$

Often we will not and cannot restrict the process $(Y(t))$ on a compact interval as $[0,1]$. Then, we have to define the values differently. Suppose e.g. that $(Y(t))_{t\in\mathbb{R}}$ is a process on the whole real line $\mathbb{R}$. Then we substitute (3.114) by

$$S^{(n)}(q) = \lim_{N\to\infty}\frac{1}{N}\sum_{k=0}^{N2^n-1} 2^{-nqs_k^{(n)}}$$

and (3.113) by

$$N^{(n)}(a,\epsilon) = \text{card}\left\{k = 0,\ldots,N2^n - 1; a - \epsilon \le s_k^{(n)} \le a + \epsilon\right\} \tag{3.117}$$

3.8 Multifractal Models and the Influence of Small Scales

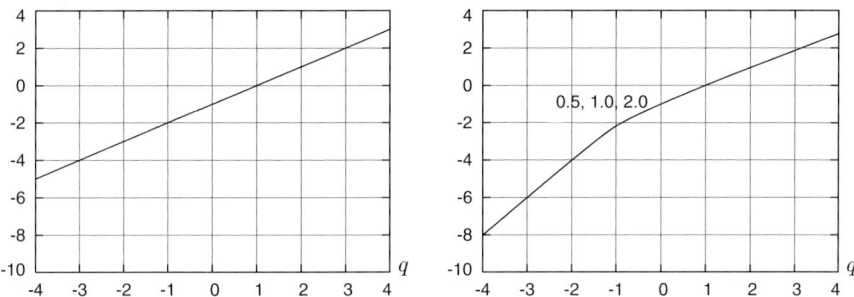

Fig. 3.23. Deterministic envelope function $T(q)$ for deterministic distribution (left) and exponential distributions with $\lambda = 0.5, 1.0, 2.0$ (right)

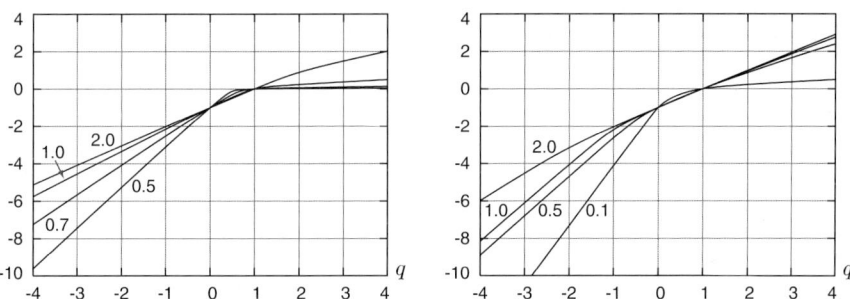

Fig. 3.24. Deterministic envelope function $T(q)$ for Pareto distributions with $k = 1$ and $\alpha = 0.5, 0.7, 1.0, 2.0$ (left) and Weibull distributions with $\beta = 1$ and $\tau = 0.1, 0.5, 1.0, 2.0$ (right)

If $(Y(t))$ is ergodic, then we obtain $S^{(n)}(q) = 2^n \mathbb{E}_\Omega(2^{-nqs_K^{(n)}})$ a.s. (resp. the uniform distribution), and thus, naturally $\mathbb{E}_\Omega(S^{(n)}(q)) = S^{(n)}(q)$ a.s. resp.

$$T(q) = \tau(q, \omega) \text{ a.s.} \tag{3.118}$$

It is evident that we cannot expect (3.118) in general (note that we assumed an ergodic process!). Nevertheless, we have (3.118) in most interesting situations and $T(q)$ can be considered as a *deterministic envelope* of $\tau(q, \omega)$. We have added ω to emphasize the dependence of the path.

Proposition 3.93. *We have with probability 1*

$$\tau(q, \omega) \geq T(q)$$

for all $q \in \mathbb{R}$, satisfying $T(q) < \infty$.

We defined τ^* and denoted the existence of the limit (3.115) in theorem 3.90 as crucial. This limit exists for the so called binomial cascades, as we will then treat as example in one of the next sections. In turn, they are used for modeling the IP traffic. But in general for a lot of applications of these processes we can state and use the inequality

$$\limsup_{n\to\infty} -\frac{1}{n}\log_2 S^{(n)}(a) \leq \underline{f}^*(a) = \inf_{q\in\mathbb{R}}\left(qa - \underline{f}(a)\right)$$

With this we obtain for $a \in \mathbb{R}$

$$f(a) = T^*(a) = \inf_{q\in\mathbb{R}}\left(qa - T(a)\right) \tag{3.119}$$

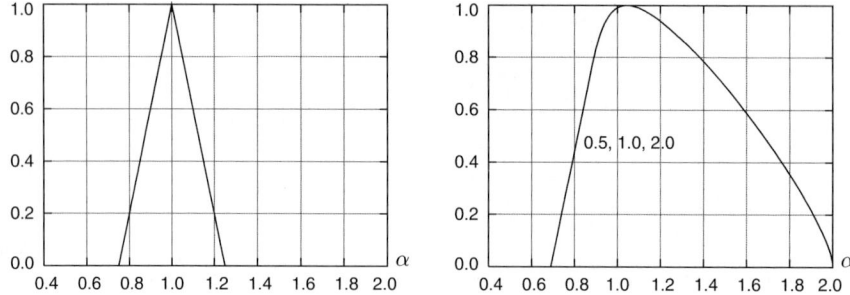

Fig. 3.25. Multifractal spectrum and the Hölder exponent T_α for deterministic distribution (left) and exponential distributions with $\lambda = 0.5, 1.0, 2.0$ (right)

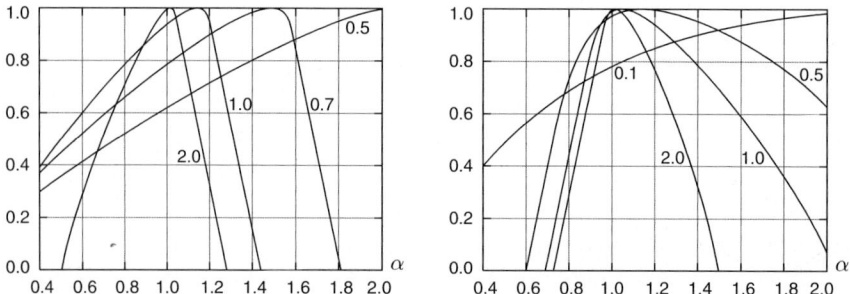

Fig. 3.26. Multifractal spectrum and the Hölder exponent T_α for Pareto distributions with $k = 1$ and $\alpha = 0.5, 0.7, 1.0, 2.0$ (left) and Weibull distributions with $\beta = 1$ and $\tau = 0.1, 0.5, 1.0, 2.0$ (right)

The transition from the partition function τ to the deterministic envelope function T consists in changing from averaging within a path over the single coarse exponents to the averaging within and over *all* paths of the process. This in turn implies for the function $f(a)$ that we change now from the probability of one path to the distribution within and over all paths. That means that

3.8 Multifractal Models and the Influence of Small Scales

we have to average over $N^{(n)}(a,\epsilon)$ in dependence of ω. Fixing n, a and ϵ, we let $\eta(\omega) = \mathbf{1}_{s_K^{(n)}(\omega) \in [a-\epsilon, a+\epsilon[}$. That means $\eta(\omega) = 1$ if $s_k^{(n)}(\omega) \in [a-\epsilon, a+\epsilon[$ and $\eta(\omega) = 0$ else. Here, $(\omega, k) \in \Omega \times \{0, 1, \ldots, 2^n - 1\}$. Clearly we obtain the equation $\mathbb{E}(\eta) = \mathbb{P}_{\Omega \times 2^n}(\eta = 1)$ ($\mathbb{P}_{\Omega \times 2^n}$ is the product measure on the product space $\Omega \times \{0, 1, \ldots, 2^n - 1\}$). This probability can be computed according to the well known rule of Fubini for product measures. Hence, it is possible to exchange the integrations: Either we fix a k and average over all paths (thus, averaging over $\mathbb{P}_\Omega([a-\epsilon, a+\epsilon[)$, or we fix an ω, i.e., a path and average first within the path (then we obtain $\frac{N^{(n)}(a,\epsilon)}{2^n}$). Subsequently we use the expectation over all paths. Summarizing we obtain

$$\mathbb{P}_{\Omega \times 2^n}\left([a - \epsilon \leq s_K^{(n)} < a + \epsilon[\right)$$
$$= 2^{-n} \sum_{k=0}^{2^n-1} \mathbb{P}_\Omega\left([a - \epsilon \leq s_k^{(n)} < a + \epsilon[\right)$$
$$= \mathbb{E}_\Omega\left(\frac{N^{(n)}(a,\epsilon)}{2^n}\right)$$

With this derivations we can define the *deterministic grain spectrum*.

Definition 3.94. *The deterministic grain spectrum of a path will be define for $a \in \mathbb{R}$ by*

$$F(a) = \lim_{\epsilon \downarrow 0} \limsup_{n \to \infty} \frac{1}{n} \log_2 \mathbb{E}_\Omega\left(N^{(n)}(a,\epsilon)\right)$$

As for the grain-based spectrum $f(a)$ we set

$$\underline{F}(a) = \lim_{\epsilon \downarrow 0} \liminf_{n \to \infty} \frac{1}{n} \log_2 \mathbb{E}_\Omega\left(N^{(n)}(a,\epsilon)\right)$$

Thus, the random variable $f(a)$, which is dependent of the path, turns into a real number. As already done for the grain-based spectrum we can formulate a main theorem for the deterministic spectrum.

Theorem 3.95. *We have for all $a \in \mathbb{R}$*

$$F(a) \leq T^*(a)$$

In particular: Assume for all $a \in \mathbb{R}$ the limit $T(a)$ exists and the function $a \longmapsto T(a)$ is concave and differentiable, then the limit

$$F(a) = \lim_{\epsilon \downarrow 0} \lim_{n \to \infty} \frac{1}{n} \log_2 \mathbb{E}_\Omega\left(N^{(n)}(a,\epsilon)\right) \tag{3.120}$$

exists and it follows

$$F(a) = \underline{F}(a) = T^*(a) \tag{3.121}$$

How does the random variable $f(a)$ behave with respect to the deterministic value $F(a)$? The upcoming theorem gives an answer.

Theorem 3.96. *For a fixed $a \in \mathbb{R}$ follows*

$$f(a, \omega) \le F(a) \text{ a.s.}$$

Suppose for all $n \in \mathbb{N}$ the set of random variables $s_k^{(n)}$, $k = 0, \ldots, 2^n - 1$ is iid and suppose $F(a) = \underline{F}(a) > 0$, then we have a.s.

$$f(a, \omega) = \underline{f}(a, \omega) = F(a) \tag{3.122}$$

Note that only the equality in the limit on both sides is required. This does not mean that the limit need to exist according to (3.120). If $(Y(t))$ has independent increments (which we want to assume in our IP models), then we can apply the fact (3.122) to the process $(Y(t))$, and not only for the coarse exponents of the increments but also for the coarse Hölder- and wavelet exponent.

We summarize the results in a corollary.

Corollary 3.97. *We assume that the limit $T(a)$ exists for all $a \in \mathbb{R}$ and that the function $a \longmapsto T(a)$ is concave and differentiable. Let further up to a finite set of $n \in \mathbb{N}$ the coarse exponents $(s_k^{(n)}, k = 0, \ldots, 2^n - 1)$ be iid, which are used for the definition. Let $\hat{a} \in \mathbb{R}$ so that $T(\hat{a}) > 0$. Then we have a.s.*

$$\underline{f}(\hat{a}, \omega) = f(\hat{a}, \omega) = \tau^*(\hat{a}, \omega) = F(\hat{a}) = T^*(\hat{a})$$

The functions T^* and F can certainly assume negative values, while this is not the case for f. Thus, F and T^* are good estimators for f, where they are positive. Finally we summarize all relations in the final theorem of this section.

3.8.4 Multifractal Formalism

We give first a short summary on the different spectra.

Theorem 3.98. *If all values are built according to the same coarse exponent $s_k^{(n)}$, then we have for all $a \in \mathbb{R}$*

$$\dim(K^a) \le \underline{f}(a) \le f(a) \le \tau^*(a) \overset{a.s.}{\le} T^*(a) \tag{3.123}$$

Here, all but the last inequalities hold pathwise. Analogously we have

$$\dim(E^a) \le f(a) \overset{a.s.}{\le} F(a) \le T^*(a)$$

3.8 Multifractal Models and the Influence of Small Scales

A short comment to the last is in order at this point. On the left side of the inequality we find values, which reflect the singularity resp. irregularity of the path best. In contrast the possibility for computations improves going to the right side, and with the help of the Legendre transformations τ^*, F^* and T^* we are able to compute the values sufficiently. The assertion of theorem 3.98 is often called the *multifractal formalism*. Nevertheless another way of expression is commonly in use: If there is equality of the partition function with its Legendre transform, then we say the *multifractal formalism holds*. In this case it holds, that means in (3.123) appears equality (see for more details [122]). Though if in some situations we can detect equality, the inequality is valid and commonly (considered pathwise). On the other hand we find a lot of interesting relationships between the grain-based spectrum and the partition function. We will state them in the next section.

We start with the main result and remark that the coarse exponents $s_k^{(n)}$ with infinite values do not contribute to the construction of the grain-based spectrum resp. partition function. Hence, we assume that all exponents are finite. Simultaneously, we know that the computation of the grain-based spectrum and partition function is difficult. Thus, the following theorem is of great importance.

Theorem 3.99. *Let a path $Y(\cdot)$ be given. Then we have:*

- *If the set $\{s_k^{(n)},\ k = 0, \ldots, 2^n - 1,\ n \in \mathbb{N}\}$ is not bounded from below and above in $\mathbb{R}$, then it follows $\tau(q) = -\infty$, $q \neq 0$.*
- *If the set $\{s_k^{(n)},\ k = 0, \ldots, 2^n - 1,\ n \in \mathbb{N}\}$ is bounded from below and not bound from above in $\mathbb{R}$, it holds*

$$\tau(q) = \begin{cases} f^*(q) & \text{for all } q > 0 \\ -\infty & \text{for all } q < 0 \end{cases}$$

  (Leftsided multifractal)
- *If the set $\{s_k^{(n)},\ k = 0, \ldots, 2^n - 1,\ n \in \mathbb{N}\}$ is not bounded from below, but bounded from above in $\mathbb{R}$, then we obtain*

$$\tau(q) = \begin{cases} -\infty & \text{for all } q > 0 \\ f^*(q) & \text{for all } q < 0 \end{cases}$$

  (Rightsided multifractal)
- *If the set $\{s_k^{(n)},\ k = 0, \ldots, 2^n - 1,\ n \in \mathbb{N}\}$ is in $\mathbb{R}$ bounded, then we get*

$$\tau(q) = f^*(q),\ q \in \mathbb{R}$$

  (Bothsided multifractal)

Which properties does the partition function τ have? The preceding theorem tells us that the different cases and its properties can be determined by the

Legendre transform. Thus, τ is always concave and, depending on the four cases in theorem 3.99, we observe continuity on the sets $\mathbb{R}$, $\{q > 0\}$ or $\{q < 0\}$. In addition, it is up to countable many points, differentiable. We deduce from the definition that τ is non decreasing as long as the exponents $s_k^{(n)}$ are positive up to finite set. This stems from the case $s_k^{(n)} = h_k^{(n)}$ for paths with bounded variation or in the case $s_k^{(n)} = w_k^{(n)}$ if almost all paths are square integrable. But the function could be decreasing, provided the process exists only in the distributive sense. In any acse τ is more robust than f, since τ is constructed using expectation operator and no double limits are involved. Is it possible to obtain similar result as for τ in theorem 3.99 for f, too? We try it and will note first a simple to prove assertion.

Proposition 3.100. *The grain-based spectrum f is lower semi-continuous, that means, if (a_m) is a sequence converging to a, then it follows*

$$f(a) \geq \limsup_{m \to \infty} f(a_m)$$

We do not obtain equality in theorem 3.99 with exchanged rôles. But twice application of the Legendre transform will help us over this obstacle.

Theorem 3.101. *(Central multifractal formalism).* *We have for all $a \in \mathbb{R}$*

$$f(a) \leq f^{**}(a) = \tau^*(a) \tag{3.124}$$

where f^{**} is the Legendre transform of f^*. In addition we have equality, if the derivative from the right side (resp. from the left side) $\tau'(q+) = \lim_{x \downarrow q} \tau'(x)$ (resp. $\tau'(q-) = \lim_{x \uparrow q} \tau'(x)$) exists in $\mathbb{R}$. That means if $a = \tau'(q+)$ (resp. $a = \tau'(q-)$, then it follows

$$f(a) = f^{**}(a) = \tau^*(a) = q\tau'(q+) - \tau'(q+)$$

(analogously for $a = \tau'(q-)$).

The formulation in (3.124) is only determined by local properties of the partition function. From classical analysis we know that a concave function is almost everywhere differentiable (up to countable many point). Thus, we can also define $\tau'(q+)$ for all those q, if τ is differentiable for all sequences $x_n \downarrow q$ provided it is differentiable at x_n ($n \in \mathbb{N}$). The only problem consists whether the limit $\tau'(q+)$ exists in $\mathbb{R}$. In this sense we have to understand the right sided (resp.) leftsided limit in the above theorem 3.101. In the light of theorem 3.99 and proposition 3.100 we either can formulate results for the whole of $\mathbb{R}$, for the set $\{q < 0\}$, for the set $\{q > 0\}$ or for no set.

How do some of the values of τ look like and what shape does f have? For this we remember that $S^{(n)}(0)$ is the number of the finite values of $s_k^{(n)}$:

- If all $s_k^{(n)}$, $k = 0, \ldots, 2^n - 1$, $n \in \mathbb{N}$ are finite then we have $\tau(0) = -1$. In general, $-\tau(0)$ indicates the dimension of the support of the path $Y(\cdot)$.

3.8 Multifractal Models and the Influence of Small Scales

- The number $S^{(n)}(0)$ counts per definition the number of the finite $s_k^{(n)}$, $k = 0, \ldots, 2^n - 1$. Hence, we have $S^{(n)}(0) \geq 0$ and thus, $\tau(0) \leq 0$. Furthermore, we deduce $S^{(n)}(0) \geq N^{(n)}(a, \epsilon)$ for $a \in \mathbb{R}$ and $\epsilon > 0$. This implies

$$f(a) \leq -\tau(0) \tag{3.125}$$

- In some cases the singular exponents are monotone. That means e.g. if $s_k^{(n)} = h_k^{(n)}$. Then we get

$$2^{-n s_k^{(n)}} \geq \max\left(2^{-n s_{2k}^{(n+1)}}, 2^{-n s_{2k+1}^{(n+1)}}\right)$$

Thus, we deduce

$$S^{(n)}(1) \geq 2 S^{(n+1)}(1) \quad \text{and} \quad \tau(1) \geq -1$$

(the $h_k^{(n)}$ are bounded from below). In the case $s_k^{(n)} = \alpha_k^{(n)}$, we have

$$S^{(n)}(1) = S^{(n+1)}(1) \quad \text{and} \quad \tau(1) = 0$$

Hence, we obtain

$$f_\alpha(a) \leq a$$

Thus, we get equality in the case $a = \tau'(1+)$ and $a = \tau'(1-)$.

- If the process is not monotone (the cascades are monotone for example), then we do not have (3.125) necessarily. For some monotone processes we can identify $\tau'(1)$ as the dimension of the information of the support [82].
- We can define the *support* of a function as the closure of the set, where the function is not constant and the *kernel* is the smallest set on which the functions reveals its whole variability, i.e. that all values are assumed by $s_k^{(n)}$. In the special case of a distribution function the support consists of all open intervals with positive probability. The kernel indicates the smallest set with probability 1.

We finally give some comments on the deterministic partition function.

Proposition 3.102. *If $T(q)$ is finite for a $q > 0$, then the set of the finite exponents $s_k^{(n)}$, $k = 0, \ldots, 2^n - 1$, $n \in \mathbb{N}$, is bounded from below for almost all paths. In addition, we have for all $q > 0$ with finite $T(q)$*

$$T(q) = F^*(q)$$

If $T(q)$ is finite for a $q < 0$, then the set of the finite exponents $s_k^{(n)}$, $k = 0, \ldots, 2^n - 1$, $n \in \mathbb{N}$, is bounded from above for almost all paths. In addition, it holds for all $q < 0$ with finite $T(q)$

$$T(q) = F^*(q)$$

For all $q \in \mathbb{R}$, where T is differentiable, i.e., if $a = T'(q) \in \mathbb{R}$ exists, we deduce

$$F(a) = T^*(a) = qT'(q) - T(q) \tag{3.126}$$

3.8.5 Construction of Cascades

The concept of binomial cascades has greater importance in modeling TCP based traffic. Hence, we deal in the following in more detail the so called *monotone* processes, and will apply the results to irregularity. The construction is two-fold: First we will construct geometrically a path, which is modeled in the sequel stochastically. We explain this heuristical view more precisely.

Each path of a *binomial cascade* is a distribution function $\mathcal{M}_b(t)$. In particular, each $\mathcal{M}_b(t)$ possesses the property of a distribution function: It is continuous from right and monotone increasing. But almost everywhere $\mathcal{M}_b(\cdot)$ is not differentiable, i.e. it does not have a representation of the form $\mathcal{M}_b(t) = \int_{-\infty}^{t} \mathcal{M}'_b(s)ds$. As known from probability theory, every distribution function defines a measure on $\mathbb{R}$. We denote the corresponding measure by $\mu_b(]-\infty, t[) = \mathcal{M}_b(t)$. In fact, it is easier to define the measure μ_b first and construct in the sequel the process $\mathcal{M}_b$. We can denote μ_b as *binomial measure* and $\mathcal{M}_b$ as *binomial cascade*.

We proceed to the geometric construction. For this we successively divide the interval $[0,1]$. In this way we construct a tree, where each branch consists of an 'interval' and in each level the preceding intervals will be divided. Formally we can express this as follows.

For every $t \in [0,1]$ we select uniquely a sequence $k_1, k_2, \ldots$ such that $t \in I_{k_n}^{(n)} = [k_n 2^{-n}, (k_n+1)2^{-n}[$ holds for all $n \in \mathbb{N}$. Explicitly, we have $k_n = [t2^n]$ (remember the definition (3.108)). With this we uniquely determine for every $t \in [0,1]$ a sequence (k_n). We define this fact.

Definition 3.103. *We call a sequence $(k_n)_{n \in \mathbb{N}}$, for which $I_{k_{n+1}}^{(n+1)}$ is a subinterval of $I_{k_n}^{(n)}$, an embedded sequence.*

Vice versa it is evident that averaging over a interval sequence of half open intervals, belonging to an embedded sequence, defines exactly one point $\{t\}$. How can we define for an ω a binomial measure μ_b? For this we consider an interval $I_{k_n}^{(n)}$ on the level n. We divide it into the subintervals $I_{2k_n}^{(n+1)}$ and $I_{2k_n+1}^{(n+1)}$.

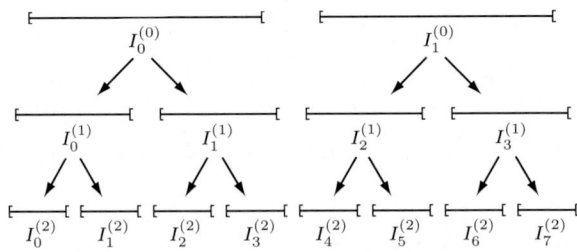

Fig. 3.27. Scheme of the embedded intervals constructing the binomial measure

3.8 Multifractal Models and the Influence of Small Scales

Considering both intervals in the classical Lebesgue measure, both have length $2^{-(n+1)}$. We want to weight each subinterval with positive numbers $M^{(n+1)}_{2k_n}$ and $M^{(n+1)}_{2k_n+1}$, in such a way that the union of both subintervals has the mass of $I^{(n)}_{k_n}$, i.e. preserving the mass. Hence, we stress that $M^{(n+1)}_{2k_n} + M^{(n+1)}_{2k_n+1} = 1$. The $M^{(n)}_{k_n}$ are called *multipliers*, if they fulfill in addition the three properties, described A) to C) below. But before we define the mass of μ_b for a particular path.

Definition 3.104. *For given multipliers $M^{(n)}_{k_n}$ we define the* binomial measure *according to*

$$\mu_b(I^{(n)}_{k_n}) = M^{(n)}_{k_n} \cdot M^{(n-1)}_{k_{n-1}} \cdots M^{(1)}_{k_1} \cdot M^{(0)}_0 \qquad (3.127)$$

and the binomial cascade *is given by $\mathcal{M}_b(0) = 0$ and*

$$\mathcal{M}_b\left((k_n+1)2^{-n}\right) - \mathcal{M}_b(k_n 2^{-n}) = \mu_b\left(I^{(n)}_{k_n}\right) \qquad (3.128)$$

From measure theory we know that the half open dyadic interval form a intersection stable generator of the Borel-σ-algebra. Hence, the definition in (3.128) suffices for determining the whole measure. Integrals and expected values are of great importance. We show this by an example for a continuous function g (it is true for a general class of functions g)

$$\int_{-\infty}^{\infty} g(t) d\mu_b(t) = \lim_{n \to \infty} \sum_{k=0}^{2^n - 1} g(k 2^{-n}) d\mu_b\left(I^{(n)}_k\right) \qquad (3.129)$$

$$= \lim_{n \to \infty} \sum_{k=0}^{2^n - 1} g(k 2^{-n}) \left(\mathcal{M}_b\left((k+1)2^{-n}\right) - \mathcal{M}_b(k 2^{-n})\right)$$

$$= \int_{-\infty}^{\infty} g(t) d\mathcal{M}_b(t)$$

We defined $\mathcal{M}_b$ only at dyadic points. Since $\mathcal{M}_b$ is continuous from right as distribution function and the dyadic numbers lie dense in the interval $[0,1]$, the definition in (3.128) suffices.

We defined *one* path up to now. To generate a process, we have to stress that the multipliers $M^{(n)}_{k_n}$ are random variables, which fulfill the following conditions. As initial value, the so called basic mass, we choose a positive random variable $M^{(0)}_0$:

A) *Preserving of the mass*: One chooses $M^{(n)}_k > 0$ almost surely everywhere and stresses

$$M^{(n+1)}_{2k} + M^{(n+1)}_{2k+1} = 1 \text{ a.s., for all } n, k \qquad (3.130)$$

Thus, the $\mathcal{M}_b$ is uniquely determined. But it is possible to weaken A) with the alternative condition A').

A′) *Preserving of the mean:*

$$\mathbb{E}(M_0 + M_1) = 1$$

This condition enables more flexibility in modeling. But one has to be more careful in the definition (3.127) and modify it in a certain direction (see [127]).

B) *Embedded independence:* For easier computations in applications (and because it often can be assumed), we stress that for an embedded sequence (k_n) the sequence of multipliers is stochastic independent. We obtain (similar as for all existing moments)

$$\mathbb{E}_\Omega\left(M_{k_n} \cdots M_0^{(0)}\right) = \mathbb{E}_\Omega\left(M_{k_n}^{(n)}\right) \cdots \mathbb{E}_\Omega\left(M_0^{(0)}\right) \qquad (3.131)$$

C) *Identical distribution:* For all n, k we choose the following equation in distributional sense

$$M_k^{(n)} = \begin{cases} M_0^{(1)} = M_0 & \text{for } k \text{ is even} \\ M_1^{(1)} = M_1 & \text{for } k \text{ is odd} \end{cases}$$

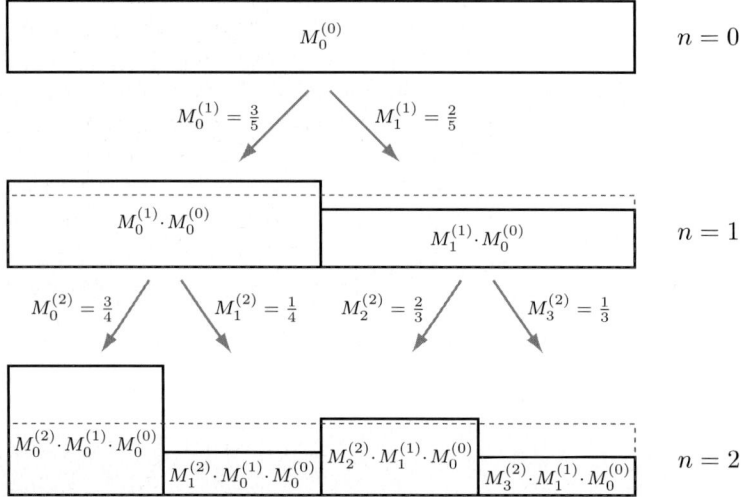

Fig. 3.28. Binomial cascade under the condition A) – preserving the mass (dashed line)

We should remark that under the assumption A′) we cannot expect $\mathcal{M}_b(1) = M_0^{(0)}$ anymore. Similar the chosen identity (3.127) for the construction for $\mu_b(I_k^{(n)})$ has to be understood in distributional sense because of the iid property of the multipliers and not as pure identity

3.8 Multifractal Models and the Influence of Small Scales

$$\mu_b\left(I_{k_n}^{(n)}\right) \stackrel{d}{=} M_{k_n}^{(n)} \cdot M_{k_{n-1}}^{(n-1)} \cdots M_{k_1}^{(1)} \cdot M_0^{(0)} \qquad (3.132)$$

Because we have $\mathbb{E}(\mathcal{M}_b(1)) = \mathbb{E}(M_0^{(0)})$, we get for the conditional expectation

$$\mathbb{E}\left(\mu_b\left(I_{k_n}^{(n)}\right) \mid M_{k_n}^{(n)}, \ldots, M_{k_1}^{(1)}\right) = M_{k_n}^{(n)} \cdots M_{k_1}^{(1)} \mathbb{E}\left(M_0^{(0)}\right) \qquad (3.133)$$

and it follows in particular $\mathbb{E}(\mu_b(I_{k_n}^{(n)})) = \frac{1}{2^n}\mathbb{E}(M_0^{(0)})$. This implies for simulations that in the process of simulations one does not receive the value of $\mu_b(I_{k_n}^{(n)})$ using (3.133), but only the *expected* value. The decisive advantage of the alternative definition A') lies in the fact that one can use unbounded multipliers as e.g. the lognormal distribution for the random variables M_0 and M_1. Then the marginal distribution of $\mu_b(I_{k_n}^{(n)})$ turn out lognormal. In general we can deduce under the conditions A'), B) and C) that $\mu_b(I_{k_n}^{(n)})$ is according to the central limit theorem at least *asymptotical lognormal* distributed, of course under the assumption that the variance in (3.127) is finite.

After considering the construction of the binomial cascades we want to comment on the irregularity of the paths. Here, a problem concerning the measure μ_b arises. Since we did define it only in the distributional sense and since it does not possesses paths in realizations, we have to proceed differently. Thus we restrict ourselves to the treatment of the multifractal analysis of the distribution function $\mathcal{M}_b$. The following proposition shows us that both the coarse increment exponent $\alpha_k^{(n)}$ as the coarse Hölder exponent leads to the same (deterministic) partition function. The consideration of the wavelet exponent will be done separately.

Proposition 3.105. *For a binomial cascade* $(Y(t)) = (\mathcal{M}(t))$ *which is almost surely increasing we have*

$$T_\alpha(q) = T_h(q), \text{ for all } q \in \mathbb{R}$$

In addition, we get for almost all paths

$$\tau_\alpha(q,\omega) = \tau_h(q,\omega), \text{ for all } q \in \mathbb{R}$$

We want to compute the envelope resp. the partition function precisely. For this we select the coarse exponent of the increments $\alpha_k^{(n)}$. Because of proposition 3.105 we also obtain the result for the coarse Hölder exponent. We have $\mathcal{M}_b((k_n+1)2^{-n}) - \mathcal{M}_b(k_n 2^{-n}) = \mu_b(I_{k_n}^{(n)})$. According to (3.127) (resp. (3.132)) we get by the independence condition B) (see also (3.131)) and condition C)

$$\mathbb{E}\left(S_{\alpha,\mathcal{M}_b}^{(n)}\right) = \sum_{k_n=0}^{2^n-1} \mathbb{E}\left(\left(M_{k_n}^{(n)}\right)^q \cdots \left(M_{k_1}^{(1)}\right)^q\right) \mathbb{E}\left((\mathcal{M}_b(1))^q\right) \qquad (3.134)$$

$$= \mathbb{E}\left((\mathcal{M}_b(1))^q\right) \cdot \sum_{j=0}^{n} \binom{n}{j} \mathbb{E}(M_0^q)^j \mathbb{E}(M_1^q)^{n-j}$$

$$= \mathbb{E}\left((\mathcal{M}_b(1))^q\right) \cdot (\mathbb{E}(M_0^q) + \mathbb{E}(M_1^q))^n$$

In case A) we have $\mathcal{M}_b(1) = M_0^{(0)}$ almost surely. We summarize this fact in a theorem.

Theorem 3.106. *Let $\mathcal{M}_b$ be a binomial cascade, for which A'), B) and C) hold. It follows $T(q) = -\infty$, if one of the values $\mathbb{E}(M_0^q)$, $\mathbb{E}(M_1^q)$ or $\mathbb{E}(\mathcal{M}_b(1)^q)$ is infinite. Otherwise we have*

$$T_{\alpha, \mathcal{M}_b}(q) = T_{h, \mathcal{M}_b}(q) = -\log_2 \mathbb{E}\left(M_0^q + M_1^q\right) \tag{3.135}$$

If the multipliers M_0 and M_1 have at least one existing negative moment (i.e. $q < 0$), then it holds in the case of the coarse exponent of the increment $\alpha_k^{(n)}$ for all $a \in \mathbb{R}$ with $T^*(a) > 0$

$$\dim(K^a) = \dim(E^a) = \underline{f}(a) = f(a) = \tau^*(a) = F(a) = T^*(a) \tag{3.136}$$

Assuming further that the random variables M_0 and M_1 are bounded away from 0 almost surely, then we get for the coarse Hölder exponent $h_k^{(n)}$ and all $a \in \mathbb{R}$ with $T^*(a) > 0$ the equality (3.136) almost surely.

We should present some examples of binomial cascade.

Example 3.107. Beta-binomial cascade: The multipliers M_0 and M_1 are identical beta distributed. As known the density of the distribution is $g(x) = c_p x^{p-1}(1-x)^{p-1}$, for $x \in [0, 1]$ and $g(x) = 0$ for $x \notin [0, 1]$. Here, $p > 0$ is a given parameter, describing the tail behavior (see (2.32)), and c_p is a constant (to obtain a probability density g). Because of the condition (3.130) we get a symmetric distribution, since M_0 and $M_1 = 1 - M_0$ are uniformly distributed. The β distribution has a finite moment for all $q > -p$, i.e. $\mathbb{E}(M_0^q) < \infty$ for $q > -p$. Thus, we deduce for the binomial cascade (with (3.135) and theorem 3.106)

$$T_\alpha(q) = -1 - \log_2 \frac{\Gamma(p+q)\Gamma(2p)}{\Gamma(2p+q)\Gamma(p)}, \quad q > -p \tag{3.137}$$

and $T_\alpha(q) = -\infty$ for $q \leq -p$. We considered this fact when investigating binomial cascades in the TCP traffic.

Example 3.108. Assuming in example 3.107 explicitly $p = 1$, then we obtain the uniform distribution. We conclude $T_\alpha(q) = -1 + \log_2(1+q)$, if $q > -1$. Applying the Legendre transform, then the explicit form

$$T_\alpha^*(a) = 1 - a + \log_2(e) + \log_2\left(\frac{a}{\log_2(e)}\right)$$

for $a > 0$ and $T^*(a) = -\infty$ if $a \leq 0$ follows.

Example 3.109. A very interesting case is revealed by the *lognormal binomial cascades*. We use the lognormal distribution for the multipliers M_0 and M_1 (see e.g. for the lognormal distribution [38]). In this case we have to substitute

3.8 Multifractal Models and the Influence of Small Scales

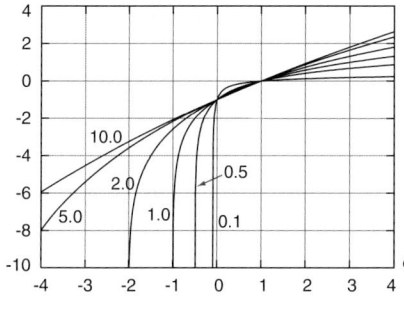
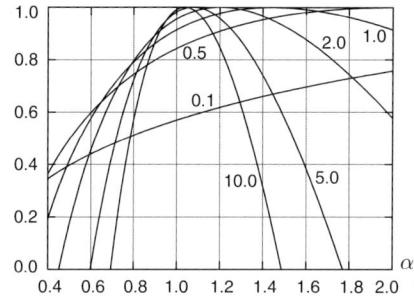

Fig. 3.29. Deterministic envelope function $T(q)$ (left) and multifractal spectrum and the Hölder exponent T_α (right) for the MWM model with $p = 0.1, 0.5, 1.0, 2.0, 5.0, 10.0$

A) by A'). Lognormal means that we used a $\mathcal{N}(m, \sigma^2)$-distributed Gaussian variable ξ and considered $M = \exp(\xi)$. Then we have $\mathbb{E}(M^q) = \mathbb{E}(\exp(\xi)^q) = \exp(\frac{qm+q^2\sigma^2}{2})$. Assuming again M_0 and M_1 as identical distributed, then (according to A')) the respective mean has to be $\frac{1}{2}$, and we obtain the equality $\frac{m+\sigma^2}{2} = \log(2)$. From (3.135) we get for the lognormal binomial cascade

$$T_\alpha(q) = (q-1)\left(1 - \frac{\sigma^2}{2\log(2)}q\right) \tag{3.138}$$

for all $q \in \mathbb{R}$, for which $\mathbb{E}(\mathcal{M}_b(1)^q)$ is finite. The parabola in (3.138) has two points of value 0: 1 and $q_{\text{krit}} = \frac{2\log(2)}{\sigma^2}$. To obtain a non degenerated cascade, we have to have $T'(1) > 0$, which is equivalent to $q_{\text{krit}} > 1$. Then, we conclude $\mathbb{E}(\mathcal{M}_b(1)^q) < \infty$ for all $q < q_{\text{krit}}$ (see [127]). Indeed, we can describe the tail behavior of $\mathcal{M}_b(1)$: It is $\mathbb{P}(\mathcal{M}_b(1) > x) \sim x^{q_{\text{krit}}}$ for large $x \to \infty$ (see [105]). We omit the details and present the results:

- From the representation we can detect that T is differentiable for all $q < q_{\text{krit}}$. Then we can compute the Legendre transform according to (3.126) for $a = T'(q)$, $q < q_{\text{krit}}$, i.e. for $a > a_{\text{krit}} = T'(q_{\text{krit}}) + \frac{\sigma^2}{2\log(2)}$

$$T^*(a) = 1 - \frac{\log(2)}{2\sigma^2}\left(a - 1 - \frac{\sigma^2}{2\log(2)}\right)^2, \quad a > a_{\text{krit}}$$

If $a \leq a_{\text{krit}}$, then we have $T^*(a) = a \cdot q_{\text{krit}}$. Hence, at a_{krit} the parabola of T^* turns into its tangent, going through the origin with slope q_{krit}. Since $q_{\text{krit}} > 1$, we have $a_{\text{krit}} < 0$ and the spectra for negative a possesses a finite but negative value $T^*(a)$.
- Deterministic spectra: It holds $T^*(a) = F(a)$ for all $a \in \mathbb{R}$ with $a = T'(q)$ according to 3.102.
- Partition function: All path of the cascade are increasing, thus, the coarse exponent of the increments $\alpha_{k_n}^{(n)}$ has to turn positive and hence, $\tau_\alpha(q)$ is

monotone increasing. On the other hand we know that the deterministic partition function $T_\alpha(q)$ is a parabola, where $T_\alpha(q)$ attains finite values. Thus, consequently we have $\tau(q) > T(q)$ for $q > \frac{1+q_{\text{krit}}}{2}$, and hence, q lies on the right of the maximum of T.

- Negative (real) exponents: It seem surprising on the first glimpse that in the case of lognormal binomial cascades negative values of $a < 0$ can occur, for those with finite $T^*(a)$. Negative exponents show that the increment within one subinterval can be larger as in the preceding one, which is impossible for increasing paths. The contradiction can be solved by the reasoning that the exponent is not directly responsible for the single ingredients of the measure but for the $M_{k_i}^{(i)}$, and here, what should be expected, not for virtual paths.

Remark 3.110. Finally we want to give some remarks on the shape of the deterministic partition function or envelope. We start with the Legendre transform T^* and consider first the part $T^*(a) > 0$. This reveals the properties of the path. In case of the cascade we have, according to the theorem 3.106, that $T^*(a)$ is the value of the Hausdorff dimension $\dim(E^a)$, thus, reflecting the local structure of the path using the coarse exponents $s_k^{(n)}$. Similar the value $T^*(a)$ indicates the values of the grain-based spectral $f(a)$ resp. the Legendre transform of the partition function $\tau^*(a)$. In general T^* consists as an upper bound (see theorem 3.98). The positive part of T^* lies between its roots, which are in turn representing the corresponding slope of the tangents to $T(q)$, which run through the origin. As disadvantage one cannot interpret the negative part of T^* as dimension. Let's consider the values a, which occur in the spectrum. As already discussed those a make trouble for which $T^*(a) < 0$ hold. They are called the *latent* exponents. The *virtual* are those with $a < 0$. For the coarse exponents $\alpha_k^{(n)}$ resp. $h_k^{(n)}$ of monotone processes all virtual exponents are latent ones. We can realize this by the fact that $T(1) = 0$ and hence, $T^*(a) \leq a$. Of course no increasing path can show a negative coarse exponent of the increments and virtual exponents are rarely. But this observation will change for the wavelet exponents.

Remark 3.111. Stationary increments: Because of property C) of the binomial cascades the increments $\mathcal{M}_b((k+1)2^{-n}) - \mathcal{M}_b(k2^{-n})$ are identical distributed, if – and this can be assumed – the M_0 and M_1 are identical distributed. In this case we have stationary increments of the first order, but we will not get those of second order.

Remark 3.112. One can extend the binomial structure to the so called c-adic cascades Then, the geometric representation is not given by a tree since in each knot of a level we have c-subtrees. In addition we could alter the deterministic or the stochastic numbers $c(n)$ for each levels n. Then we call these objects *multinomial cascades*. On this very sophisticated structure we will go into details and refer on the literature (e.g. [218]).

3.8 Multifractal Models and the Influence of Small Scales

The coarse Hölderexponent resp. the exponent of the increments could only be defined for $\mathcal{M}_b$. To present in addition a multifractal analysis for μ_b, we have to use the wavelet exponents.

For this we first use the Haar wavelet. The wavelet coefficients are

$$2^{-\frac{n}{2}} \mathcal{D}_{n,k_n}(\mu_b) = \mu_b\left(I_{2k_n}^{(n+1)}\right) - \mu\left(I_{2k_n+1}^{(n+1)}\right) \qquad (3.139)$$

$$= \prod_{j=0}^{n} M_{k_j}^{(j)} \left(M_{2k_n}^{(n+1)} - M_{2k_n+1}^{(n+1)}\right)$$

The scaling coefficients turn out to be

$$\mathcal{C}_{n,k}(\mu_b) = \int_{-\infty}^{\infty} \psi_{n,k}(t) d\mu_b(t) = 2^{\frac{n}{2}} \int_{k2^{-n}}^{(k+1)2^{-n}} d\mu_b(t) = 2^{\frac{n}{2}} \mu_b\left(I_k^{(n)}\right) \qquad (3.140)$$

Changing from the Haar wavelet to a general one with arbitrary support in $[0,1]$, we apply the substitution $s = 2^n t - k_n$, to gain from (3.139) the equation

$$2^{-\frac{n}{2}} \mathcal{D}_{n,k_n}(\mu_b) = \int_{I_{k_n}^{(n)}} \varphi(2^n t - k_n) d\mu_b(t) \qquad (3.141)$$

$$= M_{k_n}^{(n)} \cdots M_{k_1}^{(1)} \int_0^1 \varphi(s) d\mu_b^{n,k_n}(s)$$

The measure μ_b^{n,k_n} is formed for every embedded sequence (k_m) at the knot k_n starting with length m and depending of the interval. We give a more precise definition.

Definition 3.113. *For a given embedded sequence $k_1, \ldots, k_n$ we set*

$$\mu_b^{n,k_n}\left(I_{i_m}^{(m)}\right) = \hat{M}_0^{(0)} M_{2k_n+i_1}^{(n+1)} M_{4k_n+i_2}^{(n+2)} \cdots M_{2^m k_n + i_m}^{(n+m)}$$

where $(i_1, i_2, \ldots, i_m)$ is an embedded sequence defined depending on the interval $I_{i_m}^{(m)}$. Furthermore $\hat{M}_0^{(0)}$ is an independent copy of $M_0^{(0)}$. Finally we set $\mathcal{M}_b^{n,k_n}(t) = \mu_b^{n,k_n}([0,t[)$.

We define the wavelet coefficients for binomial cascade measures. Now we will transfer (3.141) to the distribution function $\mathcal{M}_b$. For this let $t \in [0,1]$ and $k_n = k_n(t) = [2^n t]$. We compute

$$\mathcal{M}_b(t) - \mathcal{M}_b(k_n 2^{-n}) = \mu_b\left([k_n 2^{-n}, t]\right) = \mathcal{M}_b^{n,k_n}(2^n t - k_n) M_{k_n}^{(n)} \cdots M_{k_1}^{(1)}$$

How do the particular wavelet coefficients look like? Again we pick a mother wavelet φ with support in $[0,1]$. This implies that $\varphi_{n,k} = \varphi(2^n \cdot - k)$ has a support in $I_k^{(n)}$. Choosing $s = 2^n t - k$ it follows

$$\int_{I_k^{(n)}} \varphi_{n,k} \mathcal{M}_b(t) dt = \int_{I_k^{(n)}} \varphi(2^n t - k_n) \left(\mathcal{M}_b(t) - \mathcal{M}_b(k_n 2^{-n}) \right) dt$$

$$= 2^{-n} \cdot M_{k_n}^{(n)} \cdots M_{k_1}^{(1)} \cdot \int_0^1 \varphi(s) \mathcal{M}_b^{n,k_n}(s) ds$$

To obtain a closed representation, we define the following random variables:

$$\eta_{n,k}(\mu_b) = \int_0^1 \varphi(t) d\mu_b^{n,k_n}(t) \quad \text{resp.} \quad \eta_{n,k}(\mathcal{M}_b) = \int_0^1 \varphi(t) \mathcal{M}_b^{n,k_n}(t) dt$$

We are now able to summarize the results in the following proposition.

Proposition 3.114. *Suppose φ is a mother wavelet with support in $[0,1]$ and μ_b satisfies the conditions A) to C). Then we have for the wavelet coefficients (random variables!)*

$$\mathcal{D}_{n,k_n}(\mu_b) = 2^{\frac{n}{2}} M_{k_n}^{(n)} \cdots M_{k_1}^{(1)} \eta_{n,k_n}(\mu_b)$$
$$\mathcal{D}_{n,k_n}(\mathcal{M}_b) = 2^{-\frac{n}{2}} M_{k_n}^{(n)} \cdots M_{k_1}^{(1)} \eta_{n,k_n}(\mathcal{M}_b)$$

In addition, the random variables η_{n,k_n} and $M_{k_i}^{(i)}$, $i = 0, \ldots, n$ are, as for μ_b and for $\mathcal{M}_b$ pairwise independent and

$$\eta_{n,k_n} \stackrel{d}{=} \eta_{0,0} = \mathcal{D}_{0,0} \tag{3.142}$$

Remark 3.115. Proposition 3.114 has a great impact for modeling the TCP traffic. Starting with the data, obtained by the wavelet analysis, one models with the help of the result (3.142) a binomial cascade. For this suitable assumptions to the starting distributions of M_0 and M_1 are made. Consequently, one receive with the help of the data, thus, the wavelet coefficients, a reliable multifractal model. We discussed this in more detail in section 3.8.2.

How can we determine the irregularity of binomials cascades using the wavelet coefficients? As already mentioned they are particularly useful for the treatment of binomial measures μ_b. Again we start with the computation of the deterministic envelope. The proposition 3.105 resp. the computation in (3.134) can be transferred and we obtain

$$\mathbb{E}\left(S_{w,\mu_b}^{(n)}(q)\right) = 2^{nq} \mathbb{E}\left(|\mathcal{D}_{0,0}(\mu_b)|^q\right) \cdot \left(\mathbb{E}_\Omega(M_0^q) + \mathbb{E}(M_1^q)\right)^n$$

Similar we can determine $\mathbb{E}(S_{w,\mathcal{M}_b}^{(n)}(q))$. We summarize

Proposition 3.116. *Assuming the condition that the moments of the wavelet coefficients $\mathbb{E}(|\mathcal{D}_{0,0}(\mu_b)|^q)$ resp. $\mathbb{E}(|\mathcal{D}_{0,0}(\mathcal{M}_b)|^q)$ and the moments $\mathbb{E}(M_0^q)$, $\mathbb{E}(M_1^q)$ and $\mathbb{E}(\mathcal{M}_b(1)^q)$ are all finite, we have*

$$T_{w,\mu_b}(q) + q = T_{w,\mathcal{M}_b}(q) = T_{\alpha,\mathcal{M}_b}(q)$$
$$T_{w,\mu_b}^*(a-1) = T_{w,\mathcal{M}_b}^*(a) = T_{\alpha,\mathcal{M}_b}^*(a) \tag{3.143}$$

3.8 Multifractal Models and the Influence of Small Scales

We considered the moments for the coarse wavelet exponents. How can we determine in particular the $w_k^{(n)}$? For this we have to pose some additional conditions on the initial distribution of M_0 resp. M_1, so that the wavelet coefficient $\mathcal{D}_{0,0}$ will not turn out to be too small. Hence, it suffices to stress $|\mathcal{D}_{0,0}| \geq \epsilon$ for a sufficient small $\epsilon > 0$ (for the Haar wavelet we have $\mathcal{D}_{0,0} = 2M_0 - 1$ a.s.). With this condition we get for all $t \in [0, 1]$

$$\lim_{n \to \infty} \frac{1}{n} \log_2 \left(\int_0^1 \varphi(s) d\mu_b^{n,k_n}(s) \right) = 0$$

Taking in addition (3.141) into account, then it follows

$$w_{\mu_b}(t) = \liminf_{n \to \infty} -\frac{1}{n} \log_2 \left(2^{\frac{n}{2}} |\mathcal{D}_{n,k_n}| \right) = \alpha_{M_b}(t) - 1$$

Here, we have a coincidence with (3.143), and we realize that the wavelet exponent of the measure differs from the exponent of the increment α of the distribution function by the value 1, assuming the cascade is sufficiently 'nice' (expressed by the boundedness of the wavelet coefficient from below). In case this condition is not satisfied that simple observation may fail. This is also expressed by considering the coarse Hölder exponent. Using the above proposition we can formulate the following corollary.

Corollary 3.117. *Suppose μ_b satisfies the conditions A) to C) and the random variables $\eta_{n,k}(\mu_b)$ resp. $\eta_{n,k}(M_b)$ are uniformly bounded away from 0. In this case we have equality in the multifractal formalism, i.e.*

$$dim(K^a_{w,\mu_b}) = f_{w,\mu_b}(a) = \tau^*_{w,\mu_b}(a) = T^*_{w,\mu_b}(a) = T^*_{\alpha,M_b}(a+1) \qquad (3.144)$$

resp.

$$dim(K^a_{w,M_b}) = f_{w,M_b}(a) = \tau^*_{w,M_b}(a) = T^*_{w,M_b}(a) \qquad (3.145)$$

Here, all equations in (3.145) have to be understood almost surely (as in (3.144) with exception of the last one).

We have for the Haar wavelet $\eta_{n,k} = M_{2k}^{(n+1)} - M_{2k+1}^{(n+1)} = 2M_{2k}^{(n+1)} - 1$. The restriction being bounded away from 0 seems just for applications quite unrealistic. Looking more closely to the scene we can provide some technique to avoid this unwished restriction

- Being bounded away from 0 is needed, whenever we divide by the random variable or simply if we have to consider negative moments, as done by the multifractal formalism. Thus, it is possible to substitute the stronger assumption being bounded away from 0 by the weaker condition of the existence of all negative moments of $\eta_{n,k}$. If we drop condition C) and demand that the distribution of the multipliers depend on the level n, then we can skip the boundedness on finite many levels $\eta_{n,k}$, i.e. $\eta_{n,k}$ is only bounded away from 0 for large n. We realize just in the case of

the treatment of network traffic that on the fine scales (small n) discrete distributions on $[-1,1]$ are used, which have a large variance and no mass around 0.
- There is another, maybe more crude method, to avoid small wavelet coefficients. For this we use the maximum of the wavelet coefficient for all t in its neighborhood and define a new wavelet exponent $\tilde{w}$. For this coefficient we can derive a multifractal formalism. More detailed conditions can be found in [28, 123].

With the help of an example we want to investigate, if we can apply the simple rule in all cases – we loose one degree of Hölder continuity after differentiating. We already remarked before the last one that this cannot be done as simple.

Example 3.118. For this let $\mathcal{M}_b(t) = 5t + t^3 \sin(\frac{1}{t})$. Its derivative reads as $\mathcal{M}'_b(t) = 5 + 3t^2 \sin(\frac{1}{t}) - t\cos(\frac{1}{t})$ (it exists for all $t \in \mathbb{R}$) and it is strictly positive on $[0,1]$. Hence, $\mathcal{M}_b$ is monotone increasing. The local Hölder continuity at $t=0$ is 3, since $|\mathcal{M}_b(t) - 5t| \leq |t|^3$ is the best polynomial approximation. But at $t=0$ we have only $|\mathcal{M}'_b(t) - 5| \leq |t|$ and thus, the Hölder exponent $H_{\mathcal{M}'_b} = 1$. The regularity decreases rate 2 and this happens because $\mathcal{M}_b$ oscillates very heavy in a neighborhood of 0.

We want to examine the relationship between the Hölder exponent of the path $\mathcal{M}_b$ and the derivative $\mathcal{M}'_b$. Since we are not interested in the orthogonality of the wavelets, we use for our purposes different mother wavelets than the Haar wavelet. We consider the so called Gaußkernel $\exp(-\frac{x^2}{2})$ and its derivatives (we need differentiable wavelets). To study the relationship between the particular wavelet coefficients, we apply the partial integration in the following step:

$$\int_0^1 g(t)d\mu_b(t) = \lim_{n\to\infty} \sum_{k=0}^{2^n-1} g(k2^{-n})\left(\mathcal{M}_b\left((k+1)2^{-n}\right) - \mathcal{M}_b(k2^{-n})\right)$$

$$= \lim_{n\to\infty} \sum_{k=0}^{2^n-1} g(k2^{-n})\mathcal{M}_b(k2^{-n})\left(g\left((k-1)2^{-n}\right) - g(k2^{-n})\right)$$

$$+ \mathcal{M}_b(1)g(1-2^{-n}) - \mathcal{M}_b(0)g(-2^{-n})$$

$$= \mathcal{M}_b(1)g(1) - \mathcal{M}(0)g(0) - \int_0^1 g'(t)\mathcal{M}_b(t)dt$$

Hence

$$\int_0^1 g(t)d\mu_b(t) = \mathcal{M}_b(1)g(1) - \mathcal{M}_b(0)g(0) - \int_0^1 g'(t)\mathcal{M}_b(t)dt \qquad (3.146)$$

Here, we have used (3.129), integrated partially and permutated some terms. If $\mathcal{M}_b$ allows a derivative, then $d\mu_b(t)$ turns to $\mathcal{M}'_b(t)dt$.

We define $g(t) = 2^{\frac{n}{2}}\varphi(2^n t - k)$ for a differentiable mother wavelet φ and obtain

3.8 Multifractal Models and the Influence of Small Scales

$$g'(t) = 2^{\frac{3n}{2}} \varphi'(2^n t - k)$$

Since we have $\mathcal{M}_b(1) = 1$ and $\mathcal{M}_b(0) = 0$ ($\mathcal{M}_b$ is the distribution function with support in $[0, 1]$), it follows by (3.146)

$$\mathcal{D}_{n,k}(\varphi, \mu_b) = 2^{\frac{n}{2}} \varphi(2^n - k) - 2^n \mathcal{D}_{n,k}(\varphi', \mathcal{M}_b) \quad (3.147)$$

We can estimate the argument of φ in (3.147) for large t asymptotically. It holds because of $k_n = [t2^n]$

$$2^n - k_n = 2^n - [t2^n] \sim 2^n - t2^n = (1-t)2^n, \text{ for large } t$$

Because of the choice of the mother wavelet $\varphi(t)$ it decays for $t \to \infty$ exponentially (see [74] for the definition of the Schwarz space) and thus, we get

$$w_{\varphi, \mu_b} = -1 + w_{\varphi', \mathcal{M}_b}(t)$$

Summarizing we have the relationship given in the following theorem.

Theorem 3.119. *For a differentiable mother wavelet φ with exponential decay if $t \to \infty$ it follows*

$$E^{a+1}_{\varphi, \mu_b} = E^a_{\varphi', \mathcal{M}_b}, \quad f_{\varphi, \mu_b}(a) = f_{\varphi', \mathcal{M}_b}(a+1), \quad \tau^*_{\varphi, \mu_b}(a) = \tau^*_{\varphi', \mathcal{M}_b}(a+1) \quad (3.148)$$

To understand the result in (3.148) more precisely, we want to discuss the following observation. The derivative φ' has a vanishing moment less than the mother wavelet (simple computation with partial integration). As we already remarked, it is $d\mu_b(t) = \mathcal{M}'_b(t)dt$, i.e., the measure is the derivative of the distribution function and hence, its Taylor expansion has one degree less than that of $\mathcal{M}_b(t)$. As we already stated in section 3.8.2, the wavelet should be 'blind' to polynomials. The assertion (3.148) is also true, if the distribution function oscillates heavily at some points. Then the wavelet analysis is not suitable, to discover those spots. It could be said that the multifractal analysis of $\mu_b = \mathcal{M}'_b$ using φ is equivalent to that of $\mathcal{M}_b$ using φ'.

Remark 3.120. It seems to be surprising that the Haar wavelet, though it does not have vanishing moments, can describe and analyze the binomial distribution function $\mathcal{M}_b$ as its derivative μ_b. The regularity decays and thus, we could not describe $\mathcal{M}_b$. The illumination is given by the structure of the cascades themselves. In the dyadic points the polynomials are constant and thus, the property $\int \varphi(t)dt = 0$ suffices. The demand for $\int t^k \varphi(t)dt = 0$ for larger k is not necessary, even if there is a regularity of $\alpha(t) > 1$. Finally we can state that the wavelets cannot describe singularities at those points for which the degree of the approximating polynomial is higher than the regularity.

3.8.6 Multifractals, Self-Similarity and Long-Range Dependence

As above we follow the line of the survey article of Riedi [218]. We remark that as in the whole chapter of multifractal analysis we only give a short survey on the theory, which can be studied in more details in papers like [218]. In addition, there the theory is presented more mathematical formally. Up to now we sketched the concept of multifractals from the observed data material resp. signal. We saw that with the help of MWM resp. cascade models one can describe the multifractal character of the IP-based traffic. How is the multifractality connected to the concept of self-similarity and long-range dependence? We repeat the definition of self-similarity and state some properties.

Definition 3.121. *A process $(X(t))$ is called self-similar, if there exists an exponent (Hurst exponent) $H > 0$, such that*

$$X(at) = a^H X(t), \text{ for all } a > 0 \tag{3.149}$$

in the sense of equality in the marginal distribution.

We say that $(X(t))$ has *stationary* increments if for all $s, t > 0$

$$X(t+s) - X(t) \stackrel{d}{=} X(s) - X(0)$$

A self-similar process with stationary increments is called an H-sssi process. For more details we refer to the section 3.1.2. Since naturally the increments of an H-sssi process are of importance, we define

$$Z(t) = X(t+1) - X(t)$$

or the discrete version

$$Z(k) = X(k+1) - X(k)$$

Example 3.122. The most prominent example of an H-sssi process is the FBM: The only H-sssi process ($0 < H \leq 1$) with Gaussian marginal distributions is the fractional Brownian motion $(B_t^{(H)})$, as we know from section 3.2.2. The process $(Z(t))_{t \in]-\infty,\infty[} = (B^{(H)}(t+1) - B^{(H)}(t))_{t \in]-\infty,\infty[}$ is called Gaussian white noise (in continuous time) (FGN). Instead of the drift with value 1, one can choose a fixed $\delta > 0$. If $H = \frac{1}{2}$, we get the Wiener process or Brownian motion and the FBN turns into the white noise. We want to describe a representation of the FBM using the Itō integral

$$B^{(H)}(t) = \frac{1}{\Gamma\left(H + \frac{1}{2}\right)} \int_{-\infty}^{\infty} \left((t-s)_+^{H-\frac{1}{2}} - (-s)_+^{H-\frac{1}{2}}\right) dB^{(\frac{1}{2})}(s) \tag{3.150}$$

($x_+ = \max(x, 0)$). This representation goes back to Mandelbrot and Van Ness [174]. For further results we refer to the section 3.2.2. In particular the FBM has a autocovariance function, since the finite marginal distributions have a second moment.

3.8 Multifractal Models and the Influence of Small Scales

Example 3.123. There are processes, as seen in section 3.2.3, whose marginal distributions do not have necessarily a second moment. These processes were introduced, motivated by the generalization of the central limit theorem, through the symmetric α-stable distribution ($0 < \alpha \leq 2$). The characteristic function of the random variable $X(t)$ reads e.g as

$$\mathbb{E}\left(\exp(ixX(t))\right) = \exp\left(-C(t)^\alpha |x|^\alpha + i\beta x\right)$$

The H-sssi processes, whose marginal distribution are symmetric α-stable, form the next class for which the q-moments exists only for $q \leq \alpha$. In the case $\alpha = 2$ we obtain the Gaussian distribution. We denote a symmetric α-stable H-sssi process with $L_{H,\alpha}$. We know that the random variable $L_{H,\alpha}(t) - L_{H,\alpha}(s)$ is distributed according to $C(t) = |t-s|^H C_{L_{H,\alpha}(1)}$ if the increments are independent then is equivalent to $H = \frac{1}{\alpha}$ and these processes are called α-stable Lévy process. They are denoted by $\tilde{L}_{H,\frac{1}{H}} = L_H$.

Example 3.124. For an H-sssi process it must hold $H \leq \max(\frac{1}{\alpha}, 1)$, as we know already. If $H \neq \frac{1}{\alpha}$, then we replace in (3.150) the exponent $H - \frac{1}{2}$ by $H - \frac{1}{\alpha} \neq 0$ and the Brownian motion by an α-stable Lévy process. Then we obtain a general linear fractional stable motion. The special case $\alpha = 2$ gives us the FBM back, and this is the only case that the H-sssi is a 2-stable process uniquely determined and has continuous paths, i.e. the FBM. In other cases the H-sssi process is uniquely determined and possesses almost surely not continuous paths.

After these introductory remarks we turn to the multifractal analysis of self-similar processes.

Deterministic Envelope and Self Similarity

In the sequel we consider an H-sssi process $(X(t))$ and get

$$2^{-n\alpha_k^{(n)}} = \left|X\left((k+1)2^{-n}\right) - X(k2^{-n})\right| \quad (3.151)$$
$$= \left|X(2^{-n})\right| = 2^{-nH}|X(1)| = 2^{-nH}2^{-\alpha_0^{(0)}}$$

The two equation in the middle are valid in distributional sense. The second equation is a result of the stationary increments and the third follows, since we have a self-similar process. If we apply on both sides of (3.151) the logarithm, then we conclude: $-n\alpha_k^{(n)} = -nH - \alpha_0^{(0)}$ and, thus

$$\alpha_k^{(n)} = H + \frac{1}{n}\alpha_0^{(0)}$$

We can formulate this more generally.

Proposition 3.125. *Suppose $(X(t))$ is an H-sssi process. Then the singular exponents $\alpha_k^{(n)}, h_k^{(n)}$ and $w_k^{(n)}$ satisfy the scaling equation*

$$s_k^{(n)} = H + \frac{1}{n}s_0^{(0)} \quad \text{in distribution, for } s \in \{\alpha, h, w\} \tag{3.152}$$

The mean $\mathbb{E}_\Omega(2^{-ns_0^{(0)}})$ is exactly for all $Q_u < q < Q_o$ finite, where for $s_k^{(n)} = \alpha_k^{(n)}$ resp. $s_k^{(n)} = w_k^{(n)}$ we have

$$(Q_u, Q_o) = \begin{cases} (-1, \infty) & \text{for the FBM} \\ (-1, \alpha) & \text{for an s}\alpha\text{s process} \end{cases} \tag{3.153}$$

In the case $s = h$ we get

$$\begin{aligned} Q_o &= \infty \quad \text{for a FBM} \\ Q_o &\leq \alpha \quad \text{for a s}\alpha\text{s motion} \\ Q_u &= -\infty \quad \text{for a Lévy stable process or the Brownian motion} \end{aligned} \tag{3.154}$$

Using (3.152) we obtain

$$\mathbb{E}_\Omega\left(S^{(n)}(q,\omega)\right) = \mathbb{E}_\Omega\left(\sum_{k=0}^{2^n-1} 2^{-nqs_k^{(n)}}\right) \stackrel{(3.152)}{=} 2^n 2^{-nqH} \mathbb{E}_\Omega\left(2^{-qs_0^{(0)}}\right)$$

Combining (3.153) and (3.154), we can conclude

$$T_s(q) = \lim_{n \to \infty} \frac{\log_2 \mathbb{E}_\Omega\left(S^{(n)}(q)\right)}{-n} = \begin{cases} qH - 1 & \text{for } Q_u < q < Q_o \\ -\infty & \text{else} \end{cases} \tag{3.155}$$

We see that the deterministic partition function is linear. How does th deterministic grain spectrum $F(a)$ looks like? The derivation is left to the reader as exercise. Here is the result.

Proposition 3.126. *The Legendre transform T^* of the deterministic partition function of an H-sssi process is for the singular exponents $s_k^{(n)}$, $s \in \{\alpha, h, w\}$ (the upper resp. lower bound Q_o resp. Q_u are given according to proposition 3.125):*

$$T^*(a) = \begin{cases} 1 + Q_u(a - H) & \text{for } a > H \\ 1 & \text{for } a = H \\ 1 + Q_o(a - H) & \text{for } a < H \end{cases}$$

According to (3.121) it is $F(a) = \underline{F}(a)$ and thus it follows

$$F_\alpha(a) = F_w(a) = T_\alpha^*(a) = T_w^*(a) \quad \text{and} \quad F_h(a) = T_h^*(a)$$

3.8 Multifractal Models and the Influence of Small Scales

Up to now we were concerned with deterministic functions, which approximate the actual spectrum. How does the Hausdorff dimension of K^a and E^a look like? As seen in section 3.2.2 using a result of Adler, we realize that for the FBM it is $h(t) = H$. Hence we can formulate for the fractional Brownian motion

$$\dim(E_h^a) = f_h(a) = \tau^*(a) = T^*(a) = \begin{cases} 1 & \text{for } a = H \\ -\infty & \text{for } a < H \end{cases}$$

Hence, we also call the FBM *monofractal*. We should remind that a certain multifractal structure can be introduced by the FBM as we have sketched in section 3.5.1 and 3.8.1. But there we used the approach of several densities and a so called envelope function. Jaffard [124] computed for a Lévy stable motion ($L_{\frac{1}{\alpha},\alpha}(t)$) with $\alpha < 2$ its Hausdorff dimension

$$\dim(E_h^a) = \begin{cases} \frac{a}{H} & \text{for } 0 < a < H \\ -\infty & \text{else} \end{cases}$$

We proceed to the grain-based spectrum. Since the Brownian motion as the Lévy stable processes have independent increments, we obtain for a Lévy process and the singular exponents $s \in \{\alpha, h, w\}$ with proposition 3.126, theorem 3.96 and theorem 3.98

$$\underline{f}_s(a) = f_s(a) = \tau_s^*(a) = F_s(a) = T_s^*(a) \tag{3.156}$$

where $a \in \mathbb{R}$ with $T^*(a) \geq 0$ or $T^*(a) = -\infty$ holds. Here, the corresponding mother wavelet for the singular exponent w has compact support. For a FBM with $H \neq \frac{1}{2}$ there are only numerical results, which raise some hope to find theoretical verification for the FBM according to the strong decorrelation property of the wavelets. We close the consideration for H-sssi processes with the analysis of the partition function. We know, that according to (3.155) the deterministic partition function T is linear. Thus, we have to assume that there are positive as negative q, with $T(q) < \infty$. The definition tells us that $S^{(n)}(0,\omega)$ counts the increasing jumps of order 2^n. Thus, it follows $\tau(0) = -1$ (see the definition of τ). In addition we get $T(0) = -1$. Furthermore we know that τ is concave and has to lie above T (see proposition 3.93). Hence, τ has to be linear as well. Thus, we have for an H-sssi process

$$\tau(q,\omega) = qH - 1 \tag{3.157}$$

for all $q \in\,]Q_u, Q_o[$ and for almost all $\omega \in \Omega$.

Considering both for the FBM as for a Lévy stable process the values $\dim(E^a)$ and the deterministic envelop (Legendre transform of the deterministic partition function) T_s^* for $s = \alpha$ or $s = h$, then one detects decisive differences. In the case of the FBM we have $\dim(E^a)$ is only one point $a = H$, while $T_w^*(a) = T_\alpha^*(a)$ is a decreasing line, starting at $a = H$ and with the value $T_s^*(H) = 1$ and slope -1. Similar facts can be stated for the Lévy stable process. Hence, we can say that $T_\alpha^* = T_w^*$ does not correctly reflects the Hölder continuity. The reason is that $\alpha_k^{(n)}$ as $w_k^{(n)}$ are random variables, which are Gaussian or α-stable distributed and whose distribution is concentrated around the origin. Nevertheless (3.156) and (3.157) provide a good deal of information for applications. At those spots where the line has a slope of -1, we can detect for $\alpha_k^{(n)}$ resp. for $w_k^{(n)}$ a larger value, hence there, where the increments resp. the wavelet coefficients are smaller than 2^{-nH}.

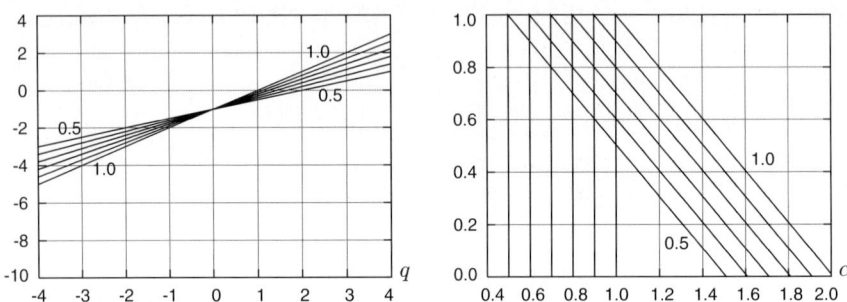

Fig. 3.30. Deterministic envelope function $T(q)$ (left) and multifractal spectrum and the Hölder exponent T_α (right) for the WIG model with $H = 0.5, 0.6, 0.7, 0.8, 0.9, 1.0$

After the treatment of the self-similarity using multifractal analysis we turn to the *long-range dependence*.

Multifractal Analysis and Long-Range Dependence

The correlation structure of the white noise in (3.18) is not sufficient for appropriate modeling. In addition the structure depends on the constant length of the steps. It is easy to realize that $r_Z(k) \sim k^{2H-2}$ holds for the FBN. For $\frac{1}{2} < H < 1$ the correlation decays as slow that it is not summable over k. In the section 3.1.2 we already discussed different approaches. We shortly review them.

Definition 3.127. *We say that the process of the increments $(Z(t) = X(t + \delta) - X(t))_{t \in \mathbb{R}}$ ($\delta > 0$ fixed) is long-range dependent, if the sum $\sum_{k=1}^{\infty} r_Z(k) = \infty$ is not convergent (set $t = k \in \mathbb{Z}$). This definition is equivalent to the long-range dependence given for aggregated processes*

$$Z^{(m)}(k) = \frac{1}{m} \sum_{i=km}^{(k+1)m-1} Z(i)$$

This means

3.8 Multifractal Models and the Influence of Small Scales

$$r_Z(k) \sim k^{2H-1}, \text{ for } k \to \infty \tag{3.158}$$
$$\Leftrightarrow \mathbb{V}ar(Z(k)) \approx m^{2-2H} \mathbb{V}ar\left(Z^{(m)}(i)\right), \text{ for } m \to \infty$$

for fixed $i \in \mathbb{N}$. Solving for H this leads to the definition of the self-similarity of second order. We say that $(Z(t))$ is self-similar of second order *for a parameter H_{Var}* if

$$H_{\text{Var}} = 1 + \frac{1}{2} \lim_{m \to \infty} \frac{\log\left(\frac{\mathbb{V}ar(Z^{(m)})}{\mathbb{V}ar(Z^{(1)})}\right)}{\log m} \tag{3.159}$$

(here the averages $Z^{(m)}$ and $Z^{(i)}$ are defined according to (3.1)).

Definition 3.128. If we have an H-sssi process $(X(t))_{t \in \mathbb{R}}$ with variance structure then we get for $Z(k) = X(k+1) - X(k)$ the equality $H_{\text{Var},Z} = H$ according to (3.149). In addition we already described the equivalence of a certain behavior of the correlation function and the spectral representation in section 3.1.2. Applying the Fourier transform on the autocorrelation r_Z then we see that

$$r_Z(k) \sim k^{2H-2}, \text{ for } k \to \infty$$
$$\Leftrightarrow \mathcal{F}(r_Z)(\lambda) \sim \lambda^{-(2H-1)}, \text{ for } \lambda \to 0$$

We can use the wavelet transform instead of the Fourier analysis and receive a similar representation (for details see [7])

$$\mathbb{V}ar(\mathcal{D}^Z_{j,k}) = \mathcal{O}\left(2^{-j(2H-1)}\right) \tag{3.160}$$

In (3.160) the factor 2^j takes over the rôle of the frequency λ. Since the wavelet coefficients are less correlated as the subordinated FBM, the wavelet methods for estimating the Hurst parameter are getting more into the center of interest. We will come back to this topic in section 4.2.7. Hence, in accordance to (3.159) we give the following definition.

Definition 3.129. *A process $(Z(t))$ is asymptotical wavelet self-similar for a parameter H_{wave}*, if the following limit exists

$$H_{\text{wave}} = \frac{1}{2} - \frac{1}{2} \lim_{j \to \infty} \frac{1}{j} \log_2 \mathbb{V}ar\left(\mathcal{D}^Z_{j,k}\right) \tag{3.161}$$

Deterministic Partition Function and Long-Range Dependence

The value $T(2)$ is closely related to the Hurst parameter H, since both measure have the scaling behavior and the statistic of second order: $T(2)$ describes the scaling behavior of the second moments of the paths (on small scales), while H indicates the decay of the correlation (long-range dependence on large scales). We have the following proposition.

Proposition 3.130. If a process $(X(t))$ has increments with vanishing first moments (that means e.g. $\mathbb{E}(Z(t)) = \mathbb{E}(X(t+1) - X(t)) = 0$), then it holds

$$H_{\mathrm{Var}} = \frac{T_{\alpha,Z}(2) + 1}{2} \qquad (3.162)$$

Considering processes, which are constructed by multipliers, such as the binomials cascades, then we do not have stationarity of second order. Hence, we cannot introduce the LRD via its covariance. In section 3.1.2 we discussed already alternative approaches. But we can define values like the variance of the aggregated traffic and the wavelet coefficients, which are, as we know, equivalent to the stationarity of second order. We can give the following computation, since the binomials cascades have positive increments $Z(k)$

$$2^{-2i}2^{(i-n)(1+T(2))} \simeq \mathbb{E}\left(|Z^{(m)}|^2\right) = \mathrm{Var}\left(Z^{(m)}\right) + \mathbb{E}\left(Z^{(m)}\right)^2$$
$$\simeq \mathrm{Var}(Z)2^{i(2H-2)} + \mathbb{E}(Z)^2$$

Because of the stationary increments $\mathbb{E}(X^{(m)})$ is independent of the scaling m. Hence we assume that for small m and i at least the limit offers a similar result as (3.162). Using (3.158) for estimating the Hurst parameters (e.g. with the variance method, see section 4.2.2), then the method is not very reliable in contrast to the wavelet method (see 4.2.7). The value $T_{w,Z}(2)$ can be considered as 'wavelet energy', i.e. as variance of the wavelet coefficient of Z. We have $\mathrm{Var}(\mathcal{D}_{j,k}) = 2^{-j}\mathbb{E}(2^j|\mathcal{D}_{j,k}|^2) \simeq 2^{-j}2^{-j(1+T(2))} = 2^{-j(2+T(2))}$ because of the stationarity of $\mathcal{M}$. Thus, we have for multiplicative processes (as the binomial cascades)

$$H_{\mathrm{wave}} = \frac{T_{w,Z}(2) + 3}{2}$$

We consider more closely the particular relationships for two important groups of processes.

Example 3.131. Parameter for the binomial cascades: If $X = \mathcal{M}_b$ a binomial cascade, then it follows using (3.143)

$$H_{\mathrm{wave},\mathcal{M}_b} = \frac{T_{w,\mu_b}(2) + 3}{2} = \frac{T_{\mathcal{M}_b}(2) + 1}{2} = H_{\mathrm{Var},\mathcal{M}_b} \qquad (3.163)$$

Example 3.132. Linear processes – FARIMA: Up to now we computed in example the parameters H and $T(q)$ with maximal two scales. This does not suffice for applications. We can establish the multifractal parameters resp. the long-range dependence property only as limit process over very small scales or very large ones. A typical example are the linear processes as the FARIMA sequences, which describe both the small time correlation and the long-range dependence. The value $\frac{T(2)+1}{2}$ is not sufficient to describe all time scales, as we see later, in particular if the small scales are characterized by a large mean. In th following we will consider a binomial cascade $M_k^{(n)}$, and will assume A),

3.8 Multifractal Models and the Influence of Small Scales

B) and instead of the identical distribution assumption C) the following condition due to [218]: The cascade possesses within a level n the distribution, satisfying the equation

$$\text{Var}\left(M_k^{(n)}\right) = \frac{2^{2-2H}\text{Var}\left(M_k^{(n-1)}\right)}{4\text{Var}\left(M_k^{(n-1)}\right)+1} \tag{3.164}$$

To determine the distribution completely we assume that $M_k^{(n-1)}$ is symmetric beta distributed on $[0,1]$. Simultaneously let $\mathbb{E}((M_k^{(1)})^2) \simeq \frac{1}{4}$, but not equal $\frac{1}{4}$, or let us assume $\text{Var}(M_k^{(1)}) \simeq 0$. The significance of this presumption can be seen in the treatment of the network traffic. This low variance is observed in a far over one minute aggregated network traffic. Iterating the condition (3.164) over the first n terms, we receive $\mathbb{E}((M_k^{(n)})^2) \simeq \frac{1}{4}$ and $\text{Var}(M_k^{(n)}) \simeq \sigma^2(2^{2-2H})^n$, which is small for small n, if $2-2H > 0 \Leftrightarrow H < 1$. Only on very small scales (large n) we have the convergence

$$\mathbb{E}\left((M_k^{(n)})^2\right) = \text{Var}\left(M_k^{(1)}\right) + \frac{1}{4}, \text{ towards the limit } 2^{-2H}$$

This influences the long-range dependence by the expression $\frac{T(2)+1}{2}$. Because of (3.134) we can use the term $-\frac{1}{n}\log_2(2^n \prod_{i=1}^n \mathbb{E}((M_k(i))^2))$ as estimator for $T(2)$. Having only measurements available for small n, we get $T(2) \simeq -\frac{1}{n}\log_2(2^n 4^{-n}) = 1$, since the convergence is slowly and we have $\frac{T(2)+1}{2} \simeq 1$. If the expression converge for $n \to \infty$, then we obtain in limit $T(2) \simeq -\frac{1}{n}\log_2(2^n 2^{-2nH}) = 2H-1$, and the true value $\frac{T(2)+1}{2} \simeq H$ appears. How can we detect this with the help of the wavelet coefficients? At least we will guess that already for small n we obtain a good estimation for H. It can be shown ([217]), that (3.161) is exact for small n. We want to motivate this in more detail. Because of the definition of binomial cascades we get $M_{2k_n}^{(n+1)} - M_{2k_n+1}^{(n+1)} = 2M_{2k_n}^{(n+1)} - 1$ and with (3.140) it turns out

$$\mathbb{E}\left(\mathcal{D}_{n,k}^2\right) = 2^n \mathbb{E}\left((M_k^{(1)})^2\right) \cdots \mathbb{E}\left((M_k^{(n)})^2\right) \cdot 4\text{Var}\left(M_k^{(n+1)}\right)$$
$$\simeq 2^n 4^{-n} 4\left(2^{2-2H}\right)^n = 4\left(2^{1-2H}\right)^n$$

which was to prove. The so called wavelet energy $\text{Var}(\mathcal{D}_{n,k})$ immediately indicates (independent of n) $H_{\text{wave}} = H$. The reason for the different estimations of $T(2)$ lies in the fact that $T(2)$ is based on the second moments, which are not centralized. The scaling behavior for coarse scales of the estimator based on $T(2)$ is hidden in the mean of the process. But also for small n we can obtain the approximation $H_{\text{Var}} \simeq H$. For this we use $\mathbb{E}((M_k^{(i)})^2) \simeq \frac{1}{4}$, $i = 1,\ldots,n-1$, $\mathbb{E}((M_k^{(n)})^2) = \frac{1}{4} + \text{Var}(M_k^{(N)})$, which leads to

$$\mathrm{Var}\left(Z^{2^{-n}}\right) = \mathbb{E}\left((M_k^{(1)})^2\right)\cdots\mathbb{E}\left((M_k^{(n)})^2\right) - \left(\frac{1}{2^n}\right)^2$$

$$\simeq \frac{1}{4^{n-1}}\mathrm{Var}\left(M_k^{(n)}\right) \simeq \frac{\sigma^2}{4}\cdot 2^{-2nH}$$

and reveals the scaling $H_{\mathrm{Var}} = H$. With an alternative initial condition $\mathbb{E}((M_k^{(1)})^2) = 2^{-2H}$ we have an exact scaling, which also gives us (3.163).

α-β Models

Based on the well-known on-off model from section 3.4.3, which was designed to explain the long-range dependence and self-similarity on the large scales in an adequate fashion, below the RTT the protocols have much more influence and the pure on-off model fails to capture the spikiness of the short-range behavior. In [227] it is shown that small rate sessions are well described by independent duration and rate, while the large rate sessions are better explained by independent file size and rate. As [227] put it:

> The patience of users is limiting the small bandwidth connections, while users with large bandwidth freely choose their file size.

This is incorporated in a two factor on-off model called α-β *on-off model*. The α-β models were first introduced by [228] to describe the different behavior of the traffic on different time scales as pointed out in section 3.5.1. The basic idea of the model is that especially in TCP based traffic on two significant scales key values as interarrival times or transmitted data size show different behavior, one for small time scales (below RTT), we call them α *scales* and consequently one for large scales, called β *scales*. The reason for this notation lies in the observation that the small scales dominate the traffic behavior and result in a very bursty character, as in the animal kingdom, where they are called α animals and in correspondence the other β animals. So the α component describes aggressive behavior given by high rates and large file transfer, while in turn the β one is the passive part. The model is a first step to give an understanding of both burstiness and long-range dependence. The model is based on the connection level consisting of all packets with the same source and destination IP addresses, port numbers and protocols.

For the model we simply describe by the process (Y_t) a certain traffic value, like transmission time or transmitted data volume, then we note by (Y_t^α) the contribution of traffic below a certain given threshold time T_0 and call it α *traffic part*). In addition, define the 'rest' $Y_t^\beta = Y_t - Y_t^\alpha$ as the β part.

3.9 Summary of Models for IP Traffic

In table 3.2 we summarize key properties of IP traffic models described in this chapter. We compare the major models, presented in the previous sections and indicate the respective key formulas. Finally, we present in the right column the tools for estimation.

Further Literature

Self-Similarity and Long-Range Dependence

At the moment there is no book on the market, dealing with the topic of self-similarity and long-range dependence. A good while short introduction to self-similarity can be found in the monograph of Embrechts and Maejima [77], though the reader should like mathematical formalism. Special literature is presented in [155, 156, 242, 243, 246]. The α−stable processes are investigated in many articles as e.g. [173]. Self-similarity and long-range dependence in others than IP traffic modeling is treated in a lot of articles, covering e.g. hydrology, economics medicine [183, 251, 101]. For those, who like to have a glimpse at the starting point of self-similar processes we recommend the original literature of Mandelbrot [172, 173, 174].

Standard and Fractional Brownian Motion The topic of the standard Brownian motion is presented in a vast variety of books. We just mention some of some which deal especially with the stochastic analysis in general like [199, 18, 134, 208]. More literature though more mathematical challenging are [121, 270, 271, 272]. A description of Gaussian as well as non-Gaussian processes is presented in the several time cited monograph of Samorodnitsky and Taqqu [226]. The relationship between the heavy-tailed distributions and the so called α-stable processes can be found in [76], which we used as a guideline for our presentation. Special literature is given by [69, 172, 67, 121]. The more and more important becoming topic of the fractional Brownian motion attracts a great number of authors, which include for the general topic [45, 65, 33, 34, 35, 205, 206, 207, 199]

FARIMA Time Series

FARIMA time series are dealt with in a number of original papers. In the monograph [68] one can find several collected papers on this topic. Again [226] gives a good introduction to the time series. In the book [200] edited by Park and Willinger several article deal with the connection of FARIMA series and network performance.

Norros Model

The Norros model is one of the striking one, since it gives a very intrinsic description of the network traffic and simultaneously is quite simple to understand. There are a number of original papers, which formed the basis of our given representation. The article [190, 192] are a good starter for a first approach to the area, while [193, 195] deal with more advance topics. The original model is used for advanced modeling like the differentiated traffic [195, 197]. Further literature, where also other aspects like multifractal are considered, can be found e.g. in [163, 176].

Heavy-Tailed Distributions and the Classical Models in IP Traffic Theory

The heavy-tailed distribution for modeling inter arrival times and the file sizes is done in several original articles [58, 59, 61]. They especially deal with the question, why in the Word Wide Web the transmitted files are heavy-tailed. The performance in connection with the heavy-tailed distribution is presented e.g. in [94, 212, 29, 245]. The connection between the classical theory and the IP traffic can be read in some monographs like [253, 20] or in the original literature [104, 49, 50, 152, 151, 201].
Nevertheless, one should give a physical justification, why the aggregated traffic leads to the use of the self-similar processes, starting with the classical traffic theory and its M/G/n approach. This is extensively done e.g. in [264, 250, 86, 262, 263, 152, 85, 41].

Multifractal Processes

This area is one of the newest areas in the IP traffic theory, which was introduced according to the well known reasons treated in the sections 3.5.1,3.6,3.7 and 3.8. A detailed survey is given by the article [218]. The connection to IP traffic is presented in a vast variety of papers, where we only cite some for a first glance [3, 6, 112, 177, 180, 213, 219, 220]. The article [84] the different transmission protocols as TCP/IP and UDP are considered and its influence concerning the open as well as closed loop algorithm on the traffic modeling investigated. The article [217] shows the advantages of the non-Gaussian fractals in comparison to the Gaussian WIG models. Theoretical results as well as the introduction to the topics of cascades are given in [14, 17, 83, 128, 129]. In [48] the MWM are transformed to the more advanced cascade models. More on the telecom process is given in [131].

3.9 Summary of Models for IP Traffic   319

Table 3.2. Summary of different IP traffic models

Mathematical Model	Key Ingredients	Crucial Estimations
1. FBM	Traffic amount A_t up to time t modeled by $A_t = mt + \sqrt{am} Z_t$	mean rate m, traffic variance a, Hurst exponent H for FBM (Z_t)
2. Fractional Levy	Traffic amount A_t up to time t modeled by $A_t = mt + (am)^{\frac{1}{\alpha}} Z_t$	mean rate m, traffic variance a, Hurst exponent H for FLM (Z_t) and α-stable distribution
3. Envelope process	easy approach for a multi-FBM from estimation finding a Hurst function $H(x)$; increments of multi-FBM can be estimated by a upper bound process $\hat{A}_t = \int_0^t \overline{m} + \gamma \sigma H(x) x^{H(x)-1} dx$ where $\overline{m}$ is the average traffic rate, σ^2 its variance and $H(\cdot)$ a Hölder function or Hurst function	idea is to use a multifractal approach as done e.g. in [217, 219] and find a $H(\cdot)$ in connection to the Hölder exponent function $T(q)$; remember there is a connection between the Hurst exponent and the function $T(q)$ for multifractal processes (see section 3.8.6); mostly H is a quadratic function
4. Multi-FBM	as in model 1. but with different FBM on different scales; formal definition with standard Brownian motion according to 3.82	as in model 3.
5. α-β model	partition of the traffic into an α- and β-component; α-traffic is highly non-Gaussian, β-traffic is purely Gaussian; α-traffic below RTT, β-traffic is the difference of all traffic and α-part	determination of a threshold time scale, as e.g. RTT; α-traffic estimation via the on-off approach, hence estimation of the interarrival time resp. traffic load using heavy-tail distribution (Pareto); β-traffic with normal distribution
6. Multifractal	according to MWM approach and a tree structure; starting at initial distribution $p_{0,0}$ with the expected value $\mathbb{E}(V_{0,0}) = 1$ on the coarsest level and the Hurst exponent; it is possible to simulate the traffic using $M_j \sim \beta(p_j, p_j)$-distributed multipliers; the p_j fulfill the second order equations (see (4.286)); p_{-1} will be computed with the help of expectation and variance on the coarsest level; $T(q)$ is responsible for the p_j	maximal scale depth j; mean of the traffic differences at reference time points i, weighted by Hölder exponent q; calculation of $T(q)$ and $T^*(\alpha)$; initial β-distribution for the coarsest time interval

4
Statistical Estimators

> *It remains to investigate, if, fixed by the increase in observations, the probability grows for that amount that the number of the good observation in relation to the bad ones attains the true relationship, and this in such a way that the probability surpasses finally every degree of certainty, ...*
>
> *Jakob Bernoulli (18th century)*

4.1 Parameter Estimation

A broad variety of approaches exists to estimate an appropriate distribution and its parameters, when given a sample of data. They contain among others: unbiased estimators, regression methods or maximum likelihood estimator (see e.g. [118, 98, 158]). In this chapter we introduce relevant estimators for characteristic properties of IP traffic. We start with some standard estimators and later introduce appropriate approaches with regard to IP traffic. Further details on basic concepts of parameter estimation are summarized in monographs like [208].

Like in most models IP traffic is measured in careful experiments and described by appropriate distributions with *optimal* parameters. In general, we apply all estimators under the premise of optimal parameters. Therefore, we have to specify the term 'optimal' at first. As described e.g. in [261] we start with a statistical experiment $(\mathcal{X}, (W_\theta)_{\theta \in \Theta})$ whereas $\mathcal{X}$ denotes the *sample space*. In our context a statistical experiment is represented by the measurement of characteristic properties of IP traffic. These experiments are repeated N times, which leads to N values of the random variable ξ. We describe these measurements as experiments with *sample size N*.

The decision space is denoted by $\mathbb{D} = \mathbb{R}$ and the estimator function is a measurable map $g : \mathcal{X} \longrightarrow \mathbb{D}$. This is necessary, since we usually apply a composition with random variable ξ. In most cases we estimate a certain value or a part $\gamma(\theta)$ (e.g. conditional distribution), whereas $\gamma : \Theta \longrightarrow \mathbb{R}$ represents the value of dependence of parameter θ. We denote the estimator as *point estimator*. To determine, i.e. to estimate the expectation $\gamma(\theta) = \mu_\theta$ of a random variable ξ, the point estimator

$$\bar{\xi} = \frac{1}{N} \sum_{i=1}^{N} \xi_i$$

is used, whereas $\xi_1, \ldots, \xi_N$ represent (iid) independent and according to ξ distributed random variables. If $x_1, \ldots, x_N$ are realizations of this random vector $(\xi_1, \ldots, \xi_N)$, then

$$\overline{x} = \frac{1}{N} \sum_{i=1}^{N} x_i$$

is a *point estimation* of μ. With this

$$g(x) = \overline{x}, \quad x_1, \ldots, x_N \in \mathcal{X}$$

and

$$g(\xi) = \overline{\xi}$$

is the corresponding point estimator. In a similar form we consider

$$\mathcal{S}^2 = \frac{1}{N-1} \sum_{i=1}^{N} (\xi_1 - \overline{\xi})^2$$

as point estimator of variance σ^2.

A common approach to measure the optimality consists in minimizing the risk, not having determined the correct model. To avoid the effects of alternating signs, an appropriate measure is the difference calculated as quadratic loss function

$$R(\theta, g) = (\gamma(\theta) - g)^2 \tag{4.1}$$

In the remainder of this section we will examine this term more closely, although without formal derivation.

4.1.1 Unbiased Estimators

With the expressions above we describe the risk of the estimator g by

$$R(\theta, g) = \int_{\mathcal{X}} (\gamma(\theta) - g(x))^2 \, dF_\theta(x) = \mathbb{E}_\theta \left((\gamma(\theta) - g(\xi))^2 \right) \tag{4.2}$$

The F_θ are distributions, which depend on the assumption of parameter θ. The relation (4.2) can be written in the form

$$R(\theta, g) = (\gamma(\theta) - \mathbb{E}_\theta (g(\xi)))^2 + \mathrm{Var}_\theta (g(\xi))$$

The term

$$\gamma(\theta) - \mathbb{E}_\theta (g(\xi)) \tag{4.3}$$

represents the *systematic error* that occurs with the use of estimator g and the existence of parameter θ. In the literature the systematic error is often denoted as *bias*. We will use this term as from now. The second term in 4.3 denotes the variation of the estimator. With our estimation we want to minimize the risk, i.e. we want to:

- avoid a systematic error, i.e. the bias,
- achieve a variance as low as possible.

To fulfill the first demand, we select an estimator g with

$$\mathbb{E}_\theta\left(g(\xi)\right) = \gamma(\theta)$$

and denote it by *unbiased estimator*. With such an (unbiased) estimator the risk is

$$R(\theta, g) = \mathrm{Var}_\theta\left(g(\xi)\right)$$

We give two elementary examples.

Example 4.1. We consider a binomial distributed random variable ξ with parameters N and θ. Then we can show that $\frac{\xi}{N}$ is an unbiased estimator for $\gamma(\theta) = \theta$: we know for a binomial distribution that $\mathbb{E}(\xi) = N\theta$ and with this it follows

$$\mathbb{E}\left(\frac{\xi}{N}\right) = \frac{1}{N}\mathbb{E}(\xi) = \frac{1}{N}N\theta = \theta$$

Thus $g(\xi) = \frac{\xi}{N}$ is an unbiased estimator for θ.

The next example illustrates that not every estimator is unbiased.

Example 4.2. We consider an iid sequence $\xi, \ldots, \xi_N$ of realizations of the random variable ξ, which is distributed with density

$$f(x) = \begin{cases} e^{-(x-\delta)} & \text{for } x > \delta \\ 0 & \text{else} \end{cases}$$

We will show that $g(\xi) = \overline{\xi}$ is not an unbiased estimator for $\gamma(\delta) = \delta$. The mean of ξ is given by

$$\mu = \int_0^\infty x e^{-(x-\delta)} dx = 1 + \delta$$

However,

$$\mathbb{E}(\overline{\xi}) = \frac{1}{N}\sum_{i=1}^N \mathbb{E}(\xi_i) = \mu = 1 + \delta \neq \delta$$

With this follows that $\overline{\xi}$ is obviously a biased estimator of δ.

Now the question arises, whether $g(\xi) = \mathcal{S}^2$ is an estimator of σ^2? We give the result by a remark and a proof.

Remark 4.3. If the point estimator is unrestricted, then $\mathcal{S}^2$ is an unbiased estimator of the variance $\gamma(\theta) = \sigma_\theta^2$ of a random variable ξ.

Proof. We build a sequence of realizations of the random variable ξ by iid random variables $\xi_1, \ldots, \xi_N$

$$S^2 = \frac{1}{N-1} \sum_{i=1}^{N} (\xi_i - \bar{\xi})^2$$

With this we get

$$\begin{aligned}
\mathbb{E}(S^2) &= \mathbb{E}\left(\frac{1}{N-1} \sum_{i=1}^{N} (\xi_i - \bar{\xi})^2\right) \\
&= \frac{1}{N-1} \mathbb{E}\left(\sum_{i=1}^{N} \left((\xi_i - \mu) - (\bar{\xi} - \mu)\right)^2\right) \\
&= \frac{1}{N-1} \left(\sum_{i=1}^{N} \mathbb{E}\left((\xi_i - \mu)^2\right) - n\mathbb{E}\left((\bar{\xi} - \mu)^2\right)\right) \\
&\stackrel{1)}{=} \frac{1}{N-1} \left(\sum_{i=1}^{N} \sigma^2 - n\frac{\sigma^2}{n}\right) = \sigma^2
\end{aligned}$$

$\square$

This fact is one reason, why the empirical variance is calculated with the factor $\frac{1}{N-1}$ instead of $\frac{1}{N}$.

We now address the problem to fulfill equation '1)' above. At first we quote the so called *Cramér-Rao Inequality*

$$\mathbb{V}\mathrm{ar}(g) \geq \frac{1}{N\mathbb{E}\left(\left(\frac{\partial \log f(\xi)}{\partial \theta}\right)^2\right)}$$

whereas f denotes the density of a random variable ξ and N the number of iid realizations of ξ. With this we get the following property.

Property 4.4. If g is an unbiased estimator for θ and

$$\mathbb{V}\mathrm{ar}(g) = \frac{1}{N\mathbb{E}\left(\left(\frac{\partial \log f(\xi)}{\partial \theta}\right)^2\right)} \tag{4.4}$$

then g is an unbiased estimator that minimizes the variance.

The denominator of (4.4) is often described as *degree of information*. With this we see that the larger the information the smaller the variance is.

Consistency

By an example we will demonstrate that the previous criteria for the determination of an optimal estimator do not necessarily lead to a suitable approach. We consider a set of sample data that depends on the parameter θ and is distributed with density

$$f(x) = r \frac{1}{\sigma\sqrt{2\pi}} \exp\left(-\frac{1}{2}\left(\frac{x-\theta}{\sigma}\right)^2\right) + (1-r)\frac{1}{\pi}\frac{1}{1+(x-\theta)^2}$$

whereas $-\infty < x < \infty$ and $0 < r < 1$. This expression is a combination of a normal distribution with mean θ and variance σ^2 and a Cauchy distribution with $\alpha = \theta$ and $\beta = 1$. We remind that α-stable distributions play an important rôle and that normal and Cauchy distributions are two prominent representatives of this class (see section 3.2.3). If we choose r near 1, e.g. $r = 1 - 10^{-1000}$ and $\sigma = 10^{-1000}$ very small, then we may assume that (because of small variance) ξ assumes the value θ and with this the estimator is precise for θ. However, the variance of a Cauchy distribution does not exist and hence not the variance of the estimator. Consequently the method of minimizing the variance fails. From a mathematical perspective this argumentation is inconclusive, but it clearly shows that we have to draw our attention to the distribution itself. This is the reason for an alternative approach which considers the probability that the estimators takes values near θ. A particular tool helpful for this approach is the Chebyshev inequality (see e.g. [38, 208]). With this we define a *consistent estimator*.

Definition 4.5. *An estimator g is* consistent *to the parameter θ if and only if*

$$\lim_{N \to \infty} \mathbb{P}\left(|g(\xi) - \theta| < c\right) = 1$$

where N denotes the sample size.

Theorem 4.6. *If $g(\xi)$ is an unbiased estimator of parameter θ and if the variance $\mathbb{V}ar(g(\xi)) \to 0$ for an infinite number of realizations, i.e. $N \to \infty$, then the estimator $g(\xi)$ is consistent to θ.*

Example 4.7. We show that for a normal distributed sample the estimator S^2 is consistent for the variance σ^2. According to theorem 4.3, S^2 is an unbiased estimator of σ^2. Hence, we can apply theorem 4.6 and have to show that $\mathbb{V}ar(S^2) \to 0$, if the sample size tends towards infinite, i.e. $N \to \infty$. With

$$\mathbb{V}ar(S^2) = \frac{2\sigma^4}{N-1}$$

it follows immediately that $\mathbb{V}ar(S^2) \to 0$ for $N \to \infty$.

Example 4.8. To discuss the following example we have to introduce the concept of *order statistic* first. For this purpose we randomly select a sample of size N with a steady distribution function. We order the values x, i.e. the smallest value is assigned to the random variable Y_1, the next value is assigned to the random variable Y_2 and so on. The random variables $Y_1, \ldots, Y_N$ are described as *order statistic*. Y_1 denotes the first order statistic, Y_2 the second order statistic etc. In case of $N = 2$ we have

$$y_1 = x_1 \text{ and } y_2 = x_2, \text{ for } x_1 < x_2$$
$$y_1 = x_2 \text{ and } y_2 = x_1, \text{ for } x_2 < x_1$$

The following theorem about the distribution applies.

Theorem 4.9. *Given a random sample of size N from an infinite set with the value $f(x)$ at x. The probability density g_r of the r-th order statistic Y_r is*

$$g_r(y_r) = \tag{4.5}$$
$$\frac{N!}{(r-1)!(N-r)!} \cdot \left(\int_{-\infty}^{y_r} f(x)dx\right)^{r-1} \cdot f(y_r) \cdot \left(\int_{y_r}^{\infty} f(x)dx\right)^{N-r}$$

if $-\infty < y_r < \infty$.

We consider example 4.2 again. We now want to show that the first order statistic Y_1 of a $f(x)$-distributed random variable is a consistent estimator of δ. We deduce this as follows: we substitute f in (4.5) and obtain as first order statistics

$$g_1(y_1) = N \exp\left(-(y_1 - \delta)\right) \left(\int_{y_1}^{\infty} e^{-(x-\delta)} dx\right)^{N-1}$$
$$= N \exp\left(-N(y_1 - \delta)\right)$$

if $y_1 > \delta$ and $g_1(Y_1) = 0$ otherwise. Now, we immediately get

$$\mathbb{E}(Y_1) = \delta + \frac{1}{N}$$

With this follows that $\mathbb{E}(Y_1) - \delta \to 0$ if $N \to \infty$. Hence, Y_1 is *asymptotic unbiased*. Furthermore, it is

$$\mathbb{P}(|Y_1 - \delta| < c) = \mathbb{P}(\delta < Y_1 < \delta + c)$$
$$= \int_{\delta}^{\delta+c} Ne^{-N(x-\delta)} dx$$
$$= 1 - \exp(-Nc)$$

which leads to $\lim_{N \to \infty}(1 - \exp(-Nc)) = 1$. With this result and the definition 4.5 it follows that Y_1 is a consistent estimator of δ.

Sufficient Statistic

Definition 4.10. *An estimator $g(\xi)$ is exactly sufficient for the parameter θ, if for every realization of $g(\xi)$ the conditional density $f(x_1, \ldots, x_N \mid g(\xi))$ of the realization vector $(x_1, \ldots, x_N)$ under the condition $g(\xi) = \theta$ is independent of θ (see e.g. [98, 38, 208] about the term 'conditional density').*

In practice, it is hard to verify this. This is, why we use the *factorization theorem* according to Neymann.

Theorem 4.11. *The estimator $g(\xi)$ of parameter θ is a sufficient estimator, if and only if the joint distribution function of the realization vector $(\xi_1, \ldots, \xi_N)$ and θ can be written in the form*

$$f(x_1, \ldots, x_N; \theta) = h_1(g(x), \theta) h_2(x_1, x_2, \ldots, x_N) \qquad (4.6)$$

Here $h_1(x, \theta)$ depends only on the realizations of $g(\xi)$, i.e. $g(x)$ and θ. h_2 is independent of θ.

To prove that an estimator is sufficient we apply theorem 4.11. To prove that an estimator is not sufficient we better consult the definition.

Example 4.12. We consider a set of samples ξ_1, ξ_2, ξ_3 and show that

$$\eta = \frac{1}{6}(\xi_1 + 2\xi_2 + 3\xi_3)$$

is not a sufficient estimator for the parameter θ of a Bernoulli distribution: because of (4.6), we have to show that

$$f(x_1, x_2, x_3 \mid y) = \frac{f(x_1, x_2, x_3, y)}{g(y)}$$

is for some values of the random variables ξ_1, ξ_2 and ξ_3 not independent of θ. For this we assume the realizations $x_1 = 1$, $x_2 = 1$ and $x_3 = 0$. Then for the realizations y of η we get

$$y = \frac{1}{6}(1 + 2 \cdot 1 + 3 \cdot 0) = \frac{1}{2}$$

Thus it follows

$$f(1, 1, 0 \mid \eta = \frac{1}{2}) = \frac{\mathbb{P}(\xi_1 = 1, \xi_2 = 1, \xi_3 = 0, \eta = \frac{1}{2})}{\mathbb{P}(\eta = \frac{1}{2})}$$

$$= \frac{f(1, 1, 0)}{f(1, 1, 0) + f(0, 0, 1)}$$

whereas

$$f(x_1, x_2, x_3) = \theta^{x_1 + x_2 + x_3}(1 - \theta)^{3 - (x_1 + x_2 + x_3)}$$

with $x_i = 0$ or $x_i = 1$ for $i = 1, 2, 3$. Because of $f(1, 1, 0) = \theta^2(1-\theta)$ and $f(0, 0, 1) = \theta(1-\theta)^2$ follows

$$f\left(1, 1, 0 \mid \eta = \frac{1}{2}\right) = \frac{\theta^2(1-\theta)}{\theta^2(1-\theta) + \theta(1-\theta)^2} = \theta$$

and thus, is not independent of θ. Consequently, the estimator is not sufficient. We remark that for every n $\overline{\xi} = \frac{\xi_1 + \ldots + \xi_n}{n}$ is a sufficient estimator of a Bernoulli distributed random variable.

Example 4.13. We show that for a $\mathcal{N}(\mu, \sigma^2)$-distributed random variable, the empirical mean $\overline{\xi}$ is a sufficient estimator of μ: for the realizations $x_1, \ldots, x_n$ we have

$$f(x_1, \ldots, x_n; \mu) = \left(\frac{1}{\sigma\sqrt{2\pi}}\right)^n \cdot \exp\left(-\frac{1}{2}\sum_{i=1}^n \frac{(x_i - \mu)^2}{\sigma^2}\right)$$

Furthermore, we have

$$\sum_{i=1}^n (x_i - \mu)^2 = \sum_{i=1}^n ((x_i - \overline{x}) - (\mu - \overline{x}))^2$$

$$\overset{!}{=} \sum_{i=1}^n (x_i - \overline{x})^2 + \sum_{i=1}^n (\overline{x} - \mu)^2$$

$$= \sum_{i=1}^n (x_i - \overline{x})^2 + n(\overline{x} - \mu)^2$$

We notice that for '!' it applies $\sum_{i=1}^n (x_i - \overline{x})(\mu - \overline{x}) = 0$. With this follows

$$f(x_1, \ldots, x_n; \mu) = \left(\frac{\sqrt{n}}{\sigma\sqrt{2\pi}} \cdot \exp\left(-\frac{1}{2}\left(\frac{\overline{x} - \mu}{\frac{\sigma}{\sqrt{n}}}\right)^2\right)\right)$$

$$\cdot \left(\frac{1}{\sqrt{n}} \cdot \left(\frac{1}{\sigma\sqrt{2\pi}}\right)^{n-1} \cdot \exp\left(-\frac{1}{2}\sum_{i=1}^n \frac{(x_i - \mu)^2}{\sigma^2}\right)\right)$$

The first factor on the right depends only on the estimated value $\overline{x}$ and mean μ, while the second factor is independent of μ. Following theorem 4.11 $\overline{\xi}$ is therefore a sufficient estimator of μ under the condition of known variance σ^2.

Robustness

It is difficult to give a rigorous definition of the term *robustness*, since uniform criteria are missing. The term robustness describes generally the quality of an estimator to be hardly affected by changes of the assumptions. We shortly outline the idea of robustness. If we assume e.g. exponential distributed interarrival times, the estimator shall find the missing data by means of the observed realizations. But, if it turns out that the initial data is Weibull distributed, then the property of the estimator to indicate mean or variance correctly should not be affected. In this case, we say that the estimator is robust. Certainly, other properties may change and quantitative changes will occur, but their degree (not specified exactly as already mentioned) should be low.

4.1.2 Linear Regression

Linear regression and especially regression lines are classified under the generic term of linear methods. Since we will often apply these methods for estimating characteristic properties of IP traffic, we give a short introduction here. Further details are given in several monographs [118, 158].

Straight Line Fitting

Again we start with N realizations $\mathcal{X} = \{\xi_1, \ldots, \xi_N\}$ of the random variable ξ with the assumed distribution $F_\theta, \theta \in \Theta$. The result of the random experiment depends on some independent parameters $t_1, \ldots, t_N$. We give a detailed example with regard to World Wide Web traffic.

Example 4.14. We examine the size of 50 objects of mimetype text in World Wide Web traffic. Table 4.1 summarizes the sizes in ascending order. Apparently each size occurs only once. Therefore, we can assign each object size the discrete probability $\mathbb{P}(\xi = x_i) = \frac{1}{50} = 0.02$ with x_i for $i = 1, \ldots, 50$ representing the different object sizes. With this we calculate the empirical mean $\mu = \mathbb{E}(\xi) = 160,201$ and the empirical variance $\text{Var}(\xi) = 562,011,106,276$.

Table 4.1. Size (in byte) of text objects in world wide web traffic.

No.	Size	No.	Size	No.	Size	No.	Size	No.	Size
1.	155	11.	353	21.	691	31.	2,250	41.	35,056
2.	165	12.	385	22.	807	32.	2,386	42.	40,590
3.	177	13.	438	23.	856	33.	3,549	43.	49,353
4.	208	14.	463	24.	991	34.	5,899	44.	54,624
5.	221	15.	472	25.	997	35.	6,367	45.	68,397
6.	233	16.	571	26.	1,060	36.	6,440	46.	116,751
7.	236	17.	623	27.	1,223	37.	7,828	47.	395,856
8.	237	18.	635	28.	1,467	38.	10,521	48.	751,309
9.	242	19.	649	29.	2,003	39.	11,066	49.	1,263,871
10.	301	20.	659	30.	2,108	40.	12,806	50.	5,145,481

Figure 4.1 illustrates the measured data by an *empirical distribution*. The x-axis represents the object sizes and the y-axis the respective cumulative

distribution $F_\xi(x) = \mathbb{P}(\xi \leq x)$. The probability of object sizes between two measured data points is calculated by linearization.

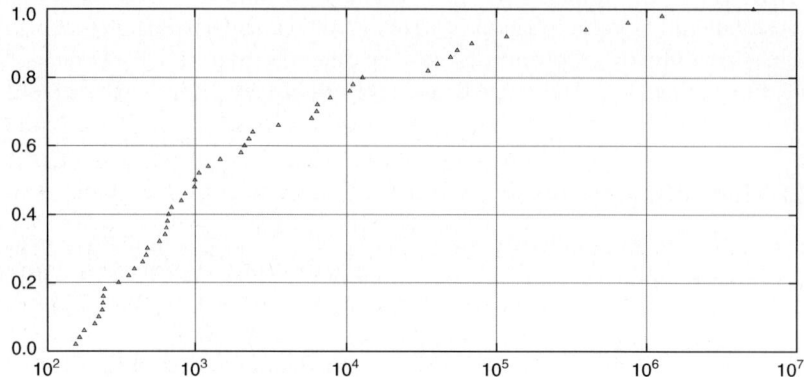

Fig. 4.1. (Empirical) probability distribution of object sizes

Because we consider ξ_i iid to ξ (same assumption applies to all experiments), we now formulate the equations

$$\xi_1 = a_1 + a_2 t_1 + \epsilon_1$$
$$\xi_2 = a_1 + a_2 t_2 + \epsilon_2$$
$$\vdots = \vdots$$
$$\xi_{N-1} = a_1 + a_2 t_{N-1} + \epsilon_{N-1}$$
$$\xi_N = a_1 + a_2 t_N + \epsilon_N$$

With these equations we assume that the measured data behaves linearly to the intervals (expressed by the straight line $y = a_1 + a_2 t$) and that the stochastic fluctuation can be indicated by the respective random variables ϵ_i. Since we assume with this model that no preference of difference exists below or above the straight line we have $\mathbb{E}(\epsilon_i) = 0$, $i = 1, \ldots, N$. For $i = 1, \ldots, N$ we now obtain

$$\mathbb{E}(\xi_i) = a_1 + a_2 t_i$$

With this we do not postulate a strictly valid linear relation, but only a possible description of the trend of the measured data, which can be of use for further analysis or forecasts.

In our example we selected quantities only instead of the exact distribution, and therefore, we have to choose a specific approach. Since we assume a heavy-tail distribution, we start with a Pareto distribution (see also because of corollary 3.46)

$$F_\xi(x) = 1 - \left(\frac{k}{x}\right)^\alpha \tag{4.7}$$

However, this distribution represents no straight line and is therefore not suitable as a regression line. For this we have equivalently to reorder equation (4.7) and apply the complementary distribution function $F_\xi^c(x) = 1 - F_\xi(x)$

$$\log\left(F_\xi^c(x)\right) = \log\left(\left(\frac{k}{x}\right)^\alpha\right) = \alpha \log\left(\frac{k}{x}\right)$$
$$= \alpha\left(\log k - \log x\right)$$
$$= -\alpha \log x + \alpha \log k$$

We substitute

$$\log(F_\xi^c(x)) = \tilde{y} = a_2 \tilde{x} + a_1$$

with $\tilde{x} = \log x$, $a_2 = -\alpha$ and $a_1 = \alpha \log k$. With the regression line $a_2 \tilde{x} = a_1$ we now get α and k according to $\alpha = -a_2$ and $\log k = \frac{a_1}{\alpha}$. Finally, we obtain k by

$$\log k = \frac{a_1}{-a_2} \Leftrightarrow k = e^{-\frac{a_1}{a_2}}$$

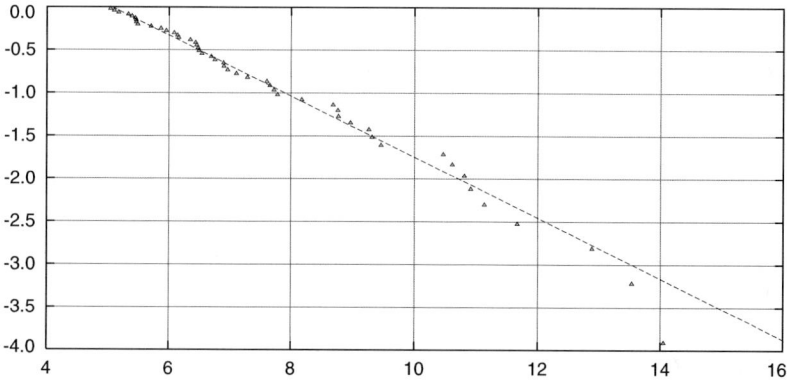

Fig. 4.2. Linearization and regression line of an (empirical) complementary cumulative distribution

With the regression line in figure 4.2 we obtain

$$\alpha = 0.3545 \quad \text{and} \quad k = 159.8178$$

Generally, this approach can be transferred to multiple parameters (denoted as *regressors*), for example

$$X_i = a_1 + a_2 t_i^{(1)} + a_3 t_i^{(2)} + a_4 t_i^{(3)} + \epsilon_i, \; i = 1, \ldots, N$$

How can we determine a_1, a_2 which indicate the trend? For this, we formulate the *method of least squares*. This method proves that depending on the samples $x = (x_1, \ldots, x_N)$ we have to choose the values $\hat{a}_1(x)$ and $\hat{a}_2(x)$, such that

$$\sum_{i=1}^{N}(x_i-(\hat{a}_1(x)+\hat{a}_2(x)t_i))^2 = \inf_{(a_1,a_2)\in\mathbb{R}^2}\sum_{i=1}^{N}(x_i-(a_1+a_2t_i))^2$$

is fulfilled. In practice, we partially derive the right term (without 'inf') with respect to a_1 resp. a_2. The minima $(\hat{a}_1(x),\hat{a}_2(x))$ have to fulfill the equations

$$\frac{\partial}{\partial a_1}\sum_{i=1}^{N}(x_i-(a_1+a_2t_i))^2 = 0 \quad\text{and}\quad \frac{\partial}{\partial a_2}\sum_{i=1}^{N}(x_i-(a_1+a_2t_i))^2 = 0$$

This yields to

$$\sum_{i=1}^{N}(x_i-a_1-a_2t_i) = 0 \quad\text{and}\quad \sum_{i=1}^{N}t_i(x_i-a_1-a_2t_i) = 0$$

Substituting $\bar{x}=\frac{1}{N}\sum_{i=1}^{N}x_i$ and $\bar{t}=\frac{1}{N}\sum_{i=1}^{N}t_i$, we get

$$\hat{a}_2(x) = \frac{\sum_{i=1}^{N}(t_i-\bar{t})(x_i-\bar{x})}{\sum_{i=1}^{N}(t_i-\bar{t})^2} \quad\text{and}\quad \hat{a}_1(x) = \bar{x}-\hat{a}_2(x)\bar{t}$$

The pair $(\hat{a}_1(x),\hat{a}_2(x))$ is denoted the *least square estimation*, the projection

$$(\hat{a}_1,\hat{a}_2):\mathbb{R}^N\longrightarrow\mathbb{R}^2$$

is denoted the *least square estimator (LSE)*. With $y=\hat{a}_1(x)+\hat{a}_2(x)t$ the *regression line* is defined.

Linear Model

A *linear model* is given, if for a N dimensional random variable Υ the form applies

$$\Upsilon = \mathcal{A}\theta + \epsilon$$

where $\Upsilon=(x_1,\ldots,x_N)^T$, $\tilde{\theta}=(\theta_1,\ldots,\theta_p)^T$ and $\epsilon=(\epsilon_1,\ldots,\epsilon_N)^T$. $\mathcal{A}$ is a known $N\times p$ matrix and $\tilde{\theta}$ the parameter vector to be estimated in the parameter space $\Theta\subset\mathbb{R}^p$. For the random vector ϵ, which describes the fluctuations, we have

- $\mathbb{E}(\epsilon_1)=\ldots=\mathbb{E}(\epsilon_N)=0$
- $\mathrm{Var}(\epsilon_1)=\ldots=\mathrm{Var}(\epsilon_N)=\sigma^2$ with unknown $\sigma^2>0$
- $\mathrm{Cor}(\epsilon_i,\epsilon_j)=0$ for $i\neq j$

These three conditions mathematically express that with random fluctuations no systematic distortion, no mutual influence and no uniform conditions exist. Obviously, the latter two conditions are fulfilled with iid realizations, i.e. with random variables $\xi_1,\ldots,\xi_N$. Since, apart from the unknown parameter $\tilde{\theta}$, σ^2 is also unknown, we altogether have the unknown parameter vector

$$\theta = \left(\tilde{\theta}, \sigma^2\right) \in \Theta \subset \mathbb{R}^p \times [0, \infty[$$

In general, this vector cannot describe the entire model, since the special structure of the fluctuation vector ϵ is not estimated. However, this does not affect the following results. The matrix $\mathcal{A}$ is often denoted as *design matrix* and describes the observation conditions or the experimental setup, respectively. In the ordinary linear case the design matrix is

$$\mathcal{A} = \begin{bmatrix} 1 & t_1 \\ 2 & t_2 \\ \vdots & \vdots \\ N & t_N \end{bmatrix}$$

As in the one-dimensional case the least square estimator is used for estimating $\tilde{\theta}$, too. Assuming the linear model $\Upsilon = \mathcal{A}\tilde{\theta} + \epsilon$, then for the sample $x = (x_1, \ldots, x_N) \in \mathbb{R}^N$ the vector

$$\hat{\theta}(x) \in \mathbb{R}^p$$

is denoted as *least square estimated value* if the equation

$$\left(x - \mathcal{A}\hat{\theta}(x)\right)^T \left(x - \mathcal{A}\hat{\theta}(x)\right) = \inf_{\tilde{\theta} \in \mathbb{R}^p} (x - \mathcal{A}\tilde{\theta})^T (x - \mathcal{A}\tilde{\theta})$$

is fulfilled. The corresponding estimator

$$\hat{\theta} = \begin{bmatrix} \theta_1 \\ \vdots \\ \theta_N \end{bmatrix} : \mathbb{R}^N \longrightarrow \mathbb{R}^p$$

assigns to each sample a least square estimator value that is denoted as least square estimator. As we can easily deduce from the definition of the least square estimator, the solution can be obtained by geometrical interpretation. Hence, the solution is given in the form of the *normal equation*.

Theorem 4.15. *Given that $\Upsilon = \mathcal{A}\tilde{\theta} + \epsilon$ is a linear model with $\tilde{\theta} \in \mathbb{R}^p$. $x \in \mathbb{R}^N$ is a sample. Then $\hat{\theta}(x)$ is a least square estimated value if*

$$\mathcal{A}^T \mathcal{A} \hat{\theta}(x) = \mathcal{A}^T x \tag{4.8}$$

To obtain a least square estimated value, we have to solve the *normal equation* (4.8). In doing so we will consider explicit and definite solutions only. If

$$\text{rang}\mathcal{A} = p \quad \text{and} \quad \tilde{\Theta} = \mathbb{R}^p$$

we denote the linear model a model with *full rank*. The inequality $p \leq N$ is necessary, because the maximum number of linear independent columns, i.e. p, cannot be larger than the number of linear independent columns, i.e. N. The significance of the definition is formulated in the following theorem (see e.g. [223] about the term rank of a matrix).

Theorem 4.16. *In a linear model* $\Upsilon = \mathcal{A}\tilde{\theta} + \epsilon$ *with full rank the least square estimator is given by*

$$\hat{\theta}(x) = \left(\mathcal{A}^T \mathcal{A}\right)^{-1} \mathcal{A}^T x, \ x \in \mathbb{R}^N \tag{4.9}$$

We remark that we developed plausible approaches to determine a least square estimator so far. But what about the risk of having determined a 'bad' estimator? We give a definition to determine that risk.

Definition 4.17. *We consider a sample space* $\mathcal{X} = \mathbb{R}^N$ *with square loss function according to (4.1) and denote an estimator of the form*

$$g : \mathbb{R}^N \longrightarrow \mathbb{R}, \ g(x) = a^T x, \ a \in \mathbb{R}^N$$

as linear estimator. *A linear estimator $\hat{g}$ is denoted as* uniform best linear unbiased estimator, *if:*

- *$\hat{g}$ is unbiased and*
- *for every linear unbiased estimators g*

$$R(\theta, \hat{g}) \leq R(\theta, g), \ \theta \in \Theta$$

For unbiased estimators the second condition can be reduced to

$$\mathbb{V}\mathrm{ar}_\theta\left(\hat{g}(\xi)\right) \leq \mathbb{V}\mathrm{ar}_\theta\left(g(\xi)\right), \ \theta \in \Theta$$

The least square estimator is given with the following result.

Theorem 4.18. *Given a linear model* $\Upsilon = \mathcal{A}\tilde{\theta} + \epsilon$ *with full rank. According to (4.9) the least square estimator is* $\hat{\theta}(x) = (\mathcal{A}^T \mathcal{A})^{-1} \mathcal{A}^T x$. *For $\alpha \in \mathbb{R}^N$ we have to estimate $\gamma(\theta) = \alpha^T \tilde{\theta}$. Then $\alpha^T \hat{\theta}$ represents a uniform best linear unbiased estimator with the risk*

$$R\left(\theta, \alpha^T \hat{\theta}\right) = \sigma^2 \alpha^T \left(\mathcal{A}^T \mathcal{A}\right)^{-1} \alpha, \ \text{for all } \theta \in \Theta$$

To estimate the single components θ_i, we substitute $\alpha = e_i$ the i-th unit vector in $\mathbb{R}^p$ (each component is 0 apart from the i-th), i.e. $e_i^T \tilde{\theta} = \theta_i$ and conclude that $\hat{\theta}_i$ is an uniform best linear unbiased estimator for θ_i.

Finally, we have to estimate as remaining parameter the variance σ^2. If we consider the model of linear regression and calculate the least square estimator $\hat{\theta}(x) = (\hat{\theta}_1(x), \hat{\theta}_2(x))$ for an observed realization $x = x_1, \ldots, x_N$, then the corresponding regression line $\hat{\theta}_1(x) + \hat{\theta}_2(x)t$ takes at t_i the values $\hat{\theta}_1(x) + \hat{\theta}_2(x)t_i$. Thus the quadratic distance is given by

$$SSE : \mathbb{R}^N \longrightarrow [0, \infty[, \ SSE(x) = \sum_{i=1}^N \left(x_i - \hat{\theta}_1(x) - \hat{\theta}_2(x)t_i\right)^2$$

We denote SSE as *sum of squares error*. It is obvious that by SSE we can describe the variance of the error, because it defines the distance between a single measurement and the regression line and builds the sum over all quadratic differences. The larger the difference is the larger SSE and the more distinctive the error variable ϵ_i becomes. We illustrate this with the following definition.

Definition 4.19. *Given a linear model* $\Upsilon = \mathcal{A}\tilde{\theta} + \epsilon$. *We define the sum of the error squares according to*

$$SSE : \mathbb{R}^N \longrightarrow [0,\infty[, \ SSE(x) = \left(x - \mathcal{A}\hat{\theta}(x)\right)^T \left(x - \mathcal{A}\hat{\theta}(x)\right)$$

Is SSE an unbiased estimator for σ^2? The next theorem gives the answer for a system with full rank.

Theorem 4.20. *In a linear model* $\Upsilon = \mathcal{A}\tilde{\theta} + \epsilon$ *with full rank we have*

$$\mathbb{E}_\theta\left(SSE(\xi)\right) = (N-p)\sigma^2, \ \text{for all } \theta = (\tilde{\theta}, \sigma^2)$$

To estimate the unknown parameter $p < N$ is required, because for e.g. $p = N = 2$ two points $(t_1, x_1), (t_2, x_2)$ are necessarily on the regression line and thus $SSE(x) = 0$. But the variance of ϵ is not necessarily identical to 0.

Proposition 4.21. *If* $p < N$ *in a linear model* $\Upsilon = \mathcal{A}\tilde{\theta} + \epsilon$ *with full rank, then* $\frac{SSE}{N-p}$ *represents an unbiased estimator of* σ^2.

4.1.3 Estimation of the Heavy-Tail Exponent α

The estimation of the exponent in complementary distribution functions of the form $F^c(x) \sim cx^{-\alpha}$ is of great importance for IP traffic modeling. With this distribution we describe e.g. on-off phases as well as the size of transferred files in good approximation. Several methods exist to calculate or estimate the decay of a heavy-tail distributed random variable representing measured sample data. We will introduce four selected approaches:

- Regression method,
- Hill Estimator,
- Mean excess function $CME(x)$,
- Scaling estimators according to Crovella and Taqqu.

At first we remind that most considered random variables exhibit a behavior of the form

$$\mathbb{P}(\xi > x) = F^c_\xi(x) \sim x^{-\alpha}L(x), \ x \to \infty, \ 0 < \alpha < 2$$

with slowly varying function L. Here '$\sim$' denotes that

$$\lim_{x \to \infty} \frac{F^c(x)}{x^{-\alpha}L(x)} = 1$$

We now consider $L(x) = c = $ const. With the logarithm of x and $F^c(x)$ and considering only larges values of x we get

$$\frac{d \log F^c(x)}{d \log x} \sim -\alpha \qquad (4.10)$$

The relation (4.10) reveals that applying $\log F^c(x)$ over $\log x$ values for large x results in a straight line with slope $-\alpha$.

Regression Line

We draw a diagram with measured data according to (4.10). The x axis represents the values under investigation e.g. size of transferred files, duration of transfers, number of packets per transfer or duration of connections. These values occur with a certain frequency, which is represented by the *empirical* probability distribution F. According to the complementary distribution function we write down the logarithm of x against the logarithm of frequencies in a table and draw the corresponding data in a diagram. The slope of a regression line in the logarithmic diagram denotes $-\alpha$. We remark that we have to define x_0 carefully above, which the measured data provides a good approximation by the regression line. This diagram is often denoted as *CD plot*. The example given in section 4.1.2 yields to the diagrams in fig. 4.3.

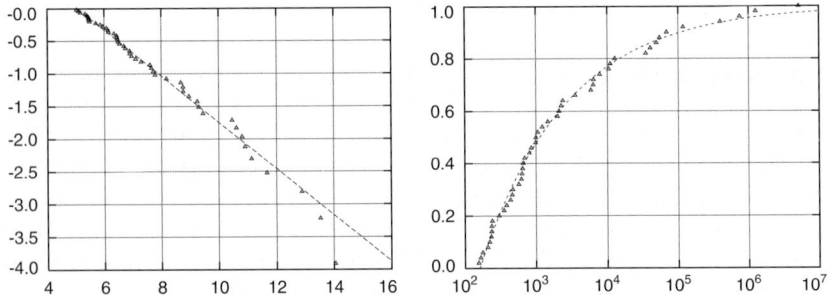

Fig. 4.3. Regression line with complementary distribution function (left) and corresponding empirical distribution (right)

Both diagrams prove that in this case the regression line yields to a good approximation and that $\tilde{x}_1 = 5$ is an appropriate starting point. However, the logarithmic diagram on the left indicates that for large x the deviation between measured data and regression line increases.

Hill Estimator

The Hill estimator gives the exponent α as a function of the k largest values. For this we reorder the sample data x_i, $i = 1, \ldots n$ by size and denote the resulting series by $x_{(i)}$, $i = 1, \ldots, n$. Now we apply the k-th order statistics

$$\mathcal{H}_{k,n} = \left(\frac{1}{k} \sum_{j=0}^{k-1} \left(\log x_{(n-i)} - \log x_{(n-k)} \right) \right)^{-1}$$

In practice we plot the estimated value $\mathcal{H}_{k,n}$ for the tail exponent α over the degree of the order statistic, i. e. k. For increasing k the curve asymptotically approaches a certain value that is regarded as tail exponent.

Mean Excess Function

With the definition $CME_\xi(x) = \mathbb{E}(\xi - x \mid \xi \geq x)$ and because of the empirical distribution function and the corresponding empirical conditional expectation $\mathbb{E}(\xi - x \mid \xi \geq x)$ we again obtain a curve that depends on x. Provided that the distribution is heavy-tailed, the slope of this curve is strictly monotone with x. In section 2.7.4 we thoroughly investigated the excess function (see e.g. (2.35) and theorem 2.49). Further details and particular applications of the excess function are given in [76].

Scaling Estimator

All approaches we treated so far show a considerable disadvantage. The CD plot as well as the Hill estimator and the excess function require a certain value x_0, above which the log-log plot behaves with slope $-\alpha$. In practice, we can rarely determine a precise value of x_0 adequately. Furthermore, the exponent significantly depends on x_0, in most cases a large x_0 leads to a larger slope. However, with a steeper slope we may neglect a relevant part of the data and eventually cannot state anything about the subexponential behavior of the distribution. This was the reason for Crovella and Taqqu to introduce a new estimator that is based on building averages, i.e. the already known blocks $X^{(m)}$.

The scaling estimator is based on the fact that the quality of the tail is determined by the scaling properties of sample data, if it is compounded. For this we again build the non-overlapping blocks of length m for $m \in \mathbb{N}$, or more precisely

$$X_i^{(m)} = \sum_{j+(i-1)m+1}^{im} X_j$$

We remind of the definition of a stable distribution. If F is a α-stable distribution, then for non-negative numbers s, t

$$s^{\frac{1}{\alpha}} X_1 + t^{\frac{1}{\alpha}} X_2 \stackrel{d}{=} (s+t)^{\frac{1}{\alpha}} X$$

with X, X_1, X_2 distributed to F. If F belongs in case of $\alpha = 2$ to the domain of normal attraction of a α-stable distribution G_α, i.e. $F \in DNA(G_\alpha)$, then

F is the normal distribution (see corollary 3.46). Generally, for $n \in \mathbb{N}$ and $F \in DNA(G_\alpha)$ we obtain

$$X_1 + \ldots + X_n \stackrel{d}{\sim} c_n X$$

Then for $F \in DNA(G_\alpha)$ we have $c_n = \frac{1}{n^{\frac{1}{\alpha}}}$ (see corollary 3.46) and call the iid series (X_i) *strictly α-stable*. As from now we denote this property *scaling property*. It indicates that the property of the tail is constant with the aggregation. We already know this fact in a similar form for subexponential distribution. Apart form the normal distribution the Pareto distribution is a well known example of this property, especially in the case of $\alpha < 2$. In principle we have two difficulties to deal with:

- Which m do we have to choose?
- Above which value of x do we have to measure the tail?

From now we consider a sequence of iid X_i strictly α-stable random variables and build $X^{(m)} = (X_i^{(m)})$ as above. For two different values $m_1 < m_2$ we determine the tail distributions $\mathbb{P}(X^{(m_i)} > x)$ and draw the respective curves in a diagram. The essential values δ and τ denote the horizontal and vertical distance respectively.

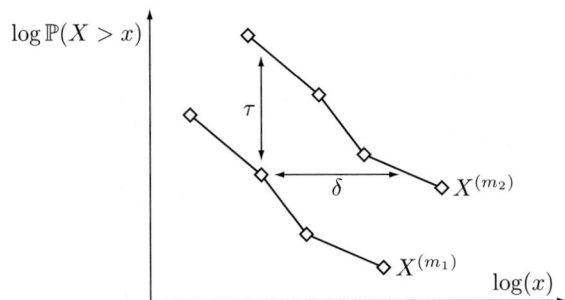

Fig. 4.4. Determination of δ and τ for the Crovella-Taqqu estimator

We will try to answer the questions above depending on δ and τ. For this we assume $F \in DNA(G_\alpha)$ stable. The problem is to estimate α. We choose $x_1 > 0$ and consider a certain point

$$\left(\log x_1, \log \mathbb{P}(X^{(m_1)} > x_1)\right)$$

whereas $\mathbb{P}$ denotes the empirical distribution. We now choose a $x_2 > 0$ with

$$\mathbb{P}\left(X^{(m_1)} > x_1\right) = \mathbb{P}\left(X^{(m_2)} > x_2\right)$$

Because of $F \in DNA(G_\alpha)$ we have approximately

4.1 Parameter Estimation

$$\frac{1}{m_1^{\frac{1}{\alpha}}} X^{(m_1)} \stackrel{d}{=} \frac{1}{m_2^{\frac{1}{\alpha}}} X^{(m_2)}$$

With this follows

$$\mathbb{P}\left(X^{(m_1)} > x_1\right) = \mathbb{P}\left(X^{(m_2)} > \left(\frac{m_2}{m_1}\right)^{\frac{1}{\alpha}} x_1\right)$$

According to the definition we find

$$\delta = \log x_2 - \log x_1 = \log\left(\left(\frac{m_2}{m_1}\right)^{\frac{1}{\alpha}} x_1\right) - \log x_1 = \frac{1}{\alpha} \log \frac{m_2}{m_1} \quad (4.11)$$

Although we can determine α by the difference δ (m_1 and m_2 are known), we do not know, which point x_1 we have to choose or which one is appropriate. Therefore, we have to determine a new parameter τ. We set

$$\tau = \log \mathbb{P}\left(X^{(m_2)} > x_1\right) - \log \mathbb{P}\left(X^{(m_1)} > x_1\right) \quad (4.12)$$

Now we have to distinguish two cases: at first let $\alpha < 2$. Because $F \in DNA(\alpha)$ it follows according to corollary 3.46

$$F^c(x) \sim cx^{-\alpha}, \text{ for } x \to \infty$$

Thus, for large x the logarithm yields to

$$\log \mathbb{P}\left(X^{(m)} > x\right) = \log c + \log m - \alpha \log x$$

Substituting in (4.12), we obtain

$$\tau = (\log c + \log m_2 - \alpha \log x) - (\log c + \log m_1 - \alpha \log x) = \log \frac{m_2}{m_1} \quad (4.13)$$

Considering (4.11) and (4.13), we see that the quotient $\frac{\tau}{\delta}$ estimates α. The asymptotic of $(X_i^{(m)})$ is true according to the general central limit theorem 3.45, after subtracting the empirical expectation $\mu = \mathbb{E}(X)$. Afterwards we have to determine x_1 such that

$$\tau \approx \log \frac{m_2}{m_1}$$

Since the values of $m_1 < m_2$ are arbitrary, we set $\Delta = \frac{m_2}{m_1}$. In most cases $\Delta = 2$ is used. Thus, x_1 is determined, such that $\tau = \log \Delta$. With this we have

$$\alpha = \frac{\log \Delta}{\delta}$$

The second case $\alpha = 2$ implies according to corollary 3.46 that F is normal distributed with μ and σ. To verify this, we have to measure a large set of sample data, because convergence is reached slowly. The same phenomena arises with simulations – in most cases the estimation of this parameter is rather vague.

The implementation to estimate the heavy-tail exponent according to the given algorithm was proposed by Crovella and Taqqu in [60]. Roughly, the underlying algorithm consists of the following steps:

- Calculate δ and τ for a set of data $m_1 < m_2$, $\Delta = \frac{m_2}{m_1}$ and subset x_1.
- Calculate for this subset $\hat{\alpha} = \frac{\log \Delta}{\delta}$.
- Compare τ with $\log \Delta$. If τ and $\log \Delta$ are almost equal, then abort the search for x_1. We now have $\hat{\alpha}$ as an estimator for α if e. g.

$$|\tau - \log \Delta| < \Theta \log \Delta$$

is true for a given Θ.

We want to give some comments on the scaling estimator. Besides the estimation of parameter α with this method we can:

a) determine the range (i.e. the values x) where heavy-tail behavior occurs,
b) determine after rescaling, whether the series $\frac{1}{m} X^{(m)}$ shows asymptotic self-similarity.

The question is how we can observe both properties? To answer this we have to distinguish between two cases:

- If F is pure α-stable (especially normal distributed), then a) is fulfilled according to the definition. Property b) follows due to the theorems about self-similarity, that is the general central limit theorem 3.45.
- For $F \in DNA(\alpha)$ we have a) for large x following (3.44) and also b) following corollary 3.45. Because this applies for large x only, we only can prove b) for small x very vaguely. However, we obtain better results for small x with a larger θ. If $F \in DNA(\alpha)$ then $F^c(x) \sim cx^{-\alpha}$ according to corollary 3.46. With this we have a good convergence for Pareto distributions already for small x. Otherwise convergence occurs slowly for α close to 2. Generally, a) can be obtained for large x, only, self-similarity in b) can be observed for all scales.

To observe convergency in particular for b), we have to subtract the empirical expectation μ according to the general central limit theorem if $\alpha > 1$.

- $\alpha > 1$: If $\mu \neq 0$ then the estimator is insufficient and we have to subtract the expectation. In doing so, the range of heavy-tail occurrence in a) may be shifted.
- $\alpha < 1$: According to the general central limit theorem we do not subtract the expectation in this case (for $\alpha < 1$ no expectation exists in the asymptotic $x \to \infty$). If we subtract the expectation, nevertheless we may obtain negative values, which distort the estimation. This is the main reason for a careful handling, especially for $\alpha \approx 1$.

4.1 Parameter Estimation 341

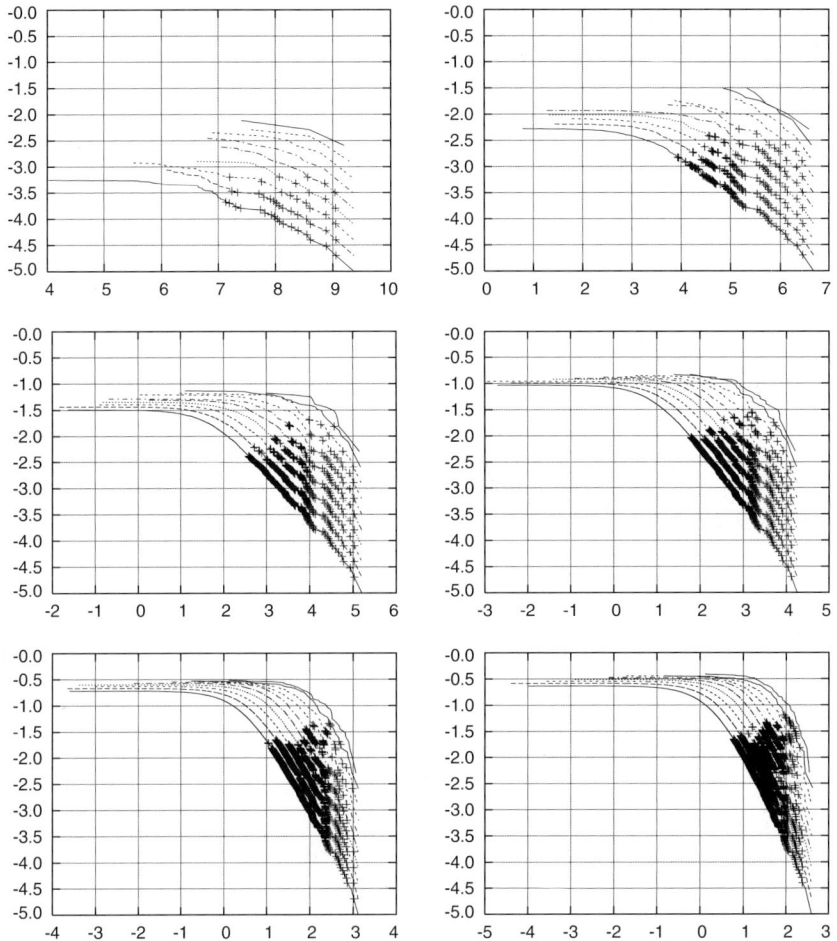

Fig. 4.5. α estimator for Pareto distribution with shifted data set

Table 4.2. Results for α estimator

fig.	function	parameter	α estimator
1.	Pareto	$k = 1, \alpha = 0.5$	$\alpha = 0.674$
2.	Pareto	$k = 1, \alpha = 0.7$	$\alpha = 0.798$
3.	Pareto	$k = 1, \alpha = 0.9$	$\alpha = 0.962$
4.	Pareto	$k = 1, \alpha = 1.1$	$\alpha = 1.126$
5.	Pareto	$k = 1, \alpha = 1.5$	$\alpha = 1.429$
6.	Pareto	$k = 1, \alpha = 1.8$	$\alpha = 1.615$

342   4 Statistical Estimators

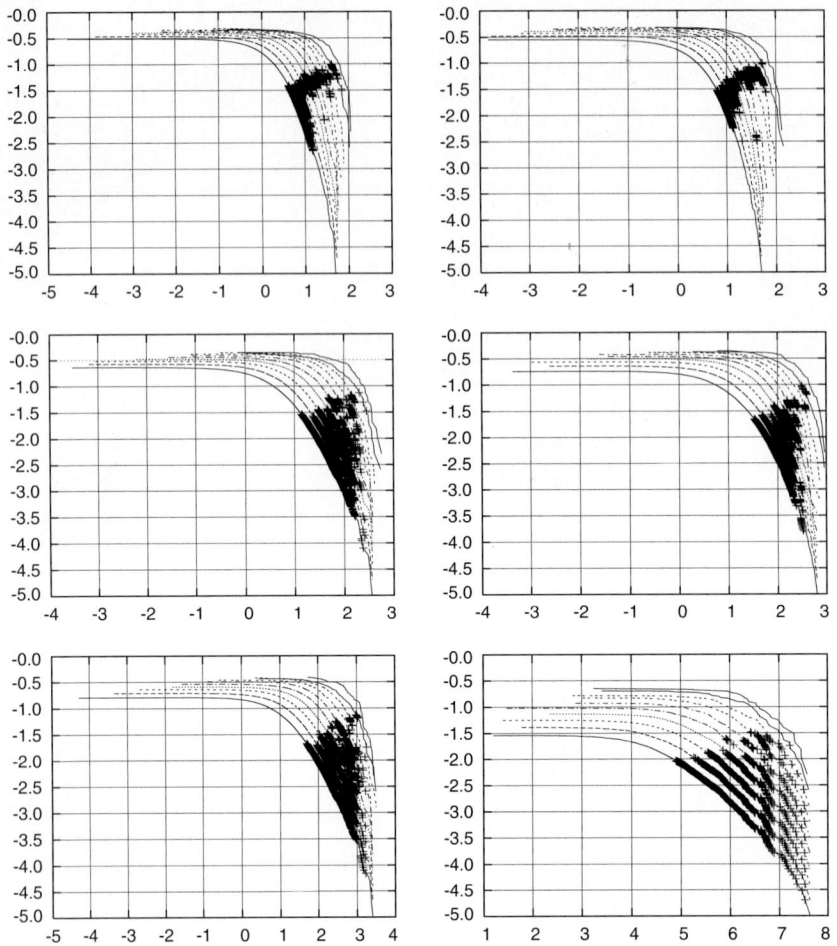

Fig. 4.6. α estimator for lognormal (left column) and Weibull (right column) distribution

Table 4.3. Results for α estimator

fig.	function	parameter	α estimator
1.	lognormal	$\mu = 1.0$, $\sigma = 1.0$	$\alpha = 2.066$
2.	Weibull	$a = 1.0$, $b = e^{-0,5}$	$\alpha = 2.266$
3.	lognormal	$\mu = 1.0$, $\sigma = 1.5$	$\alpha = 1.787$
4.	Weibull	$a = 1.0$, $b = e^{-1}$	$\alpha = 1.191$
5.	lognormal	$\mu = 1.0$, $\sigma = 2.0$	$\alpha = 1.460$
6.	Weibull	$a = 1.0$, $b = e^{-2}$	$\alpha = 0.936$

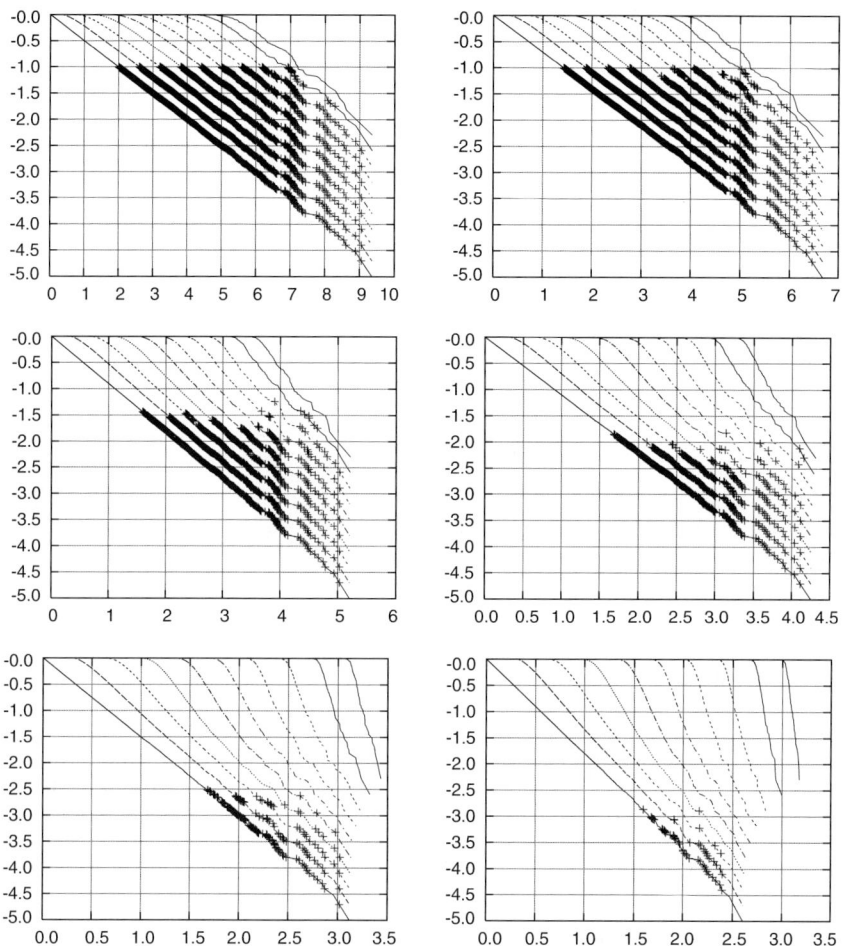

Fig. 4.7. α estimator for Pareto distribution with non-shifted data set

Table 4.4. Results for α estimator

fig.	function	parameter	α estimator
1.	Pareto	$k = 1.0, \alpha = 0.5$	$\alpha = 0.502$
2.	Pareto	$k = 1.0, \alpha = 0.7$	$\alpha = 0.689$
3.	Pareto	$k = 1.0, \alpha = 0.9$	$\alpha = 0.892$
4.	Pareto	$k = 1.0, \alpha = 1.1$	$\alpha = 1.110$
5.	Pareto	$k = 1.0, \alpha = 1.5$	$\alpha = 1.631$
6.	Pareto	$k = 1.0, \alpha = 1.8$	$\alpha = 2.031$

4.1.4 Maximum Likelihood Method

We begin our description of the well known *maximum likelihood estimator* with a short introduction to the method of moments. As we already stated in the previous section, a single parameter of a given distribution can be determined in several ways. In the remainder of this chapter we will introduce approaches to estimate the Hurst parameter. Hence, we are in the need for an appropriate approach to determine an estimator with most of the desired properties. Among these approaches are the method of moments, the maximum likelihood estimator, the Bayes estimator (see e.g. [98, 38]) and the method of least square.

Method of Moments

The method of moments is based on the mapping of moments of a random variable. Thus, we give a definition here.

Definition 4.22. *The k-th moment of N realizations $x_1, \ldots, x_N$ is the mean of the k-th power of the observations, i.e.*

$$m_k = \frac{1}{N} \sum_{i=1}^{N} x_i^k \tag{4.14}$$

To determine the parameters of an assumed distribution, we have to solve as much of the equations

$$m_k = \mu_k$$

as possible (μ_k is the k-th moment of the distribution).

Example 4.23. We consider a gamma distribution and estimate the parameters α and β. At first we introduce two variables

$$m_1 = \mu_1 \quad \text{and} \quad m_2 = \mu_2$$

whereas $\mu_1 = \alpha\beta$ and $\mu_2 = \alpha(\alpha+1)\beta^2$ (we remark that the second moment is not the variance). With this we get

$$m_1 = \alpha\beta \quad \text{and} \quad m_2 = \alpha(\alpha+1)\beta^2$$

By solving these equations, we obtain the estimated parameters

$$\hat{\alpha} = \frac{m_1^2}{m_2 - m_1^2} \quad \text{and} \quad \hat{\beta} = \frac{m_2 - m_1^2}{m_1}$$

Finally, we substitute definition (4.14), and with $\bar{x} = \frac{1}{N} \sum_{i=1}^{N} x_i$ we get

$$\hat{\alpha} = \frac{N \bar{x}^2}{\sum_{i=1}^{N}(x_i - \bar{x})^2} \quad \text{and} \quad \hat{\beta} = \frac{\sum_{i=1}^{N}(x_i - \bar{x})^2}{N \bar{x}^2}$$

If we want to estimate expectation and variance only, we do not have to assume a certain distribution.

Maximum Likelihood Estimator

The Maximum Likelihood Estimator, which was developed at the beginning of the 20th century, is a sufficient and asymptotically unbiased minimum variance estimator. The idea behind this estimator is to determine the parameters of a given sample of data, such that the observations are realized with maximal likelihood. Below we consider the estimation of a single parameter only. However, this approach can be easily generalized to the estimation of multiple parameters.

The probability that the observations $x_1, \ldots, x_N$ occur is given by

$$\mathbb{P}(\xi_1 = x_1, \ldots, \xi_N = x_N) = f(x_1, \ldots, x_N, \theta)$$

whereas the joint N dimensional distribution depends on the parameter θ. Since $x_1, \ldots, x_N$ are known, the right equation is a function of the unknown θ.

Definition 4.24. *If $x_1, \ldots, x_N$ are observed realizations of a random variable ξ, whose distribution f depends on parameter θ, then the* Likelihood Function *is given by*

$$L(\theta) = f(x_1, \ldots, x_N; \theta)$$

whereas $\theta \in D$ is in a given domain $D \subset \mathbb{R}$.

To apply the Maximum Likelihood method, we have to search for the maximum of the function L over D.

Example 4.25. Let a sample of data with exponential distributed interarrival times be given. We observe N values $t_1, \ldots, t_N$ and determine the parameter θ of the appropriate distribution. The likelihood function is given by

$$L(\theta) = f(t_1, \ldots, t_N; \theta) = \prod_{i=1}^{N} f(t_i, \theta) = \frac{1}{\theta^N} \exp\left(-\frac{1}{\theta} \sum_{i=1}^{N} t_i\right)$$

Differentiating the term $\log L(\theta)$ with respect to θ yields to

$$\frac{d \log(L(\theta))}{d\theta} = -\frac{N}{\theta} + \frac{1}{\theta} \sum_{i=1}^{N} t_i$$

Then the extremum (set derivative to 0) is

$$\hat{\theta} = \frac{1}{N} \sum_{i=1}^{N} t_i = \bar{t}$$

Example 4.26. We consider N realizations $\xi_1, \ldots, \xi_N$ of sample data that are (μ, σ^2)-distributed. We determine the values of μ and σ^2. The Likelihood function of both parameters is given by

$$L(\mu, \sigma^2) = \prod_{i=1}^{N} \mathcal{N}(x_i, \mu, \sigma^2)$$

$$= \left(\frac{1}{\sigma\sqrt{2\pi}}\right)^N \exp\left(-\frac{1}{2\sigma^2}\sum_{i=1}^{N}(x_i - \mu)^2\right)$$

Partial differentiation of $\log L(\mu, \sigma^2)$ with respect to μ and σ leads to

$$\frac{\partial (\log L(\mu, \sigma^2))}{\partial \mu} = -\frac{1}{\sigma^2}\sum_{i=1}^{N}(x_i - \mu)$$

$$\frac{\partial (\log L(\mu, \sigma^2))}{\partial \sigma^2} = -\frac{N}{2\sigma^2} + \frac{1}{2\sigma^4}\sum_{i=1}^{N}(x_i - \mu)^2$$

We set both equation to 0 again, and it follows

$$\hat{\mu} = \frac{1}{N}\sum_{i=1}^{N} x_i = \overline{x} \quad \text{and} \quad \hat{\sigma}^2 = \frac{1}{N}\sum_{i=1}^{N}(x_i - \overline{x})^2$$

Example 4.27. We consider a uniform distribution in the interval $[0, \alpha]$. We determine from a sample $x_1, \ldots, x_N$ of size N of a uniform distributed random variable the maximum likelihood estimator for α. We build the likelihood function according to

$$L(\alpha) = \prod_{i=1}^{n} f(x_i, \alpha) = \left(\frac{1}{\alpha}\right)^n$$

if α is greater than or equal to the largest value of x_i. The likelihood function is set to 0 otherwise. Since the likelihood function increases exactly with decreasing α, we have to make α as small as possible. Hence, Y_n is the n-th order statistic, a maximum likelihood estimator for α.

With the examples above gave the structure of distribution assumptions in product form. On the other hand we assumed that the parametric density family f_θ is differentiable with respect to θ. We summarize this with the following definition.

Definition 4.28. *We call a statistical experiment $(\mathcal{X}, (F_\theta)_{\theta \in \Theta})$ with the densities f_θ differentiable, if $\Theta \subset \mathbb{R}$ is an open interval and for all observed realizations $x = (x_1, \ldots, x_N) \in \mathcal{X}$ the mapping*

$$\theta \longmapsto f_\theta(x) \text{ is differentiable with respect to } f_\theta(x) > 0$$

In particular for a differentiable experiment we have

$$\frac{\partial}{\partial \theta} \log(f_\theta(x)) = \frac{\frac{\partial}{\partial \theta} f_\theta(x)}{f_\theta(x)}$$

We denote the mapping

$$L(\theta, x) = \log(f_\theta(x))$$

as *loglikelihood function* and

$$\frac{\partial}{\partial \theta} L(\theta, x) = 0 \qquad (4.15)$$

as *likelihood equation*.

By selecting an observed realization $x = (x_1, \ldots, x_N) \in \mathcal{X}$, it follows for a Maximum Likelihood estimate $\hat{\theta}(x)$

$$L\left(\hat{\theta}(x), x\right) = \sup_{\theta \in \Theta} L(\theta, x)$$

and hence,

$$\frac{\partial}{\partial \theta} L\left(\hat{\theta}(x), x\right) = 0 \qquad (4.16)$$

However, (4.15) is a necessary constraint only. Solutions of the likelihood equations may exist which are no maximum likelihood estimators. In all cases the maximality has to be proved.

Finally we want to address the product form of parametrised densities in examples 4.25 and 4.27. Again, we consider in general a differentiable experiment and repeat it N times, i.e. we have an iid sequence $\xi_1 \ldots, x_N$ with observed realizations $x = (x_1, \ldots, x_N)$. Then, the joint density of $(\xi_1, \ldots, \xi_N)$ with respect to the estimator parameter θ is given by

$$f_{\theta, N} = \prod_{i=1}^{N} f_\theta(x_i)$$

The likelihood function is

$$L_N(\theta, x) = \log(f_{\theta, N}(x)) = \sum_{i=1}^{N} L(\theta, x_i)$$

Here, we observe the advantage of the likelihood function: if we execute subsequent experiments, we obtain a density distribution in product form. With this, the likelihood function generates a sum that is easy to handle according to theorems on stochastic limits (see [38, 208]).

If we determine a maximum likelihood estimator $\hat{\theta}(x)$ for N observed values $x = (x_1, \ldots, x_N)$, then it follows according to (4.16)

$$L_N\left(\hat{\theta}(x), x\right) = \sup_{\theta \in \Theta} L_N(\theta, x)$$

and

$$\frac{\partial}{\partial \theta} L_N\left(\hat{\theta}(x), x\right) = \sum_{i=1}^{N} \frac{\partial}{\partial \theta} L\left(\hat{\theta}(x), x_i\right) = 0$$

The equation

$$\sum_{i=1}^{N} \frac{\partial}{\partial \theta} L(\theta, x_i) = 0$$

is often denoted as *likelihood equation of sample size N*.

Example 4.29. We consider an experimental sequence that is described by iid random variables $\xi_1, \ldots, \xi_N$. The distribution F_θ is unknown and depends on parameter θ. We estimate the unknown expectation $\gamma(\theta) = \mathbb{E}_\theta(\xi_1)$. For this we assume that the variance $\sigma_\theta^2 = \text{Var}(\xi_1)$ is finite. As consistent estimator we already know

$$g_N(x_1, \ldots, x_N) = \frac{1}{N} \sum_{i=1}^{N} x_i$$

For every θ we have

$$\frac{1}{\sqrt{N}\sigma_\theta} \left(g_N(x_1, \ldots, x_N) - \gamma(\theta)\right) = \frac{1}{\sqrt{n\text{Var}(\xi_1)}} \sum_{i=1}^{N} (\xi_i - \mathbb{E}_\theta(\xi_i))$$

and with the central limit theorem we conclude

$$\frac{1}{\sqrt{N\text{Var}(\xi_1)}} \sum_{i=1}^{N} (\xi_i - \mathbb{E}_\theta(\xi_1)) \longrightarrow \mathcal{N}(0,1) \text{ in distribution}$$

We say that the sequence of estimations $(g_N(x))_{N \in \mathbb{N}}$ is asymptotically normal with $\gamma(\theta) = \mathbb{E}_\theta(\xi_1)$ and variance σ_θ^2 of the observations.

Confidence Interval

Our findings were focused on point estimators so far. The goal was to determine or estimate certain parameters under the assumption of given distributions. However, we omitted the estimation of possible errors or, in other words, of confidence in the results. We give a short answer to this. Assuming a point estimator $\hat{\theta}$ of a parameter θ, then the deviation can be described with the variance $\text{Var}(\xi)$ or other informations about the distribution of the random variable ξ. We consider an *interval estimation* here. For this, we assume an interval $]\theta_1, \theta_2[$ with θ inside. For θ_1 or θ_2 we select an appropriate random variable ξ_1 or ξ_2, respectively. For a given bound γ and probability $\mathbb{P}$ this leads to

$$\mathbb{P}(\xi_1 < \xi < \xi_2) = \mathbb{P}(\theta_1 < \theta < \theta_2) = 1 - \gamma$$

Hence, the interval $\theta_1 < \theta < \theta_2$ is denoted as $(1-\gamma 100)\%$ *confidence interval*. The value $1 - \gamma$ is denoted as *grade of confidence*, whereas θ_1 and θ_2 are denoted confidence limits. The challenge is to determine for a small γ (i.e. for a high percentage $1 - \gamma$) a small interval $]\theta_1, \theta_2[$.

4.2 Estimators of Hurst Exponent in IP Traffic

Most estimators follow the statistical reproduction of the self-similarity that is determined by autocorrelation, spectral density f or its reciprocal $1/f$ (see also [7, 81]). The best choice depends on the regarded application and the grade of complexity. Generally, we distinguish between two basic types of models:

- The first model is the classical traffic theory with its distinction between arrival and service processes. We apply the models M/G/1 or GI/G/1, respectively, and estimate the appropriate model for arrival or service processes based on the measured data. The underlying formulas give quantitative or asymptotical evidence about certain characteristics, such as queue length, distribution of waiting times and loss probability (see e.g. sections 2.7.5 and 2.8.2). The essential point is estimating the parameters of the regarded distributions, e.g. of exponential distributions for interarrival times or of Pareto and Weibull distributions for connection and general service times. We presented approaches for these estimations in the above section.
- Models with heavy-tail distributions share the phenomena of long-range dependence or self-similarity respectively. Here, the challenge is to determine under the premise of these models the parameters that characterise the respective models best. With this, we have to distinguish between discrete models (time series, especially FARIMA) a time continuous processes like FBM, FBN and α-stable motions.

Methods that estimate the parameters of already compound and with this in most cases time continuous traffic appear to be interesting but complex. We will present and discuss some established methods and comment their application. We begin with pure *parametric estimators* and later introduce the improved and more precise *semiparametric estimators* which presume a certain form of spectral density. In the last case, we assume that long term dependence depends on the Hurst exponent H according to our definition and that the covariance structure is unknown. With semiparametric estimators we determine the Hurst parameter H of stationary process (X_t) (e.g. amount of data, waiting time) with a spectral representation of the form

$$f(\lambda) = \hat{f}(\lambda) \big|1 - \exp(-i\lambda)\big|^{1-2H} \qquad (4.17)$$

In most cases we have $\frac{1}{2} < H < 1$ and $\hat{f}$ is a steady function with $\hat{f}(0) \neq 0$. We find that $f(\lambda) \sim \hat{f}(0)|\lambda|^{1-2H}$ for $\lambda \to 0$. With this we have LRD phenomena if $H \in \,]\frac{1}{2}1[$.

4.2.1 Absolute Value Method (AVM)

With the introduction to asymptotic self-similarity we already built blocks over certain intervals of the form $(X_t^{(m)})$ according to the aggregated traffic (see also section 3.4.4). We now consider the first moment of the particular blocks $(X_t^{(m)})$. Afterwards we apply the expectation $\log \mathbb{E}(X_t^{(m)})$ over time $\log t$ in a log-log diagram. The slope of the line of best fit represents the Hurst parameter H, e.g. if the traffic is LRD, the slope is equal to the value of $H-1$. To give a more formal description of this, we consider a FARIMA$[p,d,q]$ time series and define the mean

$$\mathrm{AV}_X(m) = \frac{1}{\frac{N}{m}} \sum_{k=1}^{\frac{N}{m}} \left| X_t^{(m)} - \mu \right| \quad (4.18)$$

N denotes the length of the sample data, which is divided into block of length m. The mean over the whole period is represented by $\mu = \frac{1}{N}\sum_{i=1}^{N} X_i$. In case of asymptotic self-similarity, the centralized time series $X_t^{(m)} - \mathbb{E}(X_t^{(m)})$ behaves as $m^{H-1} Y_{\alpha,d}$ with $(Y_{\alpha,d})$ an α-stable time series ($0 < \alpha \leq 2$). In case of finite variance, we have fractional Gaussian noise $Y_{\alpha,d}$ according to (3.2.3). With this, $\mathrm{AV}_X^{(m)}$ behaves as m^{H-1} for large m. We already deduced in section 3.3.3 that in the case of $\alpha = 2$ we obtain $H = d + \frac{1}{2}$ and with this, $m^{H-1} = m^{d-\frac{1}{2}}$. In case of $\alpha < 2$, we have $H = d + \frac{1}{\alpha}$ (see [226, S. 382] and (3.37), respectively) and with this, $m^{H-1} = m^{d+\frac{1}{\alpha-1}}$. Thus, the estimation according to (4.18) yields to the parameter H and not d, because the term still contains the unknown factor α.

To determine H we have to draw the results for different values of m in a log-log diagram. The AVM stands in a straight line of the method of least squares, where regression lines could provide support for the AVM. We remark that the amount of sample data N and the number of blocks have to be sufficiently large to obtain reasonably good results.

We also find that in case of no long-range dependence the slope is $-\frac{1}{2}$, because of $H = \frac{1}{2}$. Another important observation is that small values of m (i.e. building the mean over very small time scales) as well as large values of m (i.e. building the mean over the whole data) yield to no significant information. The fractality for small values is too fine grained and for large values too coarse grained to obtain reliable results; so we have to avoid these extreme scales.

An alternative estimator was proposed by Higuchi[111]:

$$L(m) = \frac{N-1}{m^3} \sum_{i=1}^{m} \left[\frac{N-i}{m} \right] \sum_{k=1}^{\left[\frac{N-i}{m}\right]} \left| \sum_{j=i+(k-1)m+1}^{i+km} X_j \right|$$

whereas $[x]$ denotes the Gaussian bracket, i.e. the largest integer g with $g \leq x$. Asymptotically, we get

$$\mathbb{E}(L(m)) \sim \mathrm{const} \cdot m^{H-2}$$

with which we again obtain H, represented by the slope of the straight line in a log-log diagram. We express the difference compared to the legacy AVM method in the following way: we do not build disjunct blocks but 'observation windows' which slide over the whole set of sample data. The computational effort is larger, but leads with huge sets of data (more than 10,000 entries) to more precise results.

Quality of Estimation

We will conclude each of the following sections with an empirical study to reflect the properties of the respective Hurst estimator. The underlying algorithms follow the methods of analysis given by Taqqu. For this we generated three different series (FGN and FARIMA(1,d,1) with $\alpha = 1,8$ and $\alpha = 2,0$), each with six different Hurst parameters $H = 0.5, 0.6, 0.7, 0.8, 0.9, 0.9\overline{6}$. Each series consists of 65,536 values and was generated 200 times. We applied the different estimators as presented in the respective section to each of these $3 \cdot 6 \cdot 200 = 3,600$ series. To illustrate the results, the boxplots include the following information:

- middle 50% of data (boxes),
- median (black line),
- 95% confidence interval about the median (shaded grey area),
- mean (black dot),
- whiskers encompassing 95% of the data (dashed lines),
- outliers that fell beyond the whiskers.

The absolute value method is among the better ones for Gaussian traffic, especially for Gaussian time series. However, it exhibits unsatisfying behavior with increments of infinite variance. With α-stable distributions the behavior gets insufficient for decreasing α. Hence this estimator is preferred for traffic whose increments are Gaussian distributed or of finite variance. This method estimates H for FARIMA time series and not d.

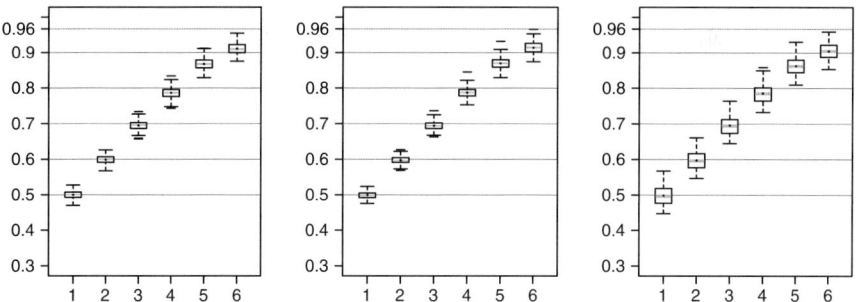

Fig. 4.8. Absolute Value method for FGN, Gaussian FARIMA series and FARIMA series with 1.5-stable generator (from left to right) with respective Hurst parameter $H = 0.5, 0.6, 0.7, 0.8, 0.9, 0.9\overline{6}$

4.2.2 Variance Method

The approach of the Variance Method is quite similar to the Absolute Value Method (AVM). However, instead of the expectation we now draw the variance line $\mathrm{Var}(X_t^{(m)})$. Again, the slope β of the regression line in the log-log diagram denotes the longe range dependence. With $-1 < \beta < 2$ we find the Hurst parameter $H = 1 - \frac{\beta}{2} \in\,]0, 1\frac{1}{2}[$. Because for $\mathrm{Var}(X_t^{(m)}) \sim cm^{2H-2}$, the slope β in the log-log diagram represents the value of $2H - 2$. With this, we have $H = 1 - \frac{\beta}{2}$.

For a more formal description let $X^{(m)}$ denote the aggregated traffic, and instead of (4.18) we now obtain the variance according to

$$\widehat{\mathrm{Var}}(X^{(m)}) = \frac{1}{\frac{N}{m}} \sum_{k=1}^{\frac{N}{m}} \left(X_k^{(m)} - \mu\right)^2$$

whereas μ represents again the mean over the entire sample data, i.e. $\mu = \frac{1}{N}\sum_{i=1}^{N} X_i$. As already stated in section 4.2.1 the centralized process $X_t^{(m)} - \mathbb{E}(X_t^{(m)})$ behaves asymptotically to m^{H-1}. In case of time series with Gaussian increments or with increments of finite variance, we find asymptotical behavior of $m^{2H-2} = m^{2d-1}$ for large $\frac{N}{m}$ and m.

For infinite variance we get a more complex solution and give a short derivation here. Because of the convergence of $(X^{(m)})$ towards an α-stable process or to a linear fractional stable noise respectively, we have

$$\widehat{\mathrm{Var}}(X^{(m)}) \sim \widehat{\mathrm{Var}}\left(m^{H-1} S_{\alpha,d}\right)$$

$$= m^{2H-2} \left(\frac{1}{\frac{N}{m}} \sum_{k=1}^{\frac{N}{m}} S_{\alpha,d}(k) - \left(\frac{1}{\frac{N}{m}} \sum_{k=1}^{\frac{N}{m}} S_{\alpha,d}(k)\right)^2 \right)$$

Because we assume (X_t) asymptotically self-similar, it follows that for large m the compounded blocks $(X^{(m)})$ behave like $m^{H-1} S_{\alpha,s}$, distributed whereas $S_{\alpha,d}$ is a FARIMA$[p,d,q]$ series with α-stable increments. With this follows in distribution

$$\widehat{\mathrm{Var}}\left(X^{(m)}\right) \sim m^{2H-2} \left(\left(\frac{N}{m}\right)^{\frac{2}{\alpha}-1} Z_{\frac{\alpha}{2}} - C^2\right)$$

whereas $Z_{\frac{\alpha}{2}}$ is a $\frac{\alpha}{2}$-stable random variable and C a constant. We obtain the asymptotic in distribution

$$\widehat{\mathrm{Var}}\left(X^{(m)}\right) \sim C(N)m^{2H-2+1-\frac{2}{\alpha}}Z_{\alpha,d}$$
$$= C(N)m^{2d-1}Z_{\alpha,d}$$

whereas $C(N)$ depends on the length N of the sample data.
If we select N very large compared to the possible compositions m, we again obtain the value of d as the slope of a straight line in the log-log diagram, namely in case of existing as well as in case of infinite variance. If the series of means exhibit leaps or a smooth decrease, this may indicate instationarity. To distinguish between this phenomena and the behavior of long-range dependence, we consider

$$\widehat{\mathrm{Var}}\left(X^{(m_i+1)}\right) - \widehat{\mathrm{Var}}\left(X^{(m_i)}\right)$$

for subsequent values m_i. This may prove to be an effective way to determine nonstationarity, and therefore should be considered apart from any common method. However, we state that for α-stable FARIMA$[1,d,1]$ series the variation increases, such that reasonable estimations cannot be guaranteed.

Quality of Estimation

We treat the Variance Estimator as our second traditional graphical estimator. It exhibits distinct biased behavior, especially for $d > 0.3$ with an arbitrary Gaussian FARIMA$[p,d,q]$ time series. This behavior does not change significantly for non-Gaussian increments. Nevertheless, the variance estimator is very robust but less accurate than the Whittle estimator described below (see 4.2.6).

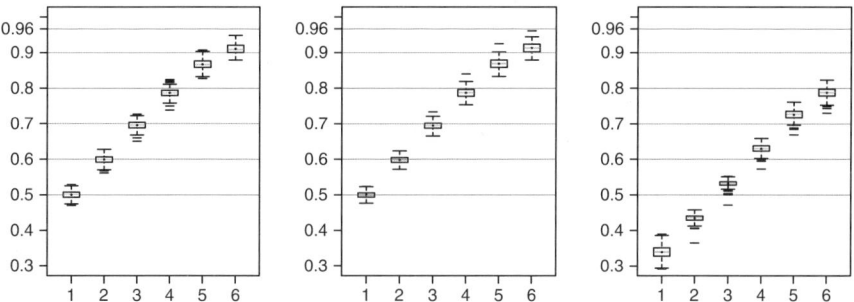

Fig. 4.9. Variance estimator of FGN, Gaussian FARIMA series and FARIMA series with 1.5-stable generator (from left to right) with respective Hurst parameter $H = 0.5, 0.6, 0.7, 0.8, 0.9, 0.9\overline{6}$

The variance estimator clearly leads to weak results for FARIMA series with 1.5-stable generator.

4.2.3 Variance of Residuals

The idea of the Variance of Residuals method is to linearise the aggregated traffic and to examine the resulting distortion, i.e. the remainder, with the variance estimator. We illustrate some details on this approach, which was introduced by Peng et al. [202].

We divide the series X_i into blocks of size m and remind of the definition of compounded traffic. Within each block we build the partial sum (in favor of a simple illustration we consider only the first block)

$$Y_t = \sum_{k=1}^{t} X_k, \quad t = 1, \ldots, m$$

Finally, we fit a line $at - b$ to this series by applying the least square estimator or the regression method respectively (see 4.1.2), i.e. we consider the residual $\sum_{t=1}^{m}(Y_t - at - b)$. For this 'residual' we now estimate the variance

$$\frac{1}{m} \sum_{t=1}^{m}(Y_t - at - b)^2 \qquad (4.19)$$

In this equation we may regard Y_t as sum of all observed traffic until time t. How does this estimator behave dependent on H and d? For an answer we first define two random variables

$$Z_1 = \int_0^1 L_{\alpha,H}(t)dt \quad \text{and} \quad Z_2 = \int_0^1 t L_{\alpha,H}(t)dt$$

whereas $(L_{\alpha,H}(t))_{t \in \mathbb{R}}$ is a α-stable fractional motion (for a definition see section 3.2.3) with $H = d + \frac{1}{\alpha}$. According to this, we obtain for large m due to the self-similarity (note that $0 < \alpha \leq 2$ and $\alpha = 2$ in case of fractional Brownian motion) in distribution

$$Y_{mt} \sim m^H L_{\alpha,H}(t)$$

This yields to the approximations

$$\sum_{t=1}^{m} Y_t \sim \int_0^m Y_t dt \stackrel{d}{\sim} m^{H+1} Z_1$$

$$\sum_{t=1}^{m} tY(t) \sim \int_0^t Y_t dt \stackrel{d}{\sim} m^{H+2} Z_2$$

We determine the coefficients with (4.19) by calculating

$$a = \frac{\sum_{t=1}^{m} tY_t \frac{1}{m} \sum_{t=1}^{m} Y_t \sum_{t=1}^{m} t}{\sum_{t=1}^{m} t^2 \frac{1}{m} \left(\sum_{t=1}^{m} t\right)^2} \stackrel{d}{\sim} \frac{m^{H+2} Z_2 - m^{H+2} \frac{Z_1}{2}}{\frac{m^3}{3} - \frac{m^3}{4}}$$

$$= m^{H-1}(12Z_2 - 6Z_1)$$

$$b = \frac{1}{m}\sum_{t=1}^{m} Y_t - \frac{1}{m}\sum_{t=1}^{m} at \overset{d}{\sim} m^H Z_1 - m^H \frac{12Z_2 - 6Z_1}{2}$$
$$= m^H(4Z_1 - 6Z_2)$$

In distribution we have

$$\frac{1}{m}\sum_{t=1}^{m}(Y_t - at - b)^2 \overset{d}{\sim} m^{2H} \int_0^1 (L_{\alpha,H} - 6t(2Z_2 - Z_1) - (4Z_1 - 6Z_2))^2 dt \quad (4.20)$$

The integral is independent of m and thus, the variance of the distortion in (4.19) behaves as m^{2H} apart of a constant. We remark that we presented a rough derivation only. A mathematically rigorous illustration for finite as well as for infinite variance is given in [248].

We have two options to estimate the Hurst parameter. The first is to apply the approach as described above to each block $\frac{N}{m}$ and calculate the mean over all blocks. The results drawn in log-log diagram versus m should give a straight line with slope $2H$.

The second option is to obtain a new random variable by building averages over all blocks. However, we have to be careful, because according to (4.20), the integral behaves as $\frac{\alpha}{2}$-stable random variable. If the original sample data is of infinite variance, then the new random variable shows none either – this behavior is often true for Internet traffic. With averaging these $\frac{\alpha}{2}$-stable random variables of $\frac{N}{m}$ blocks, we get an asymptotic of $m^{1-\frac{2}{\alpha}}$. The corresponding variance behaves as $m^{2H+1-\frac{\alpha}{2}} = m^{2d+1}$. This option is feasible, if we are interested in d instead of H. However, for a large set of sample data the variation is too large, so that we favor the first option in practice.

Quality of Estimation

Apart from the absolute value method the variance of residuals is the only estimator for FARIMA time series that estimates H instead of d.

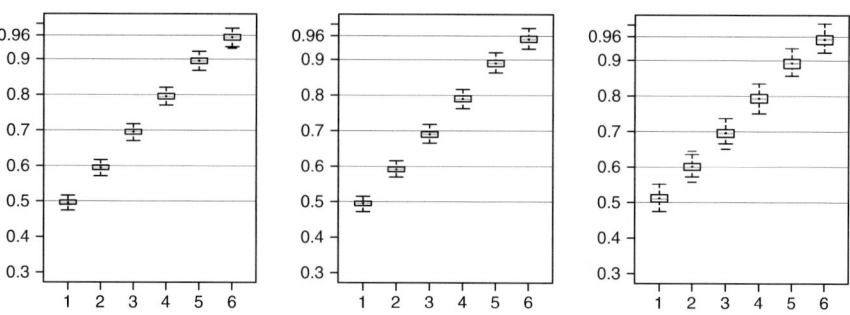

Fig. 4.10. Variance of Residuals estimator of FGN, Gaussian FARIMA series and FARIMA series with 1.5-stable generator (from left to right) with respective Hurst parameter $H = 0.5, 0.6, 0.7, 0.8, 0.9, 0.9\overline{6}$

The variance of residuals exhibits excellent results for all simulated processes, especially for FARIMA series with 1.5-stable generator. For Gaussian and stable FARIMA$[0, d, 0]$ series it leads to better results compared to analytical estimators, like Whittle estimator or wavelet analysis (see below). Here, LSE shows the smallest value of all graphical estimators. However, the behavior worsens for time series with increasing short-range dependence of p and q. Although its robust behavior for traditional distributions, the bias of this estimator rises significantly for increments in time series of Pareto or lognormal distributions, as well as for all kinds of FARIMA$[1, d, 1]$ time series.

4.2.4 R/S Method

The R/S method belongs to the traditional estimators and is still in broad use today. This approach was first investigated by Hurst [115]. We want to give a detailed description and start with a formal definition.

Definition 4.30. *Let (X_k) be a time series or simply a series of random variables. Similar to the previous section we set $Y_t = \sum_{k=1}^{t} X_k$ for $t \in \mathbb{N}$ and may regard Y_t as the sum of load at time t, here. We denote*

$$R(l, k) = \max_{1 \leq j \leq k} \left(Y_{l+j} - Y_l - \frac{j}{k}(Y_{l+k} - Y_l) \right)$$
$$- \min_{1 \leq j \leq k} \left(Y_{l+j} - Y_l - \frac{j}{k}(Y_{l+k} - Y_l) \right)$$

the adjusted target region *and define*

$$S(l, k) = \left(\frac{1}{k} \sum_{j=l+1}^{l+k} \left(X_j - \hat{X}_{l,k} \right)^2 \right)^{\frac{1}{2}}$$

whereas $\hat{X}_{l,k} = \frac{1}{k} \sum_{j=k+1}^{k+l} X_j$ represents the mean of the l-th block.

The fraction

$$Q(l, k) = \frac{R(l, k)}{S(l, k)}$$

is denoted as *rescaled adjusted target region*. The value $S(l, k)$ represents the variance estimator of the time series. The index l ensures that we move one block to the right in each step. In favor of a simple notation we write $Q(k) = Q(0, k)$ in case of $l = 0$ from now on. For e.g. stationary increments, we omit the notion of l, because $Y_{l+j} - Y_l$ distributed like $Y_j - Y_0$. For H-sssi random series (X_i) the term $Q(k) = \frac{R(k)}{S(k)}$ thus specifies an estimator of H for $k \to \infty$. With theorem 4.32 and 4.34 we will give further details on this below.

How can we interpret the specific parameters? $R(k)$ exhibits a behavior of the form $k^H = k^{d+\frac{1}{\alpha}}$ because of the compound random variables Y and their asymptotical behavior. Otherwise, $S(k)$ is, as already described, the square root of the variance of a data series, which is in turn proportional to $k^{\frac{2}{\alpha}-1}$. With this it follows $S(k) \sim k^{\frac{1}{\alpha}-\frac{1}{2}}$. Then the joint convergence for $R(k)$ and $S(k)$ yields for $Q(k)$ to

$$Q(k) = \frac{R(k)}{S(k)} \sim k^{d+\frac{1}{2}}, \text{ for } k \to \infty$$

Hence, drawing the quotient $Q(k)$ in a log-log diagram leads to the value of parameter d again. A more rigorous illustration of this heuristic derivation is given in several monographs (see [173, 23, 32]).

To estimate the Hurst parameter with this method we assume long-range dependence (this can be proved by other estimators) and draw $\log Q(k)$ versus $\log k$ as above. For each $k \in \mathbb{N}$ we have $n - k + 1$ copies, namely $Q(0,k), \ldots, Q(n-k, k)$. This implies that for large k we have to consider a large number of realizations.

Theorem 4.31. *Let (X_k) be a stationary time series in the strict sense such that (X_k^2) is ergodic and $\frac{1}{n}\sum_{j=1}^{[nt]} X_j$ converges weakly to the Brownian motion for $n \to \infty$, i.e. convergence in a functional sense. Then for $k \to \infty$ the following asymptotic applies*

$$k^{\frac{1}{2}} Q(k) \stackrel{d}{\sim} \xi$$

whereas ξ is a non-degenerated random variable.

If the well known central limit theorem applies, then the term $k^{\frac{1}{2}} Q(k)$ converges also to a random variable. In that case we obtain the Hurst parameter with value $\frac{1}{2}$. The result is different, if we assume convergence to the fractional Brownian motion.

Theorem 4.32. *Let (X_k) be a stationary time series in the strict sense, such that (X_k^2) is ergodic and $\frac{1}{n}\sum_{j=1}^{[nt]} X_j$ converges weakly to the fractional Brownian motion with Hurst parameter H for $n \to \infty$, i.e. is convergence in a functional sense. Then, for $k \to \infty$ the following asymptotic applies*

$$k^H Q(k) \stackrel{d}{\sim} \xi$$

whereas ξ is a non-degenerated random variable.

How can we determine the distinct behavior? We draw the respective realizations $\log Q(k)$ (as described above $n - k + 1$ times) versus $\log k$. If we observe a scattered plot around a straight line with slope $\frac{1}{2}$, then we have the result according to theorem 4.31. But if the scattered plot is of slope $H > \frac{1}{2}$, i.e. a straight line of the form $H \log k$ with dispersion $\log \xi$, then we have a fractional Brownian motion according to theorem 4.32. The particular property of this method is formulated in the next theorem that specifies the transfer on non-existing variance.

Theorem 4.33. *Let (X_k) be a iid time series such that $\mathbb{E}(X_k^2) = \infty$. Furthermore, let the series be in attraction to an α-stable distribution (see section 3.2.3) with $0 < \alpha < 2$. Then, for $k \to \infty$ the following asymptotic applies*

$$k^{\frac{1}{2}} Q(k) \overset{d}{\sim} \xi$$

whereas ξ is a non-degenerated random variable.

With this the asymptotic of the R/S statistic also applies to heavy-tail marginal distributions. However, proofs and propositions about convergence are rather difficult. Further details are given in an article by [23]. To provide more insight in theorems 4.31 to 4.33, we cite a result that goes back to B. Mandelbrot.

Theorem 4.34. *Let (X_k) be a stationary time series in the strict sense such that*

$$\left(\frac{1}{n^{H_1} L_1(n)} \sum_{j=1}^{[nt]} X_j, \frac{1}{n^{H_2} L_2(n)} \sum_{j=1}^{[nt]} X_j^2 \right) \longrightarrow (M(t), V(t))$$

converges, whereas L_1, L_2 are slowly varying functions and the convergence has to be seen in a certain functional sense, such that building of inf and sup remain. Then, the renormalized R/S statistic

$$\left(\frac{R([kt])}{k^J L(k) S(k)}, 0 \leq t \leq 1 \right) \tag{4.21}$$

converges, whereas L is a slowly alternating function and $J = H_1 - \frac{H_2}{2} + \frac{1}{2}$.

With this theorem we also determine, if a $0 \leq J \leq 1$ and a slowly varying function exist, such that a convergence as in (4.21) is possible. As described in [23] J depends only on the Hurst exponent and not on the finite-dimensional marginal distributions of the stationary time series. If $J = d + \frac{1}{2}$ applies, then R/S is robust, whereas $d \in]0, \frac{1}{2}[$ represents the measure of the long-range dependence of the FARIMA series. According to [32], we summarize the R/S method with the following three steps:

- First, calculate $Q(l, k)$ for an adequate number of different k, l.
- Second, draw $\log Q(l, k)$ versus $\log k$ for different k with a dispersion of the respective values of l.
- Finally, we plot a line of best fit $y = a \log k + b$ which describes for large values of k our sample data. By means of the least square method we determine appropriate values of $\tilde{a}, \tilde{b}$ and then $\tilde{a} = \tilde{H}$ is an estimator of the Hurst parameter.

We give some comments on possible difficulties with the R/S method. It is hard to determine, at which value of k the asymptotic occurs. For a finite set of sample data the distribution of Q is neither symmetric nor normal

Quality of Estimation

The R/S method is rather inapplicable for accurate estimations and should be used only to get a rough idea of the behavior of a time series. This estimator exhibits clear weaknesses for all paths and fits with high variation to some degree at $H = 0.6$ only. For non-Gaussian processes like the FARIMA time series with 1.5-stable generator the R/S estimator is not applicable.

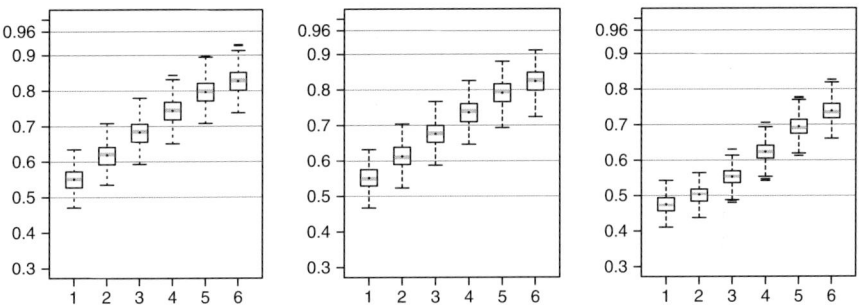

Fig. 4.11. R/S estimator of FGN, Gaussian FARIMA series and FARIMA series with 1.5-stable generator (from left to right) with respective Hurst parameter $H = 0.5, 0.6, 0.7, 0.8, 0.9, 0.9\overline{6}$

4.2.5 Log Periodogram – Local and Global

We begin with a general description of the periodogram estimator and introduce specific variants afterwards. Foundation is the *Periodogram* of a time series (X_t) with length N. Accordingly, to the frequency λ we build

$$I(\lambda) = \frac{1}{2\pi N} \left| \sum_{k=1}^{N} X_k e^{-ik\lambda} \right|^2 \qquad (4.22)$$

If the time series is of finite variance, then the spectral density is estimated herewith. As we already described in section 3.1.2 in case of long term dependence, the time series behavior as $|\lambda|^{-2d}$ at the origin. Drawing the dependence of the frequency in a log-log diagram, we again may determine the value $-2d$ as slope. However, this applies only in case of finite variance. For infinite variance the solution gets far more complicated, and we still miss a satisfying theoretical results today (see [144, 145] and [148]). However, empirical results indicate that the method can also be extended to this case.

For semiparametric methods we have to distinguish between the local and global periodogram. Roughly, the Hurst parameter is determined by the slope of the logarithm of the spectral density against the logarithm of the frequency of the applied Fourier transform. The local periodogram uses frequencies near 0 only, the global periodogram all frequencies, instead. Although, a limitation of the shape of the spectral density exists. Both methods are Gaussian estimators, i.e. they are applied to estimate Gaussian processes.

For a detailed description of the *local logperiodogram estimator* we consider equation (4.17) at first. A logarithmic regression is recommended to estimate the exponent by means of the first $m(N)$ Fourier coefficients. We define for N values the periodogram depending on the frequency of the Fourier series

$$I_N(\lambda) = \frac{1}{2\pi N} \sum_{k=1}^{N} |X_k \exp(-ik\lambda)|^2$$

With the semiparametric approach (4.17) we deduce the following estimator over the first $m(N)$ frequencies for $j = 1, \ldots, m(N)$

$$\log(I_N(\lambda_j)) = \log\left(\hat{f}(0)\right)$$
$$+ (1 - 2H) \log\left(|1 - \exp(-i\lambda_j)|\right) \quad (4.23)$$
$$+ \log\left(\frac{\hat{f}(\lambda_j)}{\hat{f}(0)}\right) + \log\left(\frac{I_N(\lambda_j)}{f(\lambda_j)}\right)$$

The local Log-Periodogram estimator is consistent for the parameter $m(N) = N^\alpha$, $0 < \alpha < 1$. Taking the series of Fourier coefficients $\lambda_j \to 0$, the third term (4.23) vanishes, because $\hat{f}$ is steady at 0 and $\log 1 = 0$. The fourth term can be regarded as error term and is not iid.

The expectation depends on j, and we remind that $I_N(\lambda_j)$ is a random variable depending on X_k. Under certain properties of the function $\hat{f}$ this term converges in distribution to a $\mathcal{N}(\mu, \sigma^2)$-distributed random variable. More details about these results are given in [116]. The complexity is in the order of the regular regression problem, i.e. with m frequencies the complexity is of order $\mathcal{O}(m)$ (see 4.1.2). The estimator is applied to Gaussian processes, where normally $\hat{f}$ is differentiable and considered locally bounded in a neighbourhood of 0.

We now consider the *global estimator*. We already pointed out that the first periodogram estimator is denoted as local, because the used frequencies converge to 0. With the global periodogram the situation is different. The model, introduced by Moulines and Soulier [182], uses the whole range of observable Fourier frequencies. However, there is a certain drawback: with this approach we have not only to determine the regarded Hurst parameter H, but we also we have to estimate the function $\log(\hat{f}(\cdot))$. For this let (e_j) be a base of the Hilbert space of $\mathcal{L}^2(\mathbb{R})$ representing the function $\hat{f} \in \mathcal{L}^2(\mathbb{R})$

4.2 Estimators of Hurst Exponent in IP Traffic

$$\log(\hat{f}) = \sum_{k=0}^{\infty} \alpha_j e_j$$

With the method described here, the so called FESM (*fractional exponential estimator method*), we apply the orthonormal base

$$e_j(\lambda_j \cdot) = \frac{1}{\sqrt{\pi}} \cos(j\lambda_j \cdot)$$

We divide the N periodogram variables $X_1, \ldots, X_N$ to K blocks of length m, whereas the last index is omitted. The same separation applies to the respective Fourier frequencies $\lambda_1, \ldots, \lambda_N$. Over each block we now build a weighted sum and renormalise to obtain the expectation 1. Then we build the logarithm and get the variables $(Z_{N,k})_{k=1,\ldots,K}$. With the LSE (see section 4.1.2) we finally obtain

$$Z_{N,k} = (1 - 2H) \log\left(|1 - \exp(-ix_k)|\right) + \sum_{j=0}^{m} \alpha_j e_j(x_k) + \mathcal{R}_{N,k}, \quad k = 1, \ldots, K$$

whereas $x_k = \frac{(2k+1)\pi}{N}$. We remark that the error terms $\mathcal{R}_{N,k}$ are neither independent nor centralized, and the bias with constant m does not converge asymptotically to 0.

The FESM is consistent and asymptotically normal, if we select blocks of length $m = N^\beta$, whereas $\frac{1}{2\delta} < \beta < \frac{2}{3}$. The factor δ describes the Hölder continuous of the function $\hat{f}$ here. Finding the right value m by using a least square estimator, is equivalent to finding the right value by using the local periodogram estimator (see [117]). The model estimates Gaussian processes, and the function $\hat{f}$ should be bounded away from 0, where we stress a decaying condition on the Fourier coefficients of the function $\log(\hat{f})$.

Quality of Estimation

All three log periodogram estimators provide good results for FGN, whereas the compound estimator becomes considerably biased for the Gaussian series. The estimation of a non-Gaussian series leads to partially wrong results, e.g. short-range dependency at $H \approx 0.4$ instead of the expected long term dependency at $H = 0.6$. This last graphical estimator is the methodical foundation of the Whittle estimator. It proves to be the best graphical estimator for Gaussian processes. Although we remark that the precision is excellent in case of FARIMA$[0, d, 0]$ series, its behavior worsens in case of increments without variance. This is the reason, why the periodogram estimator is not recommended for non-Gaussian time series.

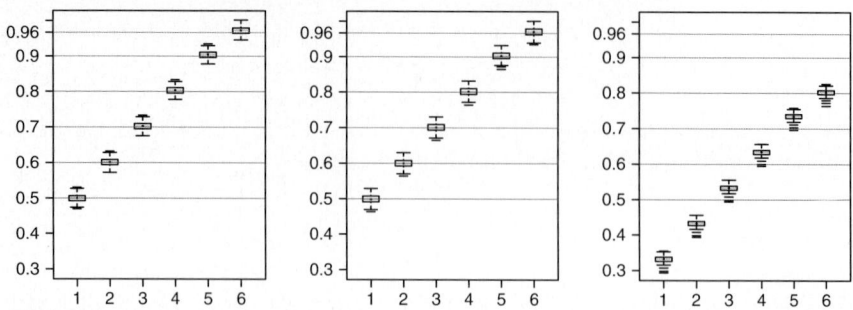

Fig. 4.12. Local Log-Periodogram estimator of FGN, Gaussian FARIMA series and FARIMA series with 1.5-stable generator (from left to right) with respective Hurst parameter $H = 0.5, 0.6, 0.7, 0.8, 0.9, 0.9\overline{6}$

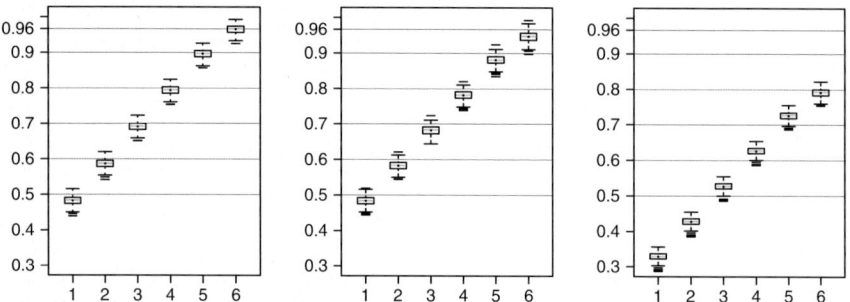

Fig. 4.13. Global Log-Periodogram estimator of FGN, Gaussian FARIMA series and FARIMA series with 1.5-stable generator (from left to right) with respective Hurst parameter $H = 0.5, 0.6, 0.7, 0.8, 0.9, 0.9\overline{6}$

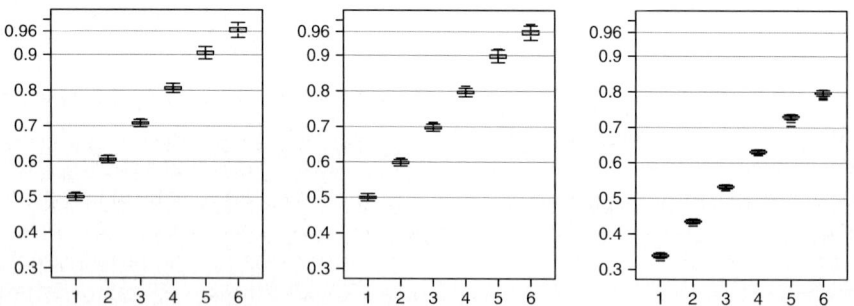

Fig. 4.14. Cumulative Periodogram estimator of FGN, Gaussian FARIMA series and FARIMA series with 1.5-stable generator (from left to right) with respective Hurst parameter $H = 0.5, 0.6, 0.7, 0.8, 0.9, 0.9\overline{6}$

4.2.6 Maximum Likelihood and Whittle Estimator

Before we turn our attention to the Whittle estimator (which is not a semi-parametric estimator), we want to express some fundamental remarks on parametric estimators. With the methods introduced so far, we determine, whether observed data exhibits long-range dependence. These approaches were based mostly on heuristic techniques, statistical theory was not applied. However, parametric attempts are useful to achieve the best possible model by appropriate selection of the parameters, i.e. in most cases by maximization. In that case we apply the correlation structure or spectral density respectively rather than evaluating the asymptotical behavior.

This led to frequent assumptions like 'for large values of k only' or 'if λ is as small as possible'. The immediate question arises at which values for k we may begin valid evaluations. We remind of the scaling estimators introduced by Crovella and Taqqu for the determination of heavy-tail distributions. There is a certain risk in considering asymptotic behavior for the wrong range in favor of better results for the Hurst exponent.

We now apply the Maximum Likelihood estimator to determine the Hurst exponent in the case of Gaussian processes. The results are also valid for more general time series with heavy-tail distributed increments [76, section 7.5.]. For our descriptions we follow the approach given by Beran [30]. We start with a finite stationary time series $X_1, \ldots, X_n$ with mean 0 (generalized by shifting) and variance σ^2. We assume an autocorrelation of the series

$$\gamma(k) \sim k^{2H-2}$$

i.e. long-range dependence occurs again if $\frac{1}{2} < H < 1$. By applying the spectral density, we examine the exact Hurst parameter and select a family of densities $\{f(\lambda, \theta); \theta \in \Theta\}$. Here, the unknown vector

$$\theta = \left(\sigma^2, H, \theta_3, \ldots, \theta_m\right)^T$$

represents the parameter to be specified with the Maximum Likelihood method. Furthermore, we assume that the time series is a FARIMA series

$$X_i = \sum_{k=0}^{\infty} b_k \epsilon_{i-k} \qquad (4.24)$$

whereas ϵ_j are a iid Gaussian series with mean 0 and variance σ^2. The coefficients have to fulfill appropriate conditions, as we already described in section 3.3.3. With the regularity condition we show that the maximum likelihood estimator $\hat{\theta}_k$ is of strong consistency and

$$\sqrt{n}(\hat{\theta}_k - \theta) \xrightarrow{d} \eta, \text{ for } k \to \infty \qquad (4.25)$$

whereas η is a m-dimensional vector of random variables with mean 0 and covariance matrix $\Gamma(\theta)$ for Gaussian distributed data. However, the calculative evaluation of the maximum likelihood estimator is difficult already for simple models of the form (4.24). Hence, we have to approximate the maximum likelihood function (see (4.15)). This yields to the well known Whittle estimators, which we will illustrate below (see [267] or [268] and [76] respectively for heavy-tail increments ϵ_j). We follow [30] again and assume that the Maximum Likelihood function is a function of $\log |\Gamma(\theta)|$ and $\Gamma(\theta)$, whereas Γ denotes the covariance matrix of the data (depending on θ).

The Maximum Likelihood estimator is based on the following requirements:

- $\lim_{n\to\infty} \frac{1}{n}|\Gamma_n(\theta)| = (2n)^{-1}\int_{-\pi}^{\pi} \log f(\lambda, \theta) d\lambda$ and
- replace the inverse matrix $\Gamma(\theta)^{-1}$ by $A(\theta) = (a(j-l))_{j,l=1,\dots,n}$, whereas

$$a_{jl} = a(j-l) = (2\pi)^{-1} \int_{-\pi}^{\pi} \frac{e^{i(j-l)\lambda}}{f(\lambda;\theta)} d\lambda \quad (4.26)$$

Minimizing the expression

$$L_W(\theta; x) = \frac{1}{2\pi} \int_{-\pi}^{\pi} \log f(\lambda; \theta) d\lambda + \frac{x^T A(\theta) x}{n}, \quad x \in \mathbb{R}^n \quad (4.27)$$

yields to the *Whittle estimator*. We can show that the Whittle estimator $\hat{\theta}_{W,n}$ is consistent in the sense of $\sqrt{n}(\hat{\theta}_{W,n} - \theta) \xrightarrow{d} \eta$ with η according to (4.25). With this, the Whittle estimator exhibits the same convergence in distribution as the exact maximum likelihood estimator. To calculate (4.26), we have to discretise first, i.e. we transform into a finite sum and then apply the (FFT). For the Whittle estimator we have to calculate the integral in (4.27) $n \cdot s$ times for each realization of θ. Because of our assumption of long-range dependence f has a singularity at $\lambda = 0$. With this, we can determine $\frac{1}{f}$ more precisely, and we replace (4.26) by the sum

$$\hat{a}_k = 2 \frac{1}{(2\pi)^2} \sum_{j=1}^{m} \frac{1}{f(\lambda_{j,m}; \theta)} e^{ik\lambda_{j,m}} \frac{2\pi}{m}$$

$$= \frac{1}{(\pi)} \sum_{j=1}^{m} \frac{1}{f(\lambda_{j,m}; \theta)} e^{ik\lambda_{j,m}}$$

whereas

$$\lambda_{j,m} = \frac{2\pi j}{m}, \quad j = 1, \dots, m^*$$

and m^* denotes the largest integer of $\frac{m-1}{2}$.

We now describe the parametric estimator again and later introduce the local Whittle estimator. The Whittle estimator is based on the periodogram. We select a vector θ of unknown parameters. The spectral density $f(\theta, \cdot)$ depends on this vector. Then we build the function

$$L_W(\theta; x) = \frac{1}{2\pi} \left(\int_{-\pi}^{\pi} \frac{I(\lambda; x)}{f(\lambda; \theta)} d\lambda + \int_{-\pi}^{\pi} \log f(\lambda; \theta) d\lambda \right) \quad (4.28)$$

whereas $I(\lambda; x)$ is the periodogram according to (4.22). Normalizing the spectral density to 1, we may set the last integral to 0. Hence, we define $f^* = cf$ and $\int_{-\pi}^{\pi} \log f^*(\lambda; \theta,) d\lambda = 0$. The basic idea is now to minimize the integral in (4.28), such that the periodogram reproduces the spectral density at best. In practice the integral is approximated by a Fourier series of different frequencies. In case of FARIMA$[0, d, 0]$ time series, θ consists of d only, in case of FARIMA$[p, d, q]$ time series θ contains also the autoregressive and moving average parts p, q. We remark that the Whittle method estimates d and not H. With this, we have the following approximation

$$\overline{L}_W(\theta; x) = 2\frac{1}{(2\pi)^2} \left(\sum_{j=1}^{[m]} \log f(\lambda_{j,m}; \theta) \frac{2\pi}{m} + \sum_{j=1}^{[m]} \frac{I(\lambda_{j,m}; x)}{f(\lambda_{j,m}; \theta)} \frac{2\pi}{m} \right) \quad (4.29)$$

Since the periodogram is easily calculated by means of the Fast Fourier transform, the discrete Whittle estimator can be determined in good approximation as well.

We give some remarks on possible disadvantages of the Whittle estimator:

- The parametric form of the spectral density has to be known in advance to apply $f(\theta, \lambda)$. If the parametric form is unknown, the estimator will become very biased and hence, is not sufficiently robust (see [93, 32, 145, 149]).
- With an increasing amount of sample data the computational efforts become rather huge, whereby this method exhibits clear shortcomings compared to the graphical estimators.

We may circumvent the problem of a missing form of the spectral density with a *compound Whittle estimator*. For this we select an appropriate amount of sample data and build

$$X_k^{(m)} = \frac{1}{m} \sum_{j=m(k-1)+1}^{mk} X_j$$

We find the definition of the compound traffic again that led to the idea of asymptotical self-similarity (see section 3.1.2). Certainly, a small number of aggregated data increases the variation. With large amounts of sample data as well as with compound data, the term $X^{(m)} - \mathbb{E}(X^{(m)})$ is near the fractional Brownian motion in case of finite variance. Additionally, the bias is decreased because we may assume the spectral density of the FBM. In case of infinite variance we asymptotically reach a linear α-stable process. However, we should assume the FGN model, as observed in the article by [149].

With the *local Whittle estimator* we now turn to another semiparametric estimator. This estimator was introduced and described first by Robinson [221]. We start with a likelihood function

$$L_N(\theta) = \frac{1}{4\pi} \int_{-\pi}^{\pi} \log(f(\lambda)) + \frac{I_N(\lambda)}{f(\lambda)} d\lambda \quad (4.30)$$

whereas $I_N(\lambda)$ represents a periodogram of the vector $(X_1, \ldots, X_N)$ of observations

$$I_N(\lambda) = \frac{1}{2\pi N} \sum_{j=1}^{N} |X_j \exp(-ij\lambda)|^2$$

With a renormalization of f, such that $\int_{-\pi}^{\pi} \log(f(\lambda))d\lambda = 0$, we obtain as estimator

$$\left(\hat{H}(N), \widehat{\hat{f}_N(\cdot)}\right) = \text{Argmin}_{(H,\hat{f})} L_N(\theta)$$

Because $\hat{f}$ is not of interest, here we calculate the Whittle estimator in a neighbourhood of $\lambda = 0$, where $\hat{f}$ is almost constant $\hat{f}(0)$. Therefore this estimator is typically denoted *local Whittle estimator*. We now replace the integral in (4.30) by a sum that is more suitable for discrete observations. The sums are build over different frequencies, where the maximum frequency converges to 0, if the number of observations converges to ∞. With the minimization we get

$$L_N(H) = \log\left(\frac{1}{m(N)} \sum_{j=1}^{m(N)} \frac{I_N(\lambda_j)}{\lambda_j^{1-2H}}\right) + \frac{1-2H}{m(N)} \sum_{j=1}^{m(N)} \log(\lambda_j)$$

whereas $\lambda_j = \frac{2j\pi}{N}, j = 1, \ldots, [\frac{N-1}{2}]$ represent the Fourier frequencies. The estimator is denoted by

$$\hat{\theta} = \text{Argmin}_{H \in \Theta} L_n(H)$$

whereas $\Theta = [H_1, H_2]$ represent the estimation interval. Even if the complexity of the calculation is of grade $\mathcal{O}(m(N) \log(N))$, the complexity may raise, because the range for minimization are unknown. If the function $\hat{f}$ is differentiable on the interval $]0, \delta[$, then $\hat{f}$ is of order $\mathcal{O}(\lambda^{-1})$ around 0. And if $m(N) \to \infty$ for $N \to \infty$, then the estimator is consistent. Under stricter differentiability at 0 and stronger conditions at m, the estimator of H is asymptotical normal with rate $\sqrt{m}$ and asymptotical variance $\frac{1}{4}$ (see [221]). Although no theoretical results exist, this estimator seems also suitable for non-Gaussian processes with infinite variance of the increments.

Apart from the semiparametric approach in (4.17) we may alternatively apply

$$f(\lambda) \sim g(d)|\lambda|^{-2d}$$

whereas g is a function of d. In principle, this method is equivalent to the Whittle estimator. However, as already remarked for the first approach, the frequencies are cut above $\frac{2\pi m}{n}$. The estimator is build by minimizing the expression (see (4.29))

$$R(d) = \log\left(\frac{1}{M} \sum_{j=1}^{M} \frac{I(\lambda_j)}{\lambda_j^{-2d}}\right) - 2d \frac{1}{M} \sum_{j=1}^{M} \log \lambda_j$$

Quality of Estimation

In principle, the Whittle estimators exhibits the best behavior of all estimators. They are preferred if the parametric form of the time series is known. However, we have to distinguish between the different Whittle estimators:

- The Whittle estimator yields to good results for the parameters ϕ_1 and ψ_1. Problems arise, if instead of an actual FARIMA[0, d, 0] a FARIMA[1, d, 1] series is selected as first attempt.
- The compounded Whittle estimator is rather unbiased, the error of estimation may become larger compared to the Whittle estimator. With FARIMA[1, d, 1] series the error is low compared to graphical methods.
- Compared to the Whittle estimator the locale Whittle estimator does not estimate ϕ_1 and ψ_1 of a FARIMA[1, d, 1] series. If the parametric form of a time series is not known, this is the best approach.

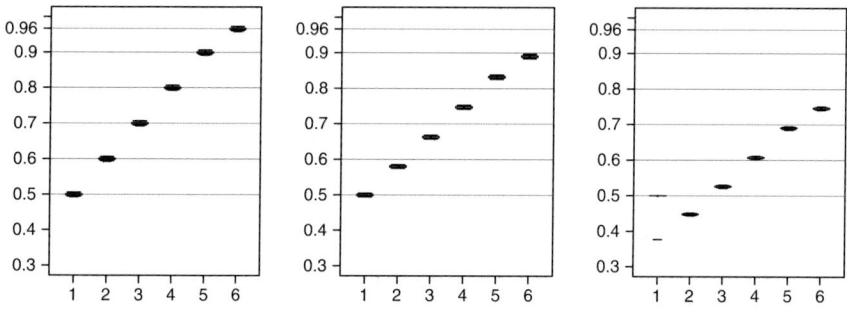

Fig. 4.15. Whittle estimator (FGN) of FGN, Gaussian FARIMA series and FARIMA series with 1.5-stable generator (from left to right) with respective Hurst parameter $H = 0.5, 0.6, 0.7, 0.8, 0.9, 0.9\overline{6}$

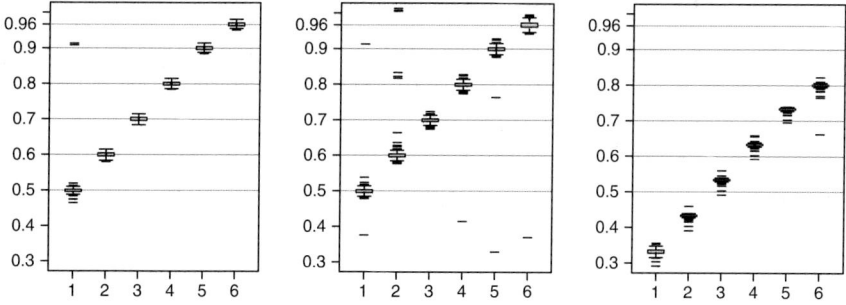

Fig. 4.16. Whittle estimator (FARIMA) of FGN, Gaussian FARIMA series and FARIMA series with 1.5-stable generator (from left to right) with respective Hurst parameter $H = 0.5, 0.6, 0.7, 0.8, 0.9, 0.9\overline{6}$

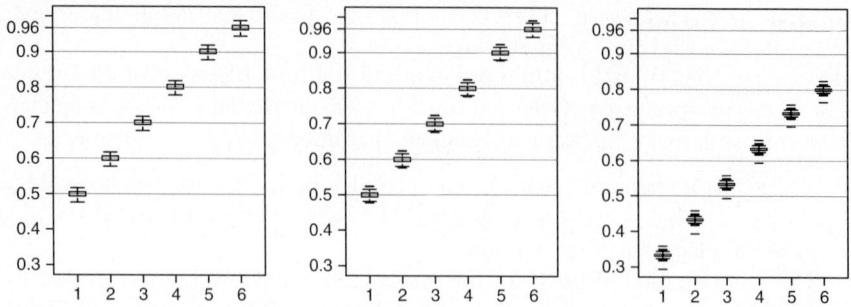

Fig. 4.17. Local Whittle estimator of FGN, Gaussian FARIMA series and FARIMA series with 1.5-stable generator (from left to right) with respective Hurst parameter $H = 0.5, 0.6, 0.7, 0.8, 0.9, 0.9\overline{6}$

All three Whittle estimators exhibit superb results for Gaussian processes. Similar to the log-periodogram estimator the results become partly wrong for for FARIMA time series with 1.5-stable generator (ϵ_t).

4.2.7 Wavelet Analysis

In contrast to the survey-like illustration of the estimators above we will give a detailed presentation of the wavelet analysis, which was developed over the past years and delivers promising results. With the wavelet transformation (here, the selection of specific wavelets is relevant) the signal or sample data under investigation is transformed into a series of wavelet coefficients. Afterwards we estimate the Hurst parameter by using this series.

We describe the approach developed by Flandrin [90] and Abry et al. [2, 4] by means of Gaussian stationary processes. At first we assume a stationary process $(X_t)_{t \in \mathbb{R}}$ with $\int_{-\infty}^{\infty} X_t^2 dt < \infty$ almost surely. Then, we select a mother wavelet φ with $M \geq 1$ vanishing moments and define the wavelet coefficients of discretization n and at point k with

$$\mathcal{D}_{n,k}^X = \frac{1}{\sqrt{n}} \int_{-\infty}^{\infty} X_t \varphi\left(\frac{t}{n} - k\right) dt \qquad (4.31)$$

It is important that (X_t) as well as the wavelet φ are quadratically integrable (this is valid e.g. if φ has compact support). The wavelet coefficients have the following characteristics:

- Since the process (X_t) is stationary, the series of wavelet coefficients $\mathcal{D}_{n,k}^X$ is also stationary for $n > 0$.
- Since (X_t) is Gaussian, the wavelet coefficients are also Gaussian.
- Since (X_t) is a LRD process that fulfils constraint (4.17), the property of M vanishing moments leads to the fact that the series of wavelet coefficients is SRD.

- Since (X_t) is Gaussian and fulfils (4.17), it applies under Hölder continuity for the exponent $\beta > 0$ of the function $\hat{f}$

$$\mathbb{E}((\mathcal{D}_{n,k}^X)^2) \sim C(\varphi, H, \hat{f}) n^{2H-1}, \text{ if } n \to \infty \text{ for all } k \in \mathbb{N} \quad (4.32)$$

with $C(\varphi, H, \hat{f}) > 0$ is constant. More details are described in [27].

With the following steps we illustrate how the calculated wavelet coefficients of the sample data (X_t) can be utilized to estimate the exponent H:

- We assume that the process was observed for all times $t \in [0, N]$. We selected a step count of n_j, $j = 1, \ldots, J$ and define

$$S_N(n_j) = \frac{1}{N_j} \sum_{k=1}^{N_j} \left(\mathcal{D}_{n_j,k}^X\right)^2$$

whereas $N_j = [\frac{N}{n_j}]$ represent the existing coefficients to the discretization n_j ($[r]$ = denotes the largest natural number that is less or equal than r). We remark that $S_N(n_j)$ is a consistent estimator for the empirical mean $\mathbb{E}((\mathcal{D}_{n_j,k}^X)^2)$ with fixed n_j and $N \to \infty$ (see the unbiased estimators in section 4.1.1 and [27]).

- With (4.32) follows the linearization

$$\log(S_N(n_j)) = (2H - 1)\log n_j + K(\varphi, H, \hat{f}) + \mathcal{R}(n_j, N)$$

whereas $K(\varphi, H, \hat{f}) = \log(C(\varphi, H, \hat{f})) \in \mathbb{R}$ and $\mathcal{R}(n_j, N) \to 0$ in probability, if n_j and $\frac{N}{n_j} \to \infty$.

- A linear regression of the vector $(\log(S_N(n_j)))_{j=1,\ldots,J}$ versus the vector $(\log(n_j))_{j=1,\ldots,J}$ yields to a slope, which represents a suitable estimator $\hat{H}(N)$ for the Hurst parameter H. The estimator is consistent in a sense that $\hat{H}(N) \xrightarrow{\mathbb{P}} H$ for $n_j \to \infty$ and $[\frac{N}{n_j}] \to \infty$.

Remark 4.35. We give two remarks on the estimation illustrated above:

- With increasing M the series of coefficients $\mathcal{D}_{n,k}^X$ converges to white noise and theoretically leads to a better estimator of H

$$\mathbb{E}\left(\mathcal{D}_{n,k_1}^X \mathcal{D}_{n,k_2}^X\right) = \mathcal{O}\left(|k_1 - k_2|^{2(H-M)-2}\right)$$

However, the asymptotic variance of the estimator $\hat{H}(N)$ increases with M.

- With multiscaling analysis the wavelet coefficients are calculated in a pyramidically structured tree. With this further restrictions apply to the mother wavelet (e.g. orthogonality) and the step count n_j. However, apart from the number of vanishing moments and a certain regularity, no properties of the mother wavelet are taken into account.

We introduced the wavelet analysis for Gaussian processes but it can be extended to general α-stable processes as well.

Wavelet Coefficients of Self-Similar and Long-Range Dependent Processes

We start with a stochastic process (X_t) and the respective wavelet coefficients $\mathcal{D}_{j,k}$. With the calculation of the coefficients we assume either realizations of the stochastic process or wavelet coefficients as random variables (we remind of the definition of a stochastic process). If the wavelet is of compact support (or the decay behavior is sufficiently large for $t \to \infty$, depending of the stochastic process), then the following properties apply, whereas the mother wavelet φ has N vanishing moments:

a) The wavelet coefficients are identically for X_t and $X_t + P(t)$, if $P(t)$ is a polynomial of $(N-1)$-th grade. Then,

$$\int_{-\infty}^{\infty} P(t)\varphi_{j,k}(t)dt = \int_{-\infty}^{\infty} P\left(2^j(s+k)\right)\varphi(s)ds = 0$$

(we remark that $s \longmapsto P(2^j(s+k))$ is a polynomial of $(N-1)$-th grade). With this, the DWT is invariant compared to distortions by polynomials of $(N-1)$-th grade. In particular, $X_t + c$ and X_t have the same wavelet coefficients.

b) In our model we assumed stationary increments of the observed processes (the fractional Brownian motion is the only Gaussian process with this property). What about the wavelet coefficients in this case? Let $(X_t)_{t \in \mathbb{R}}$ be a process with stationary increments, i.e. the finite dimensional distributions of $(X_{t+h} - X_t)_{t \in \mathbb{R}}$ are independent of t. If we consider a scaling of $j \in \mathbb{Z}$, then the series $(\mathcal{D}_{j,k})_{k \in \mathbb{Z}}$ is stationary. As a simplification we assume $j = 0$ and show that the one-dimensional distributions of $(\mathcal{D}_{j,k})_{k \in \mathbb{Z}}$ are independent of k:

$$\mathcal{D}_{0,k+k_0} = \int_{-\infty}^{\infty} X_t \varphi(t - k - k_0)dt \tag{4.33}$$

$$= \int_{-\infty}^{\infty} X_{s+k_0}\varphi(s-k)ds = \int_{-\infty}^{\infty} (X_{s+k_0} - X_s)\varphi(s-k)ds$$

$$\stackrel{d}{=} \int_{-\infty}^{\infty} (X_s - X_0)\varphi(s-k)ds = \int_{-\infty}^{\infty} X_s \varphi(s-k)ds = \mathcal{D}_{0,k}$$

The third identity in (4.33) is given by the properties of wavelets (see e.g. [126]) or [63]). The fourth identity in (4.33) is a consequence of the stationary increments (the integral is approximated with a sequence of sums here). Similarly, we can show for a selection of points $k_1, \ldots, k_n$ and real numbers $r_1, \ldots, r_n$ that $\sum_{i=1}^{n} r_i \mathcal{D}_{0,k_i+k_0} \stackrel{d}{=} \sum_{i=1}^{n} r_i \mathcal{D}_{0,k_i}$. With this follows the stationarity. In a similar way we conclude for wavelets with vanishing moments that the sequence $\mathcal{D}_{j,k}$ (j constant) has stationary increments, if (X_t) is stationary of order N (see about the term stationarity of order N e.g. [38]). In particular, $(\mathcal{D}_{j,k})$ is stationary, if (X_t) is stationary as well.

4.2 Estimators of Hurst Exponent in IP Traffic

c) What about a process (X_t) that is self-similar to a Hurst parameter H? In this case we again obtain for a constant $j \in \mathbb{Z}$

$$\mathcal{D}_{j,k} \stackrel{d}{=} 2^{j(H+\frac{1}{2})}\mathcal{D}_{0,k}, \quad k \in \mathbb{Z} \tag{4.34}$$

How can we derive this result? We deduce equation $X_{2^j s} \stackrel{d}{=} 2^{jH}X_t$ because of the self-similarity. With this we have

$$\mathcal{D}_{j,k} = \int_{-\infty}^{\infty} X_{2^j s} 2^{-\frac{j}{2}}\varphi(s-k)ds \stackrel{d}{=} 2^{j(H+\frac{1}{2})}\mathcal{D}_{0,k}$$

The summand $\frac{1}{2}$ in the exponent follows due to the normalization with $2^{-j\frac{1}{2}}$.

d) In a further step we assume that the process $(X_t)_{t \in \mathbb{R}}$ is H-sssi ($0 < H < 1$). I.e. the process is self-similar with stationary increments, mean 0 and an existing second moment, like a fractional Brownian motion. From property b) we derive that $\mathbb{E}(\mathcal{D}_{j,k}) = 0$, and with (4.34) follows

$$\mathbb{E}(\mathcal{D}_{j,k}^2) = \mathbb{E}(\mathcal{D}_{0,0}^2)2^{j(2H+1)} \tag{4.35}$$

Taking the logarithm to base 2, on both sides we get

$$\log_2\left(\mathbb{E}(\mathcal{D}_{j,k}^2)\right) = \log_2\left(\mathbb{E}(\mathcal{D}_{0,0}^2)\right) + (2H+1)j \tag{4.36}$$

The function on the right hand is linear in j and of slope $2H + 1$. We already see the relevance of the wavelet coefficients to determine the Hurst parameter and want to provide deeper insight into this method.

e) Instead of the variance we apply the covariance to estimate the Hurst parameter. We already know from property c) above that the sequence $(\mathcal{D}_{j,k})_{k \in \mathbb{Z}}$ is SRD for a constant j. Hence, the covariance decays fast to 0 although the sequence is correlated

$$\mathbb{E}(\mathcal{D}_{j,k_1}\mathcal{D}_{j,k_2}) = \text{const}(j)|k_1 - k_2|^{2(H-N)} \tag{4.37}$$

whereas with N the number of vanishing moments is selected sufficiently large and $\text{const}(j)$ denotes a constant depending on j. To detect no LRD property, i.e.

$$\sum_{k=0}^{\infty} \mathbb{E}\left(|\mathcal{D}_{j,k}\mathcal{D}_{j,0}|\right) < \infty$$

we must choose $N > H + \frac{1}{2}$, i.e. at least $N = 2$. The Haar wavelet does not have sufficient vanishing moments.

Instead of equation (4.37) we can estimate the Hurst exponent with the spectral representation. For the second moment of the wavelet coefficients we have

$$\mathbb{E}(D_{j,k}^2) = \int_{-\infty}^{\infty} f_X(s)2^j|\hat{\varphi}(2^j s)|^2 ds \tag{4.38}$$

whereas
$$\hat{\varphi}(\lambda) = \int_{-\infty}^{\infty} \exp(-i2\pi\lambda t)\varphi(t)dt$$
represents the Fourier transform of the mother wavelets and f_X the spectral density of the signal. If X is LRD with a spectral density of the form $f_X(\lambda) \sim |\lambda|^{-\gamma}$ (compare (3.22)), then it follows

$$\mathbb{E}(\mathcal{D}_{j,k}^2) \sim C_f 2^{j\gamma} \int_{-\infty}^{\infty} |\lambda|^{-\gamma}|\hat{\varphi}(\lambda)|^2 d\lambda, \text{ for } j \to \infty \qquad (4.39)$$

as estimator. The covariance of the wavelet coefficients is

$$\mathbb{E}(D_{j,k}D_{j,k'}) = \int_{-\infty}^{\infty} f_X(\lambda) 2^j |\hat{\varphi}(2^j\lambda)|^2 \exp\left(-i2\pi(k-k')2^j\lambda\right) d\lambda$$

We see that $\mathbb{E}(\mathcal{D}_{j,k}\mathcal{D}_{j,k'})$ is a function that only depends on the difference $|k-k'|$. The asymptotical behavior $|k-k'| \to \infty$ is determined by the behavior of the Fourier transform $f_X|\hat{\varphi}(2^j s)|^2$ at the origin. If the mother wavelet φ has N vanishing moments, then $\hat{\varphi}$ behaves as $|\lambda|^N$ at the origin (compare e.g. [37, 63, 126]). Therefore, the LRD property caused by the factor $|\lambda|^{-\gamma}$ in (4.39) is compensated by the factor $|\lambda|^{2N}$ (the integral (4.39) contains $|\hat{\varphi}(2^j\lambda)|^2$).

Essentially, we want to achieve that the wavelet coefficients $\mathcal{D}_{j,k}$ are decorrelated, i.e. *not* LRD. Therefore, we have the constraint $2N > \gamma$ for the vanishing moments of the wavelets. With this we obtain for the j-th octave of the wavelet a decrease of the covariance in approximative form (note that $f_X(\lambda) \sim |\lambda|^{-\gamma}$ around 0).

$$\mathbb{E}(D_{j,k}D_{j,k'}) \cong |k-k'|^{\lambda-2N-1} \qquad (4.40)$$

We see that relation (4.40) is consistent with (4.37) by the transition of a H-sssi process to the increments $Z_t = X_t - X_{t-h}$. We can prove that (Z_t) is LRD with $\gamma = 2H - 1$ as we know from (3.22). For $\tilde{H} = H - 1$ (4.37) is consistent with (4.40):

$$2(\tilde{H} - N) = 2H - 2N - 2 = \gamma - 2N - 1$$

The reason for estimating the 'exponent of the frequency' γ instead of H is, because according to (3.23) that the relation $2N > \gamma$ is equivalent to $2N > 2H - 1 \Leftrightarrow N > H - \frac{1}{2}$. With this, we can already apply the Haar wavelet here in contrast to (4.37). In practice, we use wavelets with two or three vanishing moments, because of the faster decorrelation in (4.40) and the weak localization of wavelets for higher vanishing moments.

We summarize common properties of both approaches to estimate the wavelet coefficients for H-sssi processes:

- Both approaches are stationary for constant scales.

- Both approaches exhibit short time dependency (decorrelation) (compare (4.37) and (4.40)).
- The wavelet coefficients reflect self-similarity and LRD property (compare equation (4.35)).

These properties are originated on the one hand in the invariance of scales of the processes (given by the self-similarity) and on the other hand in the multiscale analysis of the wavelet, which seems to be almost created for the IP traffic modeling. This implies that the mother wavelet has to possess at least 0 vanishing moments, and all wavelet can be deduced from it via translation and scaling. This fits exactly into the definition of H-sssi processes. This method can be transferred to other processes without characteristic on certain scales, e.g. $\frac{1}{f}$-noise (see [81]), multifractal processes (like binomial cascades) or multiplicative cascades (see for a more detailed overview [5]).

If we want to estimate the parameters of sample data, we encounter two problems: First we have to identify the basic pattern, i.e. if we face SRD or LRD. Afterwards we have to proceed to the quantitative estimation of the necessary parameter. We want to treat long-range dependence here and exemplarily consider five discrete models.

Logarithmic Scale Diagram

As already stated in our short introduction to wavelets (see section 3.8.2), we know that the values of j represent the frequencies of the signal under consideration. The lesser the value of j, the shorter the wavelength, i.e. the higher frequent the signal is. We take advantage of this fact of localization with the estimation of LRD. As we already detect with relation (4.36), the logarithmic representation of the second moment of the wavelet frequencies $\mathbb{E}(\mathcal{D}_{j,\cdot})$ reveals a possible estimation of the Hurst parameter. In detail we calculate for each scale $j \in \mathbb{Z}$ the value of

$$s_j = \log_2\left(\mathbb{E}(\mathcal{D}_{j,\cdot}^2)\right) \qquad (4.41)$$

and draw it over j. The slope of the shifted straight line represents the value of $2H + 1$. This method is referred to as *exact logarithmic scale diagram*. We have to consider the precondition that the underlying process is H-sssi.

Representations for different wavelets with different vanishing moments $N = 1, N = 3$ or $N = 6$ naturally lead to changing figures, since the constants in (4.37) depend on the choice of the wavelet. Simultaneously we will see a straight line up to the small values of j. Already a few estimated values of H reveal a strong similarity in the slope and indicate the result on convergence (3.20). The LRD property of the applied fractional white noise is reproduced clearly by the asymptotic for large time scales (i.e. large j, hence, lower frequencies). Additionally, we realize that with increasing j and N the impact of the wavelet diminishes – a characteristic of the logarithmic scale diagram. With the computation of (4.41) we also introduce an error: it is not possible to compute the second moment exactly. Thus, we have to consider a confidence interval. In addition, generating the wavelet, using the multiscale analysis has an impact on the width of the confidence interval: with increasing j the tree algorithm of the multiscale analysis of the existing coefficients is halved, which leads to an enlargement of the confidence interval.

Estimation with Logarithmic Scale Diagram

How can we compute the s_j with (4.41)? For this, we consider the unbiased estimator of the variance

$$\hat{\sigma}_j = \frac{1}{n_j} \sum_{k=1}^{n_j} |\mathcal{D}_{j,k}|^2$$

whereas n_j is the number of the existing wavelet coefficients for the scale j (see the section 4.1.1 for the estimator of the variance). The logarithm, i.e. the value $\log_2(\hat{\sigma}_j)$, is an estimator of s_j. But because of $\mathbb{E}(\log_2(\cdot)) \neq \log_2(\mathbb{E}(\cdot))$, it is biased. Thus, we introduce a correction factor r_j

$$\mathcal{D}_j = \log_2(\hat{\sigma}_j) - r_j$$

We denote by σ_j^2 the variance of $\mathcal{D}_{j,\cdot}$, and we compute the confidence interval using the Gaussian approximation of $\mathcal{D}_{j,\cdot}$. We consider the interval $[j_1, j_2]$, where we observe the LRD behavior via on a straight line. The value j_1 is denoted as *lower or small scale* or *high frequency* and correspondingly the index j_2 *higher scale* or *lower frequency*. As done by most graphical solutions to estimate the Hurst parameter, we find the interval in the region of the straight line of the diagram. Determining the exponent, we can use the weighted linear regression over the spots $\mathcal{D}_j$. More precisely, we define $S = \sum_{j=j_1}^{j_2} \frac{1}{\sigma_j^2}$, $S_1 = \sum_{j=j_1}^{j_2} \frac{j}{\sigma_j^2}$ and $S_2 = \sum_{j=j_1}^{j_2} \frac{j^2}{\sigma_j^2}$. With this, the estimator $\hat{\gamma}$ for γ reads as

$$\hat{\gamma} = \frac{\sum_{j=j_1}^{j_2} \mathcal{D}_j \frac{(jS - S_1)}{\sigma_j^2}}{S S_2 - S_1^2} = \sum_{j=j_1}^{j_2} w_j \mathcal{D}_j$$

The estimator is unbiased, provided the scale diagram indicates the interval $[j_1, j_2]$. The coefficients w_j for $j = j_1, \ldots, j_2$ are defined resp. fixed by the middle part of the expression. Several questions concerning the computation of the correction term r_j and the variance σ_j^2 remain, because we cannot exactly determine these expressions. Nevertheless, we have some approximations at hand, as the following equations express

$$r_j = \frac{\Gamma'\left(\frac{n_j}{2}\right)}{\Gamma\left(\frac{n_j}{2}\right) \log 2} - \log_2\left(\frac{n_j}{2}\right) \sim \frac{-1}{n_j \log 2}$$

$$\sigma_j^2 = \frac{\zeta\left(2, \frac{n_j}{2}\right)}{(\log 2)^2} \sim \frac{2}{n_j (\log 2)^2}$$

whereas Γ is the gamma function, Γ' its derivative and $\zeta(2, z) = \sum_{k=0}^{\infty} \frac{1}{(z+k)^2}$ the generalized Riemannian zeta function. These approximations hold in particular, if the wavelet coefficients $\mathcal{D}_{j,k}$ are Gaussian.

Under the ideal condition of independence and for large n_j, we deduce

$$\mathbb{V}\mathrm{ar}(\hat{\gamma}) = \sum_{j=j_1}^{j_2} \sigma_j^2 w_j^2 \sim \frac{1}{n}\frac{1 - 2^{-j}}{F} \qquad (4.42)$$

with $F = F(j_1, J) = (\log 2)^2 2^{1-j_1}(1 - (\frac{J^2}{2}+2)2^{-J} + 2^{-2J})$ and $J = j_2 - j_1 + 1$. The above method can be applied independently of the considered situation (LRD, H-sssi etc.). We only have to demand that the wavelet possesses sufficient vanishing moments to guarantee the decorrelation. The first important item in question is the scaling interval $[j_1, j_2]$. Having stated that we have a H-ss model, we can choose $j_1 = 1$ and j_2 as large as possible. In other situations this will get more complex. We will turn to the adequate determination of j_1 in the next section.

Estimation of Long-Range Dependence

We already stated above that long-range dependence is defined asymptotically. As already seen for the estimation of heavy-tail distributions, we encounter several problems: we cannot detect the suitable frequency (for the spectral density) resp. the scaling, where the long-range dependence starts. Similar to section 4.1.3, we have to impose an extra condition to determine the frequency resp. the scaling. We obtain this with the *estimation quality*. For each index j_1 we apply the method of least squares and hence, obtain the optimal $j_1 = j_1^{kQS}$. Thus, we define

$$kQS(\hat{\gamma}) = (\mathbb{E}(\gamma - \hat{\gamma}))^2 = ((\gamma - \mathbb{E}(\hat{\gamma}))^2 - \mathbb{V}\mathrm{ar}(\hat{\gamma}) \qquad (4.43)$$

whereas $\mathbb{V}\mathrm{ar}(\hat{\gamma})$ is computed according to (4.42). The value j_2 will be chosen very high, e.g. in practice as described in (4.44) below. If more values of j_1^{kQS} are found, then we choose naturally the smallest one to define a largest possible region. In small scales the bias decreases, but because of the low number of sample data, we get a variance of the estimator. With growing n the number j_1^{kQS} increases, since the $\mathcal{D}_j$ decreases. The bias remains constant at j, since it depends on the correct long-range dependence. Thus, $j_1(n)^{kQS}$ is a non-decreasing function in n. We may consider j_1^{kQS} as transition point between SRD and LRD. But the point j_1^{kQS} depends on the applied method – here the wavelet based estimator, which is not an absolute value.

Analysis of Discrete Data

It is important to keep in mind that the (discrete) wavelet transform is defined for *continuous* processes and not for FARIMA time series. However, the measurement of sample data always leads to time discrete series. Otherwise, FARIMA time series are an important approach for IP traffic modeling.

The continuity of the wavelet transform is required for equation (4.31) and cannot be build for discrete processes. In addition, the approach $\mathcal{D}_{0,k} = X_k$ is arbitrary and without authorization. We briefly treat an alternative approach, which is promising for the determination of γ. It was developed by Veitch, Taqqu und Abry [259].

We start with a stationary discrete process (X_k) and spectral density g_x. The decisive region is found for frequencies in the interval $]-\frac{1}{2}, \frac{1}{2}]$. Hence, we will choose a *continuous* process $(\tilde{X}_t)$ with identical spectral density in the interval $]-\frac{1}{2}, \frac{1}{2}]$. We now apply the discrete wavelet transform to the continuous process $(\tilde{X}_t)$. For application of the multiscale analysis we can combine the transition from X to $\tilde{X}$ together with the subsequent multiscale analysis in one step. We remember the discrete convolution and determine

$$\mathcal{D}_{0,k}^{\tilde{X}} = (X \star I)(k)$$

whereas I represents a discrete filter

$$I(m) = \int_{-\infty}^{\infty} \sin\left(\pi(t+m)\right) \frac{\varphi_0(t)}{\pi(t+m)} dt$$

The filter depends on the used wavelet only. An infinite support of I can be bounded, too. With this filter we can prevent errors on the first two octaves and thus, the values of $\mathcal{D}_j$ are valid on all scales.

Examples for Estimation of Short-Range and Long-Range Dependence

We now explain the method developed by Abry et al. by means of the sample data given in [4]. We start with three scenarios:

a) Exact data, i.e. theoretical data: here, the FGN and Gaussian FARIMA time series are used. All scenarios as SRD, LRD and combined SRD/LRD are covered.
b) Empirical data, modeled by FGN resp. Gaussian FARIMA time series.
c) Empirical data, modeled by arbitrary increments (with and without existing variance).

In the scenario a) we know the exact process and are aware of all numerical values. Thus, we can compute exactly and obtain an estimator for $\mathrm{Var}(\hat{\gamma})$ explicitly from equation (4.38) as variance of the wavelet coefficient. Here, f_X

is the spectral density of the H-sssi process and $\hat{\varphi}$ is the Fourier transform of the mother wavelet. We obtain the value j^{kQS} from 2 or 3 for the FGN and j^{kQS} for the FARIMA series between 1 and 5 (1 and 2 for the homogeneous FARMA$[0, d, 0]$-series, mostly defined 5 otherwise).

The scenarios b) and c) were considered by simulations of 50 realization. Here, the exact values are not known, and we have to estimate each time the time series of length $n = 10,000$. As remarked above, we can choose j_2 as large as possible. Indeed, we use the fact

$$j_2 = [\log_2 n - \text{const}] \tag{4.44}$$

Here, $[\cdot]$ is the entire function (the largest integer less than or equal to) and const $= \log_2(2N+1)$, whereas N reflects the vanishing moments of the mother wavelet. By this large number we prevent that the limiting effects of the estimation falsifies the variance. As example we note $n = 10,000$, $N = 4$ and hence, $j_2 = 14$.

We repeat briefly the models of the FARIMA time series. For this we apply the wavelet method on a FARIMA$[p, d, q]$ time series and start with a FARIMA$[0, d, 0]$ series, which is given in the form $X_t = \Delta^{-d}\epsilon_t$, $t \geq 1$, with ϵ_t iid Gaussian distributed increments with mean 0 and variance σ^2. We remind of the exact analysis in section 3.3.3 and note the backward operator $B\epsilon_t = \epsilon_{t-1}$ and $b_t(-d) = \frac{\Gamma(t+d)}{\Gamma(d)\Gamma(t+1)}$, $t = 1, 2, \ldots$ with Γ the gamma function. Then, we can write Δ in the form

$$\Delta^{-d}\epsilon_t = \sum_{i=0}^{\infty} b_i(-d) B^i(\epsilon_t)$$

The FARIMA series is LRD, if $0 < d < \frac{1}{2}$ and

$$\gamma = 2d$$

If we now consider a FARIMA$[1, d, 1]$ series, then we have an equation

$$X_t - \Phi_1 X_{t-1} = \Delta^{-d}\epsilon_t - \theta_1 \Delta^{-d}\epsilon_{t-1}$$

whereas Φ_1 and θ_1 represent the autoregressive and moving average coefficients. We obtain with the Fourier transform

$$f_X(\lambda) = \sigma^2 |1 - e^{-2i\lambda}|^{-2d} \frac{|1 - \theta_1 e^{-2i\lambda}|^2}{|1 - \Phi_1 e^{-2i\lambda}|^2}, \quad -\frac{1}{2} < \lambda < \frac{1}{2}$$

We discuss this by means of an example of a FARIMA process with significant SRD phenomena. We start with a FARIMA$[1, d, 1]$ series defined for $H = d + \frac{1}{2} = 0.7$, $\phi_1 = 0.3$ and $\psi_1 = 0.7$. Using the logarithmic scale diagram, we obtain a slope of $\gamma = 0.4$ over the first four octaves of the wavelet scales. It is again $n = 10,000$, $N = 3$ and $j_1^{kQS} = 5$ (with the estimated values, described in (4.41) until (4.43)).

The values for j_1 vary in the computations from 1 to 5 and separate clearly the regions of SRD and LRD. Comparing the scale regions, i.e. the value j_1^{kQS}, we realize that the first both scenarios are almost identical. Only for two exact values of FARIMA$[1, d, 1]$ series with $H = 0.7$ we obtain the estimated value $j_1^{kQS} = 6$, whereas we expected as exact value $j_1^{kQS} = 5$.

After all, the results show that the wavelet analysis with semiparametric approach gains comparable good results. The Whittle estimator assumes a spectral density of the form $f(\lambda) \sim C|\lambda|^{-\gamma}$ for $0 < \lambda < \frac{2\pi m}{n}$ (n is the number of the observed values, m a suitable parameter). The value $\frac{m}{n}$ describes the results for high frequencies and corresponds with the scale 2^{-j_1} in the case of the wavelet analysis (see section 4.2.6).

Simultaneously, we can extend the FARIMA time series to non-Gaussian increments. Here, on the one hand side, increments with variance (exponential and lognormal) are used and on the other hand, increments with non-existing variance (Pareto distribution and sαs distribution [$\alpha = 1.2$ and $\alpha = 1.5$]). We emphasized that with these particular distributions the resulting time series not necessarily have marginal distributions (only increments ϵ_j). Even if the variance does not exist (as for the Pareto distribution or in general for α-stable distribution with $\alpha < 2$), the wavelet analysis reveals very good results. But the variance of the estimator $\hat{\gamma}$ increases relatively fast, whereas the Bias stays very low.

Further results and detailed analysis are beyond the scope of this book. We refer to the original papers [4, 247, 119] and the indicated literature therein.

Quality of Estimation

Using the wavelet analysis for Gaussian processes we obtain almost as good results as with the Whittle estimator. The wavelet estimator reveals weak results for non-Gaussian FARIMA series, even less significant.

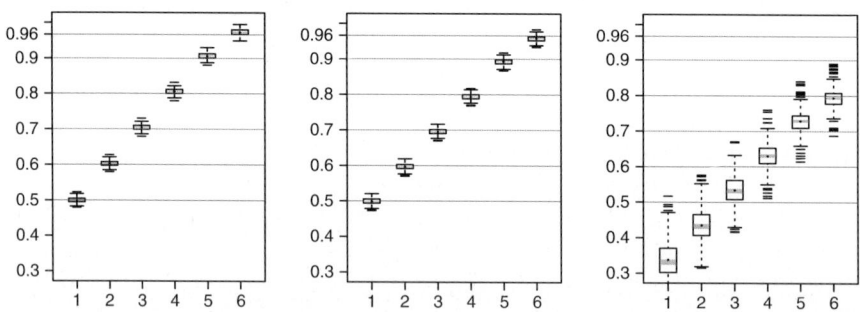

Fig. 4.18. Wavelet estimator of FGN, Gaussian FARIMA series and FARIMA series with 1.5-stable generator (from left to right) with respective Hurst parameter $H = 0.5, 0.6, 0.7, 0.8, 0.9, 0.9\overline{6}$.

4.2.8 Quadratic Variation

Before we conclude our results, we encounter again a semiparametric estimator. This method is based on the wavelet estimator for time continuous processes. Here, the process is discretised, i.e. we consider a time series. Therefore, we select discrete values $X_1, \ldots, X_N$ of the stationary (load-) process (X_t) und define the approximating wavelet coefficients

$$D^X_{n_j,k} = \frac{1}{\sqrt{n_j}} \sum_{i=1}^{N} X_i \varphi\left(\frac{i}{n_j} - k\right)$$

We 'discretise' the integral of the wavelet coefficients with a ordinary Riemann sum. The wavelet possesses not necessarily vanishing moments on the grid $\frac{i}{n_j} - k$. With this, the series $(D^X_{n_j,k})_{k \in \mathbb{N}}$ is not necessarily SRD. However, this is the case, if the number N of the selected points and with this also n_j tend towards ∞.

As with our descriptions of different approaches in sections 4.2.1 to 4.2.4, we may examine compounded traffic as well here. For this, we build $Z_n = \sum_{i=1}^{N} X_i$ and alternatively define the coefficients according to

$$D^X_{n_j,k} = \sum_{i=1}^{l_j} d_{i,j} Z_{n_j(k+i)}$$

whereas $(D_{i,j})_{i=1,\ldots,n_j}$, $n_j \in \mathbb{N}$ represents a discrete filter with M vanishing moments ($\sum_{i=1}^{n_j} i^m d_{i,j} = 0$, for all $m = 0, 1, \ldots, M-1$ and $\sum_{i=1}^{n_j} i^M d_{i,j} \neq 0$). With this, we achieve that the discretization of the integration of the wavelet transform exhibits similar properties as the time continuous wavelet transform. Indeed the $D^X_{n_j,k}$ are SRD if $M \geq 2$. We substitute the coefficients $D^X_{n_j,k}$ for $D^X_{n_j,k}$ and apply again the same estimation procedure as with the wavelet method. The *generalised quadratic variation* of the discretised process (X_k) is

$$S_{N,n_j} = \frac{1}{N} \sum_{k=1}^{N} \left(D^X_{n_j,k}\right)^2$$

This method can be applied to processes with stationary increments like FBM or general α-stable processes.

Only few results exist about the goodness of convergence and the asymptotic of the quadratic variation. In the original article by [26] several approaches with different results are discussed. Compared to the wavelet method one approach achieves the same goodness of convergence. Another approach achieves a goodness of $\sqrt{N}$ (see [24]). The complexity is of the order of n, i.e. $\mathcal{O}(n)$. More details can be found in the original articles by Bardet [24] as well as in Stoev, Papiras and Taqqu [235].

4.2.9 Remarks on Estimators

We concluded each section of our introduction into selected estimators with some remarks on their quality. We close this chapter with comments on their behavior in particular with regard to different time series.

In principle, we have to distinguish between two basic scenarios for an evaluation of the estimators:

a) pure FGN processes and FARIMA$[0, d, 0]$ time series,
b) Gaussian FARIMA$[p, d, q]$ time series and FARIMA time series with increments of infinite variance.

Generally, all methods estimate the value of d and not H (apart from the absolute value estimator and the variance of residuals). All estimators behave well in case of a). In case of b) we have to distinguish four more cases:

- Gaussian FARIMA$[p, d, q]$ series: The results essentially depend on the AR part p or the MA part q, respectively. If both parameters are negative, then the estimators are unbiased, and we obtain similar results to case a). If both parameters are positive, then the estimators are not unbiased anymore. The Whittle estimator is an exception of this rule: it is unbiased for $p = q = 1$ and the locale Whittle estimator delivers the best result.
- If the increments are exponentially or lognormal distributed, then nearly all estimators are unbiased. However, two exceptions exist: the absolute value method (for large values of d) and the variance of residuals behave extremely biased for lognormal distributions.
- Considering large areas of symmetric α-stable increments (i.e. increments of infinite variance) all estimators are unbiased for FARIMA$[0, d, 0]$ series apart from the absolute value method. However, the infinite variance causes smaller confidence intervals. In case of FARIMA$[1, d, 1]$ time series all estimators behave similar to the Gaussian case apart from the absolute value method again.
- Estimations of Pareto distributions and non-symmetric α-stable increments show similar behavior as symmetric α-stable distributions, except for the rather biased estimator of the variance of residuals.

In summary the log periodogram and the local Whittle estimator yield to the best results of all considered estimators in this chapter. The wavelet approaches – the time continuous as well as the discretized quadratic variant – are fast and easy to implement and show good performance, especially for Gaussian processes. Wavelet estimators are also suitable for non-Gaussian processes but with increasing variance. The worst results with regard to variability encounter were achieved with the local periodogram.

Situations remain difficult, if it is not possible to identify the structure of the sample data in advance. Distortions of the long-range behavior might lead to different results for all estimators.

Remark 4.36. We give some closer comments on this:

4.2 Estimators of Hurst Exponent in IP Traffic

- If the data is distorted with FGN the Whittle estimator and the periodogram lead to best results.
- There is no estimator that is universally suited for all scenarios, e.g. the Whittle estimator leads to poor results for non-LRD data.
- Diverging estimations may occur for LRD. However, the LRD phenomena may exist, even if the estimators differ as long as the result of the estimators is $\frac{1}{2} < H < 1$.
- If there is no sufficient estimation of the Hurst exponent H, then the existence of LRD behavior is unlikely.
- Separation of the sample data is recommended. Thus, multifractals and the respective wavelet analysis appear to be adequate.
- Simultaneous estimation with different methods is highly recommended. The Whittle estimators (common, local as well as compounded) lead to good results for the Hurst parameter. However, the Whittle estimator leads to poor results in case of distortions with non-LRD periodic processes.
- The existence of LRD is unlikely, if different estimators do not lead to stable Hurst exponents $H \in]\frac{1}{2}, 1[$ (e.g. with small confidence intervals).

In practice it is important to know, if some rules of thumb exist. Karagiannis, Faloutsos and Riedi already pointed out in [132] that for consequent studies the determination of a Hurst exponent is futile without stating the respective method in use. Furthermore, the confidence interval and the coefficient of correlation have to be specified precisely as an essential information about the quality of the estimator. Additionally, the estimation should consider different methods, because some approaches tend to rather optimistic results (e.g. Whittle estimator, Periodogram).

To estimate the dimensioning and performance of a network, the determination of the Hurst exponent constitutes a first and effective step. This estimation is crucial and thus, has to be robust to implement the adequate models as described in the next chapter. Only with carefully constructed models, the dimensioning of networks is carried out efficiently, such that weaknesses can be identified and adjusted in advance. However, the application on complex networks is beyond the scope of this monograph.

Remarks on Estimators in the Literature

A broad variety of literature exist about estimators of the grade α for heavy-tail distributions as well as of the Hurst exponent in the approach of LRD property. We already cited this literature in the respective sections above. Apart from that many authors carry out in-depth examination of observed data such as network traffic, e.g. [151]. The approach to estimate the exponent of heavy-tail distributions is originated from the work of Crovella and Taqqu [60]. In their work a series of example measurements is presented together with a reference implementation. At present the Whittle estimator belongs to the well established methods and thus, is described in many monographs, e.g. the already cited monograph [77] provides a well compiled number of papers and reports. The estimation of FARIMA series is described in an overview given by [247] particularly with regard to different qualities of the respective estimators and the application of standard methods to non-Gaussian increments. The articles [151] and [132] describe scenarios, where different estimators lead to varying results under miscellaneous assumptions. With this work a fundamental hint is given that not merely a single method exists for all situations, but rather selected estimators have to be applied to each data stream.

Further Literature

There are a lot of articles dealing with the existing data material and its treatment, like in [151]. A good description of the maximum likelihood estimator is given in [14]. Further literature for the Hurst exponent can be found in [36, 66, 70, 99, 184]. The estimation methods for the heavy-tailed distributions are due to Crovella and Taqqu [60]. In the monograph [77] the reader finds more literature on the topic of estimation of self-similar processes. The estimation concerning FARIMA time series well exposed in [247]. The articles [151, 132] describe very well the situation, where different the estimators lead to different results. General topics of estimation theory can be found e.g. in [76, 209].

5

Performance of IP: Waiting Queues and Optimization

> *Like as the waves make towards the pebbled shore,*
> *So do our minutes hasten to their end,*
> *Each changing place with that which goes before,*
> *In sequent toil all forwards do content.*
>
> William Shakespeare (16th century)

In this chapter we will deal with the problem of determine waiting queues in various models as well as some aspects of optimization. It is obvious that we cannot cover the full range of approaches, offered by the research community. In this respect, we will provide suitable literature for the interested reader. Our optimization approaches will be twofold: first on the basis of network flows, given by the formulas of fractional Brownian motion (and here in particular of the Norros model as introduced in the section 3.3.4), second optimization techniques, once by applying the Lyapunov function (and here for deterministic traffic description) as well as an optimal control approach using stochastic optimization, which is initiated in economics. Hence, at the end of this chapter we will enter the optimization not purely from the network view, but as optimization from the equilibrium point of view given by the offer of the network provider and the user as consumer.

5.1 Queueing of IP Traffic for Perturbation with Long-Range Dependence Processes

In this section we will return to the models of IP-based traffic, introduced in chapter 3. In particular we considered the perturbation of deterministic traffic, using FBM, multiscale FBM, multifractal FBM and fractional Lévy processes. These processes allow a fairly easy marginal distribution, though, as we will see, for the queueing distribution we will find only lower bounds. This fact reveals that we cannot expect exponential asymptotics, as seen in the short-range dependence situation in case of M/M/n models. These derived formulas, mainly from the original literature and mostly inspired by the approach of Norros, will be a starting point for the network optimization, shortly dealt with in section 5.3.2.

5.1.1 Waiting Queues for Models with Fractional Brownian Motion

In the sections 2.7.3, 2.7.5 and 2.8.2 we investigated waiting time distributions for the systems M/G/1, GI/M/n and G/G/1 and by examples indicated the impact of a heavy-tail distribution on the waiting time distribution. We look now at an alternative storage model, which is especially suitable for the traffic modeling in a LAN, using the Norros approach as starting point. In the later sections 5.1.3 and 5.1.4 this model will be transferred to the general case of Lévy processes and multifractal Brownian motion.

We recall the accumulated traffic up to time t

$$A_t = mt + \sqrt{am} B_t^{(H)}, \; t \in \,]-\infty, \infty[$$

and define the storage process with respect to the fractional Brownian motion by

$$X_t = \sup_{s \leq t}(A_t - A_s - C(t-s)), \; t \in \,]-\infty, \infty[, \; C > m \qquad (5.1)$$

Here, A_t represents the load process with parameters m, a and $H \in \,]\frac{1}{2}, 1]$. With $C > m$ we denote the capacity of the network, i.e. the rate of the outgoing data flow. We can interpret the expression (5.1), so that one has to subtract from the 'new' load $A_t - A_s$ the outgoing data amount $C(t-s)$. Afterwards the supremum (or maximum) over the differences up to time t is taken. The process (X_t) is stationary, since (A_t) is stationary, and by the theorem of Birkhoff [98, prop. 9.3.9] the supremum is finite almost everywhere. Later, we partly generalize the queueing in a slightly more complicated scenario of a network. Though (A_t) can assume negative values, the supremum will make (X_t) always positive. In addition, one can prove that for all $t \in \,]-\infty, \infty[$

$$\lim_{s \to -\infty} (A_t - A_s - C(t-s)) = -\infty \; \text{a.s.}$$

We want to determine the probability that X_t lies over a fixed threshold x. This gives

$$\epsilon = \mathbb{P}(X_t > x) \qquad (5.2)$$

We can consider x as buffer capacity. The value $\mathbb{P}(X_t > x)$ expresses the probability that the load process jumps over the capacity x. We want to indicate by ϵ a limit or *quantile* for the expression $\epsilon = \mathbb{P}(X_t > x)$, which can be regarded as upper limit for the overflow probability. Then, we will compute by $t = t_0$ the time, when the overflow probability exceeds the value ϵ at first. This value ϵ will be mainly determined by the requests of the QoS. Next we will compute the time t_0, where the relationship $\epsilon = \mathbb{P}(X_t > x)$ holds first. The self-similarity of the fractional Brownian motion (Z_t) allows a better representation of the expression (5.2).

Theorem 5.1. *With the definition (5.2) we get*

5.1 IP Traffic for Perturbation with Long-Range Dependence Processes

$$f^{-1}(\epsilon) = (C-m)(ma)^{-\frac{1}{2H}} x^{\frac{1-H}{H}} \tag{5.3}$$

where the function f is defined according to

$$f(z) = \mathbb{P}\left(\{\omega \in \Omega; \sup_{t \geq 0}(Z_t(\omega) - zt) > 1\}\right)$$

and depends only on H, but not on m, a, C, and x.

A short outline of the proof can be found in section 5.3.2. The right side of (5.3) will turn for $H = \frac{1}{2}$, i.e. the Brownian motion, to

$$\frac{1-\rho}{\rho a} = \text{const}, \quad \rho = \frac{m}{C} \tag{5.4}$$

This fact can be seen briefly as follows. Assume $\epsilon > 0$ and let the buffer capacity x be given. Then $f^{-1}(\epsilon) = \text{const}$ and it follows by (5.3)

$$\text{const} = (C-m)(ma)^{-1}x = \left(\frac{m}{\rho} - m\right)\frac{1}{ma}x = \frac{1-\rho}{\rho a}x$$

Since x is constant as well, (5.4) follows. We can consider $1-\rho$ as free capacity, and halving the capacity means a doubling of the buffer capacity (to get 'const' back). The value $\rho = \frac{m}{C}$ is the average load in the traffic model of Norros. It is equivalent to the notion of $\rho = \frac{\lambda}{\mu}$ defined for classical queueing models like M/M/n. As in section 2.4 we stress $\rho \leq 1$ to keep the system stable. The value $1 - \rho = \frac{C-m}{C}$ can be considered as the (relative) free capacity of the system.

The case is different for $H > \frac{1}{2}$. We solve first (5.3) for x and get

$$x = \left(f^{-1}(\epsilon)\right)^{\frac{H}{1-H}} \cdot a^{\frac{1}{2(1-H)}} \cdot C^{\frac{2H-1}{2(H-1)}} \cdot \frac{\rho^{\frac{1}{2(1-H)}}}{(1-\rho)^{\frac{H}{1-H}}}$$

Here, we see that, if H is close to 1, a bisection of $(1-\rho)$ requests much more storage capacity. This fact is supported by the observed behavior that in the packet switched traffic the flow cannot be increased by arbitrary enlarging the buffer. Finally we can express (5.3) for dimensioning the bandwidth

$$C = m + f^{-1}(\epsilon) \cdot a^{\frac{1}{2H}} \cdot x^{-\frac{1-H}{H}} \cdot m^{\frac{1}{2H}} \tag{5.5}$$

We can also derive by (5.5) that for $H > \frac{1}{2}$ the link capacity C grows more slowly than linear in m. Hence, a multiplex gain is achieved by bundling links of higher capacity. In figure 5.5 on page 388 we depict (5.5) for different H. The value $f^{-1}(y)$ is model-depending, i.e. only the QoS parameter ϵ and the nature of the FBM is entering. In the figures 5.1 and 5.6 we depict two key values, responsible for the behavior of the required buffer capacity (depending on the QoS value ϵ) and the service rate. With figure 5.1 it is obvious that

386   5 Performance of IP: Waiting Queues and Optimization

already a small long-range dependence changes the behavior significantly. Additionally, for high values of H we see an immediate increase near $\rho = 1.0$. In figure 5.6 the dependence of the free capacity for the service rate C shows that the service rate reacts for the standard Brownian motion only marginal compared to the FBM, provided the relative free capacity declines.

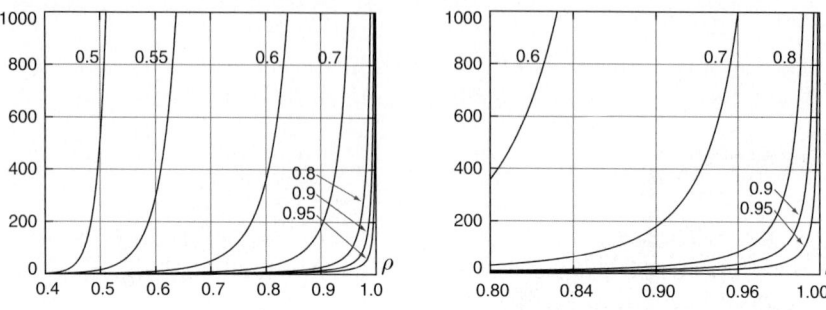

Fig. 5.1. The factor $\rho^{\frac{1}{2(1-H)}}/(1-\rho)^{\frac{H}{1-H}}$ for different Hurst exponents and scales on the x-axis

The figures 5.2 to 5.4 give under the certain selection of the Hurst parameter H the buffer requirement x for matching the given QoS rate ϵ. The buffer requirement depends on the given capacity C and the traffic load ρ. The average traffic amount m is incorporated and the parameter a is set to 2.8. The values on the vertical axes are scale free, but should indicate the relative value compared for the different Hurst parameter and respective capacity C and traffic load ρ.

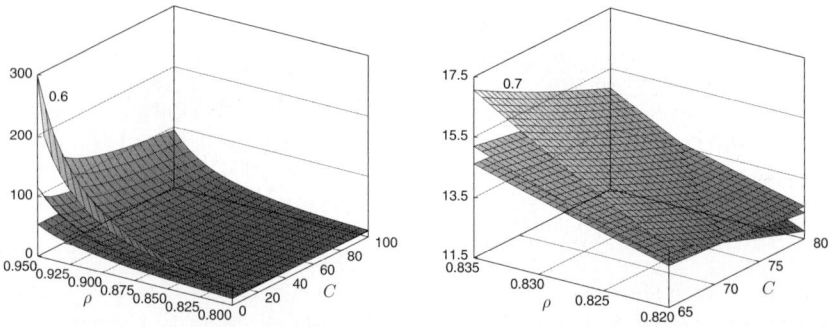

Fig. 5.2. The buffer requirement x for Hurst exponents $H = 0.5, 0.55, 0.6$ (left) and for pure long-range dependence with $H = 0.55, 0.6, 0.7$ (right)

As expected, x is higher for increasing H. In the right diagram of figure 5.2 x reaches the lowest value for $H = 0.7$, $C = 80$ and $\rho = 0.82$ and surpasses the

5.1 IP Traffic for Perturbation with Long-Range Dependence Processes

other two manifolds at $C = 70$. This reflects the fact that lowering the service rate with an simultaneous increase of the traffic load reacts more sensitive for higher long-range dependent traffic.

The left diagram of figure 5.3 indicates that for high traffic load and low values of C the buffer requirement increases rapidly for higher values of H. The right figure shows that higher long-range dependence reacts highly sensitive at lower values of C, but decreases more significantly for higher C.

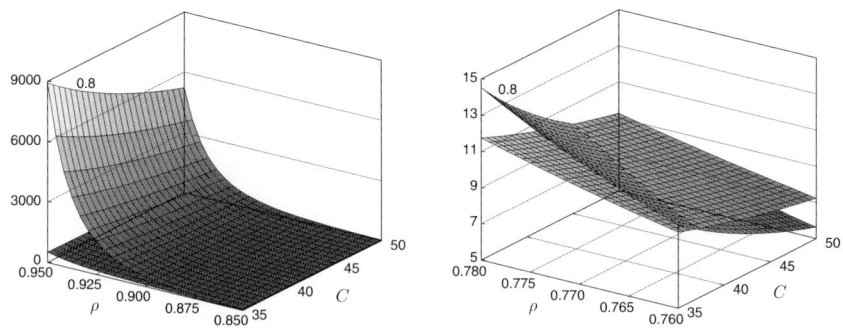

Fig. 5.3. The buffer requirement x for pure long-range dependence with Hurst exponents $H = 0.7$, 0.8 (left and right)

In figure 5.4, both diagrams reflect again that the more long-range dependent traffic is higher for low values of C and high traffic load ρ.

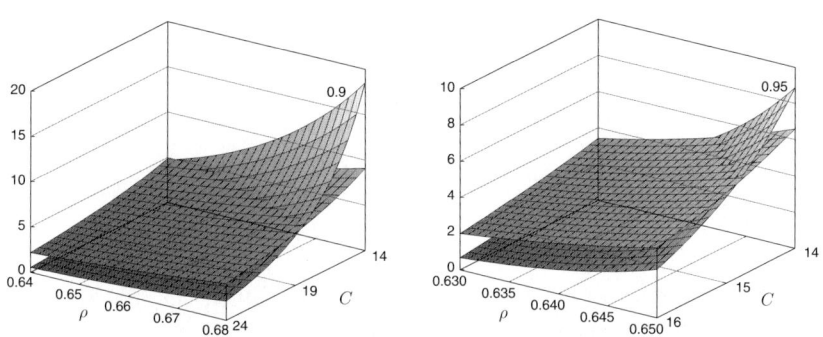

Fig. 5.4. The buffer requirement x for pure long-range dependence with Hurst exponents $H = 0.8$, 0.9 (left) and $H = 0.9$, 0.95 (left)

On the other hand, assuming an average traffic flow of m and a given buffer threshold x, we can compute the necessary service rate (or bandwidth) C, depending on the given queueing or QoS probability ϵ. For this we can use the equation (5.5) and depict the situation for different traffic situations, expressed by the Hurst exponents H ranging from 0.5 to 0.95 in figure 5.5. In the lower right diagram we selected small ranges for x and m to demonstrate that for very low values of buffer capacity x the service rate increases very slow for higher values of long-range dependence.

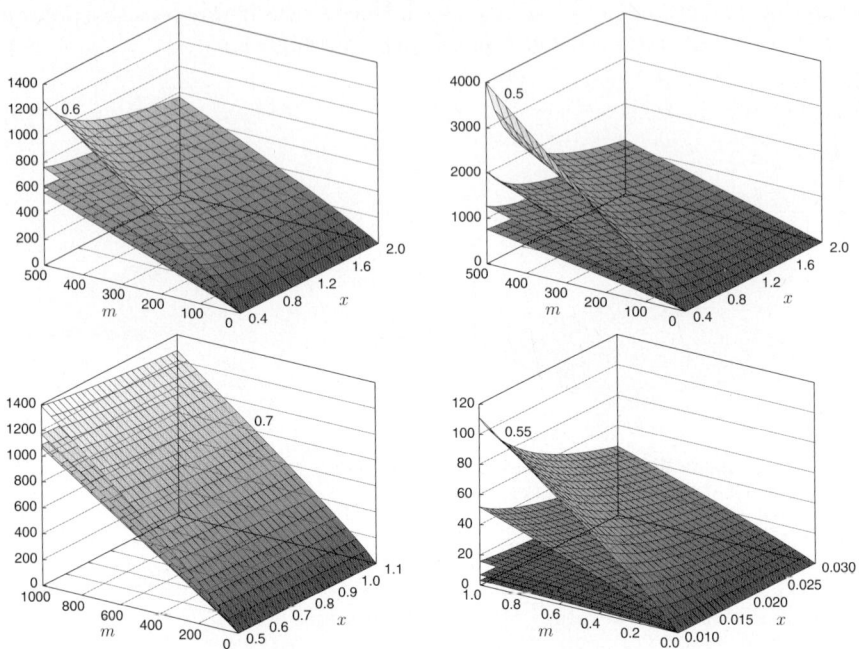

Fig. 5.5. The service rate C in dependence of the mean traffic rate m and the buffer requirement x according to equation (5.5) for Hurst parameter $H = 0.6, 0.7, 0.8, 0.9$ (upper left), $H = 0.5, 0.55, 0.6, 0.7$ (upper right), $H = 0.7, 0.8, 0.9, 0.95$ (lower left), and $H = 0.55, 0.6, 0.7, 0.8, 0.95$ (lower right)

An alternative representation of the system capacity C is given by

$$C(\rho) = \text{const} \cdot \rho^{\frac{1}{2H-1}} \cdot (1-\rho)^{-\frac{2H}{2H-1}} \qquad (5.6)$$

Equation (5.6) indicates the service rate or capacity in the network in dependence of the traffic load ρ. The crucial factor is $\rho^{\frac{1}{2H-1}} \cdot (1-\rho)^{-\frac{2H}{2H-1}}$, which depends on the Hurst exponent H and the traffic load ρ. The necessary capacity $C(\rho)$ depending on the traffic load is increasing more than linear, if the traffic load grows. This increase is higher for larger values of H, expressing the fact that higher values of the Hurst exponent means a deeper long-range dependence – the network has to build up reserve capacity for this dependence.

As seen in the left diagram of figure 5.6, the capacity increases slowlier with higher H, but then close to 1.0 rapidly. The exception is the Brownian motion

5.1 IP Traffic for Perturbation with Long-Range Dependence Processes

case $H = 0.5$. The right diagram of figure 5.6 is an equivalent representation, revealing the dependence of the relative free capacity $1 - \rho$. It shows that the expected decay for the capacity or service rate is larger and particularly strong if the Hurst parameter is high. This indicates that the lack of free capacity (low values) is much more crucial, provided the long-range dependence is more significant. Again more free capacity has to be stored for this dependence and thus, for occurring higher traffic load, resulting from the higher LRD phenomena. We see with equation (5.6) that for a high Hurst exponent $H = 0.95$, halving the free capacity for low values of ρ (0.8 to 0.4) requires already more than six times as much service rate.

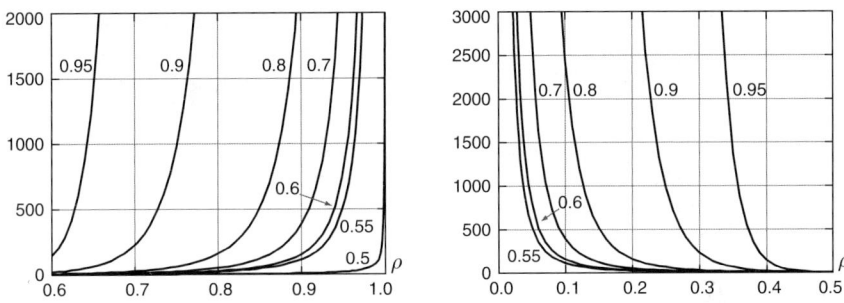

Fig. 5.6. Dependence of the service rate or capacity C on the relative traffic amount $\rho = \frac{m}{C}$ (left) and the relative free capacity $1 - \rho = \frac{C-m}{C}$ (right) for different Hurst exponents H

Of course, we are interested for important key values like computing the 'waiting probability'. But this raises the problem (it already appeared in the general case of M/G/n systems) that an explicit representation is not possible. Here we encounter the problem that the marginal distribution of (X_t) cannot be explicitly expressed. But according to Norros there exists a lower bound [190].

Theorem 5.2. *Let (X_t) be a storage process as defined in (5.1) with parameter m, a, H, and $C > m$. Then we have*

$$\mathbb{P}(X_t > x) \geq \Phi^c \left(\frac{t^H (C-m)^H x^{1-H}}{\varphi(H)\sqrt{am}} \right) \tag{5.7}$$

where $\varphi(H) = H^H(1-H)^{1-H}$, and Φ^c is the complementary standard Gaussian distribution function $\mathcal{N}(0,1)$.

We obtain (5.7) from the estimation

$$\mathbb{P}(X_t > x) = \mathbb{P}\left(\sup_{s \leq t} (A_t - A_s - C(t-s)) > x \right)$$

$$\geq \max_{t \geq 0}(A_t > t + x) = \max_{t \geq 0} \Phi^c \left(\frac{(1-m)t + x}{t^H \sqrt{am}} \right)$$

where the maximum is attained at $t_0 = \frac{Hx}{(1-H)(C-m)}$. Inserting t_0, it follows (5.7), and thus, the right side is independent of t. Now, we deduce the simple approximation

$$\Phi^c(z) \approx \frac{1}{\sqrt{2\pi}z} \exp\left(-\frac{z^2}{2}\right) \sim \exp\left(-\frac{z^2}{2}\right)$$

for large z. Hence, we get by inserting into (5.7)

$$\mathbb{P}(X_t > x) \sim \exp\left(-\frac{(C-m)^{2H}}{2\varphi(H)^2 am} x^{2-2H}\right) \quad (5.8)$$

for large values of x. In the figures 5.7 we indicated three scenarios for the complementary waiting time distribution with values $C = 480$ MBit, $m = 22$ MBit/s and $a = 2.8$ MBit/s. The first scenario reflects pure Markovian traffic with $H = 0.5$, the other scenarios illustrate LRD traffic for $H = 0.6$ and $H = 0.7$. Though, if the probability in the LRD cases lies already for small x above the Markovian case ($H = 0.5$), there exists an intersection point for relatively small values, before 10% of the overflow probability.

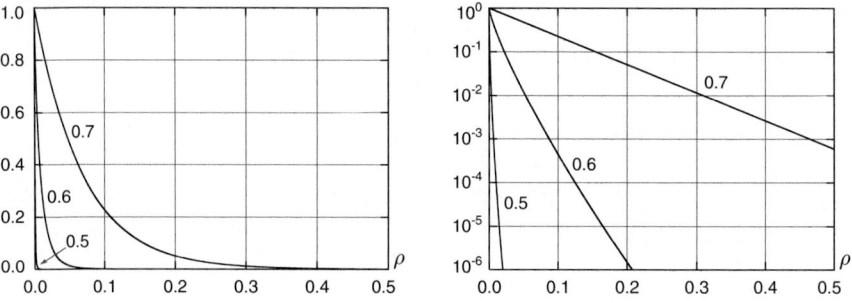

Fig. 5.7. Fractional waiting time distribution for different Hurst parameter H, linear (left) and logarithmic (right) scaling on the y-axis

In the best case scenario the waiting probability for the storage is Weibullian

$$\mathbb{P}(X_t > x) \sim \exp\left(-\kappa x^\beta\right)$$

with $\beta \leq 1$ ($\beta < 1$, if $H > \frac{1}{2}$), since in (5.7) we have '$\geq$'. In particular, we see that the Hurst exponent H has a decisive impact on the waiting time distribution. How does this look like in the case of the Brownian motion ($H = \frac{1}{2}$)? Then the load process (A_t) is Markovian, and we have ($a = 1$)

$$\mathbb{P}(X_t > x) \sim \exp\left(-2\frac{C(1-m)}{m} x\right) \quad (5.9)$$

5.1 IP Traffic for Perturbation with Long-Range Dependence Processes

which reveals the well-known asymptotic waiting time distribution of M/M/1 models. Solving (5.8) for C, we get a useful relationship for the capacity of C

$$C = m + \left(\varphi(H)\sqrt{-2\log\epsilon}\right)^{\frac{1}{H}} \cdot a^{\frac{1}{2H}} \cdot x^{-\frac{1-H}{H}} \cdot m^{\frac{1}{2H}}$$

where as above $\mathbb{P}(X_t > x) = \epsilon$ is a threshold for the waiting probability. The interesting case $H > \frac{1}{2}$ leads to a Weibullian distribution $\exp(-\kappa x^{2-2H})$. But this distribution is not very suitable for application. Though it has an expected value, it lacks of a moment generating function. Since it is a subexponential distribution, its Laplace transform shows a singularity in a neighbourhood of 0 (2.33). But this is again evidence of the long-range dependence of the IP traffic and indicates that it does not perform short-range dependence like in the classical M/M/n traffic.

Example 5.3. We want to illustrate the above results by an example according to Norros. The packet switched traffic enters a router from a variety of sources, where n links exit the router with the same capacity. All links are modeled independently by the same load process A_t with the same parameter a, m, C. They all use the same buffer. Here, the X_t^i for $i = 1, \ldots, n$ represents the storage of the aggregated IP packets or bursts at time t. In section 1.2.1 we indicated the problems for the suitable selection. The aggregated data amount is given by $X_t = \sum_{i=1}^n X_t^i$. For estimating a threshold K of the buffer capacity, we can consider the value

$$\mathbb{P}\left(X_t^i > x_i,\ i = 1, \ldots, n\right) = \prod_{i=1}^n \mathbb{P}\left(X_t^i > x_i\right)$$

with the $\sum_{i=1}^n x_i = K$. In the case of short-range dependence (i.e. Markovian with $H = \frac{1}{2}$) we get

$$\mathbb{P}\left(X_t^i > x_i,\ i = 1, \ldots, n\right) = \prod_{i=1}^n \mathbb{P}\left(X_t^i > x_i\right)$$
$$= \exp\left(-\gamma \sum_{i=1}^n x_i\right) = \exp\left(-\gamma K\right)$$

using our equation (5.9). This holds for all 'single thresholds' x_i with $\sum_{i=1}^n x_i$ thus on the simplex $\mathcal{S} = \{(x_1, \ldots, x_n);\ \sum_{i=1}^n x_i = K\}$. With this, the responsibility of an overflow is uniform distributed on the simplex. In the case $H > \frac{1}{2}$ we have

$$\mathbb{P}\left(X_t^i > x_i,\ i = 1, \ldots, n\right) = \exp\left(-\gamma \sum_{i=1}^n x_i^{2-2H}\right) \leq \exp\left(-\gamma K^{2-2H}\right)$$

Hence, the inequality turns into an equality (and thus to a maximum) in the corners of the simplex $\mathcal{S}$. That means, where for an $i = 1, \ldots, n$ we get

$x_i = K$. For increasing H the probability grows, so that for a x_i the capacity threshold is already overflown. But for a large number of links n this results is no longer true. Then the single corners x_i are no longer of importance.

Later in this chapter (see e.g. the sections 5.3.2 and 5.3.3) we will optimize a simple network with different input parameters.

5.1.2 Queueing in Multiscaling FBM

In section 3.5.1 we introduced the multiscale FBM to incorporate the traffic behavior on different scales. We consider the standard topics for the traffic performance. As we proceed for the standard FBM according to the Norros model, we compute the waiting queue length. For this we define analogously the aggregated traffic

$$A_\eta(t) = mt + \sqrt{m} X_\eta(t)$$

where m represents the average arrival rate and $\sqrt{m} X_\eta(t)$ the fluctuation around the mean, where $(X_\eta(t))$ is the multiscaling FBM. With the lower bound estimation (see (3.60)) and the fact that $A_\eta(t)$ has the mean mt and the variance $m\text{Var}(t)$, we receive

$$\mathbb{P}(Q > x) \geq \sup_{t \geq 0} \Phi^c \left(\frac{x + Ct - mt}{\sqrt{m\text{Var}(t)}} \right) \qquad (5.10)$$

where Φ^c is the complementary distribution function of the standard Gaussian. To obtain an upper estimation, i.e. to assume the worst case, we have to maximize the right side in (5.10) over t. This is difficult in the general case of a M_k-FBM. Thus, we will restrict ourself to only two different scalings, i.e. the case $K = 1$. For this we use the approximation

$$\text{Var}(X_\eta(\delta)) \approx \begin{cases} \delta^{2H_1} \frac{a_1}{C(H_1)^2} & \text{for } 0 \leq \delta < \frac{1}{\omega_1} \\ \delta^{2H_0} \frac{a_0}{C(H_0)^2} & \text{for } \frac{1}{\omega_1} \leq \delta < \infty \end{cases} \qquad (5.11)$$

where (a_1, H_1) represents the small scaling, while (a_0, H_0) the large time intervals (note that $0 = \omega_0 < \omega_1$ and thus, $\frac{1}{\omega_1} \leq \frac{1}{\omega_0} = \infty$). The value $\frac{1}{\omega_1}$ indicates the step between the small and the large scaling. With (5.11) we can maximize (5.10) and receive

$$t = t^* = \begin{cases} \frac{H_1}{1-H_1} \frac{x}{C-m} & \text{for } x < x_c \\ \frac{H_0}{1-H_0} \frac{x}{C-m} & \text{for } x \geq x_c \end{cases}$$

and hence, the distribution of the waiting queue turns into

$$\mathbb{P}(Q > x) \sim \begin{cases} \exp\left(-\kappa(a_1, H_1) x^{2-2H_1}\right) & \text{for } x < x_c \\ \exp\left(-\kappa(a_0, H_0) x^{2-2H_0}\right) & \text{for } x \geq x_c \end{cases}$$

5.1 IP Traffic for Perturbation with Long-Range Dependence Processes

where
$$\kappa(a, H) = \frac{(C-m)^{2H}}{2am(1-H)^{2-2H}H^{2H}}$$

and
$$x_c = \frac{(C-m)(1-H_1)}{H_1}$$
$$\cdot \exp\left(\frac{H_0 \log\left(\frac{H_1}{H_0}\right) + (H_0-1)\log\left(\frac{H_1-1}{H_0-1}\right) + \frac{1}{2}\log\left(\frac{a_1}{a_0}\right)}{H_0 - H_1}\right)$$

In the figure 5.8 we present the tail probability for the situation $a_0 = 0.3$ MBit/s, $a_1 = 0.07$ MBit/s, $H_0 = 0.89$, $H_1 = 0.62$, $m = 75$ MBit/s, $C_1 = 90$ MBit/s and $C_2 = 100$ MBit/s (see [95]). The critical scale is reached at $x_{c_1} = 0.1138$ for C_1 and at $x_{c_2} = 0.18974$ for C_2. In both cases the tail probability resp. the QoS scale falls abrupt at that point. However, we realize that a relatively small increase from C_1 to C_2 causes a significant change of the behavior.

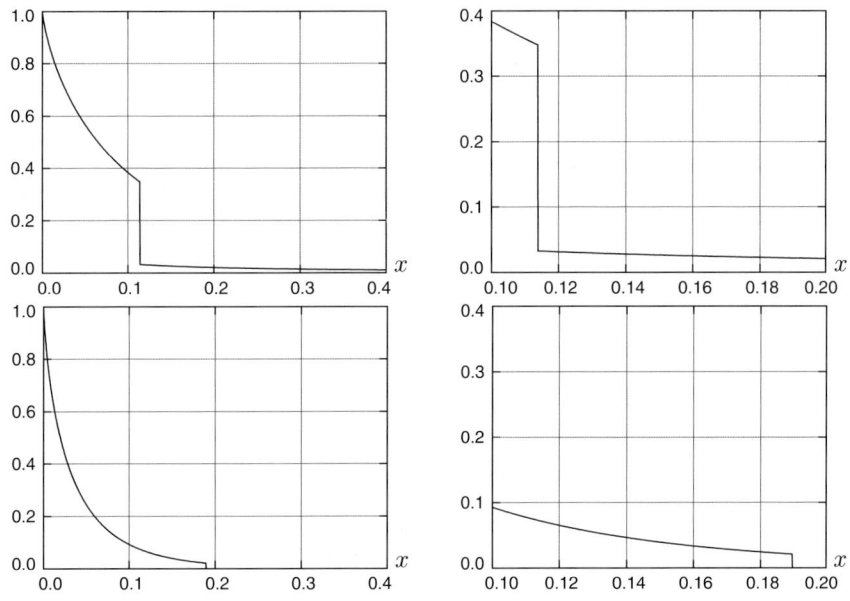

Fig. 5.8. Tail probability $\mathbb{P}(Q > x)$ for $C_1 = 90$ Mbit/s (upper left and right), $C_2 = 100$ Mbit/s (lower left and right) and different scales on x- and y-axis

The result can be extended to N independent two-scale FBM processes. Suppose m_n is the mean and $\mathbb{V}\mathrm{ar}_n(t)$ the variance of the two-scale t and the n-th flow, then we get for the distribution of the waiting queue

$$\mathbb{P}(Q > x) = \sup_{t \geq 0} \phi^c \left(\frac{x + Ct - m_n t}{\sqrt{m_n \mathbb{V}\mathrm{ar}_n(t)}} \right) \qquad (5.12)$$

But different to the pure two-scale case we cannot use the approximation (5.11), since the variance has more than two different regions. Even if we cannot solve (5.12) analytically, we may use mathematical tools like Matlab or numerical libraries for to implement appropriate algorithms.

Verification of the Model

To verify the model, we have to compare the analytically derived model above with actual data, in particular its waiting queue formula. For this a simulation program generates traffic, while putting the measured values in an infinite buffer queue with constant bit rate. But there are reasons, why the simulated results can differ from the actual data:

- There is no ideal FIFO waiting queue implemented at the router. This can happen e.g. that router information is worked in. But this delay is not significant.
- In reality, the buffer is finite and not infinite as assumed in simulations. In 2003, Fraleigh, Tobagi and Diot published traffic measurements conducted in the Sprint IP backbone [95]. They observed typical buffer sizes which correspond to 250 ms and 1 s of queueing delay. However, simulations described in the same paper reveal delays of less than 100 ms. In this case we would not observe any loss and there would be no difference between the results gained by simulations mimicking an infinite buffer and the actual routers in the network.
- The decisive difference lies, as mostly, in denying of any TCP influence, in case this protocol is used in the traffic. But we will consider situations later, where only a small amount of IP packets are influenced by this delay algorithm. In this situation the TCP exercises only a small impact.

How can we proceed? First we will determine the Hurst coefficients H_0 and H_1 using e.g. the wavelet estimators (see section 4.2.7). With the help of linear regression (method of least square) in the variance time plot we will compute the values a_0 and a_1. The decisive scale change takes place between 100 ms and 500 ms. For this we observe the scales (a_1, H_1) between 2 ms and 64 ms and (a_0, H_0) on 512 ms and 60 s. Now, we will implement this for the network A with a heavy load of user the data in the example $H_1 = 0.62$, $H_0 = 0.89$, $a_1 = 69.6$ kBit·s, $a_0 = 338$ kBit·s and $m = 75$ MBit/s. Simultaneously, we will assume a threshold $\epsilon = 0.0001$ for the probability of the delay. This is very restrictive, since in the existing networks for time sensitive traffic as VoIP a higher value is tolerated. For a bigger output the traditional model is comparable to the two scale FBM. In this region the bigger scales dominate, i.e the asymptotic of the waiting queue behavior. For smaller scales the multiscale FBM is better because the delay is described exacter.

5.1 IP Traffic for Perturbation with Long-Range Dependence Processes

Higher loads, i.e. traffic values of $\rho = \mu\lambda = 0.7$ or even $\rho = 0.9$ can lead to large delays in networks. Thus, one wants to verify the model exactly in this situations.

5.1.3 Fractional Lévy Motion and Queueing in IP Traffic Modeling

We introduced the fractional Lévy motion above on one side to incorporate the bursty character of the IP-based traffic and on the other hand as a generalization of the FBM, which we discussed in detail above and which was initially introduced by Norros for modeling the IP traffic.

Starting point is the network model due to Norros. So, we recall it from section 5.1.1 and consider a cumulated traffic at time t, denoted by $A(t)$ the amount of traffic produced in the interval $[0, t]$. Similar as in 3.3.4 we set

$$A(t) = mt + (\bar{c}m)^{\frac{1}{\alpha}} X(t, \alpha, H) \tag{5.13}$$

where as in the Norros model $m > 0$ is the mean input rate, $\bar{c} > 0$ the scaling factor (in the Norros model we denoted it by a; see also the definition in section 3.6.1) and $(X(t, \alpha, H))$ a fractional Lévy process. We briefly summarize the meanings of the different parameters used in the above model (5.13):

- The parameter $m > 0$ represents the average throughput rate.
- With α, the characteristic parameter in the stable marginal distributions, we indicate the heavy-tail distribution of the traffic load.
- We can regard

$$\bar{c} = \frac{\sigma}{\Gamma(H + 1 - \frac{1}{\alpha})^\alpha \alpha H}$$

as a kind of deviation of the traffic around its mean rate, where σ^2 is the usual variance of te traffic. Recall the meaning of $\bar{c}$ in the definition of stable distributions: in the case of $\alpha = 2$ and the Gaussian distribution we had $\bar{c} = \frac{\sigma^2}{2}$ with σ^2 its variance.
- According to theorem 3.67 the Hurst exponent H has to fulfill $H \in [\frac{1}{\alpha}, 1[$. It reflects the self-similarity of the traffic.

Our next term concerns the description of the queueing length and its distribution. Following again the approach of Norros as described in section 3.3.4, we select a single server queue with, as usual, a serving rate of $C > 0$ and an infinite buffer space, to make the model as simple as possible. The input amount follows the equation in (5.13). The parameter $\rho = \frac{m}{C}$ can be seen as traffic load resp. as the queue utilization. As in the classical traffic models we have to assume $m < C$, hence, for the traffic load $\rho < 1$, for stability reason. Depending on the service rate $C > 0$, we denote by $Q(t, C)$ the queue length at time $t > 0$. As in section 5.1.1 we have for the queueing process

$$Q(t, C) = \sup_{0 \le s \le t} (A(t) - A(s) - C(t - s))$$

which is Reich's formula [31, 210] for the virtual waiting queue length. Again as in section 5.1.1 we interpret the terms as follows:

- $A(t) - A(s)$ is the incoming traffic load during the interval $]s, t]$.
- With $C(t - s)$ is the outgoing traffic load of the single server during the same interval.
- The process $(Q(t, C))$ is stationary, since $(X(t, \alpha, H))$ has stationary increments. In addition, it is fractional stable process, which is due to the property of the driving process FLM $(X(t, \alpha, H))$.

We first describe the scaling property of the queueing process, whose proof runs basically as the one presented later for theorem 5.17.

Theorem 5.4. *The queueing process has the scaling property*

$$Q(at, C) \stackrel{d}{=} a^H \cdot Q\left(t, a^{1-H}C + \left(1 - a^{1-H}\right)m\right)$$

for all $a > 0$.

The theorem tells us basically two facts. First, the queueing process $Q(t, C)$ is 'self-similar' with the Hurst exponent of the original FLM, namely H and the new serving rate $a^{1-H} \cdot C + (1 - a^{1-H}) \cdot m$. In other words if $Q(t, C)$ is the queue length at time t for a serving rate C, then we have to apply the new serving rate $a^{1-H} \cdot C + (1 - a^{1-H}) \cdot m$ and the scale multiplier a^H. Second, if we set $\alpha = 2$, then we have in particular for $H \in [\frac{1}{2}, 1[$

$$Q(at, C) = a^H \cdot Q\left(t, a^{1-H}C + \left(1 - a^{1-H}\right)m\right)$$

This is exactly the result of theorem 5.1 (resp. in slight generalization of the one in theorem 5.17). As in the case of FBM we define for the QoS threshold

$$\epsilon = \mathbb{P}\left(Q(0, C) > x\right) = \mathbb{P}\left(\sup_{\tau \geq 0} \left(A(\tau) - C\tau\right) > x\right) \tag{5.14}$$

We can consider equation (5.14) as describing the quantitative connection between the necessary buffer size expressed by x and the QoS requirement indicated by ϵ, which shows the probability of the overflow. Like in the Norros model, we define the auxiliary function

$$q(x, z) = \mathbb{P}\left(\sup_{\tau \geq 0} \left(X(\tau, \alpha, H) - \tau z\right) > x\right)$$

As example we choose $x = 1$, and realize that the function $q(1, z)$ is strictly decreasing in z. Can we find a representation for the general function $q(x, z)$ with the help of $q(1, z)$, i.e. does there exist a kind of homogeneity resp. scaling property? The answer is given with the following proposition.

Proposition 5.5. *The function $(x, z) \longmapsto q(x, z)$ has the scaling property*

$$q(x, z) = q\left(1, x^{\frac{1-H}{H}} z\right)$$

5.1 IP Traffic for Perturbation with Long-Range Dependence Processes

As we proceeded in the case of FBM, we can express (5.14)

$$\epsilon = q\left(\frac{x}{(\bar{c}m)^{\frac{1}{\alpha}}}, \frac{C-m}{(\bar{c}m)^{\frac{1}{\alpha}}}\right) = q\left(1, \left(\frac{x}{(\bar{c}m)^{\frac{1}{\alpha}}}\right)^{\frac{1-H}{H}} \cdot \frac{C-m}{(\bar{c}m)^{\frac{1}{\alpha}}}\right)$$

We are now able to formulate two equations, expressing on the one hand the requirement for the bandwidth and on the other hand the requirement for the buffer capacity.

Proposition 5.6. *Using equation (5.14), we can formulate the bandwidth requirements by*

$$C = m + q^{-1}(1,\epsilon) \cdot \bar{c}^{\frac{1}{\alpha H}} \cdot x^{-\frac{1-H}{H}} \cdot m^{\frac{1}{\alpha H}} \qquad (5.15)$$

and for the buffer dimension

$$\frac{1-\rho}{\rho^{\frac{1}{\alpha H}}} \cdot x^{\frac{1-H}{H}} \cdot C^{\frac{\alpha H - 1}{\alpha H}} = \bar{c}^{\frac{1}{\alpha H}} \cdot q^{-1}(1,\epsilon) \qquad (5.16)$$

where $q^{-1}(1, \cdot)$ is the inverse function of $q(1, \cdot)$.

With this general results we consider some typically examples.

Example 5.7. The simplest time continuous case is the ordinary Brownian motion, i.e. $H = \frac{1}{2}$ and $\alpha = 2$. Then, inserting into (5.16), this gives, by solving for x

$$x = x(\rho) = \text{const} \cdot \rho \cdot (1-\rho)^{-1}$$

Example 5.8. We consider again a driving process having independent increments with α-stable marginal distribution and $\alpha \in]0, 2[$. This means we have an ordinary Lévy motion. Again with (5.16) we deduce a formula for the buffer dimension

$$x = x(\rho) = \text{const} \cdot \rho^{\frac{1}{1-\alpha}} \cdot (1-\rho)^{-\frac{1}{1-\alpha}}$$

As we saw for the Brownian motion, the serving rate C has disappeared because of $H = \frac{1}{2}$. Since the marginal distribution influences the traffic model in figure 5.9, we started with its incorporation. First we considered the buffer requirement x depending on the traffic load ρ and the heavy-tail exponent α. It reveals that, with falling α, the buffer overflow for a fixed QoS threshold ϵ occurs already for low values of x.

The diagrams in figure 5.9 give a more detailed description for the particular range of traffic load ρ and different range of α. Note that in all cases we have the lack of LRD, since $H = \frac{1}{2}$. We encounter Lévy stable motions. As seen, small α and high traffic load ρ results in a very low threshold x. In the right figure we realize that the threshold x falls dramatically almost to 0, if α and ρ are close to 1. The bursty traffic for small α is thus reflected by this observation.

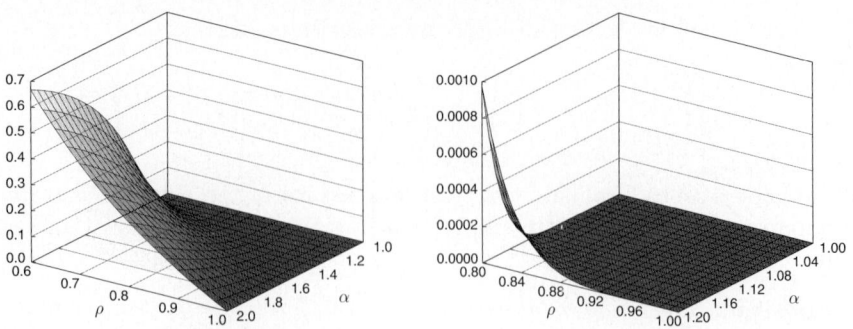

Fig. 5.9. The buffer size for which the same overflow probability ϵ occurs depending on different ranges of α and traffic load ρ

Example 5.9. Things change if we consider the case $H > \frac{1}{2}$. First, we proceed with the case $\alpha = 2$ and fix a serving rate C. Then,

$$x = x(\rho) = \text{const} \cdot \rho^{\frac{1}{2(1-H)}} \cdot (1-\rho)^{-\frac{H}{1-H}} \qquad (5.17)$$

This is exactly the result from section 5.1.1. The serving rate C is included in the constant. In opposite, if we fix a buffer size x, then it follows

$$C = C(\rho) = \text{const} \cdot \rho^{\frac{1}{2H-1}} \cdot (1-\rho)^{-\frac{H}{H-\frac{1}{2}}}$$

Here, x is part of the constant term.

Example 5.10. Finally, we generalize both parameters $H > \frac{1}{2}$ and $0 < \alpha < 2$. Again we fix a serving rate C and use (5.16). We deduce

$$x = x(\rho) = \text{const} \cdot \overline{c}^{\frac{1}{\alpha(1-H)}} \rho^{\frac{1}{\alpha(1-H)}} \cdot (1-\rho)^{-\frac{H}{1-H}} \qquad (5.18)$$

As we know from section 3.2.3, there is no arbitrary choice resp. combination of α and H for an α-stable process to be H-sssi. By theorem 3.67 we know $H \in [\frac{1}{\alpha}, [$. But on the other side we know from section 3.3, the LRD property is defined for general α-stable processes by the inequality $H > \frac{1}{\alpha}$.

Figure 5.10 reflects the buffer requirement x for satisfying the QoS threshold ϵ. We realize that a low value of α, thus, a bursty traffic, requires a larger buffer. This goes conform with the usual expectation that bursty traffic needs some 'safety' buffer space to ensure QoS guarantees. In the left diagram we see that the values of α do not significantly influence the buffer requirement as the traffic load does for high values of ρ. The reaction on the value of α is more intense for higher values of ρ as illustrated in the right diagram.

5.1 IP Traffic for Perturbation with Long-Range Dependence Processes   399

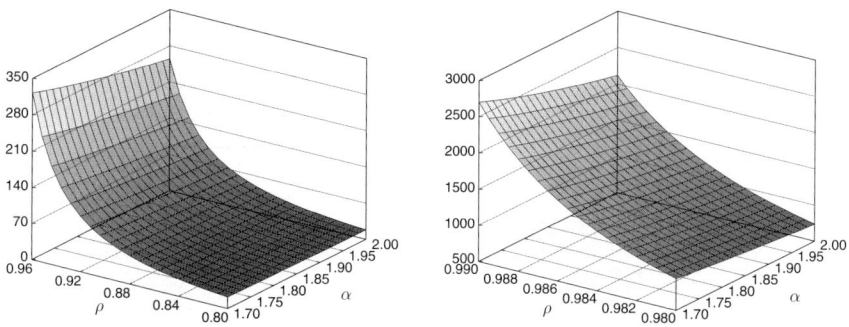

Fig. 5.10. The buffer requirement x for Hurst exponent $H = 0.6$, depending on α and different ranges of relative traffic load ρ

Figures 5.11 and 5.12 indicate the bursty traffic behavior, but this time the behavior results from the different LRD phenomena imposed by the varying Hurst exponent. Again as easily seen and expected more LRD and bursty traffic trigger a decisive impact for larger buffer requirements.

In figure 5.11, both diagrams demonstrate the dependence of the burstiness, i.e. of the parameter α in the distribution of the marginal distributions of the α-stable process. We see that the buffer requirements increases rapidly, especially depending on the Hurst parameter H.

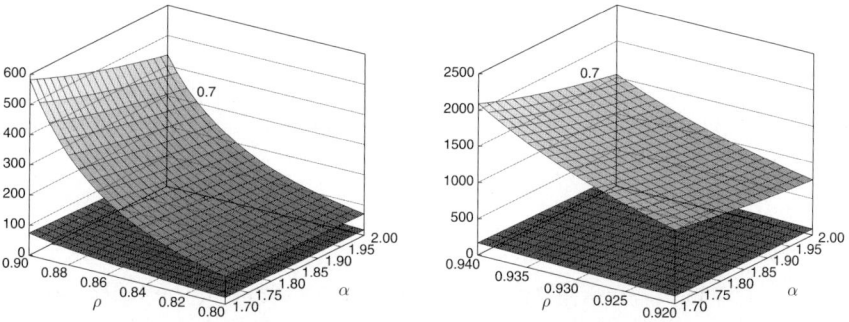

Fig. 5.11. The buffer requirement x for Hurst exponents $H = 0.7, 0.6$ (left and right), depending on α and different ranges of relative traffic load ρ

In figure 5.12 see again that the buffer requirement increases rapidly for lower values of H, bursty traffic and high traffic load ρ.

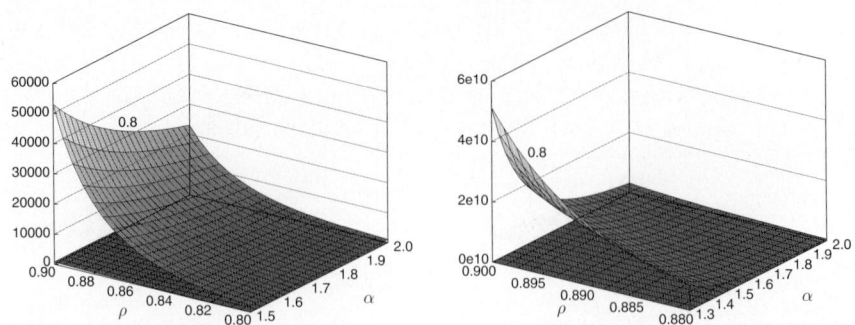

Fig. 5.12. The buffer requirement x for different Hurst exponents $H = 0.7$, 0.8 (left) and $H = 0.8$, 0.9 (right), depending on α and different ranges of relative traffic load ρ

If we set $\alpha = 2$ in (5.18), we get (5.17) back, hence a generalization of the Norros case. As in example 5.10, the serving rate is included in the constant. At last, we compute the service rate C for a fixed buffer size x

$$C = C(\rho) = \text{const} \cdot \overline{c}^{\frac{1}{\alpha H - 1}} \rho^{\frac{1}{\alpha H - 1}} \cdot (1-\rho)^{-\frac{H}{H - \frac{1}{\alpha}}}$$

Figure 5.13 depicts the necessary capacity resp. service rate for ensuring the QoS requirement in the same situation as above for the buffer requirement. As in all figures the absolute value is not important for the analysis, but the qualitative behavior of the manifolds.

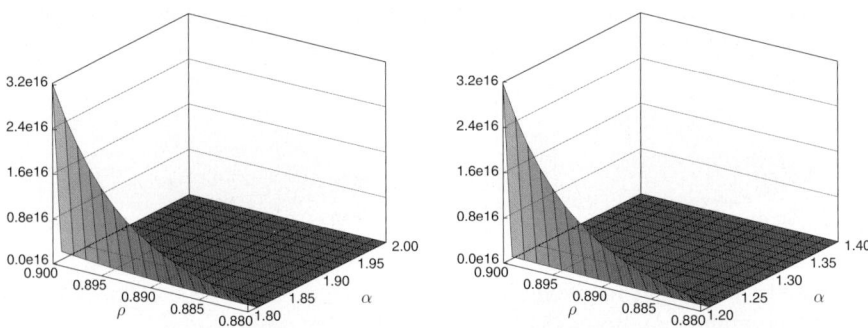

Fig. 5.13. The service rate C for the Hurst parameter $H = 0.6$ (left) and $H = 0.9$ (right), depending on the marginal distribution of the Lévy process α and the relative traffic load ρ

As seen in both diagrams shortly before low values of α no significant change is detected. The figure should demonstrate that for both values of H the service rate is heavily depending on the burstiness of the traffic, depicted by the extremely fast change for low values of α.

5.1 IP Traffic for Perturbation with Long-Range Dependence Processes

To illustrate the dependence of H, we depict in figure 5.14 similar diagrams as in figure 5.13. We show the service rate C for the same areas of α and ρ and the same Hurst exponent H. However, we also depict C for a slightly decreased value of H in each diagram and thus have to adjust the scaling on the z-axis.

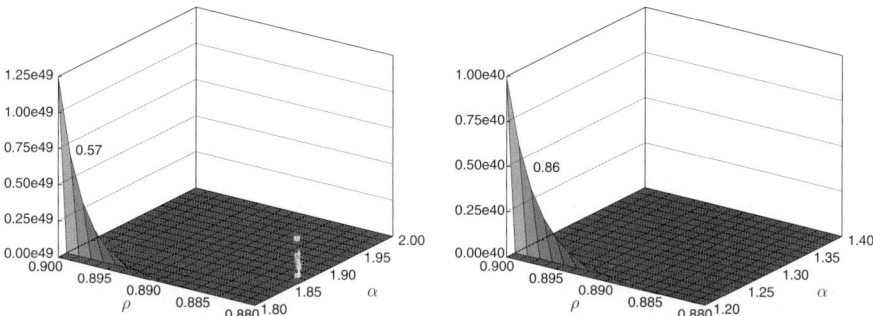

Fig. 5.14. The service rate C for the Hurst parameters $H = 0.57$, 0.60 (left) and $H = 0.86$, 0.90 (right), depending on the marginal distribution of the Lévy process α and the relative traffic load ρ

We see that the service rate C increases for a low value of the Hurst exponent very rapidly if the value of α is small and the traffic load is high. Already a small decrease for the Hurst exponent results in a very sensitive behavior for a large traffic load and small α.

In figure 5.15 we finally compare the service rate C for Hurst parameters $H = 0.8$ and 0.95 each for the same range of α.

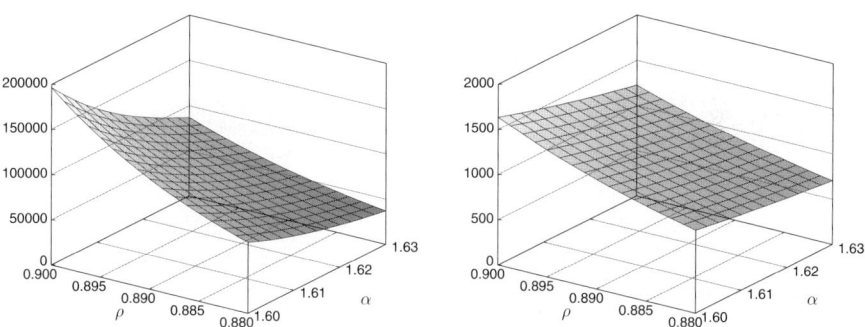

Fig. 5.15. The service rate C for the Hurst parameter $H = 0.8$ (left) and $H = 0.95$ (right), depending on the marginal distribution of the Lévy process α and the relative traffic load ρ

402   5 Performance of IP: Waiting Queues and Optimization

As it is shown the service rate is significantly higher for lower values of long-range dependence reflecting the fact that more long-range dependence requires more service rate. Obviously, the burstiness for small α in conjunction with a high traffic load results in a very fast growing value of the service rate. The increase is lower for higher values of long-range dependence.

Since we started with a QoS requirement ϵ, we would like to know, how we can detect a relationship between the buffer size x the service rate C and the QoS constraint $\epsilon > 0$. Like in the FBM case, the result is again a lower bound estimate of the following type, expressed in the succeeding theorem. We have to stress that first the estimation (5.19) is trivially fulfilled for $\alpha = 2$. In this case one should consult the already known asymptotic (5.8) for the FBM. On the other hand, as the last figure below demonstrates, the estimation should be used as asymptotic, thus not for lower values of the buffer capacity x.

Theorem 5.11. *For large buffer size, i.e. if $x \to \infty$, we have the lower estimate*

$$\epsilon = \mathbb{P}\left(Q(0,C) > x\right) \geq \Theta_\alpha x^{-\alpha(1-H)} \tag{5.19}$$

where

$$\Theta_\alpha = C_\alpha \cdot (\bar{c}m)(1-H)^\alpha \cdot \left(\frac{H}{(1-H)(C-m)}\right)^{\alpha H}$$

and

$$C_\alpha = \frac{\bar{c}}{\alpha \pi} \cdot \Gamma(\alpha+1) \cdot \sin\left(\frac{\pi \alpha}{2}\right) \tag{5.20}$$

In the figures 5.16 and 5.17 we depict the lower bound for several versions of the Hurst parameter and the range of $1 < \alpha < 2$. As expected the probability bound increases for lower α and increasing Hurst exponent. For low buffer capacity (right) we see that the exponent α has higher impact.

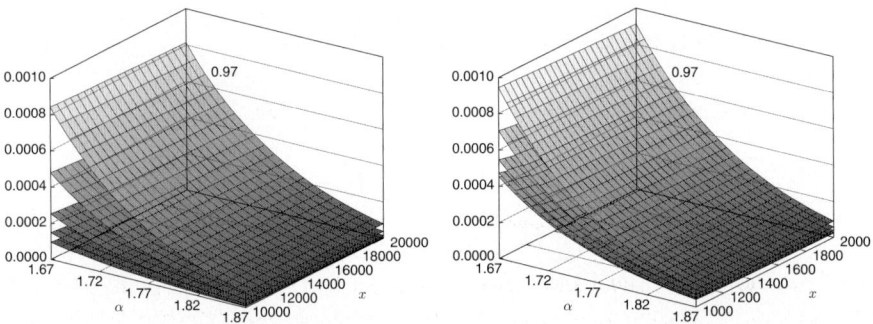

Fig. 5.16. The lower bound of equation (5.19) for Hurst exponents $H = 0.6$, 0.7, 0.8, 0.9, 0.97, depending on α and the buffer storage x

5.1 IP Traffic for Perturbation with Long-Range Dependence Processes

The left diagram in figure 5.17 demonstrates the less sensitive reaction for higher Hurst exponents at low values of α. The right diagram illustrates that the above lower bound holds only asymptotically.

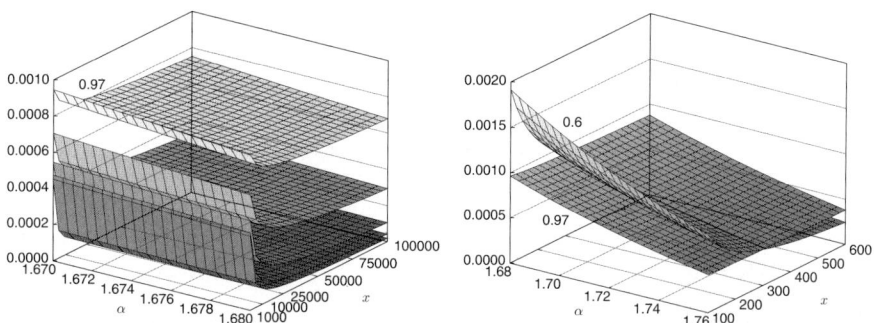

Fig. 5.17. The lower bound of equation (5.19) for Hurst exponents $H = 0.6, 0.7, 0.8, 0.9, 0.97$ (left) and $H = 0.6, 0.97$ (right), depending on α and the buffer storage x

From the above theorem we can deduce approximately the required service rate C for a given large buffer size x and a given QoS constraint by solving (5.19)

$$C = m + \left(\frac{C_\alpha}{\epsilon}\right)^{\frac{1}{\alpha H}} \cdot \overline{c}^{\frac{1}{\alpha H}} \cdot m^{\frac{1}{\alpha H}} \cdot x^{\frac{H-1}{H}} \tag{5.21}$$

Comparing (5.15) with (5.21) we see that the term $q^{-1}(1, \epsilon)$ is substituted by $\left(\frac{C_\alpha}{\epsilon}\right)^{\frac{1}{\alpha H}}$.

The result in theorem 5.11 includes several classical ones in traffic theory:

- If we consider the case of Brownian motion, i.e. $H = \frac{1}{2}$ and $\alpha = 2$, then inequality (5.20) turns into the well known Erlang formula for exponential distributed interarrival times.
- For $H = \frac{1}{2}$ and $0 < \alpha < 2$ we are in the situation of the ordinary Lévy motion, the results of which are presented in [150].
- The case of $\alpha = 2$ and $H > \frac{1}{2}$ leads to the FBM case, which lower bound was first found by Norros (see theorem 5.2).

We consider in the remainder of this section a similar situation as in the Norros case, but now with the FLM instead of the FBM. The network consists of the volume scaling parameter a, so called speed representing the bandwidth parameter x and n iid FLM driven multiplexed streams. This means:

- The workload $A(t) - A(s)$ has to be multiplied by a.
- The argument of $A(t)$ must be multiplied by x, i.e. $A(xt)$, since faster links will increase the load.

- The n links are iid, i.e. they follow the same model and the load is just the sum, since they are assumed being independent.

Hence we have the load in the interval $]s,t]$

$$\sum_{j=1}^{n} a\left(A_j(xt) - A_j(xs)\right) \tag{5.22}$$

We use the representation of $A(t)$ in our model and deduce the queueing formula

$$Q(t,a,b,n,r) = \sup_{0 \leq s \leq t} \left(\sum_{j=1}^{n} a(\bar{c}m)^{\frac{1}{2}} \left(X^{(j)}(bt,\alpha,H) - X^{(j)}(bs,\alpha,H)\right) \right.$$
$$\left. -abn(C-m)(t-s) \right)$$

where the $(X^{(j)}(t,\alpha,H))$ are iid FLM. The term $Q(t,a,b,n,C)$ describes the buffer occupancies. Thus we have by using FIFO principle

$$\Psi(a,b,n) = \frac{Q(a,b,n)}{abnr}$$

as the queueing delay. We can easily find the scaling laws for the queueing delays as

$$\Psi(a,b,n) \stackrel{d}{=} \Psi(1,b,n) \quad \text{and} \quad \Psi(a,b,n) \stackrel{d}{=} b^{-1}\Psi(a,1,n)$$

Our aim is to see, how the multiplex parameter n influences the queueing delay.

Theorem 5.12. *With the multiplex model according to equation (5.22) we have the scaling*

$$\Psi(a,b,n) \stackrel{d}{=} n^{-\frac{\alpha-1}{\alpha(1-H)}} \cdot \Psi(a,b,1)$$

In the figure 5.18 the multiplex factor $n^{-\frac{\alpha-1}{\alpha(1-H)}}$ is depicted for several Hurst exponents. As expected, the queueing probability decreases for larger numbers of lines. For higher Hurst exponents, the multiplex factor reacts more sensitive on α. The gain is smaller for higher Hurst exponents, since the decrease does not react as sensitive as for the lower Hurst exponent.

We remark that though the multiplex factor decreases with higher H, for the queueing probability we still have to encounter the factor $\Psi(a,b,1)$ as well. Over all we can see that the multiplex gain is small for small values of α, again revealing that bursty traffic is more difficult to multiplex.

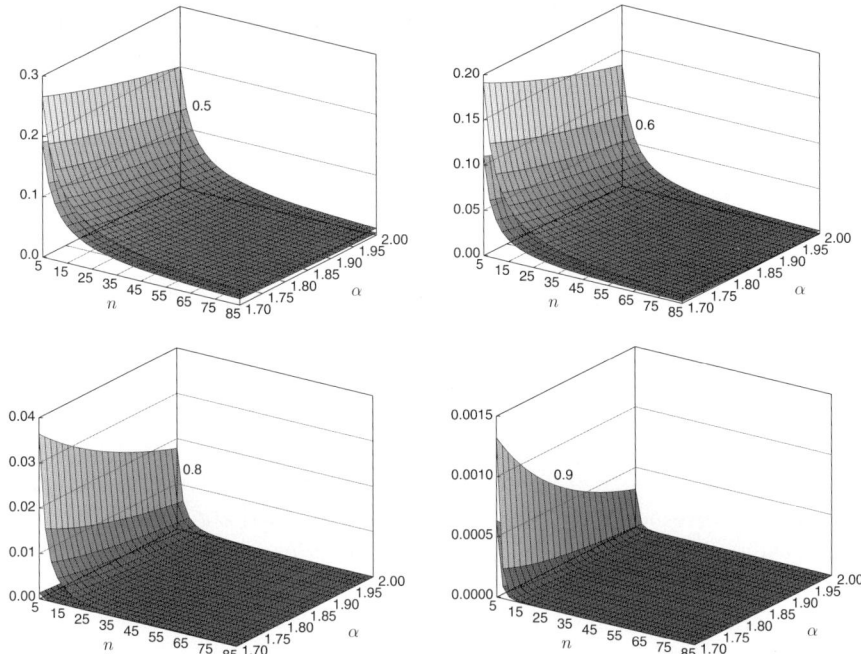

Fig. 5.18. The multiplex factor in dependence of α and the number of multiplexed lines n for Hurst parameter $H = 0.5$, 0.6 (upper left), $H = 0.6$, 0.7 (upper right), $H = 0.8$, 0.9 (lower left), and $H = 0.90$, 0.91 (lower right)

5.1.4 Queueing Theory and Performance for Multifractal Brownian Motion

In section 3.8.1 we introduced the (nonstationary) multifractal Brownian motion and the corresponding envelope process. Thus, the following considerations have to be seen in a line with section 3.8.1. We start with the computation of a queue length in multifractal IP traffic using mBm and a corresponding envelope process. Fundamental is the FIFO system, which we already considered in the classical traffic theory. We introduce the following notions:

- $A(t)$: the accumulated traffic amount, where we set $A(0) = 0$,
- $S(t)$: the served traffic amount,
- C: the capacity of the buffer.

Then the queueing length can be expressed by

$$Q(t) = A(t) - S(t) \tag{5.23}$$

and

$$S(t) = Ct + \min\left(0, \inf_{t \geq 0}\left(A(t) - Ct\right)\right) = Ct + A(\tilde{t}) - C\tilde{t} \tag{5.24}$$

where $\tilde{t} = \arg\inf_{t\geq 0}(A(t) - Ct)$, i.e. $\tilde{t}$ expresses the time at which the largest idle period in the interval $[0,t]$ ends. We define $t^* = \frac{t}{\tilde{t}}$. Then we have an expression for the serving process $S(t)$ immediately from (5.24)

$$S(t) = Ct + A\left(\frac{t}{t^*}\right) - C\frac{t}{t^*} \tag{5.25}$$

This leads with (5.23) to

$$Q(t) = A(t) - S(t) = A(t) - Ct - A\left(\frac{t}{t^*}\right) + C\frac{t}{t^*} \tag{5.26}$$

$$= A(t) - A\left(\frac{t}{t^*}\right) - Ct\left(1 - \frac{1}{t^*}\right)$$

The equation (5.26) expresses a stochastic process. We use the (deterministic) envelope process to transform the problem of (5.26) into a deterministic one and call the corresponding process $\hat{Q}(\cdot)$. First, we insert the envelope process in the expression for the service rate in (5.25). Hence, we compute $\hat{t}^* = \frac{t}{\hat{t}}$, where $\hat{t} = \arg\inf_{t\geq 0}(\hat{A}(t) - Ct)$. Thus, we define the equivalent (deterministic) envelope process for the service rate

$$\hat{S}(t) = Ct + \hat{A}\left(\frac{t}{\hat{t}^*}\right) - C\frac{t}{\hat{t}^*} \tag{5.27}$$

Hence, we obtain an upper bound for the queueing length $\hat{Q}$

$$\hat{Q}(t) = \hat{A}(t) - \hat{S}(t) = \hat{A}(t) - \hat{A}\left(\frac{t}{\hat{t}^*}\right) - Ct\left(1 - \frac{1}{\hat{t}^*}\right) \tag{5.28}$$

$$= \int_0^t m + \kappa\sqrt{am}H(x)x^{H(x)-1}dx - \int_0^{\frac{t}{\hat{t}^*}} m + \kappa\sqrt{am}H(x)x^{H(x)-1}dx$$

$$- Ct\left(1 - \frac{1}{\hat{t}^*}\right)$$

Next, we want to calculate the time t^*, where the queue length reaches its maximum. For this, we have to solve the equation

$$\hat{q}_{\max} = \max_{t\geq 0}\left(\hat{Q}(t)\right)$$

The maximal queue length is attained at time $\tilde{t}^*$, which can be computed according to an implicit given equation

$$\tilde{t}^* = \frac{\kappa\sigma\left(H(\tilde{t}^*)(\tilde{t}^*)^{H(\tilde{t}^*)} - \hat{t}H\left(\frac{\tilde{t}^*}{\hat{t}}\right)\left(\frac{\tilde{t}^*}{\hat{t}}\right)^{H\left(\frac{\tilde{t}^*}{\hat{t}}\right)}\right)}{(C-m)\left(1 - \frac{1}{\hat{t}}\right)} \tag{5.29}$$

5.1 IP Traffic for Perturbation with Long-Range Dependence Processes

Equation (5.29) is derived by computing the derivative in (5.28). Indeed we compute $\frac{dQ}{dt}$ and set $\frac{dQ}{dt} = 0$.

In the case of a the monofractal envelope process, equation (5.29) collapses to

$$\tilde{t}^* = \left(\frac{\kappa \sigma H \left(1 - (\hat{t})^{-H}\right)}{(C - m)\left(1 - \frac{1}{\tilde{t}}\right)} \right)^{\frac{1}{1-H}} \tag{5.30}$$

Equation (5.30) gives insight of the importance of multifractal modeling. The crucial time $\tilde{t}^*$, where the queue will be maximal, grows exponential with the Hurst exponent H, since it enters in the formula with the exponent $\frac{1}{1-H}$, It would overestimate the variation of the Hölder exponent.

Several Multifractal Flows

As in the classical Norros model, we consider N flows, forming an aggregated traffic. We can compute the amount of traffic by using the local asymptotically self-similarity (lass). It is straightforward that the sum of N independent FBM with the same Hurst exponent, mean m_i, $i = 1, \ldots, N$ and variance σ_i^2 is again a FBM with mean $m = \sum_{i=1}^{N} m_i$ and variance $\sigma^2 = \sum_{i=1}^{N} \sigma_i^2$ given by

$$\hat{A}^N(t) = \sum_{i=1}^{N} \hat{A}_i(t)$$

$$= \int_0^t \sum_{i=1}^{N} m_i + \kappa \left(\sum_{i=1}^{N} \sigma_i^2 H_i(s) s^{2H_i(s)-1} \right) \left(\sum_{i=1}^{N} \sigma_i^2 s^{2H_i(s)} \right)^{-\frac{1}{2}} ds$$

Inserting this into equation (5.27) reveals

$$\hat{S}^N(t) = Ct + \hat{A}^N\left(\frac{t}{\tilde{t}}\right) - C\frac{t}{\tilde{t}}$$

If we proceed as in the single queueing model, we find an upper bound for the queueing length $\hat{Q}^N$. This gives in analogy

$$\hat{Q}^N(t) = \hat{A}^N(t) - \hat{S}^N(t) = \hat{A}^N(t) - \hat{A}^N\left(\frac{t}{\tilde{t}}\right) - Ct\left(1 - \frac{1}{\tilde{t}}\right)$$

$$= \int_0^t \sum_{i=1}^{N} m_i + \kappa \left(\sum_{i=1}^{N} \sigma_i^2 H_i(s) s^{2H_i(s)-1} \right) \left(\sum_{i=1}^{N} \sigma_i^2 s^{2H_i(s)} \right)^{-\frac{1}{2}} ds$$

$$- \int_0^{\frac{t}{\tilde{t}}} \sum_{i=1}^{N} m_i + \kappa \left(\sum_{i=1}^{N} \sigma_i^2 H_i(s) s^{2H_i(s)-1} \right) \left(\sum_{i=1}^{N} \sigma_i^2 s^{2H_i(s)} \right)^{-\frac{1}{2}} ds$$

$$- Ct\left(1 - \frac{1}{\tilde{t}}\right)$$

Again, we deduce similarly

$$\hat{q}_{\max}^N = \max_{t \geq 0} \left(\hat{Q}^N(t) \right)$$

After solving, this gives the equivalent equation

$$\kappa \left(\left(\sum_{i=1}^N \sigma_i^2 H_i(t) t^{2H_i(t)-1} \right) \cdot \left(\sum_{i=1}^N \sigma_i^2 t^{2H_i(t)} \right)^{-\frac{1}{2}} \right.$$

$$+ \left(\sum_{i=1}^N \sigma_i^2 H_i \left(\frac{t}{\hat{t}} \right) \left(\frac{t}{\hat{t}} \right)^{2H_i\left(\frac{t}{\hat{t}}\right)-1} \right) \cdot \left(\sum_{i=1}^N \sigma_i^2 \left(\frac{t}{\hat{t}} \right)^{2H_i\left(\frac{t}{\hat{t}}\right)} \right)^{-\frac{1}{2}} \right)$$

$$- \left(C - \sum_{i=1}^N m_i \right) \left(1 - \frac{1}{\hat{t}} \right) = 0$$

Homogeneous Flows

In this case we have $m = m_i$ and $\sigma = \sigma_i$. This leads immediately to

$$\hat{A}^N(t) = \sum_{i=1}^N \hat{A}_i(t) = \int_0^t Nm + N^{\frac{1}{2}} \kappa \sigma H(s) s^{H(s)-1} ds$$

and for the upper bound of the queueing length

$$\hat{Q}^N(t) = \hat{A}^N(t) - \hat{S}^N(t) = \hat{A}^N(t) - \hat{A}^N \left(\frac{t}{\hat{t}} \right) - Ct \left(1 - \frac{1}{\hat{t}} \right)$$

$$= \int_0^t Nm + N^{\frac{1}{2}} \kappa \sigma H(s) s^{H(s)-1} ds - \int_0^{\frac{t}{\hat{t}}} Nm + N^{\frac{1}{2}} \kappa \sigma H(s) s^{H(s)-1} ds$$

$$- Ct \left(1 - \frac{1}{\hat{t}} \right)$$

As above we can compute the time $\tilde{t}^*$ of the largest queue

$$\tilde{t}^* = N^{-\frac{1}{2}} \frac{\kappa \sigma \left(H(\tilde{t}^*)(\tilde{t}^*)^{H(\tilde{t}^*)} - \hat{t} H \left(\frac{\tilde{t}^*}{\hat{t}} \right) \left(\frac{\tilde{t}^*}{\hat{t}} \right)^{H\left(\frac{\tilde{t}^*}{\hat{t}}\right)} \right)}{(C-m)\left(1 - \frac{1}{\hat{t}}\right)} = N^{-\frac{1}{2}} \tilde{t}_i^*$$

where $\tilde{t}_i^*$ is the time computed according to equation (5.29) with the normalized link capacity, i.e. $\tilde{C} = \frac{C}{N}$.

5.1 IP Traffic for Perturbation with Long-Range Dependence Processes

Calculation of the Equivalent Bandwidth

As already considered in the Norros model, we will deal with the following problem:

Suppose N flows with traffic mean m_i and variances σ_i are given. We model each of them according a mBm with Hölder exponents functions $H_i(\cdot)$. What is the needed link capacity C, so that the maximal queueing length $\hat{q}_{\max}^N$ occurs with probability at most $\epsilon \in \,]0,1[$?

The answer to this question consists in finding a C, such that

$$\max_{t>0} \hat{Q}^N(t) - \hat{q}_{\max}^N = 0$$

$$\Leftrightarrow \max_{t>0} \left(\hat{A}^N(t) - \hat{A}^N\left(\frac{t}{t}\right) - Ct\left(1 - \frac{1}{t}\right) \right) - \hat{q}_{\max}^N = 0$$

Here, we compute $\hat{Q}^N(t)$ according to (5.28).

Theorem 5.13. *The equivalent link capacity or bandwidth can be computed in at most*

$$n = O\left(\log(C)\right)$$

iterations, where C is the channel capacity.

Our next aim is to describe the multiplex gain, which will be analyzed in the sequel for different forms of Hölder function $H(\cdot)$. To do so, we define the equivalent bandwidth of the i-th flow by

$$eB_i = \frac{\int_0^{\tilde{t}_i^*} \tilde{m}_i + \kappa \sigma_i H_i(x) x^{H_i(x)-1} dx - \int_0^{\frac{\tilde{t}_i^*}{\tilde{t}_i}} \tilde{m}_i + \kappa \sigma_i H_i(x) x^{H_i(x)-1} dx - K}{\tilde{t}_i^*\left(1 - \frac{1}{\tilde{t}_i^*}\right)}$$

For the aggregated bandwidth we compute

$$eB(n) = \frac{\int_0^{\tilde{t}^{**}} \sum_{i=1}^n \tilde{m}_i + \kappa \frac{\sum_{i=1}^n \sigma_i^2 H_i(x) x^{2H_i(x)-1}}{\left(\sum_{i=1}^n \sigma_i^2 x^{2H_i(x)}\right)^{\frac{1}{2}}} dx}{\tilde{t}^{**}\left(1 - \frac{1}{\tilde{t}^{**}}\right)}$$

$$- \frac{\int_0^{\frac{\tilde{t}^{**}}{\tilde{t}^*}} \sum_{i=1}^n \tilde{m}_i + \kappa \frac{\sum_{i=1}^n \sigma_i^2 H_i(x) x^{2H_i(x)-1}}{\left(\sum_{i=1}^n \sigma_i^2 x^{2H_i(x)}\right)^{\frac{1}{2}}} dx - K'}{\tilde{t}^{**}\left(1 - \frac{1}{\tilde{t}^{**}}\right)}$$

Here, $eB(i)$ is the equivalent bandwidth of a single flow i and $eB(n)$ the aggregate flow of n flows. t_i^* is the time scale for queueing of the i-th flow, and

t^{**} of the aggregated system, respectively. The value K indicates the buffer size, while $K' = \frac{K}{n}$ is the relative buffer size for each connection.

The gain measure $G(n)$ for the multiplexing of n homogeneous flows is given by

$$G(n) = \frac{\sum_{i=1}^{n} eB_i}{eB(n)}$$

$$= \frac{\dfrac{\sum_{i=1}^{n} \int_0^{\tilde{t}_i^*} \tilde{m}_i + \kappa \sigma_i H_i(x) x^{H_i(x)-1} dx - \int_0^{\frac{\tilde{t}_i^*}{\tilde{t}_i}} \tilde{m}_i + \kappa \sigma_i H_i(x) x^{2H_i(x)-1} dx - K}{\tilde{t}_i^*\left(1-\frac{1}{\tilde{t}_i^*}\right)}}{\dfrac{\int_0^{\tilde{t}^{**}} \sum_{i=1}^{n} \tilde{m}_i + \kappa \frac{\sum_{i=1}^{n} \sigma_i^2 H_i(x) x^{2H_i(x)-1}}{\left(\sum_{i=1}^{n} \sigma_i^2 x^{2H_i(x)}\right)^{\frac{1}{2}}} dx - \int_0^{\frac{\tilde{t}^{**}}{\tilde{t}^{**}}} \sum_{i=1}^{n} \tilde{m}_i + \kappa \frac{\sum_{i=1}^{n} \sigma_i^2 H_i(x) x^{H_i(x)-1}}{\left(\sum_{i=1}^{n} \sigma_i^2 x^{2H_i(x)}\right)^{\frac{1}{2}}} dx - K'}{\tilde{t}^{**}\left(1-\frac{1}{\tilde{t}^{**}}\right)}}$$

It is computed as the ratio of n times the equivalent bandwidth of a flow and the equivalent bandwidth for the aggregated n homogeneous flows. If all flows are identical, we get a special case of the homogeneous flows

$$G(n) = \frac{n \cdot eB_1}{eB(n)}$$

$$= \frac{\dfrac{\int_0^{\tilde{t}_1^*} \tilde{m}_1 + \kappa \sigma_1 H_1(x) x^{H_1(x)-1} dx - \int_0^{\frac{\tilde{t}_1^*}{\tilde{t}_1}} \tilde{m}_1 + \kappa \sigma_1 H_1(x) x^{H_1(x)-1} dx - K}{\tilde{t}_1^*\left(1-\frac{1}{\tilde{t}_1^*}\right)}}{\dfrac{\int_0^{\tilde{t}^{**}} \tilde{m}_1 + n^{-\frac{1}{2}} \kappa \sigma_1 H_1(x) x^{H_1(x)-1} dx - \int_0^{\frac{\tilde{t}^{**}}{\tilde{t}^{**}}} \tilde{m}_1 + n^{-\frac{1}{2}} \kappa \sigma_1 H_1(x) x^{H_1(x)-1} dx - K'}{\tilde{t}^{**}\left(1-\frac{1}{\tilde{t}^{**}}\right)}}$$

Here, $eB(1)$ is the equivalent bandwidth of a single flow and $eB(n)$ the aggregate flow of n flows. The times $\tilde{t}_1$ and $\tilde{t}$ indicate the time scales of the effective bandwidth for the single and multiple connections, respectively. The value K indicates the buffer size, while $K' = \frac{K}{n}$ is the relative buffer size for each connection.

Example 5.14. We investigate a scenario of homogeneous flows ranging from $N = 1$ to 250 (see [179]). We use different envelope functions and different variances σ, which are, as the mean rate m, identical for all flows. The mean arrival rate is $m = 1,000$, and the lowest variance $\sigma = 10,000$. We consider the envelope functions $H(\cdot)$ from table 5.1.

In the figure 5.19 the multiplex gain for different variances and envelope functions are depicted. We chose for $a = 513.3$, $\sigma^2 = 14,400$, $\kappa = 12$, $C = 5,500$ and $K = 50$. In addition, we picked the first two Hölder functions in the above

Table 5.1. Envelope functions for different flows

Flow	Envelope function
1	$1.9x^2 - 1.9x^2 + 0.985$
2	$4.9x^3 - 7.9x^2 + 3.3x + 0.51$
3	$2.1x^4 + 1.1x^3 - 0.1x^2 + 0.8x + 0.51$
4	$\frac{\sin x}{10} + 0.61$
5	$0.5x + 0.5$

table and varied the variances each time for σ^2, $10\sigma^2$ and $100\sigma^2$. In all cases the higher the variance the higher the multiplex gain is.

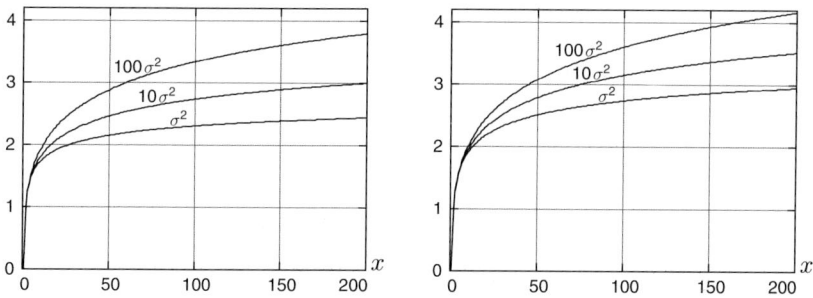

Fig. 5.19. Quadratic (left) and cubic (right) Hölder function with variances σ^2, $10\sigma^2$ and $100\sigma^2$

5.2 Queueing in Multifractal Traffic

As we did in the description of the models for IP traffic in chapter 3 we pursue an equivalent way of representation. After the view of the pure LRD phenomena and its impact on queueing we proceed to the impact on the small scales, namely how the multifractal analysis describes the queueing behavior. Though the LRD analysis does not loose its significance the small scales will perform a decisive impact on the waiting queues and its resulting impact on the QoS requirements.

5.2.1 Queueing in Multifractal Tree Models

We start with a queueing formula derived by Ribeiro, Riedi, Crouse and Baraniuk [214]. It has its origin in the classical formula for a single server due to Lindley [165]. This approach was used in the previous sections already frequently to consider the queueing behavior (see e.g. section 5.1.1, 5.1.3 resp. 5.1.4). The notation is as usual. Instead of the continuous time models, the multifractal models are constructed at discrete time spots. Thus, for $i \in \mathbb{Z}$

let $W(i)$ be the traffic per time unit that entered an infinite buffer of a single server queue with capacity or service rate $C > 0$ per time unit. As usual we denote by $Q(i)$ the queue size at time spot i. Again let $A(l)$ be the aggregated traffic between the time $-l+1$ and 0. This is expressed in

$$A(l) := \sum_{i=-l+1}^{0} W(i)$$

The traffic amount $A(l)$ is referred to the time scale r. We set as usual $A(0)=0$. By the result of Lindley we have

$$Q(0) := \max\left(Q(-1) + A(1) - C, 0\right)$$

As exercise the reader can convince herself that

$$Q(0) := \max\left(Q(-l) + A(l) - lC, A(l-1) - (l-1)C, \ldots, A(0)\right)$$

By definition it is $Q(-l) \geq 0$ for all l. This implies

$$Q(0) \geq \sup_{l \in \mathbb{N}} \left(A(l) - lC\right)$$

Suppose $-t$ was the last time spot, where the buffer was empty before time 0 ($t \geq 0$). Then we conclude

$$Q(0) \stackrel{\text{by def.}}{=} A(t) - tC \leq \sup_{l \in \mathbb{N}} \left(A(l) - lC\right)$$

So, if the queue was empty in the past at some time, we have

$$Q(0) = \sup_{l \in \mathbb{N}}(A(l) - lC) \tag{5.31}$$

For the sake of simplicity we will consider the queue at time $t = 0$. According to (5.31) we have a direct link between the queue size and the aggregated traffic $A(l)$ for multiple time scales l. This is the basic secret behind the queueing formula for multifractal traffic, since we have explicit formulas for $A(l)$. Before deriving the formulas, we make three assumptions (A1 to A3), which are crucial for the analysis, and justify them.

A1: Tail Queue Probability of Traffic at Dyadic Time Spots Captures Best the One of the Whole Traffic

Thus, we have to show that

$$\mathbb{P}\left(Q(0) > x\right)$$

is well described using the dyadic times. For a justification, we denote by

$$Q_D(0) = \sup_{m=0,\dots,n} (A(2^m) - 2^m C) \qquad (5.32)$$

The assumption A1 can be rewritten in the form

$$\mathbb{P}(Q(0) > x) \approx \mathbb{P}(Q_D(0) > x) \qquad (5.33)$$

It is obvious that $Q(0) \leq Q_D(0)$ and hence, $\mathbb{P}(Q(0) > x) \geq \mathbb{P}(Q_D(0) > x)$. So, if we want to justify (5.33) by using (5.32), we have to verify that

- only dyadic time scales are crucial and
- no time scales bigger than 2^n are required.

It can easy be assumed that there are no busy periods longer than 2^n. Hence, the time scales in question are smaller than 2^n. A little bit more work has to be done to justify that only dyadic time scales are necessary. For this we apply some results on the *critical time scale* (CTS) as done in the papers [224, 186, 100]. With the help of the CTS we will be able to give some estimates of $\mathbb{P}(Q(0) > x)$ resp. of $\mathbb{P}(Q_D(0) > x)$. The CTS is the largest scale l, so that the probability of the network traffic, exceeding the threshold x, is maximal. This reads formally

$$l^* = \arg\sup_{l \in \mathbb{N}} \mathbb{P}(A(l) - lC > x)$$

Analogously we define the *critical time scale queue* (CTSQ) as

$$CTSQ(x) = \mathbb{P}(A(l^*) - l^* C > x)$$

According to the above mentioned papers [224, 186, 100], we have

$$CTSQ(x) \approx \mathbb{P}(Q(0) > x)$$

It is evident due to the definition that

$$CTSQ(x) \leq \mathbb{P}(Q > x)$$

for all $x > 0$. As above we introduce the *critical dyadic time scale* (CDTS) according to

$$l_D^* = \arg\sup_{m=0,\dots,n} \mathbb{P}(A(2^m) - 2^m C > x)$$

and consequently the *critical dyadic time scale queue*(CDTSQ)

$$CDTSQ(x) = \mathbb{P}(A(l_D^*) - l_D^* C)$$

The CDTSQ is very suitable to substitute the CTSQ, since it provides easy computations, due to the few statistical evaluations at the dyadic time spots. We have the following chain of inequalities

$$CDTSQ(x) \leq \mathbb{P}(Q_D(0) > x) \leq \mathbb{P}(Q(0) > x)$$

for all $x > 0$.

For the numerical evidence we follow the original literature of [214]. They observe an empirical queueing situation and construct synthetic WIG and MWM traces. WIG as MWM have FGN correlation structure with a Hurst exponent of $H = 0.8$. In both situations the reveal that CDTSQ is of the same order of magnitude as the empirical trace. For the WIG model the CDTSQ is almost identical as CTSQ, while for MWM there is no formula for CTSQ. The figure 6 in [214] reveals that CDTSQ is a suitable approximation for the queue length probability $\mathbb{P}(Q > x)$. Finally, one should remark that there is no change for this result when varying the Hurst parameter.

Mathematically the integers $\mathbb{N}$ form a tiny subset within the natural numbers $\mathbb{R}$. So, we would expect (5.31) very different to its approximation (5.32). But the dyadic intervals generate all time intervals, meaning that each time spot l^* can be sandwiched between two dyadic time spots. So, close to l^* we find l_D^*. Then we can expect $CTSQ \approx CDTSQ$.

A2: Necessary Assumption for the Joint Distribution

The second assumptions concerns the modeling joint distribution: *the joint distribution* of $A(2^m)$ is determined by the right-edge node traffic $V_{m,2^m}$. It is obvious that for computing $CDTSQ(x) = \mathbb{P}(Q_D(0) > x)$ we have to know the joint distributions of $A(2^m)$. It will turn out that this distribution is given by the right node of the tree in each level. Hence, the distributions are available due to the structure of the tree models and easy to compute. First, simplifying the representation of the distribution we assert that the appropriate time instant is the very right one of the tree, i.e. at 2^j. The advantage is that at that time spot the amount $A(2^j)$ and its distribution are available and easy to determine, which in turn helps us to find the equivalent one for $Q_D(0)$. Let's do it now!

To start we consider the queueing at time spot $t = 4k + 2$ and scale $j + 2$. This is illustrated in the figure 3.21 in section 3.8. We have $A(1) = V_{j+2,4k+2}$ and $A(2) = V_{j+2,4k+1} + V_{j+2,4k+2}$. Since $A(2)$ is not detected as a node in the tree, in the model we cannot easily figure out the dependence of $A(1)$ and $A(2)$. Now let $t = 4k + 1$. Then, we have in turn $A(1) = V_{j+2,4k+1}$, $A(2) = V_{j+2,4k} + V_{j+2,4k+1} = V_{j+1,2k}$ and $A(4) = V_{j+2,4k-2} + \ldots + V_{j+2,4k+1} = V_{j+1,2k-1} + V_{j+1,2k}$. In this construction $A(1)$ and $A(2)$ are tree nodes, but not $A(4)$. Thus, to have all necessary quantities $A(1), \ldots, A(2^m)$ modeled in the tree, we have to perform the analysis at the time spot $2^n - 1$, where 2^{-n} will be the particular time unit. Because 2^n has to incorporate all necessary time spots, we need $n \geq m$. So we get

$$A(2^j) = V_{n-j, 2^{n-j}-1}, \text{ for } j = 0, \ldots, m \qquad (5.34)$$

We will mostly work with $n = m$.

A3: Approximative Independence of Large Arrivals in the Dyadic Tree

As we just saw, the traffic amount $A(1), \ldots, A(2^M)$ are not independent. But we will show in the sequel that large arrivals in the tree, i.e. the events E_i^c, where

$$E_i = \{A(2^{n-i}) < x + 2^{n-i}C\} \qquad (5.35)$$

can be assumed to be, what is called, nearly independent of each other. In fact, the complement sets E_i^c are independent. The events E_i are highly probably, since $\mathbb{P}(E_i) \approx 1$ for most i. More precisely, they converge, as proposition 5.16 will show, exponentially towards 1. Hence, if we know that E_i has occurred, this does not give us information about the other events E_j, $j \neq i$. So, E_i^c of large queue sizes gives us the required nearly independence. The next proposition reveals the recent argument more rigorously.

Proposition 5.15. *Let E_i be events of the form $\{S_i, x_i\}$ with $S_i = R_0 + \ldots + R_{i-1}$ ($1 \leq i \leq n$), where $R_0, \ldots, R_n$ are independent but arbitrary random variables. Then it holds for $1 \leq i \leq n$*

$$\mathbb{P}(E_i | E_{i-1}, \ldots, E_0) \geq \mathbb{P}(E_i) \qquad (5.36)$$

We will see that the above defined events in (5.35) are of the form required for proposition 5.15. Using (5.36), we deduce

$$\mathbb{P}(Q_D(0) > x) = 1 - \mathbb{P}(Q_D(0) < x) = 1 - \mathbb{P}\left(\bigcap_{j=1}^n E_i\right)$$

$$= 1 - \mathbb{P}(E_0) \prod_{i=1}^n \mathbb{P}(E_i | E_{i-1}, \ldots, E_0)$$

$$\leq 1 - \prod_{i=0}^n \mathbb{P}(E_i) = MSQ(x) \qquad (5.37)$$

where the last line defines the *multiscale queueing formula*. We remark that in case of independent events $E_0, \ldots, E_n$ we will obtain the equality $MSQ(x) = \mathbb{P}(Q(0)_D > x)$. The $MSQ(x)$ is a conservative approximation of the dyadic queue tail probability. In fact,

$$\mathbb{P}(Q(0) > x) \geq \mathbb{P}(Q_D(0) > x) \leq MSQ(x)$$

Dependence of the Tree Depth

Up to now we did not say anything about the coarsest and finest time scale in question. The restriction to the coarsest is given by the fact that we have to look back as far into the past, so that we find the time spot, which ensures that the queue was empty in the covered time period with probability almost 1. This largest scale will be fixed, and we look now for the finest one, what is called the *depth of modeling*.

In this respect, we have to index the used quantities with the superscript (n), as e.g. $MSQ^{(n)}$ or $A^{(n)}(2^j)$. To make things simple, we set $MSQ^{(n)} = MSQ^{(n)}(x)$ for a fixed x. With increasing n, thus the depth of the time resolution, the number of terms in the expression of $MSQ^{(n)}$ in (5.41) and (5.37) grows. Since $\mathbb{P}(E_i^{(n)})$ is naturally bounded by 1, the value $MSQ^{(n)}$ could approach easily 1 for $n \to \infty$. Can this happen? We say no and prove it now. As we know, $E_j^{(N)}$ is an event of large arrivals at level j below the tree root. If we have fixed the largest time scale, all events are obviously the same for different depth, i.e. $E_j^{(n)} = E_j^{(m)}$ as long as $n, m \geq j$. Hence,

$$1 - MSQ^{(n)} = \prod_{i=0}^{n} \mathbb{P}\left(E_i^{(n)}\right) = \prod_{i=0}^{n} \mathbb{P}\left(E_i^{(i)}\right)$$

As asserted and proved in the next proposition 5.16, we have

$$\lim_{n \to \infty} \mathbb{P}\left(E_j^{(n)}\right) = 1$$

as fast that $MSQ^{(n)}$ does not converge to 1. As a side product we will find an error bound for neglected finer time scales. For proving proposition 5.16, we need some notations and conventions. Thus, we introduce the ideal infinite-resolution of MSQ

$$MSQ^{(\infty)} = \lim_{n \to \infty} MSQ^{(n)} = 1 - \prod_{n=1}^{\infty} \mathbb{P}\left(E_n^{(n)}\right)$$

In addition, we define the *threshold scale* N. Let N be such that

$$\mathbb{P}\left(E_i^{(i)}\right) \geq 1 - 2^{-i}, \text{ for all } i \geq N \tag{5.38}$$

$$\text{and } \max_k V_{N,k} \geq x \tag{5.39}$$

Proposition 5.16. *There exists an $N \in \mathbb{N}$ such that*

$$\mathbb{P}(E_j^{(j)}) \geq 1 - 2^j$$

for all $j \geq N$ and

$$MSQ^{(\infty)} \leq MSQ^{(N)} \leq MSQ^{(\infty)} \cdot \left(1 - 2^{-N} + 2^{-N+1}\right) \tag{5.40}$$

Under the assumptions A1 to A3 we claim the following estimation for the tail queue probability

$$\mathbb{P}(Q(0) > x) \approx \mathbb{P}\left(\sup_{m=0,\ldots,n} (A(2^m) - 2^m C) < x\right)$$
$$= \mathbb{P}(A(2^m) < x + 2^m C, \ m = 0, \ldots, n)$$
$$\approx \prod_{m=0}^{n} \mathbb{P}(A(2^m) \le x + 2^m C)$$

5.2.2 Queueing Formula

According to [214] we present a *multiscale queueing formula* (MSQ) as approximation of the tail queue probability

$$MSQ(x) = 1 - \prod_{i=0}^{n} \mathbb{P}\left(A(2^{n-i}) < x + 2^{n-i} C\right) \tag{5.41}$$

The reader should be aware that other than in most queueing formulas obtained e.g. in the Norros model, here the so called *multiscale marginal distribution* of the $A(2^i)$, $i = 0, \ldots, n$, enter (5.41) and not only their variances (see e.g. [190, 56]). Numerical simulations for bursty traffic as shown in the figure 5 c) and d) in [214] reveal that

$$\mathbb{P}(Q(0) > x) \approx MSQ(x)$$

Having selected a model for describing the multifractal traffic, we only have to choose the appropriate depth of the tree for applying the formula (5.41). We will sketch the procedure for most used models.

Multiscale Queues in WIG and MWM Models

Analyzing formula (5.41), we see that we have to compute the value of

$$\mathbb{P}(E_i) = \mathbb{P}\left(A(2^{n-i}) < x + 2^{n-i} C\right)$$

This we have to derive with the help of the marginal distributions of the respective multiple time scales, represented by the tree knots i different levels. We can achieve this for the tree models, starting from $V_{0,0}$ at level 0 and the multiscale innovations $Z_{j,k}$ and $M_{j,k}$.

As usual the WIG model is the simple case to investigate. Because of its additive structure $A(2^{n-i}) = V_{i,2^i-1}$ is the sum of independent Gaussian random variables, the so called tree root variables and the independent innovations. The sum is again Gaussian with mean μ_i and variance σ_i^2. Setting μ_i and σ_i^2 the mean resp variance of the sample means resp. variance in the nodes $V_{i,k}$, we can model the tree. An alternative way is given by starting in $V_{0,0}$ and using appropriate innovations $Z_{j,k}$ according to the observed samples and compute σ_i^2 by (3.94) resp. (3.95). Suppose Φ_{μ_i,σ_i} is the distribution function of the variable $V_{i,2^i-1}$, then we obtain

$$MSQ_{\text{WIG}}(x) = 1 - \prod_{i=0}^{n} \Phi_{\mu_i,\sigma_i}\left(x + 2^{n-i}\right)$$

Deriving a similar result for the MWM model is not as straightforward. The amount $A(s^{-i})$ is a product not a sum as in the WIG model. In fact, it is the product of the tree root $V_{0,0}$ and the multiplicative innovations, hence, the product of $i+1$ independent random variables. Choosing lognormal distribution for the innovations will end up in a lognormal distribution for $A(2^{n-i})$, which is unbounded and thus certainly not bounded by 0 and 1. As mentioned in section 3.8, when treating the multifractal modeling, we found the symmetric beta distribution suitable. Fan showed (see [82, 125]) that the product of beta distributed random variables can be approximated by another beta distributed random variable with known parameters. We get for $A(2^{n-i})$ (see (3.96) and (5.34))

$$A\left(2^{n-i}\right) \sim a\beta(d_i, e_i)$$

where a is a constant, and the parameters d_i, e_i are given according to

$$d_i = \zeta \left(\theta - \zeta^2\right)^{-1} (\zeta - \theta), \quad e_i = d_i \frac{1-\zeta}{\zeta}$$

with $\zeta = 2^{-i}$ and

$$\theta = \prod_{i=-1}^{i-1} \frac{p_i + 1}{2(2p_i + 1)}$$

The above approximation due to Fan reflects exactly the mean and variance of the product of the beta random variables and approximates very well the first 10 moments of the product variable. The parameters a and p_j are gained according to the fitting procedure in section 3.8.2. If we denote by B_{M,d_i,e_i} the distribution function of the beta distribution $\beta_{0,M}(d_i, e_i)$, then we can formulate the MSQ for MWM as follows

$$MSQ_{MWM}(x) = 1 - \prod_{i=0}^{n} B_{M,d_i,e_i}\left(x + 2^{n-i}C\right)$$

Proof to Proposition 5.15

Define

$$Y_i = S_i \,|\, E_{i-1}, \ldots, E_0 \quad \text{and} \quad U_i = S_i \,|\, E_i, \ldots, E_0, \ i \geq 1$$

It suffices for proving the proposition

$$F_{Y_i}(z) \geq F_{S_i}(z) \tag{5.42}$$

for all $z \in \mathbb{R}$ and all $i \in \mathbb{N}_0$ and set in particular $z = x_i$ to get the proof. The proof of (5.42) runs by induction over i

$$F_{U_i}(z) = \begin{cases} \frac{c}{F_{Y_i}(x_i)} & \text{for } z \leq x_i \\ 1 & \text{else} \end{cases}$$

This gives immediately

$$F_{U_i}(z) \geq F_{Y_i}(z) \qquad (5.43)$$

for all z

$$\begin{aligned}
F_{Y_{i+1}}(z) &= \mathbb{P}(U_i + R_{i+1} < z) \\
&= \int_{-\infty}^{\infty} \int_{-\infty}^{z-z_{i+1}} f_{U_i}(u_i) f_{R_{i+1}}(z_{i+1}) du_i dz_{i+1} \\
&= \int_{-\infty}^{\infty} F_{U_i}(z - z_{i+1}) f_{R_{i+1}}(z_{i+1}) dz_{i+1} \\
&\overset{\text{by (5.43)}}{\geq} \int_{-\infty}^{\infty} F_{Y_i}(z - z_{i+1}) f_{R_{i+1}}(z_{i+1}) dz_{i+1} \\
&\overset{\text{by induction}}{\geq} \int_{-\infty}^{\infty} F_{S_i}(z - z_{i+1}) f_{R_{i+1}}(z_{i+1}) dz_{i+1} \\
&= \mathbb{P}(S_i + R_{i+1} < z) \\
&= F_{S_{i+1}}(z)
\end{aligned}$$

We first want to apply proposition 5.15 to the WIG model. From the structure of the WIG model we obtain (see (3.93) in section 3.8.2)

$$A(2^{-i}) = V_{i,2^i-1} = 2^{-i} V_{0,0} - \sum_{j=0}^{i-1} 2^{j-i} Z_{j,2^j-1}$$

where the $Z_{j,k}$ are the innovations in the building block of the tree. Now, we only have to set $x_i = 2^i x + 2^n C$, $R_0 = V_{0,0}$ and $R_i = -2^{i-1} Z_{i-1,2^{i-1}-1}$. Now we apply proposition 5.15 to the MWM model. According to 3.96

$$A(2^{n-i}) = V_{i,2^i-1} = V_{0,0} \prod_{j=0}^{i-1}(1 - M_j)$$

where the M_j are the multiplicative innovations of the tree. If we apply on both sides the logarithms, we obtain the desired form, setting $x_i = \log(x + 2^{n-i}C)$, $R_0 = \log(V_{0,0})$ and $R_i = \log(1 - M_{i-1})$.

Proof to Proposition 5.16

First, we will proof the existence of the threshold scale given in (5.38). All events $E_n^{(n)}$ depend on the buffer size $x^{(n)}$, the link capacity $C^{(n)}$ at time scale 2^{-n} and the arriving workload $A^{(n)}(2^{n-i})$. Since the buffer works independent of the time scale, we have $x^{(n)} = x$. In addition, the link capacity fulfills

$C^{(n)} = 2^{-n}\overline{C}$, where $\overline{C}$ is the maximal size of bytes, which can be emptied from the queue on the coarsest level. Finally, $A^{(n)}(2^{n-i}) = V^{(n)}_{i,2^i-1}$ according to (3.96) and (3.97). Thus, $A^{(n)}(2^{n-i})$ equals in distribution the random variable $aM_{-1}\prod_{j=0}^{i-1}(1-M_j)$. We do not have to indicate the depth of the tree, since the tree grows downwards with increasing n. The multiplier at level i does not change – new ones are added successively. Since the M_i are symmetrical, we obtain

$$\mathbb{P}\left(E_n^{(n)}\right) = \mathbb{P}\left(aM_1\ldots M_{n-1} < \overline{C}2^{-n}\right)$$

We abbreviate $D_i = \log_2(M_i)$ and

$$\alpha_n = -\frac{1}{n}\log_2\left(\frac{x}{a} + \overline{C}\frac{2^{-n+1}}{a}\right) \tag{5.44}$$

For all $q > 0$ we apply the Jensen inequality and the independence of the multipliers and receive the Chernoff bound

$$1 - \mathbb{P}\left(E_n^{(n)}\right) = \mathbb{P}\left(-\frac{1}{n}(D_{-1} + \ldots + D_{n-1}) < \alpha_n\right)$$
$$= \mathbb{P}\left(2^{q(D_{-1}+\ldots+D_{n-1})} > 2^{-qn\alpha_n}\right)$$
$$\leq \frac{\mathbb{E}\left(2^{q(D_{-1}+\ldots+D_{n-1})}\right)}{2^{-qn\alpha_n}}$$
$$= 2^{n(-T^{(n)}(q)-1+q\alpha_n)}$$

According to (3.103), we have for the deterministic envelope function (see definition 3.92) at level n

$$T^{(n)}(q) = -1 - \frac{1}{n}\sum_{i=-1}^{n-1}\log_2 \mathbb{E}\left(M_i^q\right)$$

We apply the logarithm and minimize for $q > 0$, which implies

$$\frac{1}{n}\log_2\left(1 - \mathbb{P}\left(E_n^{(n)}\right)\right) \leq \inf_{q>0}\left(q\alpha_n - T^{(n)}(q)\right) - 1$$
$$= -1 + \left(T^{(n)}\right)^*(\alpha_n) \leq -1$$

provided that α_n is so small that $\left(T^{(n)}\right)^*(\alpha_n) < 1$, where $\left(T^{(n)}\right)^*$ is the Legendre transform (see 3.101 and 3.119 for the definition of the Legendre transform). We remind for the concave shape of $T^{(n)}$, where negative input values remain positive, but smaller than 1. In addition, α_n decreases towards 0 and $T^{(n)}$ converges to T, implying that the 0 of $\left(T^{(n)}\right)^*$ do not change much, once n is sufficiently large. In this respect, w.l.o.g. we can assume that $\left(T^{(n)}\right)^*(\alpha_n)$ is negative for all $n \geq N$ for a critical N. Putting things together now, we obtain for $n \geq N$ that

5.2 Queueing in Multifractal Traffic

$$1 - \mathbb{P}\left(E_n^{(n)}\right) \leq 2^{-n} \leq 2^{-N}$$

which gives us finally (5.38). We have

$$\log_2\left(\mathbb{P}\left(E_n^{(n)}\right)\right) \geq -b_0\left(1 - \mathbb{P}\left(E_n^{(n)}\right)\right)$$

forcing for b_0 the definition

$$b_0 = \frac{\log_2\left(1 - 2^{-N}\right)}{-2^{-N}}$$

for all $n \geq N$. This implies

$$\log_2\left(\prod_{n=N}^{N'} \mathbb{P}\left(E_n^{(n)}\right)\right) \geq -b_0 \sum_{n=N}^{N'} \left(1 - \mathbb{P}\left(E_n^{(n)}\right)\right)$$

$$\geq -b_0 \sum_{n=N}^{N'} 2^{-n} \geq -b_0 \sum_{n=N}^{\infty} 2^{-n}$$

$$\geq -b_0 2^{-N+1}$$

Estimating the neglected terms of $MSQ^{(N)}$ by

$$1 \geq \prod_{n=N}^{\infty} \mathbb{P}\left(E_n^{(n)}\right) \geq 2^{-b_0 2^{-N+1}}$$

we conclude

$$\begin{aligned}1 - MSQ^{(N)} &\geq 1 - MSQ^{(\infty)} \\ &\geq \left(1 - MSQ^{(N)}\right) 2^{-b_0 2^{-N+1}} \\ &\geq \left(1 - MSQ^{(N)}\right)(1 - 2{-N})^2 \end{aligned} \quad (5.45)$$

Noting that (5.40) is equivalent to (5.45), we cheer for finishing the proof. How can we find the necessary tree depth N in practice? The intensively investigated multifractal formalism helps: Choosing in (3.119) $\left(T^{(n)}\right)^*(\alpha_n) < 0$, this implies that there is no exponent α_n, since $N(\alpha_n) = 0$. The definition of the coarse Hölder exponent tells us that at scale n all $\alpha(t) = -\frac{1}{j}\log_2|V_{n,k}| \geq \alpha_n$. With the help of (5.44) we get consequently $x + 2^{-n+1}\overline{C} \geq |V_{n,k}|$. For a conservative estimation we choose $\left(T^{(n)}\right)^*(\alpha_n) < 0$ for $n \geq N$, which can easily be checked, and impose as practicable requirement

$$x \geq |V_{N,k}|$$

matching with (5.39).

Impact of the Multiscale Marginals on Queueing

Usually the queueing analysis is based for IP traffic on the LRD property as done in the previous sections (see 5.1.1,5.1.3 and 5.1.4). LRD is crucial, especially, when the traffic is modeled by FBM (resp. FGN). This is motivated by the analysis of asymptotic self-similarity, obtained by the second order statistics. In the models based on the Norros approach (the FBM as well as the FLM) we used infinite buffer (though 'x' serves as the QoS threshold) and gained the estimates

$$\mathbb{P}(Q(0) > x) \simeq \exp\left(-\gamma x^{2-2H}\right) \tag{5.46}$$

for a FBM perturbation

$$\mathbb{P}(Q > x) \geq \sup_{t \geq 0} \Phi^c \left(\frac{x + Ct - mt}{\sqrt{m\mathbb{V}\mathrm{ar}(t)}}\right) \tag{5.47}$$

for the multiscaling FBM

$$\hat{Q}(t) = \int_0^t m + \kappa\sqrt{am}H(x)x^{H(x)-1}dx - \int_0^{\frac{t}{t^*}} m + \kappa\sqrt{am}H(x)x^{H(x)-1}dx$$
$$- \int_0^{\frac{t}{t^*}} m + \kappa\sqrt{am}H(x)x^{H(x)-1}dx - Ct\left(1 - \frac{1}{t^*}\right)$$

for the multifractal FBM and finally

$$\epsilon = \mathbb{P}\left(Q(0,C) > x\right) \geq \Theta_\alpha x^{-\alpha\left(\frac{3}{2}-H-\frac{1}{\alpha}\right)} \tag{5.48}$$

for the FLM. In the asymptotic (5.46) the parameter γ (as the other constants in the equations (5.47) to (5.48)) depends on certain traffic parameters like C, m or the QoS requirement x. Sticking to the first asymptotic (5.46) (the other are more complicated), we realize that the queueing is asymptotically according to Weibullian law (recall that it is a lower bound), in contrast to the SRD expressed by the exponential function in classical traffic models (when $H = \frac{1}{2}$). If x is small, we do have as depicted in several figures a not suitable description. The LRD captures only the asymptotic variance of the traffic.

In the multifractal queueing formula we incorporate according to [214] the marginal distribution in contrast to the asymptotic formulas, using second order statistics. Traffic characteristics at CTS impacts, as recent investigations revealed, more than the Hurst parameter H on small time scales. So, at any time scale the variance of traffic affects $\mathbb{P}(Q > x)$ more than H.

The multifractal queueing formula reveals that traffic with heavier tail implies that $\mathbb{P}(A(2^i), x + C2^i)$ is smaller, thus, MSQ is larger. Since the models of MWM perform heavier tails than the WIG, consequently, the MSQ is larger for MWM. But our intuition emphasizes this as well, because large bursts for different times naturally lead to larger queues. The WIG model is more closely to the Gaussian traffic model with its averaging traffic assumption. Finally, the MSQ can express, why LRD leads to larger queues than SRD (with identical m and σ): LRD has heavier tailed marginal distributions at multiple time scales than SRD.

5.3 Traffic Optimization

In this last section we briefly give some insight in the theory of optimization for network design. We select some approaches examplarily of this topic, attacking mainly two problems. The first concerns the intrinsic network optimization using key values like queueing and service capacity. In a second step we turn to the economic aspect and optimize towards an equilibrium. In the latter case, the basic approach is based on the utility, both for the user and the network. Fundamental for the investigation is a model introduced by Kelly, Maulhoo and Tan (see [138]). There, the optimization is derived by a *deterministic* equation and stochastic equations, which incorporates small stochastic perturbation, given by short-range dependent processes, as Poisson or standard Brownian motion. But, this model will be our starting point to derive a result for the equilibrium, using the optimal control technique for a *stochastic* differential equation, derived for the transmission rate. Beside a deterministic part we will use, both a short-range dependence contribution by a Poisson perturbation and the LRD phenomena by the FBM. The technique could be transferred for other stochastic processes, for which a stochastic integration theory as for Brownian resp. fractional Brownian motion is available.

5.3.1 Mixed Traffic

Up to now we considered mainly a modeling with long-range dependent processes, especially the FBM. As we know the LAN traffic investigated and documented by the well known Bellcore experiment and its subsequent modeling by Norros using the FBM, data traffic is well modeled by the long range dependent processes, looking at large scale behavior. In 3.4.4 we showed, how the simple on-off models lead in the limit to a perturbation by FBM or special fractional Lévy processes. In practice, traffic is a mixture of several different kinds. Especially VoIP is becoming more and more important. A separated line for VoIP would keep everything to the standard Erlang formulas, while a pure data line is represented by the long-range dependent processes. In reality we prefer considering a mixture or superstition of both models. Thus, we keep to the approach of Norros and insert the Poisson process for the 'voice' part. If A_t is the whole traffic amount in connection during the interval $]-\infty, t]$, (N_t) represents homogeneous Poisson process with intensity λ, $(B_t^{(H)})$ a FBM with Hurst parameter H and scaling constants m_1, m_2, m_3 and σ, then we use the approach

$$A_t = m_1 t + m_2 N_t + \sqrt{m_3} \sigma B_t^{(H)}$$

Since we are concerned with time sensitive traffic we consider the following topics:

- Queueing theory and estimation of threshold probability
- Optimization of certain network situations

As mentioned above we consider a traffic mix of voice and data. The major assumption is that the voice traffic part is independent of the data stream. This seems to be realistic, since the telephone behavior of the user as well as data transmission are not linked in any particular way.

5.3.2 Optimization of Network Flows

We follow basically the standard idea:

Suppose Q_t represents the queueing line at a server or router. The amount of traffic in an interval $[s,t]$, $s < t$ can be represented by

$$A_{s,t} = A_t - A_s$$

If $C > 0$ denotes the capacity of the router or server, then we have for the net amount of the traffic (i.e. the queue length)

$$\tilde{Q}_{s,t} = (A_t - A_s) - C(t-s)$$

Since queues can build up, we have finally

$$Q_t = \sup_{s \leq t} \tilde{Q}_{s,t} = \sup_{s \leq t} \left((A_t - A_s) - C(t-s) \right)$$

Up to now we investigated and described the traffic and its key values like queueing behavior with the help of long-range dependence processes. In the next two sections we consider an important question for network designers or maintainers: *optimization*. Certainly, we can only present a short insight in this vast area. We start with a simple example of N buffers or servers, which are linked together in a network described in the figure 5.20.

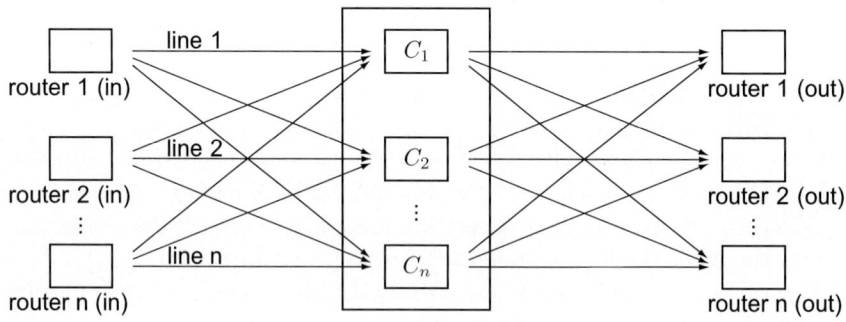

Fig. 5.20. Mesh of servers an routers

5.3 Traffic Optimization

Each router i, $i = 1, \ldots, N$, is connected with another connection determined by the pair of buffer (server) j, $j = 1, \ldots, N$. At the beginning of the network the incoming traffic is splitted by the router $\mathcal{R}$ into the N connections. We denote the accumulated traffic amount at the beginning (before $\mathcal{R}$) by A_t and the respective amount at the buffer i by $A_t^{(i)}$. Thus,

$$\sum_{i=1}^{N} A_t^{(i)} = A_t$$

Each buffer (server) i has a capacity of C_i, $i = 1, \ldots, N$. By using capacity of others buffers j, $j = 1, \ldots, N$, $j \neq i$ we can distribute the traffic amount using by partitioning the traffic according to $q_{i,j}(t) \in [0,1]$, meaning shuffling capacity of j to i. Thus, for the connection i we have a new capacity of

$$C_i(t) = C_i \left(1 - \sum_{j=1, j \neq i}^{N} q_{j,i}(t) \right) + \sum_{j=1, j \neq i}^{N} q_{i,j}(t) C_j$$

This gives restriction to the fractions $q_{i,j}(t)$

$$\sum_{i=1, i \neq j}^{N} q_{i,j}(t) = 1, \text{ for all times } t \text{ and } j = 1, \ldots, N$$

and

$$C = \sum_{i=1}^{N} C_i(t), \text{ for all times } t$$

where C is a fixed given capacity of the system. The optimization problem consists in finding the optimal fractions $q_{i,j}(t)$, such that the over all queue is minimized. We denote by $Q_i(t)$ the length of the queue at the buffer i at time t, i.e.

$$Q_i(t) = \sup_{s \leq t} \left(\left(A_t^{(i)} - A_s^{(i)} \right) - \int_s^t C_i(u) du \right)$$

Our first aim is to investigate, whether the process $(Q_i(t))$ is a kind of self-similar. We remind to the result of Norros [190], where $C_i(u) = C = \text{const}$. We assume that $C_i(\cdot)$ is continuous with a self-similar integral, i.e.

$$\text{There exists a } \tilde{H}_i \geq 0 : \int_{\alpha s}^{\alpha t} C_i(u) du = \alpha^{\tilde{H}_i} \int_s^t C_i(u) du$$

for all $i = 1, \ldots, N$. Examples of self-similar integrals are functions $C_i(u) = u^\theta$ with $\theta \geq 0$. We get the following theorem, inspired by [190, theorem 3.1].

Theorem 5.17. *Let m, H and a be given according to the Norros model (3.38). Let the capacity function $C_i(u)$ have self-similar integral w.r.t. H_i. Then for any $\alpha > 0$ the process $Q(\alpha t)$ is distributed like α^H times the corresponding process $\tilde{Q}$ with the same FBM but capacity function $C_i(u) \frac{\alpha^{\tilde{H}_i}}{\alpha^H} + m - \alpha^{1-H} m$.*

Proof. Since the proof is short and gives some insight, we transfer the Norros proof for our means. We easily conclude

$$Q(\alpha t)$$
$$= \sup_{s \leq t} \left(A(\alpha t) - A(\alpha s) - \int_{\alpha s}^{\alpha t} C_i(u) du \right)$$
$$= \sup_{s \leq t} \left(m\alpha(t-s) + \sqrt{ma} \left(B_{\alpha t}^{(H)} - B_{\alpha s}^{(H)} - \alpha^{\tilde{H}_i} \int_s^t C_i(u) du \right) \right)$$
$$\stackrel{d}{=} \alpha^H \sup_{s \leq t} \left(m\alpha^{1-H}(t-s) + \sqrt{ma} \left(B_t^{(H)} - B_s^{(H)} - \frac{\alpha^{\tilde{H}_i}}{\alpha^H} \int_s^t C_i(u) du \right) \right)$$
$$= \alpha^H \sup_{s \leq t} \left(A(t) - A(s) - \int_s^t \left(\frac{\alpha^{\tilde{H}_i}}{\alpha^H} C_i(u) + \alpha^{1-H} \right) du \right)$$

□

The process $(Q(t))$ is stationary, provided the capacity function is constant. Since we are only interested in stationary scenarios, we will assume that, after a certain time, the function, representing the assigned new capacity, $C_i(\cdot) = C_i^*$ being constant. For our needs we define certain constants, emerging from the general model due to Norros (see section 3.3.4). We assume for each user the model

$$A_t^{(i)} = m_i t + \sqrt{a_i m_i} B_t^{(H)}$$

with independent copies of FBM w.r.t. the same Hurst exponent to reflect a similar connection behavior as well as to keep the investigation as simple as possible. The mean traffic amount as well as the 'variance' parameter a_i are specific selected for user. In contrast the Hurst exponent is equal for all users to express the over all property of the considered network. In this respect we define

$$D_i = a_i^{-\frac{1}{2H}} m_i^{-\frac{1}{2H}}$$

and the function

$$g(y) = \mathbb{P}\left(\sup_{t \geq 0} \left(B_t^{(H)} - yt \right) \leq 1 \right)$$

for $i = 1, \ldots, N$. We can investigate the following four problems.

Problem 1: Minimizing the Maximal Queueing Probability

Suppose a maximal queue length x is given. Then, depending on t, we try to minimize the probability

$$\mathbb{P}\left(\max_{i=1,\ldots,N} Q_i(t) > x \right) \longrightarrow \min \tag{5.49}$$

This means, finding optimal control processes $(q_{i,j}(u))$, is minimizing (5.49) (remember that they are contained in the expression of the $(Q_t^{(i)})$). We can formulate the following theorem.

Theorem 5.18. *Consider the above described network. Then the optimal selection for the capacity distribution is given by the vector*

$$q^* = (q_{i,j}^*, \ i,j = 1, \ldots, N, \ i \neq j)$$

provided for

$$y_i^* = D_i x^{-1+\frac{1}{H}} \left(C_i \left(1 - \sum_{j=1, j \neq i}^N q_{j,i}^* \right) + \sum_{j=1, j \neq i}^N C_j q_{i,j}^* \right)$$

we have the relation for all $i,j = 1, \ldots, N$, $i \neq j$

$$\frac{D_i}{D_j} = \left(\frac{a_j m_j}{a_i m_i} \right)^{\frac{1}{2H}} = \frac{g'(y_j^*)}{g'(y_i^*)} \cdot \frac{g(y_i^*)}{g(y_j^*)}$$

where g' denote the derivative of g.

Proof. We follow basically the proof of Norros in [190]. For this recall that for all $i = 1, \ldots, N$ the stationary process $(Q_t^{(i)})$ is symmetric, i.e. $Q_0^{(i)} = \sup_{t \leq 0}(A(t) - C_i)$ is distributed as $\sup_{t \geq 0}(A(t) - C_i)$ because of the symmetry of the FBM. Thus, we define

$$p(x, \beta) = \mathbb{P}\left(\sup_{t \geq 0} \left(B(t)^{(H)} - \beta t \right) \leq x \right)$$

where we write $B(t)^{(H)}$ for the FBM with Hurst parameter H. The self-similarity of the FBM implies

$$p(\alpha x, \beta) = \mathbb{P}\left(\sup_{t \geq 0} \left(B\left(\frac{t}{\alpha^{\frac{1}{H}}} \right) - \frac{\beta}{\alpha} t \right) \leq x \right) = p\left(x, \alpha^{\frac{H}{1-H}} \beta \right)$$

and finally

$$p(x, \beta) = p\left(1, x^{\frac{1-H}{H}} \beta \right) = g\left(x^{\frac{1-H}{H}} \beta \right)$$

with the function g as defined above. Obviously, the function g is increasing for $y \geq 0$ with $g(0) = 0$ and $\lim_{y \to \infty} g(y) = 1$. We can conclude

$$\mathbb{P}\left(\max_{1 \leq i \leq N} Q_0^{(i)} > x \right) \tag{5.50}$$

$$= 1 - \mathbb{P}\left(\max_{1 \leq i \leq N} Q_t^{(i)} \leq x \right)$$

$$= 1 - \mathbb{P}\left(\max_{1 \leq i \leq N} \left(\sup_{t \geq 0} B(t)^{(H)} - \frac{C_i^* - m_i}{\sqrt{a_i m_i}} t \right) \leq \frac{x}{\sqrt{a_i m_i}} \right)$$

With the help of the independence of the copies of the FBM this results into

$$\mathbb{P}\left(\max_{1\leq i\leq N} Q_0^{(i)} > x\right) = 1 - \prod_{i=1}^{N} g\left(\left(\frac{x}{\sqrt{a_i m_i}}\right)^{\frac{1-H}{H}} \cdot \frac{C_i^* - m_i}{\sqrt{a_i m_i}}\right)$$

Hence, we have to optimize the expression

$$q \longmapsto \prod_{i=1}^{N} g_i(q)$$

where

$$g_i(q) = g\left(D_i x^{-1+\frac{1}{H}} \left(C_i \left(1 - \sum_{\substack{j=1 \\ j\neq i}}^{N} q_{j,i}\right) + \sum_{\substack{j=1 \\ j\neq i}}^{N} C_j q_{i,j} - m_i\right)\right)$$

and $q = (q_{i,j}, \ i,j = 1,\ldots,N, \ i \neq j)$. We partially differentiate w.r.t. to $q_{i,j}, \ i \neq j$ and obtain

$$\frac{\partial (\prod_{i=k}^{N} g_k)}{\partial q_{i,j}} = \prod_{\substack{k=1 \\ i\neq k\neq j}}^{N} g_k(q) \frac{\partial (g_i \cdot g_j)}{\partial q_{i,j}} \qquad (5.51)$$

$$= \prod_{\substack{k=1 \\ i\neq k\neq j}}^{N} g_k(q) \left(D_i C_j x^{-1+\frac{1}{H}} g_i'(q) g_j(q) - D_j C_j x^{-1+\frac{1}{H}} g_j'(q) g_i(q)\right)$$

$$\frac{\partial (\prod_{i=1}^{N} g_i)}{\partial q_{j,i}} = \prod_{\substack{k=1 \\ i\neq k\neq j}}^{N} g_k(q) \frac{\partial (g_i \cdot g_j)}{\partial q_{j,i}} \qquad (5.52)$$

$$= \prod_{\substack{k=1 \\ i\neq k\neq j}}^{N} g_k(q) \left(D_j C_i x^{-1+\frac{1}{H}} g_j'(q) g_i(q) - D_i C_i x^{-1+\frac{1}{H}} g_i'(q) g_j(q)\right)$$

Setting both equations (5.51) and (5.52) to 0 and dividing both by the term $\prod_{k=1, i\neq k\neq j}^{N} g_k(q)$ gives the assertion by suitable simplification. □

Our next step is to apply the above result under a certain assumption concerning the functions g, i.e. we use the lower bound and its approximation to derive a suitable expression, which could be easily computed, in particular evaluated by available programmes. We shortly outline the approach. For this we return to equation (5.8), which tells us that

$$\mathbb{P}(A_t^{(i)} - C^*t > x) \sim \exp\left(-\frac{(C_i^* - m_i)^{2H}}{2H^{2H}(1-H)^{2-2H} a_i m_i} x^{2-2H}\right)$$

We will use a similar approximation for the function $g(y)$

$$1 - g(y) \leq 1 - \max_{t \geq 0} \Phi^c\left(\frac{y+1}{t^H}\right) \sim 1 - \exp\left(-\frac{^2 y^{2H}(1-H)^{2H}}{2H^{2H}}\right)$$

Hence, it seems suitable for computation to assume that

$$g(y) = 1 - \exp\left(-\frac{(y^{2H}(1-H)^{2H}}{2H^{2H}}\right)$$

If we apply this to theorem 5.18 we get a condition for the optimal capacity distribution

$$\left(\frac{a_j m_j}{a_i m_i}\right)^{\frac{1}{2H}} = \frac{2H(y_j^*)^{2H-1}}{2H(y_i^*)^{2H-1}} \cdot \exp\left(-((y_j^*)^{2H} - (y_i^*)^{2H})\frac{(1-H)^{2H}}{2H^{2H}}\right)$$

$$\cdot \left(\frac{1 - \exp\left(-\frac{(y_i^*)^{2H}(1-H)^{2H}}{2H^{2H}}\right)}{1 - \exp\left(-\frac{(y_j^*)^{2H}(1-H)^{2H}}{2H^{2H}}\right)}\right)$$

$$= \exp\left(-((y_j^*)^{2H} - (y_i^*)^{2H})\frac{(1-H)^{2H}}{2H^{2H}}\right)$$

$$\cdot \left(\frac{1 - \exp\left(-\frac{(y_i^*)^{2H}(1-H)^{2H}}{2H^{2H}}\right)}{1 - \exp\left(-\frac{(y_j^*)^{2H}(1-H)^{2H}}{2H^{2H}}\right)}\right) \left(\frac{y_j^*}{y_i^*}\right)^{2H}$$

for $i, j = 1, \ldots, N$, $i \neq j$ since the derivative of $\exp(\cdot)$ is again $\exp(\cdot)$.

Example 5.19. We start with the example of two lines in figure 5.21. For the Norros model, we choose $a_1 = 2.8$, $m_1 = 22$, $a_2 = 1.5$, $m_2 = 18$ and $C = 500$, as initial service rate capacity $C_1 = C_2 = 250$ and $H = 0.7$.

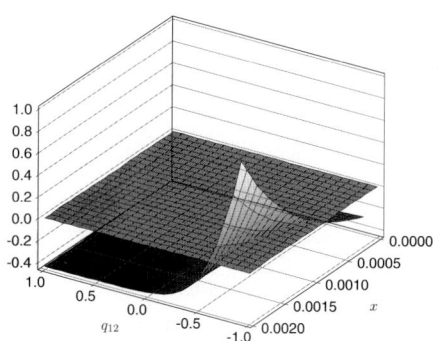

Fig. 5.21. The intersection between the horizontal flat and the manifold gives the optimal allocation

The two manifolds in figure 5.21 express the necessary service rate C_1 and C_2, depending on the buffer capacity x and the allocation q_{12}. The intersection line of both manifolds gives the optimal allocation in dependence of the values of x.

Figures 5.22 and 5.23 show the line of optimal allocations for different values of Hurst parameter H. Negative values mean more capacity contribution for line 1.

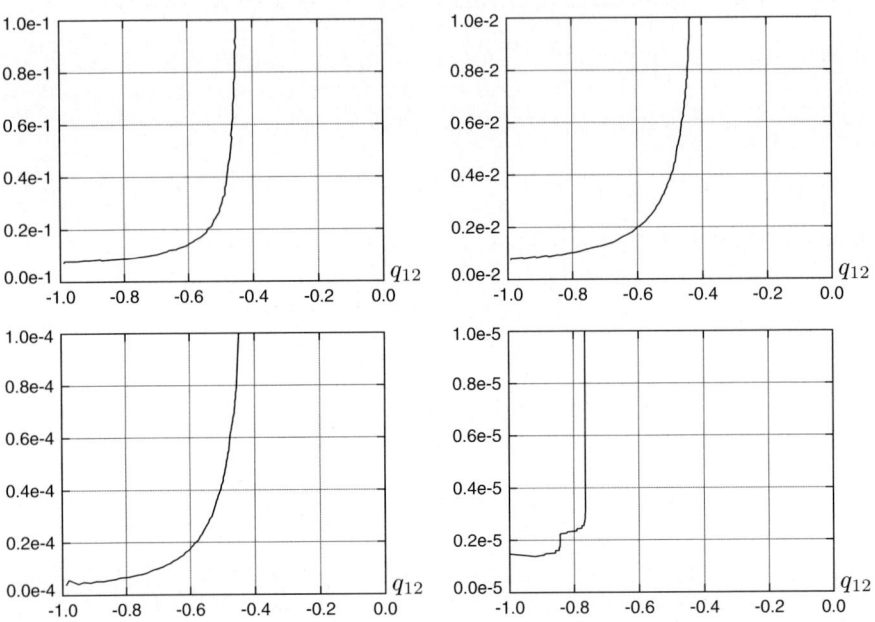

Fig. 5.22. The optimal allocation q_{12} for traffic with $H = 0.6$ (upper left), 0.7 (upper right), 0.8 (lower left), and 0.9 (lower right)

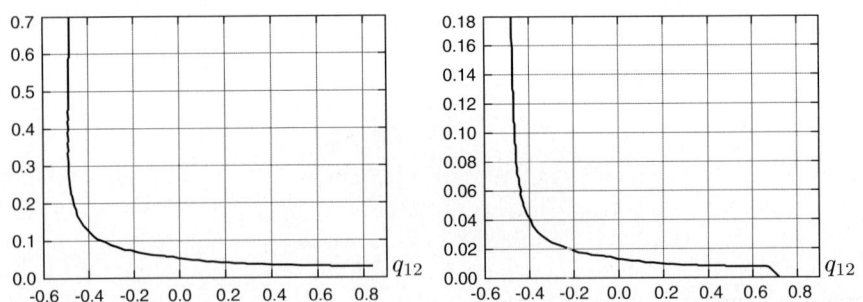

Fig. 5.23. The optimal allocation q_{12} for traffic with $H = 0.51$ (left) and 0.55 (right)

The first observation shows that with increasing value of H the buffer value x falls very fast, where the queueing probability surpasses the threshold ϵ. On the other hand, in all situations of Hurst exponent H low values of the buffer x requires more capacity given to the line 1 (expressed by the decreasing values of z_1 in the figures). This indicates that the more volatile traffic in line 1 (higher values of a) needs more capacity to match the required QoS.

Problem 2: Minimizing the Single Queueing Probabilities

If we consider different traffic requirements in different subnets, i.e. for different i, $i = 1, \ldots, N$, then (5.49) turns into: Given maximal queue lengths x_i for $i = 1, \ldots, N$ find the optimal control processes $(q_{i,j}(u))$, such that

$$\mathbb{P}(Q_i(t) > x_i) \longrightarrow \min, \text{ for all } i = 1, \ldots, N \qquad (5.53)$$

We have to remark that minimizing (5.53) indicates minimizing

$$1 - g\left(\left(\frac{x}{\sqrt{a_i m_i}}\right)^{\frac{1-H}{H}} \cdot \frac{C_i^* - m_i}{\sqrt{a_i m_i}}\right)$$

for all $i = 1, \ldots, N$ according to (5.50). If we define

$$\epsilon_i = 1 - g\left(\left(\frac{x}{\sqrt{a_i m_i}}\right)^{\frac{1-H}{H}} \cdot \frac{C_i^* - m_i}{\sqrt{a_i m_i}}\right)$$

then we have to find minimal ϵ_i. If we want to minimize on a fair level, i.e. if we choose all minimal ϵ_i equal, then we have, since g is strictly increasing, that all arguments of g are equal, i.e.

$$\left(\frac{x}{\sqrt{a_i m_i}}\right)^{\frac{1-H}{H}} \cdot \frac{C_i^* - m_i}{\sqrt{a_i m_i}} = \left(\frac{x}{\sqrt{a_j m_j}}\right)^{\frac{1-H}{H}} \cdot \frac{C_j^* - m_j}{\sqrt{a_j m_j}}$$

$$\Leftrightarrow \left(\frac{1}{\sqrt{a_i m_i}}\right)^{\frac{1-H}{H}} \cdot \frac{C_i^* - m_i}{\sqrt{a_i m_i}} = \left(\frac{1}{\sqrt{a_j m_j}}\right)^{\frac{1-H}{H}} \cdot \frac{C_j^* - m_j}{\sqrt{a_j m_j}} \qquad (5.54)$$

$$\Leftrightarrow \left(\frac{a_j m_j}{a_i m_i}\right)^{\frac{1}{2H}} = \frac{C_j^* - m_j}{C_i^* - m_i}$$

Example 5.20. Again we first illustrate with figure 5.24 the two line example. We choose for the Norros model $a_1 = 2.8$, $m_1 = 22$, $a_2 = 1.5$, $m_2 = 18$, and $C = 500$. The initial service rate capacity is set to $C_1 = C_2 = 250$. Both diagrams indicate at $H = 0.5$ the quotient $C_1/C_2 \approx 3.0$. With increasing Hurst parameter H we see that C_1 decreases and C_2 increases as expected. However, for $H = 0.9$ the relation of the service rate allocation is still $C_1/C_2 \approx 1.8$.

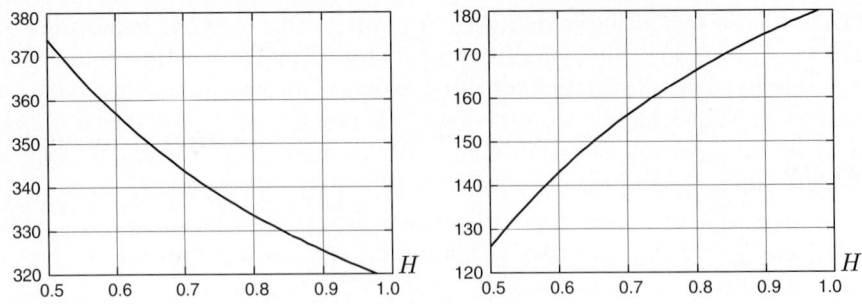

Fig. 5.24. The dependence of the optimal service rate allocation C_1 (left) and C_2 (right) for Hurst exponents $H = 0.5$ to 1.0

Example 5.21. In this example we proceed with the next step of three lines being connected. Here let $a_1 = 2.8$, $m_1 = 22$, $a_2 = 1.5$, $m_2 = 18$, $a_3 = 1.4$, $m_3 = 15$, and $C = 900$. Again the initial service rate are equal $C_1 = C_2 = C_3 = 300$. We consider two values of $H = 0.7$ and $H = 0.8$. In the figures 5.25 we see the two optimal allocations for $C2$, $C3$ and on the vertical line $C1$.

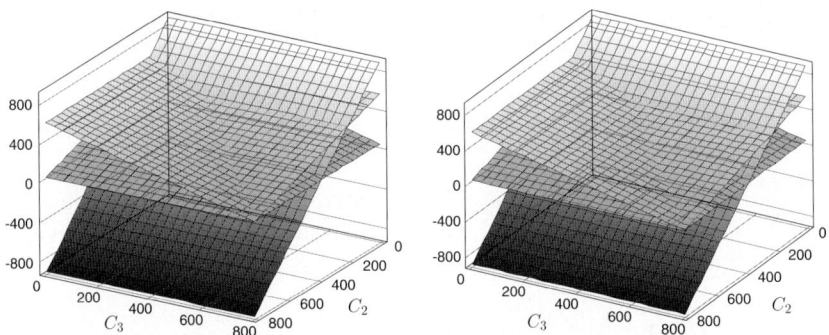

Fig. 5.25. The three manifolds intersect in the optimal point resp. optimal allocation of the three service rates C_1 (vertical line) and C_2 and C_3 in the case of $H = 0.7$ (left) and $H = 0.8$ (right)

The optimal selection of the service rates in case of $H = 0.7$ are $C_1 = 364.3$, $C_2 = 201.1$ and $C_3 = 253.3$. In case of $H = 0.8$ we get $C_1 = 427.5$, $C_2 = 212.3$ and $C_3 = 260.2$.

With this example we realize that in the Norros model an increasing Hurst parameter implies higher service rates for lower values of a and m.

Problem 3: Minimizing the Average Queueing Probability

If we allow an overflow for some subnet i and not in the average, we consider the optimal control problem: Find optimal processes $(q_{i,j}(u))$, such that

$$\mathbb{P}\left(\left(\frac{1}{N}\sum_{i=1}^{N}Q_i(t)\right) > x\right) \longrightarrow \min$$

Problem 4: Minimizing the Expected Weighted Maximal Queueing Length

Up to now we had a *local* view, considering everything for a given t. Now we look *globally*, i.e. over the whole time range. Given a maximal queue length x and a preference discount rate of $\delta > 0$ we find optimal processes $(q_{i,j}(u))$, such that

$$\mathbb{E}\left(\int_0^\infty \exp(-\delta u)\max_{i=1,\ldots,N}Q_i(u)du\right) \longrightarrow \min \quad (5.55)$$

The idea is originated from economics, where the expression in brackets represents the utility over time, thus (5.55) is minimizing the expected 'utility'. The discount factor '$\exp(-\delta u)$' evaluates future events less important than earlier ones. In our context we used a so called risk neutral utility function $u(z) = z$ to 'measure' the maximal queue length $\max_{i=1,\ldots,N} Q_i(u)$ (i.e. not a concave function $u(z) = \frac{z^\gamma}{\gamma}$ with $\gamma \in\,]-\infty,1[$ as usual in economics).

Remark 5.22. The result in problem 4 is directly linked to problem 1 in the following way, which shows that the control problem 1 can be used for solving problem 4

$$\mathbb{E}\left(\max_{i=1,\ldots,N}Q_i(t)\right) = \int_0^\infty \mathbb{P}\left(\max_{i=1,\ldots,N}Q_i(t) > x\right)dx$$

Thus using Fubini's theorem

$$\mathbb{E}\left(\int_0^\infty \exp(-\delta u)\max_{i=1,\ldots,N}Q_i(u)du\right) \quad (5.56)$$

$$= \int_\Omega \left(\int_0^\infty \exp(-\delta u)\max_{i=1,\ldots,N}Q_i(u)du\right)d\mathbb{P}(\omega)$$

$$= \int_0^\infty \exp(-\delta u)\cdot\left(\int_\Omega \max_{i=1,\ldots,N}Q_i(u)d\mathbb{P}(\omega)\right)du$$

$$= \int_0^\infty \exp(-\delta u)\cdot \mathbb{E}\left(\max_{i=1,\ldots,N}Q_i(u)\right)du$$

$$= \int_0^\infty \exp(-\delta u)\cdot\left(\int_0^\infty \mathbb{P}\left(\max_{i=1,\ldots,N}Q_i(u) > x\right)dx\right)du$$

Thus, minimizing relation (5.55) is equivalent to minimize the last double integral in (5.56). But this is in turn equivalent in find the minimal function $(u,x)\longmapsto \mathbb{P}(\max_{i=1,\ldots,N}Q_i(u) > x)$, wich is exactly problem 1. Problem 2 and Problem 3 can be transferred to the concept of minimizing the expected queueing length respectively. Since they do not incorporate new insights, we skip them. Results to problems 1 to 4 will give answer to the following questions for network design and maintenance:

- The solution to the above problems are depending on the one hand of the overall capacity $C = \sum_{i=1}^{N} C_i(t) = \sum_{i=1}^{N} C_i$. In this respect the minimal (optimal) queueing probability $\mathbb{P}\left(\max_{i=1,\ldots,N} Q_i(t) > x\right)$ will be a function on the long-range dependence parameter as well as the other traffic input values, but also the capacity C.
- In the same way as in the previous item, we have a dependence of the optimal probability as well as the optimal expected queue length on the single capacities of the subnets i.
- The optimal control processes $(q_{i,j}(u))$ can be used to implement programs to channelize the traffic into networks with better or vacant capacities. This contributes to a higher efficiency of the network.
- Since we can consider the subnets as virtual, they could represent traffic streams with different priorities. We only have to insert for different i different weights, to evaluate some virtual networks with higher priority.
- We used a relatively general framework by introducing the weight processes $(q_{i,j}(u))$. It is possible to use certain constraints on these processes to describe any other kind of network design than the above example.

Optimization Using Economic Equilibrium

We will consider a certain network structure with N connections, where, to keep things basic, all have the same priority. We sketch the approach, which we follow in detail in section 5.3.4. It should be mentioned that we outline the approach here in a broader generality than later, since the detailed investigation in 5.3.4 will be partly elusive anyhow. The situation will be the following:

1. The N lines are used in total by all consumers Γ_j, $j = 1, \ldots, M$. Each line has a capacity of $C_i(t)$, $i = 1, \ldots, N$ at time t. All consumers have the same estimation for the available bandwidth, i.e. we have for all users a utility function, which expresses the use of data minus the costs. So we have as utility

$$u(C) = u_j(C) = \frac{C^\gamma}{\gamma}$$

with $\gamma \in\,]-\infty, 1]$. The utility function indicates that unbounded increase of bandwidth does not automatically increase the utility estimation in the same way. We choose a concave utility. But it is not the capacity, which is to maximize but a compound amount:

a) First the consumer estimates her/his transferred data amount A_t up to time $T > 0$ with the utility

$$u_1(T) = \begin{cases} \frac{A_T^{\gamma_1}}{\gamma_1} & \gamma_1 \in\,]-\infty, 1],\ \gamma_1 \neq 0 \\ \log(A_T) & \end{cases}$$

b) Next she/he has to subtract the price for the guaranteed bandwidth $g(C_i(t))$ for consumer i with a continuous and concave function g expressing that more bandwidth (capacity) does not mean a linear increasing price.

$$u_2(t) = \begin{cases} \frac{g(C_i(t))^{\gamma_2}}{\gamma_2} & \gamma_2 \in \,]-\infty, 1], \; \gamma_2 \neq 0 \\ \log\left(g\left(C_i(t)\right)\right) \end{cases}$$

c) In addition the queueing length has to be subtracted, since it would cause negative judgment of the performance. For this we choose, depending on the consumer i, a threshold of length $x_i > 0$, which indicates a waiting queue $W_i(t)$ resp. waiting probability

$$u_3(t) = \begin{cases} \frac{W_i(t)^{\gamma_3}}{\gamma_3} & \gamma_3 \in \,]-\infty, 1], \; \gamma_3 \neq 0 \\ \log\left(W_i(t)\right) \end{cases}$$

For all $k = 1, 2, 3$ we have $\gamma_k \in \,]-\infty, 1] \setminus \{0\}$. Finally we have for each consumer i a optimization problem:

$$V_i(S) = \max_{C_i(t)} \left(\mathbb{E}(u_1(T)) - \mathbb{E}\left(\int_0^T u_2(u)du\right) - \mathbb{E}\left(\int_0^T u_3(u)du\right) \right)$$

$$= \max_{C_i(t)} \mathbb{E}\left(u_1(T) - \int_0^T u_2(u)du - \int_0^T u_3(u)du \right) \qquad (5.57)$$

2. The price of bandwidth is $g(C(t))$ per unit and the cost for it will be denoted by $f(C(t))$. Then we have:
   - the price for the consumer $g(C_i(t))$ and the
   - cost for the provider is $f(\sum_{i=1}^N C_i(t))$.
3. The consumer maximizes her/his expected utility over the whole time period and the provider his expected profit

$$\mathbb{E}\left(\int_0^\infty \exp(-\delta s)(g-f)\left(\sum_{i=1}^N C_i(s)\right) ds \right)$$

4. Since we also consider time sensitive applications in the traffic mixture, we have to guaranty an available traffic amount with a certain threshold as failure probability $\epsilon > 0$, i.e.

$$\mathbb{P}(Q_t > x) < \epsilon$$

This gives a constraint for the x_i of $\min_{i=1,\ldots,N} x_i \geq x$.

5.3.3 Rate Control: Shadow Prices and Proportional Fairness

Following [138] we start with a model on rate control and the so called proportional fairness. It is based on a paper of Kelly, Maulhoo and Tan. First, we will sketch the basic ideas, without going into the detail description done in the above mentioned article. Our aim is to transfer the basic idea to the traffic models, introduced in chapter 3. We start with I as the set of sources associated with a given capacity C_i. A *route* r is a non-empty subset of I and defines a users. The set of possible routes resp. users is called $\mathcal{R}$. In addition, we need a matrix $\mathcal{A} = (a_{ir})$ with $i \in I$ and $r \in \mathcal{R}$. We have $a_{ir} = 1$, if $i \in r$, i.e. the source i lies on the route of r and $a_{ir} = 0$ otherwise. Thus, $\mathcal{A}$ is a $0-1$ matrix. We assign to each route r a rate x_r. This rate gives, depending on the users a certain utility, i.e for all r let u_r be an increasing and strictly concave function on the set of all $x_r \geq 0$. A traffic, which leads to such a utility function is called *elastic*, which we already associated with the pure data traffic. In contrast to this is the time sensitive or inelastic traffic: here the utility is 0 below a certain threshold x^*, since below that rate voice or other time sensitive traffic would have no required quality. Thus, the utility function would be in that region never be strictly concave (see [229]). We assume further that the utility is additive giving an overall utility of the system by $\sum_{r \in \mathcal{R}} u_r(x_r)$. We can look at the optimization problem from different point of view: the *system optimization, user optimization* and *network optimization*.

System optimization $\mathcal{S}(u, \mathcal{A}, C)$. The aim is to find the optimal vector $\tilde{x} = (x_r, r \in \mathcal{R})$ which fulfills the following control problem

$$\sup_{\tilde{x} \in X} \sum_{r \in \mathcal{R}} u_r(x_r)$$

subject to $\mathcal{A}\tilde{x} \leq C$ and $\tilde{x} \geq 0$ where $C = (C_i, i \in I)$.

The optimization problem is mathematical tractable with the usual assumption. But it involves utility functions, which in practice is hard to specify. Hence, the problem is splitted into two subproblems: Users optimization and network optimization. For this we denote by w_r a price per unit time and receive in turn a flow of x_r, which is proportional to w_r, e.g. $x_r = \frac{w_r}{\lambda_r}$. Thus λ_r is the charge per unit flow of the user. Hence, we get a *single* optimization problem for each user:

User optimization $U_r(u_r, C)$.

$$\left(u_r\left(\frac{w_r}{\lambda_r}\right) - w_r\right) \longrightarrow \max, \text{ for } w_r \geq 0$$

Suppose now that the network knows the vector $w = (w_r, r \in \mathcal{R})$ and tries to maximize its utility function $\sum_{r \in \mathcal{R}} w_r \log x_r$ Then, we get the optimization problem for the network:

Network optimization $(\mathcal{A}, C, w)$.

$$\sum_{r \in \mathcal{R}} w_r \log x_r \longrightarrow \max$$

subject to $\mathcal{A} \leq C$ for $x \geq 0$.

We note a fact due to [137].

Theorem 5.23. *There exists always a vector* $\lambda = (\lambda_r, r \in \mathcal{R})$, *so that for* $w = (w_r, r \in \mathcal{R})$ *and* $x = (x_r, r \in \mathcal{R})$ *satisfying* $w_r = \lambda_r x_r$ *for all* $r \in \mathcal{R}$, *the vector* w *solves the User* $U_r(u_r, C)$ *problem* $(r \in \mathcal{R}$ *)and* x *solves the Network* $(\mathcal{A}, C, w)$ *problem. In particular the vector* x *is the unique solution for the system problem* $(u, \mathcal{A}, C)$.

We call a vector $x = (x_r, r \in \mathcal{R})$ *proportional fair*, if it is feasible, that is $x \geq 0$ with $Ax \leq C$, and if for any other feasible vector $\tilde{x}$ the aggregate of proportional changes is negative, i.e.

$$\sum_{r \in \mathcal{R}} \frac{\tilde{x}_r - x_r}{x_r} \leq 0 \tag{5.58}$$

We should mention that we will use a continuous version of (5.58), which we will call in the context of stochastic perturbation later, *stochastic growth rate*. In fact, in economics this factor in the non-stochastic case is well known for determine steady state growth rates.

Suppose $w_r = 1$, $r \in \mathcal{R}$, then the vector x solves Network(A, C, w), if and only if x is proportional fair. This vector x is a so called Nash bargaining solution (i.e. satisfying certain axioms of fairness, see e.g. [96]). In [178] we find this idea transferred to the context of telecommunication.

We can finally state that for a vector x the *rates per unit change* are proportionally fair, if x is feasible and if for all other feasible vectors $\tilde{x}$ we have

$$\sum_{r \in \mathcal{R}} w_r \frac{\tilde{x}_r - x_r}{x_r} \leq 0$$

Let us mention that, if we solve the network problem for a given vector of prices $w = (w_r, r \in \mathcal{R})$, then the resulting optimal rate vector x solves a slightly different system problem with weighted utility $\sum_{r \in \mathcal{R}} \alpha_r u_r(x_r)$, where $\alpha_r = \frac{w_r}{x_r u'(x_r)}$. We can say that the choice of prices w by the network corresponds implicitly to the fact that the network weights the single users utilities respectively. We decompose the optimization problem system into the two blocks of user and network. For the network problem the single utility is not required, since it is already given by the special form of the function $\sum_{r \in \mathcal{R}} w_r \log x_r$. Hence, for solving the network problem we consider the Langrangian

$$L(x, z, \mu) = \sum_{r \in \mathcal{R}} w_r \log x_r + \mu^T (C - \mathcal{A}x - z)$$

where $z \geq 0$ is a slack variable, and μ is the vector of Lagrange multipliers or shadow prices. We have as necessary optimal condition

$$\frac{\partial L}{\partial x_r} = \frac{w_r}{x_r} - \sum_{i \in r} \mu_i$$

We get for the optimal unique solution

$$x_r = \frac{w_r}{\sum_{i \in r} \mu_i} \tag{5.59}$$

where the vectors $x = (x_r, r \in \mathcal{R})$ and $(\mu_i, i \in I)$ solve

$$\mu \geq 0, \ \mathcal{A}x \leq C, \ \mu^T(C - \mathcal{A}x) = 0$$

and (5.59).

Before we consider the driving equation for the rate x_r and its solution, we introduce the so called *Dual problem*, where we optimize over the shadow prices μ:

Dual $(\mathcal{A}, C, w)$. Optimize the vector $\mu = (\mu_i, i \in I)$ such that

$$\max \sum_{r \in \mathcal{R}} w_r \log \left(\sum_{i \in I} \mu_i \right) - \sum_{i \in I} \mu_i C_i$$

under the restriction $\mu \geq 0$.

We start considering the basic system of differential equations. The underlying idea is the following. The network is offering a certain rate x_r for each user/route r, whose change $\frac{dx_r}{dt}$ is proportional to the rate itself with the constant λ_r. Hence, we first get

$$\frac{dx_r}{dt} = \kappa \lambda_r x_r = \kappa w_r$$

where κ is a still free selectable constant. Now suppose for all nodes i with $i \in r$ (for a fixed route r) the provider of this node requires a certain price p_i, depending of the rates going through this nodes, i.e. $p_i(\sum_{s: i \in s} x_s)$. This means that the price depends on the whole through traffic from other routes running through node i. In turn, this effects our user r, who tries to adjust the rate x_r by the price $x_r \sum_{i \in r} \mu_i(t)$ with $\mu_i(t) = p_i\left(\sum_{s: i \in s} x_s(t)\right)$. Hence, this results in the following system

$$\frac{d}{dt} x_r(t) = \kappa \left(\lambda_r x_r - x_r \sum_{i \in r} \mu_i(t) \right) \tag{5.60}$$

where

$$\mu_i(t) = p_i\left(\sum_{s:i\in s} x_s(t)\right) \tag{5.61}$$

But we can give an alternative explanation. This will be important, when we describe the stochastic perturbations. Each node or source i gives a feedback signal of rate $p_i(y)$ with total flow through i of y. It is assumed that a copy is sent to the user, who adjusts the rate according to the feedback signal for avoiding the congestion. So, the steady increase of the rate according to the term λx_r is partially compensated by the rate (as subtraction) $x_r \sum_{i\in r} p_i(y)$. Thus, this results in (5.60).

In the next subsection we will consider the following function $\mathcal{U}$ which will serve under some mild assumption to p_i as a Lyapunov function.

$$\mathcal{U}(x) = \sum_{r\in\mathcal{R}} w_r \log x_r - \sum_{i\in I} \int_0^{\sum_{s:i\in s} x_s} p_j(y) dy$$

where $x = (x_s, s \in \mathcal{R})$. Let's give some interpretation into this function. It shows on the right hand side first the term for maximizing the total 'utility' of the network by $\sum_{r\in\mathcal{R}} w_r \log x_r$. This is corrected by the 'feedback signal' $\sum_{i\in I} \int_0^{\sum_{s:i\in s} x_s} p_j(y) dy$. Hence, maximizing the above Lyapunov function gives an arbitrary close solution to the above mentioned network optimization problem.

Lyaponov Approach and Stochastic Perturbations

Define for certain price functions p_i under mild conditions the function

$$\mathcal{U}(x) = \sum_{r\in\mathcal{R}} w_r \log x_r - \sum_{i\in I} \int_0^{\sum_{s:i\in s} x_s} p_i(z) dz$$

As the next theorem tells us, the function $\mathcal{U}$ serves as a Lyapunov function for the system of differential equations (5.60) and (5.61). This means that an extremum of the function is a stable point of the system of differential equations. For this let p_i for $i \in I$ be non-negative, increasing and not identical 0 defined on the interval $[0, \infty[$.

Theorem 5.24. *(Kelly, Maulhoo, Tan [138]) Under the above restriction the concave function $\mathcal{U}$ provides a Lyapunov function for the system of differential equations (5.60) and (5.61). The unique value x maximising $\mathcal{U}$ is a stable point of the differential equation, where all trajectories converge in.*

We linearise the system in a neighbourhood of the stable point $x^* = (x_r^*)_{r\in\mathcal{R}}$ of the system (5.60) and (5.61) by the approach $x_r(t) = x_r^* + x^{\frac{1}{2}} y_r(t)$ according to the system of linear equations, which reads as

$$\frac{d}{dt}y_r(t) = -\kappa \left(y_r(t) \sum_{i \in r} \mu_i + x_r^{\frac{1}{2}} \sum_{i \in r} p'_i \sum_{s:i \in s} x_s^{\frac{1}{2}} y_s(t) \right)$$

$$= -\kappa \left(\frac{w_r}{x_r} y_r(t) + x_r^{\frac{1}{2}} \sum_{i \in r} \sum_{s:i \in s} p'_i a_{ir} a_{is} x_s^{\frac{1}{2}} y_s(t) \right)$$

(here p'_i is the derivative of the function p_i). In matrix form this results in

$$\frac{d}{dt}y_r(t) = -\kappa \left(WX^{-1} + X^{\frac{1}{2}} A^T P' A X^{\frac{1}{2}} \right) y(t)$$

where $X = \text{diag}(x_r, r \in \mathcal{R})$, $W = \text{diag}(w_r, r \in \mathcal{R})$ and $P' = \text{diag}(p'_i, i \in I)$. The matrix

$$WX^{-1} + X^{\frac{1}{2}} A^T P' A X^{\frac{1}{2}} \tag{5.62}$$

is real-valued, symmetric and positive definite. Hence, we can find an orthogonal matrix Γ (in particular $\Gamma \Gamma^T = Id$) and a matrix Φ, the matrix of positive eigenvalues of (5.62) in the diagonal, i.e. $\Phi = \text{diag}(\phi_r, r \in \mathcal{R})$. Then, we have

$$\Gamma^T \Phi \Gamma = WX^{-1} + X^{\frac{1}{2}} A^T P' W X^{\frac{1}{2}}$$

This results in

$$\frac{d}{dt}y_r(t) = -\kappa \Gamma^T \Phi \Gamma y(t) \tag{5.63}$$

Hence, the speed of convergence to the stable point is determined by the smallest eigenvalue of the matrix (5.62). The next step is to investigate a stochastic perturbation of the linearised equation (5.63).

$$dy_r(t) = -\kappa \left(\Gamma^T \Phi \Gamma y(t) dt + F dB^{(H)}(t) \right) \tag{5.64}$$

where F is an arbitrary $\text{card}\mathcal{R} \times \text{card}I$ matrix and $B^{(H)}(t) = (B_i^{(H)}(t), i \in I)$ is a family of independent standard fractional Brownian motions to the Hurst parameter H. As solution we get (see [35])

$$y(t) = -\kappa \int_{-\infty}^{t} \exp\left(-\kappa(t-\tau) \Gamma^T \Phi \Gamma\right) F dB^{(H)}(\tau)$$

$$\Sigma = \mathbb{E}(y(t)y(t)^T) = \kappa^2 \int_{-\infty}^{0} \exp\left(\kappa \tau \Gamma^T \Phi \Gamma\right) FF^T \exp\left(\kappa \tau \Gamma^T \Phi \Gamma\right) d\tau$$

$$= \kappa \Gamma^T \left(\int_{-\infty}^{0} \exp(\tau \Phi) \Gamma FF^T \Gamma^T \exp(\tau \Phi) d\tau \right) \Gamma \tag{5.65}$$

Define the symmetric matrix $[\Gamma F; \Phi]$ according to

$$[\Gamma F; \Phi]_{rs} = \left(\int_{-\infty}^{0} \exp(\tau \Phi) \Gamma FF^T \Gamma^T \exp(\tau \Phi) d\tau \right)_{rs} = \frac{(\Gamma FF^T \Gamma^T)_{rs}}{\phi_r + \phi_s}$$

Then, we obtain

$$\Sigma = \kappa \Gamma^T [\Gamma F; \Phi] \Gamma$$

Stochastic Version of (5.60) and (5.61)

Up to now we used a deterministic approach: the rate x_r was driven by a fixed price λ_r for the available bandwidth and a diminishing factor μ_i, depending on the node i used by route $r \in \mathcal{R}$, which is considered as a Langrange multiplier or for further interpretation as fair price. In the sequel, we want to incorporate the certainly existing random factor. The network, which we consider, is on the one hand 'deterministic' constructed by using certain routes, but the impact of the selection of the services, like WWW, FTP or Email beside the pure data transfer, requires the introduction of factors of uncertainty. We start in the next equation with the usual factor

$$dx_r(t) = \kappa w_r dt = \kappa \lambda_r x_r(t) dt \tag{5.66}$$

The rate given by equation (5.66) is reduced by an algorithm given according to congestion in node $i \in I$. Thus, we describe the individual rate of route $r \in \mathcal{R}$ by

$$dx_r(t) = \kappa \left(w_r dt - x_r(t) \sum_{i \in r} \epsilon_i dN_i \left(\epsilon_i^{-1} \int_0^t \mu_i(\tau) d\tau \right) \right) \tag{5.67}$$

with independent Poisson processes N_i and rate $\epsilon_i^{-1} \int_0^t \mu_i(\tau) d\tau$. The description resp. quantitative evaluation is more elusive than considering a limit. Thus, with $\epsilon \to 0$ we get in the limit a Brownian motion and the equation

$$dx_r(t) = \kappa \left(w_r dt - x_r(t) \sum_{i \in r} \left(\mu_i(t) dt + \epsilon_i^{\frac{1}{2}} \mu_i(t)^{\frac{1}{2}} dB_i(t) \right) \right) \tag{5.68}$$

where the $(B_i(t))_{t \geq 0}$ are a family of independent standard Brownian motions for $t \geq 0, i \in I$. Considering the linearised version according to (5.63) gives exactly (5.64), where we have $\overline{B}(t) = (B_i(t), i \in I)$ and F is a $|\mathcal{R}| \times |I|$-matrix with components

$$F_{ri} = \epsilon_i^{\frac{1}{2}} \mu_i^{\frac{1}{2}} A_{ir} x_r^{\frac{1}{2}}$$

This turns into

$$FF^T = X^{\frac{1}{2}} \cdot A^T \cdot E \cdot P \cdot A \cdot X^{\frac{1}{2}}$$

where $E = \text{diag}(\epsilon_i, i \in I)$ and $P = \text{diag}(\mu_i, i \in I)$. Finally, we obtain a stationary covariance matrix Σ, which can be calculated by (5.65). Equation (5.67) can be changed to an individual feedback, which we will consider in the next subsection in a broader view. The basic difference consists in splitting the Poisson process N_i into individual ones N_{ir}.

Remark 5.25. In this subsection we introduced the rate model of Kelly, Maulhoo and Tan, which we will use as basis for a further study of economic optimization in the succeeding subsections incorporating stochastic perturbations:

- The above approach using the Lyaponov approach to find stability results is as just mentioned investigated in details, e.g. in the papers [138, 136]. The paper [138] establishes parallel to the above Lyaponov technic for the network problem the Lyaponov-type the result for the dual algorithm, which we formulated in the equations (5.60) and (5.61). We skip the details and refer to the original literature [138].
- Also in [138] further topics are treated, like user adaptions – the user can instantly monitor her/his rate and adapt the parameter by optimized decision – and a more general optimization approach, where the restriction $Ax \leq C$ is substituted by a penalty, expressed by certain delay and loss factors. A similar idea we will follow up in the next subsection.
- In the subsection 5.76 we are concerned with a more general perturbation

$$dx_r(t) = \kappa \left(\lambda_r x_r(t) dt - x_r(t) \sum_{i \in r} \epsilon_i dN_i \left(\epsilon_i^{-1} \int_0^t \mu_i(\tau) d\tau \right) \right)$$
$$+ x_r(t) F_r dB^{(H)}(t)$$

The major difference to equation (5.67) resp. (5.68) consists in the use of the fractional Brownian motion in conjunction with a family of independent Poisson process. The latter ones should incorporate a short rate dependence and give tribute to the idea of the mixed traffic, whose model was shortly introduced in subsection 5.3.2. Here, we will investigate the impact of the used factors on the optimal rate as well as in certain allocation of network resources on the basis of economic viewpoint. The other difference to the Kelly, Maulhoo and Tan model is established by the truly existing fact of long-range dependence as widely investigated by us (and the research community), using the FBM. In fact, the relative rate $\frac{dx_r}{x_r}$ is nothing else as the accumulated traffic model due to Norros, with the slight difference of the Poisson factor. But here we are back to the model in subsection 5.3.2.

5.3.4 Optimization for Stochastic Perturbation

Different to the above approach, using Lyapunov functions, we apply the classical optimal control tools. First we start with the *network view*, i.e. we optimize the income of the network and find an optimal solution, depending on certain constraints. Next, we will optimize the constraints according to each *user* to get an equilibrium.

First, we simply describe the model and will specify mathematically the used variables. The basic scenario is transferred from the model in section 5.3.3. In general depending on a price π_r for a unit of rate x_r, the change of offer is proportional. Hence,

$$dx_r = \pi_r x_r dt$$

for each route or user r. This basic proportional rate is diminished by a certain price $\mu_{i,r}$, proportional to x_r for using the nodes $i \in I_r$. Here, I_r is the subset of resources met by the route (user) $r \in \mathcal{R}$. Expressed in the differential equation, it looks like

$$dx_r(t) = \pi_r(t)x_r(t)dt - x_r(t)\sum_{i \in I_r}\mu_{i,r}(t)dt$$

for $r \in \mathcal{R}$. On the other hand the multiplier $\mu_{i,r}$ leads to a congestion at node i which gives a feedback signal to the user to decrease the rate. The amount is depending on a certain factor ϵ_i and as a stochastic event occurring according to a stationary Poisson process $N_{i,r}$, $r \in \mathcal{R}, i \in I_r$, with rate $\lambda_i\mu_{i,r}$, which are assumed to be mutually independent, since the congestion and subsequent decision are independent of each other. Thus, we get a stochastic differential equation of the form

$$dx_r(t) = \pi_r(t)x_r(t)dt - x_r(t)\sum_{i \in I_r}(\mu_{i,r}(t)dt + \epsilon_i dN_i(\mu_{i,r}))$$

for $r \in \mathcal{R}$. In addition, we incorporate an improvement of the network. This improvement is expressed by a better performance of each node, which increases the rate for these particular bottlenecks. We model it by a family of independent Poisson processes $M_{i,r}$ and factors $\tilde{\epsilon}_r$. The rate is similar to the congestion and modeled by a rate $\tilde{\lambda}D_{i,r}$. The factor $D_{i,r} \in [0,1]$ is an investment factor of the network provider for the route r and the node or bottleneck $i \in I_r$. Thus, this factor will decrease the income of the network. Therefore we obtain the next form of the equation

$$dx_r(t) = \pi_r(t)x_r(t)dt - \sum_{i \in I_r}(\mu_{i,r}(t)x_r(t)dt + \epsilon_{i,r}(t)x_r(t)dN_{i,r}(\mu_{i,r}))$$
$$+x_r(t)\sum_{i \in I_r}\tilde{\epsilon}_r(t)dM_{i,r}(D_{i,r})$$

for $r \in \mathcal{R}$. Finally, we have to take a long-range component into account, depending on the rate x_r and an additional scaling factor a. It will be modeled by the FBM for a Hurst parameter $H \in]\frac{1}{2}, 1[$, since we want to incorporate the over all LRD property. We choose the FBM independent of the Poisson processes. Over all we obtain the stochastic differential equation for the rate x_r

$$dx_r(t) = \kappa\left(\pi_r(t)x_r(t)dt - x_r(t)\sum_{i \in I_r}(\mu_{i,r}(t)dt + \epsilon_i(t)dN_{i,r}(\mu_i))\right. \quad (5.69)$$
$$\left.+x_r(t)\sum_{i \in I_r}\tilde{\epsilon}_r(t)dM_{i,r}(D_{i,r}) + ax_r(t)dB_t^{(H)}\right)$$

The scaling parameter κ will be set 1, since we do not want to overload the solutions by constants not necessary needed for the qualitative description of the model.

Next, we select the function for maximizing the utility of the network. First, we have to specify the 'income' of the network. We choose a weight parameter $\theta \in {]}0,1[$ and define the weighted income for one route and one node, according to the *Cobb-Douglas production function*(see [10])

$$F_{i,r}(R,D) = R_{i,r}^{\theta}(1-D_{i,r})^{1-\theta} \tag{5.70}$$

where $R_{i,r} = \mu_{i,r} x_r$ for $i \in I_r$ and $r \in \mathcal{R}$. The amount $R_{i,r}$ can be interpreted as the income of the throughput of resource i generated by user or route r. This amount is multiplied by a factor of increase $1 - D_{i,r}$, where $D_{i,r}$ gives more efficiency to the throughput, but in turn decreases the income and thus, we multiply by $1 - D_{i,r}$. The factor $1 - D_{i,r}$ reflects the fact that less improvement on the net structure (small $D_{i,r}$) depending on the route r and node i, will in turn offer more income for the network. Each of these incomes is again weighted by utility. As usual, higher profit does not necessarily result in a linearly higher considered utility. Thus, we choose for the model a function $u : [0, \infty[\longrightarrow \mathbb{R}, u(x) = \frac{x^{\gamma}}{1-\gamma}$ with $\gamma \in {]}-\infty, 1[$ to illustrate that higher income will not increase the utility in the same manner. Those utility functions are called risk averse. Other functions like the logarithm are also possible. Hence, the over all utility would be

$$u(F(R,D)) = \sum_{r \in \mathcal{R}} \left(\sum_{i \in I} \left(R_{i,r}^{\theta}(1-D_{i,r})^{1-\theta} \right)^{\gamma} \right) \tag{5.71}$$

with $R = (R_{i,r}, i \in I_r, r \in \mathcal{R})$ and $D = (D_{i,r}, i \in I_r, r \in \mathcal{R})$. Obviously, the amount $R_{i,r}$ is the income received from user (route) r for using resource i as the product of rate x_r and price $\mu_{i,r}$.

Before we come to the optimal control problem, we shortly comment on the *stochastic growth rate*, which is the relative change to the rate itself in the context of (5.69)

$$\frac{dx_r}{x_r} = \kappa \left(\pi_r dt - \left(\sum_{i \in I_r} \mu_{i,r} dt + \sum_{i \in I_r} \epsilon_i dN_{i,r}(\mu_{i,r}) \right) \right.$$
$$\left. + \sum_{i \in I_r} \tilde{\epsilon}_r(t) dM_{i,r}(D_{i,r}) + a dB_t^{(H)} \right)$$

Thus, the right hand side defines a stochastic equation for a process $(Y_r(t))$. If we skip the Poisson processes, we get an equation in the from, which is basically known to us as the model of Norros from section 3.3.4

$$dY_r(t) = \kappa \left(\pi_r dt - \left(\sum_{i \in I_r} \mu_{i,r} dt + \sum_{i \in I_r} \epsilon_i dN_{i,r}(\mu_{i,r}) \right) \right.$$
$$\left. + \sum_{i \in I_r} \tilde{\epsilon}_r(t) dM_{i,r}(D_{i,r}) + a\pi_r dB_t^{(H)} \right)$$

In general, the (control-)variables $\mu_{i,r}$ and $D_{i,r}$ as well as π_r are stochastic processes. We will later specify the property to ensure an appropriate solvability of equation (5.69). In particular, we encounter a differential equation with fractional Brownian motion. As in the Norros approach, we want to obtain stability. Thus, we are interested in a stability condition

$$\mathbb{E}\left(\frac{dx_r}{x_r}\right) = \mathbb{E}\left(dY_r(t)\right) = Adt$$

with a variable $A \in \mathbb{R}$ which will be determined later as the optimal condition. This results for all initial conditions $Y_r(u) = Y_u$ in the solution of the form

$$Y_r(T) = A(T - u) + Y_u \qquad (5.72)$$

where Y_u is a given and integrable random variable. Before we proceed, we have to give some results on the stochastic calculus for the FBM.

Stochastic Calculus for the Fractional Brownian Motion

In the sequel we consider a fractional Brownian motion for $\frac{1}{2} < H < 1$, since then, we know that the covariance decays slowly. Fundamental for the stochastic calculus is the Malliavin derivative $D_{t_0}X_{t_0}$ for a stochastic process $(X_t)_{t \in \mathbb{R}}$ at time t_0 (see [73, 114, 33] for details). Further set

$$\psi(s,t) = H(2H-1)|s-t|^{2H-2}$$

$$C_H = \left(2\Gamma\left(H - \frac{1}{2}\right)\cos\left(\frac{\pi}{2}\left(H - \frac{1}{2}\right)\right)\right)^{-1}(\Gamma(2H+1)\sin(\pi H))^{\frac{1}{2}}$$

with the Gamma function $\Gamma(x) = \int_0^\infty t^{x-1}e^{-t}dt$. Then, we define the space of underlying processes for the optimal control problem.

Definition 5.26. *The space $\mathcal{L}_\psi^{1,2}(\mathbb{R})$ consists of all processes (X_t) such that*

$$\|X\|^2_{\mathcal{L}_\psi^{1,2}(\mathbb{R})} = \mathbb{E}\left(\int_{\mathbb{R}^2} X_s X_t ds dt + \left(\int_\mathbb{R}\left(\int_\mathbb{R}\psi(s,t)D_t X_t dt\right)X_s ds\right)^2\right) < \infty$$

For all following applications the condition will be satisfied, and this is the class of processes, for which one can define a suitable integration theory, such that in particular theorem 5.27 holds. If we have a deterministic process $f(t)$ (not depending on a probability space Ω), then

$$|f|^2_\psi = \|f\|^2_{\mathcal{L}_\psi^{1,2}(\mathbb{R})} = \int_{\mathbb{R}^2} f(s)f(t)\psi(s,t)dsdt$$

We start with citing without proofs some results on the integral with respect to the FBM.

Theorem 5.27. *For any process $X \in \mathcal{L}^{1,2}_\psi(\mathbb{R})^2$ we have*

$$\mathbb{E}\left(\int_\mathbb{R} X_t(\omega) dB_t^{(H)}\right) = 0$$

Theorem 5.28. *(Fractional Girsanov theorem, see [114, 35]). Let $T > 0$ and let $g : [0,T] \longrightarrow \mathbb{R}$ be continuous. Let further $K : [0,T] \longrightarrow \mathbb{R}$ satisfy the equation*

$$\int_0^T K(s)\psi(s,t)ds = g(t), \ 0 \le t \le T$$

and extend K to all of $\mathbb{R}$, such that $K(s) = 0$ for $s \notin [0,T]$. Then, define the probability measure $\hat{\mathbb{P}}$ by

$$d\hat{\mathbb{P}}(\omega) = \exp\left(-\int_0^T K(s)dB_s^{(H)} - \frac{1}{2}|K|_\psi^2\right) d\mathbb{P}(\omega)$$

Then

$$\hat{B}_t^{(H)} = \int_0^t g(s)ds + B_t^{(H)}$$

is a fractional Brownian motion in the probability space $(\Omega, \mathcal{F}, \hat{\mathbb{P}})$.

We recall the natural filtration $(\mathcal{F}_t^{(H)})$ with respect to the Brownian motion, i.e. the σ-algebra generated by the path of fractional Brownian motion up to the time t

$$\mathcal{F}_t^{(H)} = \sigma\left(B_s^{(H)}; s \le t\right)$$

We have to introduce the quasi-conditional expectation and quasi-martingales. A random variable X with formal expansion

$$X(\omega) = \sum_{n=0}^\infty \int_{[0,T]^n} f_n dB_t^{(H),\otimes n}, \ f_n \in L^2_\psi([0,T]^n) \tag{5.73}$$

is defined to lie in the space $\mathcal{G}(\mathbb{P})$, if there is a $q \in \mathbb{N}$ such that

$$\|X\|_{\mathcal{G},-q} = \sum_{n=0}^\infty n! \|f_n\|^2_{L^2_\psi([0,T]^n)} e^{-2qn} < \infty$$

Then, we define for a random variable X of the form (5.73)

$$\tilde{\mathbb{E}}\left(F|\mathcal{F}_t^{(H)}\right) = \sum_{n=0}^\infty \int_{[0,t]^n} f_n dB_t^{(H),\otimes n}$$

A process (X_t) adapted to the filtration $(\mathcal{F}_t)$ is called a *quasi-martingale*, if

$$\tilde{\mathbb{E}}\left(X_t|\mathcal{F}_s^{(H)}\right) = X_s, \text{ for all } s \le t$$

Theorem 5.29. *For any $X \in \mathcal{L}_\psi^{1,2}(\mathbb{R})$ the process*

$$M(t) = \int_0^t X_t(\omega) dB_t^{(H)}, \ t \geq 0$$

is a quasi-martingale. In particular, for all $t \geq 0$

$$\mathbb{E}(M(t)) = \mathbb{E}(M(0)) = 0$$

Theorem 5.30. *(Fractional Clark-Haussmann-Ocone (CHO) theorem, see [114]) Suppose $G \in L_2(\mathbb{P})$ is $\mathcal{F}_T^{(H)}$-measurable. Define*

$$\vartheta(t,\omega) = \tilde{\mathbb{E}}\left(D_t G | \mathcal{F}_t^{(H)}\right)$$

Then we have

$$\vartheta \in \mathcal{L}_\psi^{1,2}(\mathbb{R})$$

and

$$G(\omega) = \mathbb{E}(G) + \int_0^T \vartheta(t,\omega) dB_t^{(H)}$$

where $D_t G = \frac{dG}{d\omega}(t,\omega)$ is the Malliavin derivative or stochastic gradient of G at t, which exists almost surely in $\mathcal{G}(\mathbb{P})$ (see [114]).

Finally we come to the stochastic differential equation with fractional Brownian motion as driving term. For this, let $\alpha, \beta \in \mathcal{L}_\psi^{1,2}(\mathbb{R})$. By the differential equation

$$dX_t = \alpha_t dt + \beta_t dB_t^{(H)}$$

we mean the process in integral form

$$X_t = X_0 + \int_0^t \alpha(s) ds + \int_0^t \beta(s) dB_s^{(H)} \qquad (5.74)$$

If $\alpha(s) = \mu(s, X_s)$ and $\beta(s) = \sigma(s, X(s))$, we call (5.74) a stochastic differential equation in X. Suppose now that $\alpha(s) = \mu(s) X_s$ and $\beta(s) = \beta X_s$ with α deterministic and $\beta \in \mathbb{R}$, then it is possible to give a solution of (5.74)

Theorem 5.31. *(see [33]) Let $\alpha \in L_\psi^2(\mathbb{R})$ and $\beta \in \mathbb{R}$. Then*

$$X_t = X_0 \exp\left(\beta B_t^{(H)} + \int_0^t \alpha(s) ds - \beta t^{2H}\right)$$

is the solution of the differential equation (5.74) with initial value X_0 for $t = 0$.

In stochastic analysis the Itō formula plays a decisive rôle (see e.g. [231, 153]). How can one formulate such a formula for the fractional Brownian motion?

Theorem 5.32. *(Fractional Itô formula, see [33]) Let (X_t) be a process satisfying with initial condition X_0 the stochastic differential equation (5.74). Let $F : \mathbb{R} \longrightarrow \mathbb{R}$ be twice differentiable. Then*

$$F(X_t) = F(X_0) = \int_0^t F'(X_s)\alpha(s)ds + \int_0^t F'(X_s)\beta(s)dB_s^{(H)}$$
$$+ \int_0^t F''(X_s)D_sX_s\beta(s)ds$$

where D_sX_s is the Malliavin derivative of X at time s.

For details on the stochastic calculus of the FBM we refer to [226, 35]. Finally, we state two theorems, which give the solution of a stochastic differential equation for the perturbation of Poisson and fractional Brownian motion. The proof and further results can be found in [161, 162].

Theorem 5.33. *The following stochastic differential with constants $a, b \neq 0$ and $c > -1$*

$$dY_t = aY_t dt + bY_t dB_t^{(H)} + cY_t dN_t, \ t \geq 0$$

as shorthand notation for

$$Y_t = Y_0 + \int_0^t aY_s ds + \int_0^t bY_s dB_s^{(H)} + \int_0^t cY_s dN_s$$

is uniquely solved by the geometric fractional Brownian-Poissonian motion

$$Y_t = Y_0 \exp\left(at + bB_t^{(H)} - \frac{1}{2}b^2 t^{2H} + \ln(1+c)N_t\right)$$

Theorem 5.34. *Let $a(\cdot)$, $b(\cdot)$, $c(\cdot)$ be deterministic functions then the stochastic differential equation*

$$dY_t = a(t)Y_t dt + b(t)Y_t dB_t^{(H)} + c(t)Y_t dN_t, \ t \geq 0$$

which is a shorthand notation for

$$Y_t = Y_0 + \int_0^t a(s)Y_s ds + \int_0^t b(s)Y_s dB_s^{(H)} + \int_0^t c(s)Y_s dN_s$$

can be uniquely solved by the geometric fractional Brownian-Poissonian motion

$$Y_t = Y_0 \exp\left(\int_0^t a(s)ds + \int_0^t b(s)dB_s^{(H)} - \frac{1}{2}\int_{\mathbb{R}} \left(M_s\left(b(s)\chi_{[0,t]}(s)\right)\right)^2 ds \right.$$
$$\left. + \int_0^t \ln(1+c(s))dN_s\right)$$

5.3.5 Optimization of Network Flows Using an Utility Approach

The optimization with Poisson or Brownian motion perturbations are classical implemented in the *optimal control theory* and done in several monographs. The reader may consult e.g. [231, 162]. The problem, we are encountered, consists in a perturbation of Poisson and fractional Brownian motion. For the combined Poisson and Brownian motion case, the Itō formula is fundamental, once for the Hamilton-Jacobi-Bellman equation and second for the explicit solution. The latter one was done e.g. by [110], the first one is folklore. To transfer the technique to the fractional Brownian motion case we have to introduce the Malliavin derivative, as e.g. indicated above in theorem 5.32, called fractional Itō formula. The drawback consists in the application of the Malliavin derivative. In like paper [113] resp. in the monograph [35], the representation is not very suitable. But, the technique using Lagrange multipliers as done e.g. in [113] can be applied in certain situations to find the optimal solution in our network situation.

Before we start with finding the optimal prices and transmission rates, we introduce some notions for a general stochastic process used in economics.

Definition 5.35. *A stochastic process (Y_t) is in steady state, if the derivative of the function $t \to E[Y_t]$ is constant or $E(dY_t) = Adt$ for a suitable $A \in \mathbb{R}$.*

We return to the control problem and want to solve it by maximizing the expected utility, given according to equation (5.71) in section 5.3.4. Looking at its expectation by applying Fubini's theorem, we obtain for a solution of equation (5.72)

$$\mathbb{E}(Y_T) = \mathbb{E}(Y_u) + \mathbb{E}\left(\kappa\left(\int_u^T \pi_r dt - \left(\sum_{i \in I_r} \mu_{i,r} dt + \sum_{i \in I_r} \epsilon_i dN_{i,r}(\mu_{i,r})\right)\right.\right.$$
$$\left.\left. + \sum_{i \in I_r} \tilde{\epsilon}_r(t) dM_{i,r}(D_{i,r}) + \int_u^T a\pi_r dB_t^{(H)}\right)\right) \quad (5.75)$$
$$= \mathbb{E}(Y_u) + \kappa\left(\int_u^T \pi_r(s) ds - \int_u^T \sum_{i \in I_r}(1 + \lambda \epsilon_i)\mu_{i,r}(s) ds\right.$$
$$\left. + \int_u^T \sum_{i \in I_r} \tilde{\epsilon}_r \tilde{\lambda} \mathbb{E}(D_{i,r}(u)) ds\right)$$

since the expectation of the FBM is 0 (see theorem 5.27), $E(dN_{i,r}(s)) = \lambda_i \mu_{i,r}(s))ds$ and $E(dM_{i,r}(s)) = \tilde{\lambda} E(D_{i,r}(s))ds$ for the used Poisson processes with intensities $\lambda_i \mu_i(s)$ and $\tilde{\lambda} E(D_{i,r}(s))$ (see eg. [230, 231, 162]). In view of (5.75), we define the following process

$$\tilde{Y}_t = \mathbb{E}(Y_u) + \kappa \left(\int_u^t \pi_r(s)ds - \int_u^t \sum_{i \in I_r}(1+\lambda\epsilon_i)\mu_{i,r}(s)ds \right.$$

$$\left. + \int_u^t \sum_{i \in I_r} \tilde{\epsilon}_r \tilde{\lambda} \mathbb{E}(D_{i,r}(u))ds \right)$$

This yields for all $t \geq 0$ to

$$\mathbb{E}(Y_t) = \mathbb{E}(\tilde{Y}_t) \quad \text{and} \quad \mathbb{E}(dY_t) = \mathbb{E}(d\tilde{Y}_t)$$

Thus, we can consider the process (Y_t) as the growth rate of the data rate process $(x_r(t))$, which is driven by the fractional Brownian motion. Setting $x = (x_r, r \in \mathcal{R})$ and fixing $T > u > 0$, we are able to formulate the optimal control problem (P1) under steady state by

$$\text{(P1)} \begin{cases} V(x) = \max_{R(t),D(t)} \mathbb{E}\left(\int_u^T e^{-\delta t} u(C(t))dt \right) \\ C(t) = F(R_t, D_t) \\ dx_r(t) = \kappa x_r(t) \Big(\pi_r(t)dt - \sum_{i \in I_r}(\mu_{i,r}(t)dt + \epsilon_i(t)dN_{i,r}(\mu_i)) \\ \qquad + \sum_{i \in I_r} \tilde{\epsilon}_r(t)dM_{i,r}(D_{i,r}) + a\pi_r dB_t^{(H)} \Big), \text{ for all } r \in \mathcal{R} \\ \mathbb{E}(dY_t) = A dt \\ R(t), D(t), 1 - D(t) \geq 0, x(t) \geq 0, \text{ for all } t \geq u \end{cases}$$

where we used the notation $R(t), D(t) \geq 0$ meaning componentwise, and where we imposed the condition $1 - D(t)$, to indicate that each factor $D_{i,r}$ does not exceed 1. For simplicity we assume $\kappa = 1$. Apart from this $R_{i,r} = \mu_{i,r}x_r(t)$ is chosen to be our Markov control, where $(\mu(t)) = ((\mu_{i,r}(t), i \in I_r, r \in \mathcal{R}))$ is an adapted processes uncorrelated to (x_t) (see [230]).

To find the optimal solution, we transform this problem into an equivalent unconstraint optimization problem (P1)' given by

$$\text{(P1)}' \begin{cases} \tilde{V}_{\varphi_u}(x) = \max_{R(t),L(t)} \mathbb{E}\Big(\int_u^T e^{-\delta t}u(C(t))dt - \varphi_u\Big(\sum_{r \in \mathcal{R}}\Big(\int_u^T \pi_r(s) \\ \qquad - \int_u^T \sum_{i \in I_r}(1+\lambda\epsilon_i)\mu_{i,r}(s)ds + \int_u^T \sum_{i \in I_r}\tilde{\epsilon}_r dM_{i,r}(s)\Big)\Big)\Big) \\ C(t) = F(R(t), D(t)) \\ dx_r(t) = \kappa x_r(t)\Big(\pi_r(t)dt - \sum_{i \in I_r}(\mu_i(t)dt + \epsilon_i(t)dN_{i,r}(\mu_i)) \\ \qquad + \sum_{i \in I_r}\tilde{\epsilon}_r(t)dM_{i,r}(D_{i,r}) + a\pi_r dB_t^{(H)} \Big), \text{ for all } r \in \mathcal{R} \\ \mathbb{E}(dY_t) = A dt \\ R(t), D(t), 1 - D(t) \geq 0, x(t) \geq 0, \text{ for all } t \geq u \end{cases}$$

where φ_u is an integrable random variable, which is uncorrelated to the increments of the process $(N_{i,r}(s))_{u < s \leq T}$ and $(\mu(s))_{u < s \leq T}$. At the end, we have to

5.3 Traffic Optimization

check for the Lagrange variable φ_u, if these assumptions are fulfilled. Suppose we found for the problem (P1)' the optimal pair $((R_t^*)_{t\leq T},(D_t^*)_{t\leq T})$ and φ_u^* and $((R_t),(D_t))$ is any pair solving the original problem, then

$$\mathbb{E}\left(\int_u^T e^{-\delta t}\sum_{r\in\mathcal{R}}\sum_{i\in I_r}\frac{R_t^{\theta\gamma}(1-D_t)^{\gamma(1-\theta)}}{\gamma}dt\right)$$

$$=\mathbb{E}\left(\int_u^T e^{-\delta t}\sum_{r\in\mathcal{R}}\sum_{i\in I_r}\frac{R_{i,r}^{\theta\gamma}(t)(1-D_{i,r}(t))^{\gamma(1-\theta)}}{\gamma}dt\right.$$

$$-\varphi_u^*\left(\int_u^T\pi_r(s)ds-\int_u^T\sum_{i\in I_r}(1+\lambda\epsilon_i)\mu_{i,r}(s)ds\right.$$

$$\left.\left.+\int_u^T\sum_{i\in I_r}\tilde{\epsilon}_r dM_{i,r}(s)\right)+\varphi_u^*\left(A(T-u)\right)\right)$$

$$=\mathbb{E}\left(\int_u^T e^{-\delta t}\sum_{r\in\mathcal{R}}\sum_{i\in I_r}\frac{R_{i,r}^{\theta\gamma}(t)(1-D_{i,r}(t))^{\gamma(1-\theta)}}{\gamma}dt\right.$$

$$-\varphi_u^*\left(\mathbb{E}\left(\int_u^T\pi_r(s)ds-\int_u^T\sum_{i\in I_r}(1+\lambda\epsilon_i)\mu_{i,r}(s)ds\right.\right.$$

$$\left.\left.\left.+\int_u^T\sum_{i\in I_r}\tilde{\epsilon}_r dM_{i,r}(s)\right)+\varphi_u^*\left(A(T-u)\right)\right)\right.$$

$$\leq \mathbb{E}\left(\int_u^T e^{-\delta t}\sum_{r\in\mathcal{R}}\sum_{i\in I_r}\frac{(R_{i,r}^*)^{\theta\gamma}(t)(1-D_{i,r}^*(t))^{\gamma(1-\theta)}}{\gamma}dt\right.$$

$$-\varphi_u^*\left(\mathbb{E}\left(\int_u^T\pi_r(s)ds-\int_u^T\sum_{i\in I_r}(1+\lambda\epsilon_i)\mu_{i,r}^*(s)ds\right.\right.$$

$$\left.\left.\left.+\int_u^T\sum_{i\in I_r}\tilde{\epsilon}_r dM_{i,r}^*(s)\right)+\varphi_u^*\left(A(T-u)\right)\right)\right.$$

$$=\mathbb{E}\left(\int_u^T e^{-\delta t}\frac{(R_{i,r}^*(t))^{\theta\gamma}(1-D_{i,r}^*(t))^{\gamma(1-\theta)}}{\gamma}dt\right)$$

where we used in the second and last equation that φ_u^* is uncorrelated to $(M_{i,r}(s))_{s<T}$ and $(\mu_s)_{s<T}$, and that

$$\mathbb{E}\left(\int_u^t\pi_r(s)ds-\int_u^t\sum_{i\in I_r}(1+\lambda\epsilon_i)\mu_{i,r}(s)ds+\int_u^t\sum_{i\in I_r}\tilde{\epsilon}_r(s)dM_{i,r}(s)\right)$$

$$=\mathbb{E}\left(\int_u^t\pi_r(s)ds-\int_u^t\sum_{i\in I_r}(1+\lambda\epsilon_i)\mu_{i,r}(s)ds+\int_u^t\sum_{i\in I_r}\lambda\tilde{\epsilon}_r(s)D_{i,r}(s)ds\right)$$

$$=\mathbb{E}(Y_r(T))-\mathbb{E}(Y_r(u))=A(T-u)$$

5 Performance of IP: Waiting Queues and Optimization

This shows that an optimal solution of the original problem (P1) is implied by the optimal solution of (P1)'.
By the identity $E(Y_t) = E(\tilde{Y}_t)$, the value $\tilde{V}_{\varphi_u}(S)$ function can be rewritten as

$$\tilde{V}_{\varphi_u}(S) = \max_{R(t),D(t)} \mathbb{E}\left(\left(\int_u^T e^{-\delta s} \sum_{r \in \mathcal{R}} \sum_{i \in I_r} \frac{R_{i,r}^{\theta\gamma}(t) D_{i,r}(t)^{\gamma(1-\theta)}}{\gamma} ds \right.\right. \quad (5.76)$$

$$- \varphi_u \left(\sum_{r \in \mathcal{R}} \left(\int_u^T \pi_r ds - \sum_{i \in I_r} \int_u^T (1+\lambda\epsilon_i)\mu_{i,r}(s) ds \right.\right.$$

$$\left.\left.\left.\left. + \sum_{i \in I_r} \int_u^T \tilde{\lambda\epsilon}_r(s) D_{i,r}(s) ds \right)\right)\right)\right)$$

The corresponding solution is given by finding the optimum of the function

$$G(R,D)$$
$$= e^{-\delta s} \sum_{r \in \mathcal{R}} \sum_{i \in I_r} \frac{R_{i,r}^{\theta\gamma}(s) D_{i,r}(s)^{\gamma(1-\theta)}}{\gamma}$$
$$+ \varphi_u \left(\sum_{r \in \mathcal{R}} \left(\pi_r(s) - \sum_{i \in I_r} (1+\lambda\epsilon_i)\mu_{i,r}(s) + \sum_{i \in I_r} \tilde{\lambda\epsilon}_r(s) D_{i,r}(s)\right)\right)$$
$$= e^{-\delta s} \sum_{r \in \mathcal{R}} \sum_{i \in I_r} \frac{R_{i,r}^{\theta\gamma}(s) D_{i,r}(s)^{\gamma(1-\theta)}}{\gamma}$$
$$+ \varphi_u \left(\sum_{r \in \mathcal{R}} \left(\pi_r(s) + \sum_{i \in I_r} \tilde{\lambda\epsilon}_r(s) D_{i,r}(s)\right) - \sum_{r \in \mathcal{R}} \sum_{i \in I_r} (1+\lambda\epsilon_i)\frac{R_{i,r}(s)}{x_r}\right)$$

for each $s \in \mathbb{R}$ $s \in [u,T]$.
Computing the partial derivatives for the components $r \in \mathcal{R}$, $i \in I_r$ and setting to 0 gives for $t \in [u,T]$ (we skip the argument of $R_{i,r}$, $D_{i,r}$, x_r and their subscript, remember that $R_{i,r} = \mu_i x_r$)

$$\frac{\partial G}{\partial R_{i,r}} = e^{-\delta t} \theta \gamma \frac{R^{\theta\gamma-1}(1-D)^{\gamma(1-\theta)}}{\gamma} - \varphi_u(1+\lambda\epsilon_i)\frac{1}{x} = 0 \quad (5.77)$$

$$\frac{\partial G}{\partial D_{i,r}} = -e^{-\delta t}(1-\theta)\gamma \frac{R^{\theta\gamma}(1-D_{i,r})^{\gamma(1-\theta)-1}}{\gamma} + \varphi_u(\tilde{\lambda\epsilon}_r) = 0 \quad (5.78)$$

Solving both equations and dividing (5.78) by (5.77) gives

$$R = \frac{\theta}{1-\theta} \frac{\tilde{\lambda\epsilon}_r}{1+\lambda\epsilon_i}(1-D)x$$

Thus, we obtain for R and D

$$R = \left(\frac{\theta}{1-\theta}\tilde{\lambda}\tilde{\epsilon}_r\right)^{\frac{\gamma(\theta-1)}{1-\gamma}} \cdot (1+\lambda\epsilon_i)^{\frac{\gamma(\theta-1)+1}{\gamma-1}} \cdot \left(\frac{\theta}{\varphi e^{\delta t}}\right)^{\frac{1}{1-\gamma}} \cdot x^{\frac{\gamma(\theta-1)+1}{1-\gamma}} \quad (5.79)$$

$$1 - D = \left(\frac{\theta}{1-\theta}\tilde{\lambda}\tilde{\epsilon}_r\right)^{\frac{\gamma\theta-1}{1-\gamma}} \cdot (1+\lambda\epsilon_i)^{\frac{\gamma\theta}{\gamma-1}} \cdot \left(\frac{\theta}{\varphi e^{\delta t}}\right)^{\frac{1}{1-\gamma}} \cdot x^{\frac{\gamma\theta}{1-\gamma}} \quad (5.80)$$

To compute $\varphi = \varphi_u$ depending on the variable u, we use the following constraint

$$A(T - u) \quad (5.81)$$

$$= \mathbb{E}\left(\int_u^T \pi_r(s)ds + \sum_{i \in I_r}\left(\int_u^T \tilde{\lambda}\tilde{\epsilon}_r(s)D_{i,r}(s)ds - \int_u^T (1+\lambda\epsilon_i)\mu_{i,r}(s)ds\right)\right)$$

$$= \mathbb{E}\left(\int_u^T \pi_r(s)ds + \sum_{i \in I_r}\left(\int_u^T \tilde{\lambda}\tilde{\epsilon}_r(s)D_{i,r}(s)ds - \int_u^T (1+\lambda\epsilon_i)\frac{R_{i,r}(s)}{x_r(s)}ds\right)\right)$$

$$= \mathbb{E}\left(\int_u^T \pi_r(s)ds + \sum_{i \in I_r}\left(\int_u^T \tilde{\lambda}\tilde{\epsilon}_r(s)D_{i,r}(s)ds\right.\right.$$
$$\left.\left. - \int_u^T \frac{\theta\tilde{\lambda}\tilde{\epsilon}_r(s)}{1-\theta}(1-D_{i,r}(s))ds\right)\right)$$

$$= \mathbb{E}\left(\int_u^T \pi_r(s)ds + |I_r|\int_u^T \tilde{\lambda}\tilde{\epsilon}_r(s)ds - \sum_{i \in I_r}\int_u^T \frac{\tilde{\epsilon}_r(s)}{1-\theta}\tilde{\lambda}(1-D_{i,r}(s))ds\right)$$

where we inserted (5.79) for R_s, did some arrangements and used the notation $|I_r| = \#\{i \in I_r\}$ for the number of elements in I_r. Using (5.81) we get

$$A(T - u) - \int_u^T \pi_r(s)ds - |I_r|\int_u^T \tilde{\lambda}\tilde{\epsilon}_r(s)ds$$

$$= -\mathbb{E}\left(\int_u^T \sum_{i \in I_r}\left(\frac{1}{1-\theta}\tilde{\lambda}\tilde{\epsilon}_r(s)\right) \cdot \left(\frac{\theta}{1-\theta}\tilde{\lambda}\tilde{\epsilon}_r(s)\right)^{\frac{\gamma\theta-1}{1-\gamma}}\right.$$
$$\left.\cdot (1+\lambda\epsilon_i)^{\frac{\gamma\theta}{\gamma-1}} \cdot \left(\frac{\theta}{\varphi e^{\delta t}}\right)^{\frac{1}{1-\gamma}} \cdot x^{\frac{\gamma\theta}{1-\gamma}}ds\right)$$

Applying Fubini's theorem, differentiating with respect to u and setting $\Lambda(u) = \pi_r(u) + |I_r|\tilde{\lambda}\tilde{\epsilon}_r(u) - A$, this yields to

$$\Lambda(u) = \mathbb{E}\left(\sum_{i \in I_r}\left(\frac{1}{1-\theta}\tilde{\lambda}\tilde{\epsilon}_r(s)\right) \cdot \left(\frac{\theta}{1-\theta}\tilde{\lambda}\tilde{\epsilon}_r(s)\right)^{\frac{\gamma\theta-1}{1-\gamma}}\right.$$
$$\left.\cdot (1+\lambda\epsilon_i)^{\frac{\gamma\theta}{\gamma-1}} \cdot \left(\frac{\theta}{\varphi e^{\delta t}}\right)^{\frac{1}{1-\gamma}} \cdot x^{\frac{\gamma\theta}{1-\gamma}}(u)\right)$$

Hence, we obtain as candidate for φ_u

$$\varphi_u^{-\frac{1}{1-\gamma}} = x_r^{\frac{-\gamma\theta}{1-\gamma}}(u) \cdot (e^{\delta u})^{\frac{1}{1-\gamma}} \cdot \Lambda(u) \cdot \frac{1-\theta}{\tilde{\lambda\epsilon}_r(u)} \cdot \left(\frac{\theta}{1-\theta}\tilde{\lambda\epsilon}_r(u)\right)^{\frac{\gamma\theta-1}{\gamma-1}}$$

$$\cdot \theta^{-\frac{1}{1-\gamma}} \cdot \left(\sum_{i \in I_r}(1+\lambda\epsilon_i(u))^{\frac{\gamma\theta}{\gamma-1}}\right)^{-1}$$

Defining $\Delta_r(u) = \sum_{i \in I_r}(1+\lambda\epsilon_i(u))^{\frac{\gamma\theta}{\gamma-1}}$ and inserting it back into (5.80), reveals finally

$$1 - D_{i,r}(u) = \frac{x_r^{\frac{\gamma\theta}{1-\gamma}}(u)}{x_r^{\frac{\gamma\theta}{1-\gamma}}(u)} \cdot \Lambda(u) \cdot \frac{1-\theta}{\tilde{\lambda\epsilon}_r(u)} \cdot (1+\lambda\epsilon_i(u))^{\frac{\gamma\theta}{\gamma-1}} \cdot \Delta_r^{-1}(u) \quad (5.82)$$

$$= \left(\pi_r(u) + |I_r|\tilde{\lambda\epsilon}_r(u) - A\right) \cdot \frac{(1-\theta)(1+\lambda\epsilon_i(u))^{\frac{\gamma\theta}{\gamma-1}}}{\tilde{\lambda\epsilon}_r(u)} \cdot \Delta_r^{-1}(u)$$

and for R we get

$$R_{i,r}(u) = \left(\pi_r(u) + |I_r|\tilde{\lambda\epsilon}_r(u) - A\right) \cdot \frac{\theta(1+\lambda\epsilon_i(u))^{\frac{\gamma(\theta-1)+1}{\gamma-1}}}{\Delta_r(u)} \cdot x_r(u) \quad (5.83)$$

Thus, we conclude for the prices μ_i

$$\mu_{i,r}(u) = \left(\pi_r(u) + |I_r|\tilde{\lambda\epsilon}_r(u) - A\right) \cdot \frac{\theta(1+\lambda\epsilon_i(u))^{\frac{\gamma(\theta-1)+1}{1-\gamma}}}{\Delta_r(u)} \quad (5.84)$$

Since u and T are arbitrary, we have (5.82) to (5.84) for all $u \geq 0$. Finally, we have to check the assumptions on φ_u. The process $(\mu_{i,r}(u))$ is uncorrelated to $(x_r(u))$, because $\mu_{i,r}(u)$ is a deterministic function. The random variable φ_u depends only on the value of $x_r(u)$ and is therefore uncorrelated to the increments of $(\mu_{i,r}(t))_{u<t\leq T}$, $(M_{i,r}(t))_{u<t\leq T}$ and $(N_{i,r}(t))_{u<t\leq T}$, because of the independent increments of the Poisson process, and since the Poisson processes $(M_{i,r}(t))_{u<t\leq T}$ and $(N_{i,r}(t))_{u<t\leq T}$ are independent of the fractional Brownian motion.

By the next theorem we are able to solve the optimization problem (P1). We will fix the basic prices π_r, the time preference δ, the concavity of the utility γ, the weighting of income by the resource fees $\mu_{i,r}$, the multiplier $D_{i,r}$ and the 'congestion reply' $\epsilon_{i,r}$ with its intensity λ. The LRD property will come in as well by the scaling factor a as by the crucial Hurst exponent H. Since we start with this facts, we will optimize w.r.t. the steady state rate A. Choosing this optimized slope A, we will see, how the solution of the rate x_r depends on the other parameter just mentioned. Before starting with the theorem, we briefly comment on the above optimal income $R_{i,r}$ and its multipliers $\mu_{i,r}$ and $D_{i,r}$.

Interpretation

The multipliers $1 - D_{i,r}$ can be considered as the part of the company (resp. network provider) responsible for maintenance of the network. Thus, in principle, $D_{i,r}$ represents the development part, enabling the increase of network performance. The direct input for the network is modeled as independent Poisson processes for the nodes or bottleneck. Basically, we have two parts on the right hand side of (5.69):

- $\pi_r(t)x_r(t)dt - \epsilon_i(t)\mu_{i,r}dN_{i,r}(\mu_{i,r}) + a\pi_r x_r dB_t^{(H)}$ reflects the pure network behavior, i.e. the change of transmission rate proportional to the price together with the congestion control expressed in the 'Poisson part' of $N_{i,r}$ and the LRD property of the network incorporated by the factor $ax_r dB_r(t)^{(H)}$.
- the economic part $x_r(t\sum_{i \in I_r}(\mu_{i,r}(t)dt+x_r(t)\sum_{i \in I_r}\tilde{\epsilon}_i(t)dM_{i,r}(D_{i,r})$ which reflects the tendency to use less bandwidth according to the price of nodes (as routers or servers), whose performance is increased by the efforts of the network randomly according to the independent Poisson processes $dM_{i,r}$.

Some more comments on the parameters are in order.

- We consider the multiplier $D_{i,r}$. It is depending on the term $A - \pi_r(u)$, which indicates that a high basic price π_r does not necessary increase the surplus, expressed by the term $\sum_{i \in I_r, r \in \mathcal{R}} R_{i,r} D_{i,r}$.
- The optimal 'price' and multiplier for the congestion rate $\mu_{i,r}(u)$ given by the equation (5.84) indicates, that enlarging the traffic rate A_r will in turn result in a lower price: this indicates that given fixed price π_r and improvement rates $\tilde{\epsilon}_r$, the price $\mu_{i,r}$ have to decrease for higher rates A_r.
- If we increase the improvement rate $\tilde{\epsilon}_r$, then the network can impose in turn higher prices for using the better performing nodes.
- If the user accepts a high price for the overall network use given by π_r, the network can in turn again impose higher prices for the use of each nodes, provided the stationary rate A_r is not changed.

Theorem 5.36. *The optimal income rate* $(R^*(t))$, *the optimal part of the multiplier* $(D^*(t))$ *and the optimal growth rate* A_r^*, *depending on the given parameters* δ, γ, θ, λ *and the basic price* π_r *of the constraint problem (P1) is given by*

$$R_{i,r}^*(t) = \left(\pi_r(t) + |I_r|\tilde{\lambda\epsilon}_r(t) - A_r\right) \cdot \frac{\theta(1+\lambda\epsilon_i(t))^{\frac{\gamma(\theta-1)+1}{1-\gamma}}}{\Delta_r(t)} \cdot x_r(t)$$

$$D_{i,r}^*(t) = 1 - \left(\pi_r(t) + |I_r|\tilde{\lambda\epsilon}_r(t) - A_r\right) \cdot \frac{(1-\theta)(1+\lambda\epsilon_i(t))^{\frac{\gamma\theta}{\gamma-1}}}{\tilde{\lambda\epsilon}_r(t)\Delta_r(t)}$$

where A_r *is the steady state rate of user* r. *We have for the development of the data rate*

$$x_t^* = x_0 \exp\left(\int_0^t \pi_r(s)ds - \frac{1}{2}a^2 t^{2H} + aB_t^{(H)} \right. \tag{5.85}$$

$$- \sum_{i \in I_r} \left(\int_0^t \left(\pi_r(s) + |I_r|\tilde{\lambda}\tilde{\epsilon}_r(s) - A\right) \cdot \frac{\theta(1+\lambda\epsilon_i(s))^{\frac{\gamma(\theta-1)+1}{1-\gamma}}}{\Delta_r(s)} ds \right)$$

$$\left. - \sum_{i \in I_r} \int_0^t \log(1+\epsilon_i(s))dN_i(s) + \sum_{i \in I_r} \int_0^t \log(1+\tilde{\epsilon}_r(s))dM_i(s) \right)$$

If in particular the prices π_r and the rates $\tilde{\epsilon}_r$ and ϵ_i are time independent (i.e. constant), then the optimal rate A_r^* is the unique solution of the equation

$$\int_0^\infty \exp\left(-\delta t + \gamma\theta A_r^* t - \gamma\theta\frac{1}{2}a^2 t^{2H} + \frac{1}{2}\gamma^2\theta^2 a^2 t^{2H} \right)$$

$$\left(\theta t\left(\pi_r + |I_r|\tilde{\lambda}\tilde{\epsilon}_r - A_r^*\right) - 1 \right) dt = 0 \tag{5.86}$$

Proof. The formula for R_t^* and D_t^* is a consequence of equation (5.82) and (5.83). To verify x_t^*, we consider the following stochastic differential equation

$$dx(t) = \pi_r(t)x(t)dt$$

$$- \sum_{i \in I_r} \left(\left(\pi_r(t) + |I_r|\tilde{\lambda}\tilde{\epsilon}_r(t) - A_r\right) \frac{\theta(1+\lambda\epsilon_i(t))^{\frac{\gamma(\theta-1)+1}{1-\gamma}}}{\Delta_r(t)} x(t)dt \right)$$

$$- \sum_{i \in I_r} \epsilon_i(t)x(t)dN_i(t) + \sum_{i \in I_r} \tilde{\epsilon}_r(t)dM_i(t) + ax(t)dB_t^{(H)}$$

Using theorem 5.34 (note that the function in the fractional Brownian motion part is constant a and thus the Malliavin operator turns in to the identy), we obtain (5.85). Finally, we have to show (5.86). In doing so, we insert (R_t) and (L_t) in the value function of (P1) and obtain

$$\int_u^\infty \frac{1}{\gamma}\exp(-\delta t) \sum_{r \in \mathcal{R}} \sum_{i \in I_r} R_t^{\gamma\theta}(1-D)_t^{\gamma(1-\theta)} dt$$

$$= \int_u^\infty \frac{1}{\gamma}\exp(-\delta t)$$

$$\cdot \sum_{r \in \mathcal{R}} \sum_{i \in I_r} \left(\left(\pi_r(t) + |I_r|\tilde{\lambda}\tilde{\epsilon}_r(t) - A_r\right) \cdot \frac{\theta(1+\lambda\epsilon_i(t))^{\frac{\gamma(\theta-1)+1}{\gamma-1}}}{\Delta_r(t)} \right)^{\gamma\theta} x_r(t)^{\gamma\theta}$$

$$\cdot \left(\left(\pi_r(t) + |I_r|\tilde{\lambda}\tilde{\epsilon}_r(t) - A_r\right) \cdot \frac{(1-\theta)(1+\lambda\epsilon_i(u))^{\frac{\gamma\theta}{\gamma-1}}}{\tilde{\lambda}\tilde{\epsilon}_r(t)} \Delta_r^{-1}(t) \right)^{\gamma(1-\theta)} dt$$

$$= \int_u^\infty \frac{1}{\gamma} \exp(-\delta t)$$

$$\cdot \sum_{r \in \mathcal{R}} \sum_{i \in I_r} \left(\frac{\pi_r(t) + |I_r|\tilde{\lambda}\tilde{\epsilon}_r(t) - A_r}{\Delta_r(t)} \right)^\gamma x_r(t)^{\gamma\theta} \theta^{\gamma\theta} (1-\theta)^{\gamma(1-\theta)}$$

$$\cdot (1 + \lambda\epsilon_i(u))^{\frac{\gamma\theta}{\gamma-1}} (\tilde{\lambda}\tilde{\epsilon}_r(t))^{\gamma(\theta-1)} dt$$

$$\cdot \int_u^\infty \frac{1}{\gamma} \exp(-\delta t) \sum_{r \in \mathcal{R}} \left(\pi_r(t) + |I_r|\tilde{\lambda}\tilde{\epsilon}_r(t) - A_r \right)^\gamma \Delta_r(t)^{1-\gamma}$$

$$\cdot x_r(t)^{\gamma\theta} \theta^{\gamma\theta} (1-\theta)^{\gamma(1-\theta)} (\tilde{\lambda}\tilde{\epsilon}_r(t))^{\gamma(\theta-1)} dt$$

because of the definition of Δ_r and using the constant stochastic growth rate A of the transmitted data. We maximize according to A_r. The function

$$(A_r) \to V(S, (A_r))$$

$$= \mathbb{E}\left(\int_u^\infty \frac{1}{\gamma} \exp(-\delta t) \left(\sum_{r \in \mathcal{R}} \left(\pi_r(t) + |I_r|\tilde{\lambda}\tilde{\epsilon}_r(t) - A_r \right)^\gamma \right. \right.$$

$$\left. \left. \cdot \Delta_r(t)^{1-\gamma} x_r(t)^{\gamma\theta} \theta^{\gamma\theta} (1-\theta)^{\gamma(1-\theta)} (\tilde{\lambda}\tilde{\epsilon}_r(t))^{\gamma(\theta-1)} \right) dt \right)$$

is concave (note that $\lim_{A_r \to -\infty} V(S, A_r) = 0$ and $V(S, \tilde{A}_r) = 0$ for $\tilde{A}_r = \pi_r(t) + |I_r|\tilde{\lambda}\tilde{\epsilon}_r(t)$). We differentiate it with respect to A

$$\frac{\partial}{\partial A_r}(V(S, (A_r)))$$

$$= \frac{1}{\gamma} \int_u^\infty \frac{\partial}{\partial A_r} \left(\exp(-\delta t) \left(\sum_{r \in \mathcal{R}} \left(\pi_r(t) + |I_r|\tilde{\lambda}\tilde{\epsilon}_r(t) - A_r \right)^\gamma \right. \right.$$

$$\cdot \Delta_r(t)^{1-\gamma}$$

$$\left. \left. \cdot \mathbb{E}\left(x_r(t)^{\gamma\theta} \right) \theta^{\gamma\theta} (1-\theta)^{\gamma(1-\theta)} \left(\tilde{\lambda}\tilde{\epsilon}_r(t) \right)^{\gamma(\theta-1)} \right) \right) dt$$

$$= \frac{\theta^{\gamma\theta}(1-\theta)^{\gamma(1-\theta)}}{\gamma} \int_u^\infty \exp(-\delta t) \Delta_r(t)^{1-\gamma} \left(\tilde{\lambda}\tilde{\epsilon}_r(t) \right)^{\gamma(\theta-1)}$$

$$\cdot \frac{\partial}{\partial A_r} \left(\left(\pi_r(t) + |I_r|\tilde{\lambda}\tilde{\epsilon}_r(t) - A_r \right)^\gamma \cdot \mathbb{E}\left(x_r(t)^{\gamma\theta} \right) \right) dt$$

By using the product rule, follows

$$\frac{\partial}{\partial A_r}(V(S,A)) = \int_u^\infty \frac{\theta^{\gamma\theta}(1-\theta)^{\gamma(1-\theta)}}{\gamma} \exp(-\delta t)\Delta_r(t)^{1-\gamma}\left(\tilde\lambda\tilde\epsilon_r(t)\right)^{\gamma(\theta-1)}$$
$$\cdot \left(\mathbb{E}\left(x_r^{\gamma\theta}(t)\right)\frac{\partial}{\partial A_r}\left(\pi_r(t)+|I_r|\tilde\lambda\tilde\epsilon_r(t)-A_r\right)^\gamma\right. \quad (5.87)$$
$$\left. + \left(\pi_r(t)+|I_r|\tilde\lambda\tilde\epsilon_r(t)-A_r\right)^\gamma \frac{\partial}{\partial A_r}\mathbb{E}\left(x_r^{\gamma\theta}(t)\right)\right)dt$$
$$= \int_u^\infty \frac{\theta^{\gamma\theta}(1-\theta)^{\gamma(1-\theta)}}{\gamma} \exp(-\delta t)\Delta_r(t)^{1-\gamma}\left(\tilde\lambda\tilde\epsilon_r(t)\right)^{\gamma(\theta-1)}$$
$$\cdot \left(\mathbb{E}\left(x_r^{\gamma\theta}(t)\right)\left(-\gamma\left(\pi_r(t)+|I_r|\tilde\lambda\tilde\epsilon_r(t)-A_r\right)^{\gamma-1}\right)\right.$$
$$\left. + \left(\left(\pi_r(t)+|I_r|\tilde\lambda\tilde\epsilon_r(t)-A_r\right)^\gamma\right)\frac{\partial}{\partial A_r}\mathbb{E}\left(x_r^{\gamma\theta}(t)\right)\right)dt$$

Thus, we need to compute $\mathbb{E}(x_r^{\gamma\theta}(t))$. We now assume constant rates $\epsilon_i, \tilde\epsilon_r$ and price π_r. Because of theorems (5.29) and (5.33) we conclude with (5.85)

$$\mathbb{E}\left(x_r^{\gamma\theta}(t)\right) = x_0 \exp\left(\gamma\theta t A_r - \gamma\theta\frac{1}{2}a^2 t^{2H} + \frac{1}{2}\gamma^2\theta^2 a^2 t^{2H}\right)$$

Then, we get finally by (5.87)

$$\frac{\partial}{\partial A_r}(V(S,A)) = \theta^{\gamma\theta}(1-\theta)^{\gamma(1-\theta)}$$
$$\cdot \int_u^\infty \exp(-\delta t)\Delta_r(t)^{1-\gamma}\left(\tilde\lambda\tilde\epsilon_r(t)\right)^{\gamma(\theta-1)}$$
$$\cdot \left(x_0 \exp\left(\gamma\theta A t - \gamma\theta\frac{1}{2}a^2 t^{2H} + \frac{1}{2}\gamma^2\theta^2 a^2 t^{2H}\right)\right.$$
$$\cdot \left(\theta t\left(\pi_r(t)+|I_r|\tilde\lambda\tilde\epsilon_r(t)-A_r\right)-1\right)$$
$$\left. \cdot \left(\left(\pi_r(t)+|I_r|\tilde\lambda\tilde\epsilon_r(t)-A_r\right)^{\gamma-1}\right)\right)dt$$

Since the price functions $\pi_r, \tilde\epsilon_r$ and congestion control function ϵ_i are deterministic, we can conclude after rearrangements for the maximum A^* with $\frac{\partial}{\partial A_r}V(S,A_r^*)=0$

$$\int_0^\infty \exp\left(-\delta t + \gamma\theta A_r^* t - \gamma\theta\frac{1}{2}a^2 t^{2H} + \frac{1}{2}\gamma^2\theta^2 a^2 t^{2H}\right)$$
$$\cdot \left(\theta t\left(\pi_r + |I_r|\tilde\lambda\tilde\epsilon - A_r^*\right) - 1\right)dt = 0$$

□

5.3 Traffic Optimization

Remark 5.37. We give a short comment of the optimal rate A_r^*. It is independent of the fixed congestion rate ϵ_i. An increasing ϵ_i would on a first glance result in a lower rate x_r. But in turn, the lower price $\mu_{i,r}$ according to (5.84) will balance this effect. A decrease on the other hand has the same effect: better performance is responsible for higher prices $\mu_{i,r}$.

Up to now we know the network provider optimizes the price for using the nodes respectively the optimal investment for development. As open parameters the users have to pay for bandwidth and the surplus of performance at the bottlenecks. Thus, we consider the following utility problem for the users (resp. the route) $r \in \mathcal{R}$.

$$u_r(x_r) = w_1 \frac{x_r^{\tilde{\gamma}}}{\tilde{\gamma}} - w_2 \frac{\left(\pi_r^\theta (|I_r|\tilde{\epsilon}_r)^{1-\theta}\right)^{\tilde{\gamma}} x_r^{\tilde{\gamma}}}{\tilde{\gamma}} \tag{5.88}$$

with $w_1, w_2 \in]0,1[$, $w_1 > w_2$. Some comments should be made on (5.88). The utility approach described in equation (5.57) is modified, since a computable expression for $\mathbb{P}(Q > x)$, as queueing probability is even in the simple case of the FBM not available. Thus, this part was taken tribute as diminishing factor in the driving stochastic equation of dx_r expressed by the Poisson processes $dN_{i,r}$ as control cycle. So basically the utility of the performance expressed by $\frac{x_r^{\tilde{\gamma}}}{\tilde{\gamma}}$ reflects the network performance.

The economic part $(\pi_r^\theta(|I_r|\tilde{\epsilon}_r)^{1-\theta})^{\tilde{\gamma}} \cdot x_r^{\tilde{\gamma}}/\tilde{\gamma}$ indicates the price of the bandwidth in conjunction with the surplus factor for a better bottleneck performance. Both ingredients are weighted by factors w_1, w_2, where naturally we assume $w_1 > w_2$. We again pursue the technique described above for optimizing the network and obtain for the optimal price vector $(\pi_r, \tilde{\epsilon}_r)$ the equations

$$\pi_r = \frac{-\theta}{1-\theta} \tilde{\lambda} |I_r| \tilde{\epsilon}_r \left(\sum_{i \in I_r}(1 - D_{i,r}) + \frac{\theta}{1-\theta} \tilde{\lambda} \tilde{\epsilon}_r |I_r|^2 \right)$$

$$\tilde{\epsilon}_r = \frac{A_r^*}{\tilde{\lambda}} \left(\theta \frac{|I_r|^2}{1-\theta} + |I_r| - \frac{\theta |I_r| \sum_{i \in I_r}(1 - D_{i,r})}{1-\theta} - \frac{\sum_{i \in I_r}(1 - D_{i,r})}{1-\theta} \right)^{-1}$$

Thus, we can formulate the final theorem.

Theorem 5.38. *Let the optimal control problem (P1) be given. Then there exists a equilibrium price vector and an optimal distribution of development investment for each node $i \in I$, which is given by*

$$\tilde{\epsilon}_r{}^*(t) = \frac{A_r^*}{\tilde{\lambda}} \frac{\theta \Delta_r(t) - \frac{1-\theta}{1+\theta}}{2\theta \Delta_r(t) - \frac{1}{1+\theta}} \quad (r \in \mathcal{R})$$

$$\pi_r^*(t) = \frac{\theta}{1-\theta} \tilde{\lambda} \tilde{\epsilon}_r{}^*(t) \sum_{i \in I_r} D_{i,r}^*$$

$$= \frac{\theta}{1-\theta} |I_r| \tilde{\lambda} \tilde{\epsilon}_r{}^*(t) - \frac{\theta}{1-\theta} \tilde{\lambda} \tilde{\epsilon}_r{}^*(t) \sum_{i \in I_r} (1 - D_{i,r}^*)$$

$$\mu_{i,r}^*(t) = \left(\pi_r^*(t) + |I_r| \tilde{\lambda} \tilde{\epsilon}_r{}^*(t) - A^* \right) \frac{\theta(1 + \lambda \epsilon_i(t))^{\frac{\gamma(\theta-1)+1}{1-\gamma}}}{\Delta_r(t)} \quad (i \in I)$$

$$D_{i,r}(t)^* = 1 - \left(\pi_r^*(t) + |I_r| \tilde{\lambda} \tilde{\epsilon}_r{}^*(t) - A_r^* \right)$$
$$\cdot \frac{(1-\theta)(1 + \lambda \epsilon_i(t))^{\frac{\gamma\theta}{\gamma-1}}}{\tilde{\lambda} \tilde{\epsilon}_r{}^*(t) \Delta_r(t)} \quad (i \in I_r, \ r \in \mathcal{R})$$

The optimal A_r^* is given by the equation (5.86) for all $r \in \mathcal{R}$.

Remark 5.39. The optimal vector $(A_r^*, \tilde{\epsilon}_r(t), \pi_r^*(t), \mu_{i,r}^*(t), D_{i,r}(t)^*)$ should be commented:

1. If the rate A_r^* increases, then in turn the rate $\tilde{\epsilon}_r$ will, too. This can be interpreted that it is more efficient for the network provider to invest more into the improvement of the net, since it will, according to (5.84), enable to receive a higher price for the nodes.
2. If the congestion rates ϵ_i are constant, this implies the same for $\tilde{\epsilon}_r$ as well.
3. Again if the rates ϵ_i are constant, then we get

    π_r is constant $\Leftrightarrow D_{i,r}^*$ is constant $\Leftrightarrow \mu_{i,r}^*$ is constant.

    Thus, if we fix an optimal price π_r^* at time $t = 0$, then all the other optimal solutions will be constant, too. We have a stationary system and can skip the time parameter t.

Impact of Network Structure

The network structure enters in two decisive parts. One consists in the congestion mechanism, expressed in the term '$\lambda \epsilon_i$,' giving the amount of data being send back or denied for transmission and the rate of occurrence. This is basically a network implemented procedure and is triggered in a more predictable way. The second part reflects the long-range dependence and is incorporated in the Hurst parameter of the FBM. Thus, in some way only implicitly we observe this impact via the optimal parameter A_r^*, which is computed out of the equation (5.86). In the succeeding figures we show, how the parameter A_r^* changes, if the Hurst exponent differs between 0.5 (Brownian motion) and 0.925. We selected the free parameters $\gamma = 0.5$, $\theta = 0.7$, $\lambda = 0.1$, $\tilde{\lambda} = 0.1$, $\epsilon_i = 0.1$, $i \in I$, and $|I_r| = 20$. First, we compute the optimal A_r^* in dependence of the Hurst parameter. In the figure 5.26 we see that the optimal A_r^*

given according to (5.86) as intersection value with the horizontal axes increases with increasing H. Hence, the curves represent the partial derivatives $\frac{\partial}{\partial A_r}(V(S, A))$. The figures shows the function defined as the left hand side of (5.86).

In the figure 5.26 we compare the solutions of the different optimal A_r^*, the left diagram for 20 nodes and right diagram for 5. Clearly, the optimal A_r^* increase, while the number of nodes I_r for the route r decreases. This can be interpreted in the following way: more nodes means that the steady state rate of $\frac{dx_r}{x_r}$ is lower. On the other hand, as seen in the figures 5.27 and 5.28, there is for lower optimal A_r^* a higher number of nodes which is used.

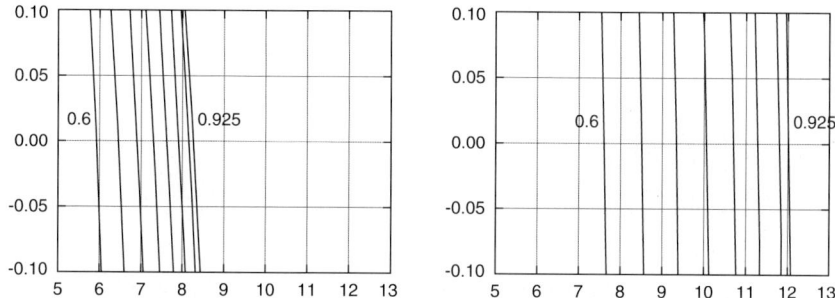

Fig. 5.26. Solutions for the optimal A_r according to equation (5.86) for different Hurst exponents $H = 0.6, 0.65, 0.7, 0.75, 0.8, 0.85, 0.9, 0.925$ and different number of nodes $|I_r| = 20$ (left) and $|I_r| = 5$ (right). The intersection with the x-axis (zero line) is the desired value of A_r.

In the figure 5.27 we present the optimal solutions of the price vector $\mu_{i,r}^*$, the development investment as part of all employees and the optimal price vectors for π_r^* and $\tilde{\epsilon}_r^*$.

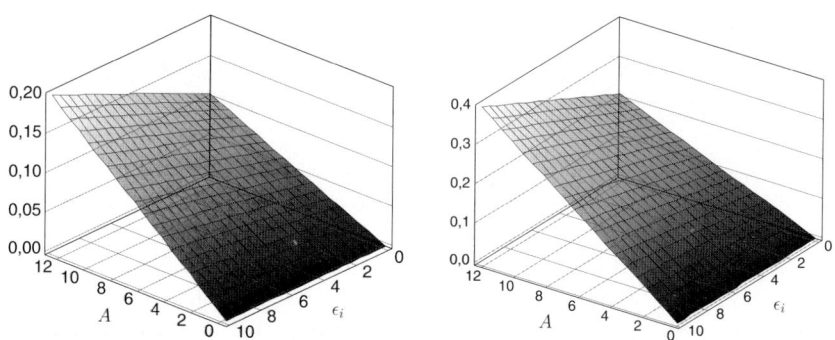

Fig. 5.27. The price μ_i depending on the steady state rate A_r and the parameter ϵ_i for different number of nodes $|I_r| = 20$ (left) and $|I_r| = 5$ (right)

Since no user is specially marked, we just choose as representative user r. For this we used first $|I_r| = 20$ and then $|I_r| = 5$ otherwise common parameters. We see that a higher number of nodes indicates in turn a lower price per rate unit. On the other hand, in the left figure we use 4 times as much nodes than in the right figure. For $H = 0.6$ we obtain as optimal rate A_r for 20 nodes 5.95 and for 5 nodes 7.86. Choosing in both cases $\epsilon_i = 6$ we get as income for the provider 1.48 in the case of 20 nodes and 0.975 in the case of 5 nodes. Thus, the cost for using one node is decreasing with more complicated network, while the overall costs are increasing.

As the figure 5.28 indicate, more nodes mean that the investment for development is even more efficient for lower steady state rate A_r^*. This is induced by the fact that it is economical reasonable to invest labor for development and increase the performance, if the number of nodes is higher.

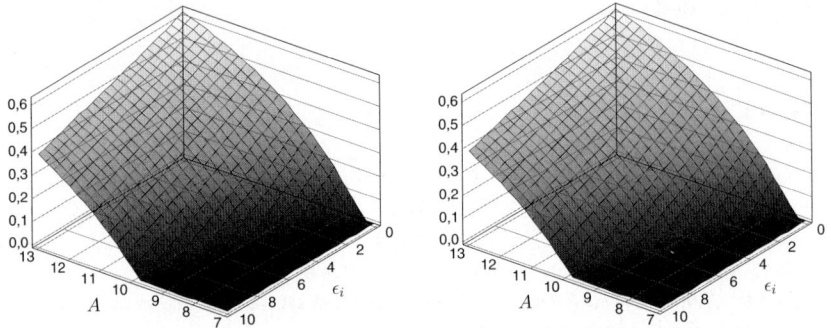

Fig. 5.28. Part of development investment $D_{i,r}$ depending on the steady state rate A_r and the parameter ϵ_i for the number of nodes $|I_r| = 20$ (left) and $|I_r| = 5$ (right)

Again we can realize that in the first case of a higher number of nodes we find that a lower steady state rate A_r^* and even a higher congestion control parameter gives rise to lower prices for each node. The comparison for the investment of development indicates that a bigger part for the development is efficient for a lower rate of A_r^* and a higher rate of congestion parameter. This is confirmed by the fact that it is reasonable to invest more for a better performance of the bottlenecks resp. nodes.

Economic Interpretation

We used a twofold approach to find an equilibrium for prices and investment of development:

- The Cobb-Douglas production function, expressed by the equation (5.70), is standard in economic where the ingredients are the prices $\mu_{i,r}^*$ (resp. $R_{i,r} = \mu_{i,r}^* x_r$) for the nodes and the multiplier $1 - D_{i,r}^*$ as part of the maintenance. The utility approach is used for finding decisions.

- For finding fair price we have chosen a similar utility approach for each user, where the utility concavity is assumed to be equal to the network provider. The decision is depending on the price for bandwidth π_r^* and the price $\tilde{\epsilon}_r{}^*$ for the improvement of the performance of the nodes. It can also be regarded as a price for priority handling.
- The result is an existing equilibrium price vector $(\pi_r^*, \mu_i^*, \tilde{\epsilon}_r{}^*)$ and optimal selection for development investment $D_{i,r}^*$.
- A higher number of bottleneck indicates that it is more efficient to induce a lower price $\mu_{i,r}^*$, which will then result in a lower optimal steady state rate A_r^*. On the other hand the prices π_r^* are lower and it is more efficient to invest in the improvement of the node performance, which is given by higher values of $D_{i,r}$.
- A higher value of $\tilde{\lambda}$ (which we did not reflect in additional figures) indicate a better use of the investment and thus, a lower value or a smaller part of $D_{i,r}^*$ is needed.

Further Literature

Queueing theory for the Norros model in done in several articles, where we consulted the original literature [190, 192]. The fractional Lévy processes and the queueing behavior is e.g. investigated in [155]. The reader can look up a general description of the queueing theory in [103, 239]. The multifractal models and the queueing results are originated in the articles of [214, 215]. For the analytical treatment of the FBM and its application to queueing and stochastic optimization a survey on the integrals with respect to the FBM are given on [64, 236]. In [71] the economical part of the queueing is treated. A more functional analytic approach using differential equations is presented in [16].

References

1. Abramowitz, M. and Stegun, I. A. (eds.) *Handbook of Mathematical Functions with Formulas and Mathematical Tables*, Dover Publications (1965).
2. Abry, P. and Veitch, D. *Wavelet Analysis for Long Range Dependent Traffic*, IEEE Transactions on Information Theory, **44**(1), pp. 2–15 (1998).
3. Abry, P., Baraniuk, R., Flandrin, P., Riedi R. H. and Veitch, D. *The Multiscale Nature of Network Traffic: Discovery Analysis and Modelling*, IEEE Signal Processing Magazine, **19**(3), pp. 28–46 (2002).
4. Abry, P., Flandrin, P., Taqqu, M. S. and Veitch, D. *Self-Similarity and Long-Range Dependence through the Wavelet Lens*, In: Doukhan, P., Oppenheim, G. and Taqqu, M. S. (eds.) Long-Range Dependence: Theory and Application, Birkhäuser, pp. 527–556 (2003).
5. Abry, P., Flandrin, P., Taqqu, M. S. and Veitch, D. *Wavelets for the analysis, estimation and synthesis of scaling data*, In: Park, K. and Willinger, W. (eds.) Self-similar network traffic and performance evaluation, John Wiley & Sons, pp. 39–88 (2000).
6. Abry, P. and Sellan, F. *The wavelet-based synthesis for the fractional Brownian motion proposed by F. Sellan and Y. Meyer: Remarks and implementation*, Applied and Computational Harmonic Analysis, **3**(3), pp. 377–383 (1996).
7. Abry, P., Gonçalves, P. and Flandrin, P. *Wavelets, spectrum analysis and $1/f$ processes*, In: Antoniadis, A. and Oppenheim, G. (eds.) Wavelets and Statistics. Lecture Notes in Statistics, **103**, Springer, pp. 15–29 (1995).
8. Ackroyd, M. *Computing the waiting time distribution for $G/G/1$ queue by signal processing methods*, IEEE Transactions on Communications, **28**(1), pp. 52–58 (1980).
9. Addie, R. G., Mannersalo, P. and Norros, I. *Performance formulae for queues with Gaussian input*, Proc. 16th International Teletraffic Congress, Edinburgh, pp. 1169–1178 (1999).
10. Aghion, P. and Howitt, P. *Endogenous growth theory*, MIT Press (1998).
11. Alsmeyer, G. *Erneuerungstheorie*, Teubner (1999).
12. Andersen, A. and Nielsen, B. *A Markovian approach for modeling packet traffic with long range dependence*, IEEE Journal on Selected Areas in Communications, **16**(5), pp. 719–732 (1998).

13. Andersson, S. and Rydén, T. *Maximum Likelihood estimation of a structured MMPP with application to traffic modeling*, Proc. 13th ITC Specialist Seminar on IP Traffic Measurement, Modeling and Management, Monterey (2000).
14. Arbeiter, M. *Random recursive construction of self-similar fractal measures. The non-compact case*, Probability Theory and Related Fields, **88**(4), pp. 497–520 (1991).
15. Arbeiter, M. and Patzschke, N. *Random self-similar multifractals*, Mathematische Nachrichten, **181**, pp. 5–42 (1996).
16. Armbruster, D., Degond, P. and Ringhofer, C. *A model for dynamics of large queuing networks and supply chains*, SIAM Journal on Applied Mathematics, **66**(3), pp. 896–920 (2006).
17. Arneodo, A., Bacry, E. and Muzy, J. F. *Random cascades on wavelet dyadic trees*, Journal of Mathematical Physics, **39**(8), pp. 4142–4164 (1998).
18. Arnold, L. *Stochastische Differentialgleichungen. Theorie and Anwendung*, Oldenbourg (1973).
19. Arvidsson, A. and Karlsson, P. *On traffic models for TCP/IP*, Proc. 16th International Teletraffic Congress, Edinburgh, pp. 455–466 (1999).
20. Assmussen, S. *Stochastic simulation with a view towards stochastic processes*, book manuscript (1999).
21. Assmussen, S. *Applied Probability and Queues*, Springer (2003).
22. Assmussen, S., Klüppelberg, C. and Sigman, K. *Sampling at subexponential times, with queueing applications*, Stochastic Processes and their Applications, **79**(2), pp. 265–286 (1999).
23. Avram, F. and Taqqu, M. S. *Robustness of the R/S statistic for fractional stable noises*, Statistical Inference for Stochastic Processes, **3**(1–2), pp. 69–83 (2000).
24. Bachelier, L. *Théorie de la spéculation*, Annales Scientifiques de l'École Normale Supérieure, **3**(17), pp. 21–86 (1900).
25. Bardet, J.-M.,Lang, G., Oppenheim, G., Philippe, A. and Taqqu, M. S. *Generators of Long-Range Dependent Processes*, In: Doukhan, P., Oppenheim, G. and Taqqu, M. S. (eds.) Long-Range Dependence: Theory and Application, Birkhäuser pp. 579–623 (2003).
26. Bardet, J.-M., Lang, G., Oppenheim, G., Philippe, A., Stoev, S. and Taqqu, M. S. *Semi-Parametric Estimators of the Long-Range Dependence Parameter*, In: Doukhan, P., Oppenheim, G. and Taqqu, M. S. (eds.) Long-Range Dependence: Theory and Application, Birkhäuser, pp. 557–577 (2003).
27. Bardet, J.-M., Moulines, E. and Soulier, P. *Wavelet estimator of long-range dependent processes*, Statistical Inference for Stochastic Processes, **3**(1–2), pp. 85–99 (2000).
28. Bacry, E., Muzy, J. and Arneodo, A. *Singularity spectrum of fractal signals from wavelet analysis: Exact results*, Journal Statistical Physics, **70**(3–4), pp. 635–674 (1993).
29. Barford, P. and Crovella, M. *A performance evaluation of Hyper Text Transfer Protocols*, Proc. ACM SIGMETRICS International Conference on Measurement and Modeling of Computer Systems, Atlanta, pp. 188-197 (1999).
30. Beichelt, F. *Stochastische Prozesse für Ingenieure*, Teubner (1997).
31. Benes, V. E. *General Stochastic Processes in the Theory of Queues*, Addison-Wesley, 1963.
32. Beran, J. *Statistics for Long-Memory Processes*, Monographs on Statistics and Applied Probability, **61**, Chapman & Hall/CRC (1994).

33. Biagini, F. and Øksendal, B. *Forward integrals and an Itō formula for the fractional Brownian motion*, preprint, Department of Mathematics, University Oslo (2004).
34. Biagini, F., Øksendal, B., Sulem, A. and Wallner, N. *An introduction to white noise theory and Malliavin calculus for fractional Brownian motion*, Royal Society Proceedings, **460**, pp. 347–372 (2004).
35. Biagini, F., Hu, Y., Øksendal, B. and Zhang, T. *Stochastic Calculus for Fractional Brownian Motion and Applications*, Springer (2008).
36. Bhansali, R. J. and Kokoszka, P. S. *Prediction of long-memory time series*, In: Doukhan, P., Oppenheim, G. and Taqqu, M. S. (eds.) Long-Range Dependence: Theory and Application, Birkhäuser, pp. 354–367 (2003).
37. Blatter, Ch. *Wavelets. Eine Einführung*, Vieweg (2003).
38. Billingsley, P. *Probability and Measure*, John Wiley & Sons (1995).
39. Black, F. and Scholes, M. *The pricing of options and corporate liabilities*, Journal of Political Economy, **81**(3), pp. 637–654 (1973).
40. Borovkov, A. A. *Stochastic Processes in Queueing Theory*, Springer (1976).
41. Brichet, F., Roberts, J., Simonian, A. and Veitch, D. *Heavy Traffic Analysis of a Storage Model with Long Range Dependent On/Off Sources*, Queueing Systems, **23**, pp. 197–215 (1996).
42. Brockwell, P. J. and Davis, R. A. *Time Series: Theory and Methods*, Springer (1991).
43. Bronstein, I. A. and Semendjajew, K. A. *Taschenbuch der Mathematik*, Harri Deutsch (1972).
44. Bahadur, R. R. and Zabell, L. S. *Large deviation of the sample mean in general vector spaces*, Annals of Probability, **7**(4), pp. 587–621 (1979).
45. Carmona, P. and Coutin, L. *Fractional Brownian motion and Markov property*, Electronic Communications in Probability, **3**, pp. 95–107 (1998).
46. Carmona, P. and Coutin, L. *Intégrale stochastique pour le mouvement brownien fractionnaire*, Comptes Rendus de l'Académie des Sciences – Series I – Mathematics, **330**, pp. 231–236 (2000).
47. Cavanaugh, J. E., Wang, Y. and Davis, J. W. *Locally Self similar processes and their wavelet analysis*, Handbook of Statistic, **21**: Stochastic processes: Modeling and simulation, Elsevier, pp. 93–135 (2003).
48. Chainais, P., Riedi, R. H. and Abry, P. *On Non-Scale-Invariant Infinitely Divisible Cascades*, IEEE Transactions on Information Theory, **51**(3), pp. 1063–1083 (2005).
49. Charzinski, J. *Activity polling and contention in media access control protocol*, Journal on Selected Areas in Communications, **19**(9), pp. 1562–1571 (2000).
50. Charzinski, J. *Good news about heavy tails*, Proc. IEEE Conference on High Performance Switching and Routing, Heidelberg, June (2000).
51. Charzinski, J. *Internet client traffic measurement and characterisation results*, Proc. International Symposium on Services and Local Access, Stockholm (2000).
52. Chitour, Y. and Piccoli, B. *Traffic circles and timing of traffic lights for cars flow*, Discrete and Continuous Dynamical Systems. Series B, **5**(3), pp. 599–630 (2005).
53. Choi, H.-K., Limb, J. O. *A Behavioral Model of Web Traffic*, Proc. 7th International Conference on Network Protocols, Toronto (1999).
54. Christoph, G. and Hackel, H. *Starthilfe Stochastik*, Teubner (2002).

55. Coclite, G. M., Garavello, M. and Piccoli, B. *Traffic flow on a road network*, SIAM Journal on Mathematical Analysis, **36**(6), pp. 1862–1886 (2005).
56. Cox, D. R. *Long-range dependence: a review*, In: David, H. A. and David, H. T. (eds.) Statistics: An Appraisal, Iowa State University Press, pp. 55–74 (1984).
57. Cox, D. and Isham, V. *Point processes*, Monographs on Statistics and Applied Probability, **12**, Chapman & Hall/CRC (1980).
58. Crovella, M. E. and Bestavros, A. *Self-Similarity in World Wide Web Traffic Evidence and Possible Causes*, IEEE/ACM Transactions on Networking, **5**(6), pp. 835–846 (1997).
59. Crovella, M. E. and Bestavros, A. *Explaining World Wide Web traffic self-similarity*, Technical Report (Revised), **TR-95-015**, Boston University Department of Computer Science (1995).
60. Crovella, M. E. and Taqqu, M. S. *Estimating the Heavy-Tail Index from Scaling Properties*, Methodology and Computing in Applied Probability, **1**(1), pp. 55–79 (1999).
61. Crovella, M. E., Taqqu, M. S. and Bestavros, A. *Heavy-tailed Probability Distribution in the World Wide Web*, In: Adler, R., Feldman, R. and Taqqu, M. S. (eds.) A practical Guide to Heavy-tails: Statistical Techniques and Applications, Birkhäuser, pp. 3–25 (1998).
62. D'Apice, C., Manzo, R. and Piccoli, B. *Packets Flow on Telecommunication Networks*, SIAM Journal of Mathematical Analysis, **38**, pp. 717–740 (2006).
63. Daubechies, I. *Ten Lectures on Wavelets*, SIAM Philadelphia (1992).
64. Decreusefond, L. *Stochastic integration with respect to fractional Brownian motion*, In: Doukhan, P., Oppenheim, G. and Taqqu, M. S. (eds.) Long-Range Dependence: Theory and Application, Birkhäuser, pp. 203–226 (2003).
65. Decreusefond, L. and Üstünel, A. S. *Stochastic Analysis of the fractional Brownian motion*, Potential Analysis, **10**(2), pp. 177–214 (1999).
66. Deo, R. S. and Hurvich, C. M. *Estimation of long memory in volatility*, In: Doukhan, P., Oppenheim, G. and Taqqu, M. S. (eds.) Long-Range Dependence: Theory and Application, Birkhäuser, pp. 313–324 (2003).
67. Deuschel, J.-D. and Strook, D. W. *Large Deviations*, Pure and Applied Mathematics, **137**, Academic Press (1988).
68. Doukhan, P., Oppenheim, G. and Taqqu, M. S. (eds.) *Long-range Dependence: Theory and Application*, Birkhäuser (2003).
69. Doukhan, P. *Limit Theorems for stationary sequences*, In: Doukhan, P., Oppenheim, G. and Taqqu, M. S. (eds.) Long-Range Dependence: Theory and Application, Birkhäuser, pp. 43–100 (2003).
70. Doukhan, P., Khezour, A. and Lang, G. *Nonparametric estimation for long-range dependent sequences*, In: Doukhan, P., Oppenheim, G. and Taqqu, M. S. (eds.) Long-Range Dependence: Theory and Application, Birkhäuser, pp. 302–311 (2003).
71. Duffield, N. *Economies of scale for long-range dependent traffic in short buffers*, Telecommunication Systems, **7**(1–3), pp. 267–280 (1997).
72. Duffield, N. G. and O'Connell, N. *Large deviation and overflow probabilities for the general single-server queue with applications*, Mathematical Proceedings of the Cambridge Philosophical Society, **118**, pp. 363–374 (1995).
73. Duncan, T. E., Hu, Y. and Pasik-Duncan, B. *Stochastic calculus for fractional Brownian motion. I: Theory* SIAM Journal on Control and Optimization, **38**(2), pp. 582–612 (2000).

74. Dunford, N. and Schwartz, *Linear Operators Part I*, Interscience (1957).
75. Durrett, R. *Stochastic Calculus: A Practical Introduction*, CRC Press (1996).
76. Embrechts, P., Klüppelberg, C. and Mikosch, T. *Modelling Extremal Events for Insurance and Finance*, Springer (1998).
77. Embrechts, P. and Maejima, M. *Selfsimilar Processes*, Princeton University Press (2002).
78. Embrechts, P. and Veraverbeke, N. *Estimates for the probability of ruin with special emphasis on the possibility of large claims*, Insurance: Mathematics and Economics, **1**(1), pp. 55–72 (1982).
79. Engel, M., Fülleborn, U., Lauterbach, D., Nalewski, H. and Stahl, A., *Rainer Maria Rilke. Werke*, Insel (1996).
80. Evans, L. C. *Partial differential equations*, American Mathematical Society (1998).
81. Fadili, M. J. and Bullmore, E. T. *Wavelet-Generalized least squares: A new BLU estimator of linear regression models with $\frac{1}{f}$ errors*, NeuroImage, **15**(1), pp. 217–232 (2002).
82. Falconer, K. *Fractal Geometry Mathematical Foundation and Application*, John Wiley & Sons (1993).
83. Falconer, K. *The multifractal spectrum of statistically self-similar measures*, Journal of Theoretical Probability, **7**(3), pp. 681–702 (1994).
84. Feldmann, A., Gilbert, A. C., Huang, P. and Willinger, W. *Dynamics of IP Traffic: A study of the role of variability and the impact of control*, Proc. ACM SIGCOMM 99, Boston, Computer Communication Review, **33**(2), pp. 301–313 (1999).
85. Feldmann, A., Gilbert, A. C., Willinger, W. and Kurtz, T. G. *The Changing Nature of Network Traffic: Scaling Phenomena*, Computer Communication Review, **28**(2), pp. 5–29 (1998).
86. Feldmann, A. *On-line call admission for high-speed networks*, Ph.D. Thesis, School of Computer Science, Carnegie Mellon University, Pittsburgh (1995).
87. Feller, W. *An introduction to probability theory and its applications, Volume I*, John Wiley & Sons (1971).
88. Feller, W. *An introduction to probability theory and its applications, Volume II*, John Wiley & Sons (1971).
89. Fischer, W. and Meier-Hellstein, K. S. *The Markov-modulated Poisson process (MMPP) cookbook*, Performance Evaluation, **18**(2), pp. 149–171 (1992).
90. Flandrin, P. *Wavelet analysis and synthesis of fractional Brownian motion*, IEEE Transactions on Information Theory, **38**(2), pp. 910–917 (1992).
91. Fonseca, N. L., Mayor, G. S. and Neto, C. A. V. *On the equivalent bandwidth of self-similar source*, ACM Transactions on Modeling and Computer Simulation, **10**(2), pp. 104–124 (2000).
92. Forster, O. *Analysis I*, Vieweg (1976).
93. Fox, R. and Taqqu, M. S. *Large-sample properties of parameter estimates for strongly dependent stationary Gaussian time series*, Annals of Statistics, **14**(2), pp. 517–532 (1986).
94. Feldmann, A. and Whitt, W. *Fitting Mixtures of Exponentials to Long-Tail Distributions to Analyze Network Performance Models*, Performance Evaluation, **31**, pp. 245–279, (1998).
95. Fraleigh, C., Tobagi, F. and Diot, C. *Provisioning IP Backbone Networks to Support Latency Sensitive Traffic*, Proc. IEEE INFOCOM 2003, San Francisco (2003).

96. Gardner, R. *Games for Business and Economics*, John Wiley & Sons (1995).
97. Gaver, D. P., Jacobs, P.A. and Latouche, G. *Finite birth-and-death models in randomly changing environments*, Advances in Applied Probability, **16**, pp. 715–731 (1984).
98. Gänssler, P. and Stute, W. *Wahrscheinlichkeitstheorie*, Springer (1980).
99. Giraitis, L. and Robinson, P. M. *Parametric estimation under long-range dependence*, In: Doukhan, P., Oppenheim, G. and Taqqu, M. S. (eds.) Long-Range Dependence: Theory and Application, Birkhäuser, pp. 229–249 (2003).
100. Glossglauser, M. and Bolot, J.-C. *On the relevance of long range dependence in network traffic*, IEEE/ACM Transactions on Networking, **7**(5), pp. 629–640 (1999).
101. Granger, C. W. *Long memory relationship and the aggregation of dynamic models*, Journal of Econometrics, **14**(2), pp. 227–238 (1980).
102. Grimm, C. and Schlüchtermann, G. *Verkehrstheorie in IP-Netzen*, Hüthig (2005).
103. Gross, D. and Harris, C. M. *Fundamentals of Queueing Theory*, John Wiley & Sons (1985).
104. Greiner, M, Jobmann, M. and Klüppelberg, C. *Telecommunication traffic, queueing models and subexponential distributions*, Queueing Systems, **33**(1–3), pp. 125–152 (1999).
105. Guivarc'h, Y. *Remarques sur les solutions d'une équation fonctionelle non-linéaire de Benoît Mandelbrot*, Comptes Rendus Academie de France Paris, **305**(4), pp. 139–141 (1987).
106. Heyde, C. C. and Yang, Y. *On defining long-range dependence*, Journal of Applied Probability, **34**(4), pp. 939–944 (1997).
107. Heyman, D. P. and Lakshman, T. V. *What are the implications of long-range dependence for VBR-video traffic engineering?*, IEEE/ACM Transactions on Networking, **4**(3), pp. 301–317 (1996).
108. Heyman, D. P. and Lucantoni, D. *Modeling multiple IP traffic streams with rate limits*, IEEE/ACM Transactions on Networking, **11**(6), pp. 948–958 (2003).
109. Hogg, R. V. and Craig, A. T *Introduction to Mathematical Statistics*, Macmillan (1978).
110. Holden, H. and Øksendal, B. *A white noise approach to stochastic differential equations driven by Wiener and Poisson processes*, In: Grosser, M., Hörmann, G., Kunzinger, M. and Oberguggenberger, M. (eds.) Nonlinear theory of generalized functions, Chapman & Hall/CRC, pp. 293–313 (1999).
111. Higuchi, T. *Approach to an irregular time series on the basis of the fractal theory*, Physica D, **31**(2), pp. 277–283 (1988).
112. Huang, P., Feldmann, A. and Willinger, W. *A non-intrusive, wavelet-based approach to detecting network performance problems*, Proc. ACM SIGCOMM Workshop on Internet Measurement, San Francisco, pp. 213–227 (2001).
113. Hu, Y., Øksendal, B., Sulam, A. *Optimal consumption and portfolio in a Black-Scholes market driven by fractional Brownian motion*, preprint (2000).
114. Hu, Y. and Øksendal, B. *Fractional white noise calculus and application to finance*, Infinite Dimensional Analysis, Quantum Probability and Related Topics, **6**, pp. 1–32 (2003)
115. Hurst, H. E. *Long-term capacity of reservoirs*, Transactions of the American Society of Civil Engineers, **116**, pp. 770–808 (1951).

116. Hurvich, C. M., Deo, R. and Brodsky, J. *The mean squared error of Geweke and Porter-Hudak's estimator of the memory parameter of a long-memory time series*, Journal of Time Series Analysis, **19**(1), pp. 19–46 (1998).
117. Iouditsky, A., Moulines, E. and Soulier, P. *Adaptive estimation of the fractional differencing coefficient*, Bernoulli, **7**(5), pp. 699–731 (2001).
118. Irle, A. *Wahrscheinlichkeitstheorie and Statistik*, Teubner (2001).
119. Istas, J. and Lang, G. *Quadratic variations and estimation of the local Hölder index of a Gaussian process*, Probabilité et Statistiques, **33**(4), pp. 407–436 (1997).
120. Itō, K. *Multiple Wiener integral*, Journal of the Mathematical Society of Japan, **3**, pp. 157–169 (1951).
121. Jacod, J. and Shiryaev, A. N. *Limit Theorems for Stochastic Processes*, A Series of Comprehensive Studies in Mathematics, **288**, Springer (1987).
122. Jaffard, S. *On the Frisch-Parisi conjecture*, Journal des Mathématiques Pures et Appliquées, **79**(6), pp. 525–552 (2000).
123. Jaffard, S. *Multifractal formalism for functions, Part I: results valid for all functions*, SIAM Journal of Mathematical Analysis, **28**(4), pp. 944–970 (1997).
124. Jaffard, S. *The multifractal nature of Lévy-processes*, Probability Theory and Related Fields, **114**(2), pp. 207–228 (1999).
125. Johnson, N. L., Kotz, S. and Balakrishnan, N. *Continuous Univariate Distributions, Volume 2*, John Wiley & Sons (1995).
126. Kaiser, G. *A Friendly Guide to Wavelets*, Birkhäuser (1994).
127. Kahane, J.-P. and Peyrière, J. *Sur certaines martingales de Benoit Mandelbrot*, Advances in Mathematics, **22**, pp. 131–145 (1976).
128. Kahane, J.-P. *Sur le chaos multiplicatif*, Annales Sciences Mathematique Québec, **9**, pp. 105–150 (1985).
129. Kahane, J.-P. *Positive martingales and random measures*, Chinese Annals of Mathematics, **8b**, pp. 1–12 (1987).
130. Kahane, J.-P. *Random covering and multiplicative processes*, In: Bandt, C., Graf, S. and Zähle, M. (eds.) Fractal Geometry and Stochastics II, Progress in Probability, **46**, Birkhäuser, pp. 125–146 (2000).
131. Kaj, I. and Taqqu, M. S. *Convergence to fractional Brownian motion and to the telecom process: the integral representation approach*, Technical Report, **2004:16**, Uppsala University, Department of Mathematics (2004).
132. Karagiannis, T., Faloutsos, M. and Riedi, R. H. *Long-Range Dependence: Now you see it, now you don't*, Proc. GLOBECOM 2002, Taipei, pp. 2165–2169 (2002).
133. Karamata, J. *Sur un mode de croissance régulière. Théorèmes fondamentaux*, Bulletin de la Société mathématique de France, **61**, pp. 55–62 (1933).
134. Karatzas, I. and Shreve, S. E. *Brownian motion and stochastic calculus*, Springer (1988).
135. Kaufmann, J. *Blocking in a shared resource environment*, IEEE Transactions on Communications, **29**(10), pp. 1474–1481 (1981).
136. Kelly, F. P. *Notes on effective bandwidths*, In: Kelly, F. P., Zachary, S. and Ziedins I. B. (eds.) Stochastic Networks: Theory and Applications, Royal Statistical Society Lecture Notes Series, **4**, Oxford University Press, pp. 141–168 (1996).
137. Kelly, F. P. *Charging and rate control for elastic traffic*, European Transactions on Telecommunications, **8**, pp. 33–37 (1997).

138. Kelly, F. P., Maulhoo, A. K. and Tan, D. K. H. *Rate control for communication networks: shadow prices, proportional fairness and stability*, Journal of the Operational Research Society, **49**, pp. 237–318, (1998).
139. Kemeny, J. G. and Snell, J. L. *Finite Markov chains*, Springer (1983).
140. Kesten, H. and Spitzer, F. *A limit theorem related to a new class of self similar processes*, Probability Theory and Related Fields, **50**(1), pp. 5–25 (1979).
141. Kleinrock, L. *Queueing Systems, Volume 1*, John Wiley & Sons (1975).
142. Kleinrock, L. *Queueing Systems, Volume 2*, John Wiley & Sons (1976).
143. Kleinrock, L. and Tobagi, F. *Packet Switching in Radio Channels: Part I – Carrier Sense Multiple-Access Modes and Their Throughput-Delay Characteristics*, IEEE Transactions on Communications, **23**(12), pp. 1400–1416 (1975).
144. Klüppelberg, C. and Mikosch, T. *The integrated periodogram for stable processes*, Annals of Statistics, **24**(5), pp. 1855–1879 (1996).
145. Klüppelberg, C. and Mikosch, T. *Self-normalized and randomly centered spectral estimates*, In: Robinson, P. M. and Rosenblatt, M. (eds.) Athens Conference on Applied Probability and Time Series Analysis. Volume II: Times Series Analysis in Memory of E. J. Hannan, Lecture Notes in Statistics, **115**, pp. 259–271 (1996).
146. Kobayashi, H. *Stochastic Modelling: Queueing Models; Discrete-Time Queueing Systems*, In: Louchard, G. and Latouche, G., Probability Theory and Computer Science, Academic Press (1983).
147. Konheim, A. *An elementary solution of the queueing system GI/G/1*, SIAM Journal of Computing, **4**, pp. 540–545 (1979).
148. Kokoszka, P. and Mikosch, T. *The integrated periodogram for long-memory processes with finite or infinite variance*, Stochastic Processes and their Application, **66**(1), pp. 55–78 (2000).
149. Kokoszka, P. and Taqqu, M. S. *Parameter estimation for infinite variance fractional ARIMA*, Annals of Statistics, **24**(5), pp. 1880–1913 (1996).
150. Konstantopoulos, T. and Lin, S.-J. *Macroscopic models for long-range dependent network traffic*, Queuing Systems: Theory and Applications, **28**(1–3), pp. 215–243, (1998).
151. Krunz, M. and Matta, I. *Analytical Investigation of the Bias Effect in Variance-Type Estimators for Inference of Long-range Dependence*, Computer Networks, **40**(3), pp. 445–458 (2002).
152. Kurtz, T. G. *Limit theorems for workload input models*, In: Kelly, F. P., Zachary, S. and Ziedins, I., Stochastic Networks: Theory and Applications, Oxford Science Publications, pp. 119–139 (1996).
153. Lamperti, J. W. *Stochastic Processes*, Springer (1977).
154. Lamperti, J. W. *Semi-stable stochastic processes*, Transactions of the American Mathematical Society, **104**, pp. 62–78 (1962).
155. Laskin, N., Lambadaris, I., Harmantzis F. and Devetsikiotis, M. *Fractional Lévy Motion and its application to Network Traffic Modeling*, Computer Networks, **40**, pp. 363–375 (2002).
156. Lau, W.-C., Erramilli, A., Wang, J. L. and Willinger, W. *Self-similar traffic generation: The random midpoint displacement algorithm and its properties*, Proc. IEEE International Conference on Communications '95, Seattle, pp. 466–472 (1995).
157. Lee, Y. D., van de Liefvoort, A. and Wallace, V. L. *Modeling Correlated Traffic with Generalized IPP*, Performance Evaluation, **40**(1–3), pp. 99–114 (2000).

158. Lehn, J. and Wegmann, H. *Einführung in die Statistik*, Teubner (2000).
159. Legendre, A.-M. *Mémoire sur l'intégration de quelques équations aux différence partielles*, Mémoires de l'Académie des Sciences, pp. 309–351 (1787).
160. Leland, W. E., Taqqu, M. S., Willinger, W. and Wilson, D. V. *On the self-similar nature of Ethernet traffic (extended Version)*, IEEE/ACM Transactions on Networking, **2**(1), pp. 1–15 (1994).
161. Leupolz, M. *Geometric Poisson and fBm processes, basic results, applications and estimations*, Diploma Thesis, Mathematical Department, Ludwig-Maximilians-Universität, Munich (2006).
162. Leupolz, M. and Schlüchtermann, G. *Optimal allocations for resources and research*, preprint (2008).
163. Lévy Véhel, J. and Riedi, R. H. *Fractional Brownian motion and data traffic modeling: the other end of the spectrum*, In: Lévy Véhel, J., Lutton, E. and Tricot, C. (eds.) Fractals in Engineering, Springer, pp. 185–202 (1997).
164. Lévy Véhel, J. and Vojak, R. *Multifractal analysis of Choquet capacities: Preliminary results*, Advances in Applied Mathematics, **20**(1), pp. 1–43 (1998).
165. Lindley, D. V., *The theory of queues with a single server*, Cambridge Philosophical Society, **48**, pp. 277–289 (1952).
166. Lucantoni, D. M. *New results on the single server queue with a batch Markovian arrival process*, Stochastic Models, **7**(1), pp. 1–46 (1991).
167. Lucantoni, D. M. *The BMAP/G/1 queue: A tutorial*, In: Donatiello L. and Nelson R. (eds.) Models and Techniques for Performance Evaluation of Computer and Communication Systems, Springer (1993).
168. Lucantoni, D. M. *After long range dependency (LRD) discoveries, what are the lessons learned so far to provide QoS for Internet advanced applications*, Presentation 17th International Teletraffic Congress Panel Discussion, Salvador da Bahia (2001).
169. Maejima, M. *A remark on self-similar processes with stationary increments*, Canadian Journal of Statistics, **14**(1), pp. 81–82 (1986).
170. Maejima, M. *Limit theorem for infinite variance sequences*, In: Doukhan, P., Oppenheim, G. and Taqqu, M. S. (eds.) Long-Range Dependence: Theory and Application, Birkhäuser, pp. 157–164 (2003).
171. Mah, B. *An empirical model of HTTP network traffic*, Proc. IEEE INFOCOM '97, Kobe, pp. 592–600 (1997).
172. Mandelbrot, B. B. *Fractals and Scaling in Finance*, Springer (1997).
173. Mandelbrot, B. B. *Limit theorems on the self-normalized range for weakly and strongly dependent processes*, Zeitschrift für Wahrscheinlichkeitstheorie und verwandte Gebiete, **31**(4), pp. 271–285 (1975).
174. Mandelbrot, B. B. and Van Ness, J. W. *Fractional Brownian motions, fractional noises and applications*, SIAM Review, **10**(4), pp. 422–437 (1968).
175. Mannersalo, P. and Norros, I. *Multifractal analysis of real ATM traffic: A first Look*, COST257TD, VTT Information Technology (1997).
176. Mannersalo, P. and Norros, I. *A most probable path approach to queueing systems with general Gaussian input*, Computer Networks, **40**(3), pp. 399–412 (2002).
177. Mannersalo, P., Norros, I. and Riedi, R. H. *Multifractal products of stochastic processes: Construction and some basic properties*, Advances in Applied Probability, **34**(4), pp. 888–903 (2002).

178. Mazumdar, R., Mason, L. G. and Douligeris, C. *Fairness in network optimal flow control: optimality of product forms*, IEEE Transactions on Communications, **39**(5), pp. 775–782 (1991).
179. Melo, C. A. V. and da Foncesca, N. L. S. *Envelope process and computation of the equivalent bandwidth of multifractal flows*, Computer Networks, **48**(3), pp. 351–375 (2005).
180. Meyer, Y., Sellan, F. and Taqqu, M. S. *Wavelets, generalized white noise and fractional integration: the synthesis of fractional Brownian motion*, Journal of Fourier Analysis and Applications, **5**(5), pp. 465–494 (1999).
181. Mikosch, T., Resnick, S., Rootzén, H. and Stegeman, A. *Is network traffic approximated by stable Lévy motion or fractional Brownian motion?*, Annals of Applied Probability, **12**(1), pp. 23–68 (2002).
182. Mikosch, T. and Starica, C. *Long-range dependent effects and ARCH modeling*, In: Doukhan, P., Oppenheim, G. and Taqqu, M. S. (eds.) Long-Range Dependence: Theory and Application, Birkhäuser, pp. 439–459 (2003).
183. Montanari, A. *Long-range dependence in hydrology*, In: Doukhan, P., Oppenheim, G. and Taqqu, M. S. (eds.) Long-Range Dependence: Theory and Application, Birkhäuser, pp. 461–472 (2003).
184. Moulines, E. and Soulier, P. *Semiparametric spectral estimation for fractional processes*, In: Doukhan, P., Oppenheim, G. and Taqqu, M. S. (eds.) Long-Range Dependence: Theory and Application, Birkhäuser, pp. 251–301 (2003).
185. Muzy, J., Bacry, E. and Arneodo, A. *Multifractal formalism for fractal signals: The structure function approach versus the wavelet transform modulus-maxima method*, Physical Review E, **47**, pp. 875–884 (1993).
186. Neidhardt, A. L. and Wang, J. L. *The concept of relevant time scales and its application to queuing analysis of self-similar traffic (or is Hurst naughty or nice?)*, ACM SIGMETRICS Performance Evaluation Review, **26**(1), pp. 222–232 (1998).
187. Neuts, M. F. *Structured Stochastic Matrices of M/G/1 Type and Their Applications*, Dekker (1989).
188. Neuts, M. F. *Models based on the Markovian arrival process*, IEICE Transactions on Communications, **E75-B**(12), pp. 1255–1265 (1992).
189. Neuts, M. F. *Modeling data traffic streams*, Proc. 13th International Teletraffic Congress, Copenhagen, pp. 1–6 (1991).
190. Norros, I. *A storage model with self-similar input*, Queueing Systems, **16**(16), pp. 387–396 (1994).
191. Norros, I. *Busy periods of fractional Brownian storage: a large deviations approach*, Advances in Performance Analysis, **2**(1), pp. 1–19 (1999).
192. Norros, I. *On the use of fractional Brownian motion in the theory of connectionless networks*, IEEE Journal on Selected Areas in Communication, **13**(6), pp. 953–962 (1995).
193. Norros, I. and Kilpi, J. *Gaussian traffic modelling for differentiated services*, Proc. 15th Nordic Teletraffic Seminar, Lund, pp. 49–61 (2000).
194. Norros, I. *Four approaches to the fractional Brownian storage*, Fractals in Engineering, pp. 154–169 (1997).
195. Norros, I. *Most probable path techniques for Gaussian systems*, Proc. Networking 2002, Pisa (2002).
196. Norros, I., Valkeila, E. and Virtamo, J. *An elementary approach to a Girsanov formula and analytical results on fractional Brownian motions*, Bernoulli, **5**(4), pp. 571–587 (1999).

197. Norros, I. and Pruthi, P. *On the applicability of Gaussian traffic models*, Proc. Thirteenth Nordic Teletraffic Seminar, Trondheim, pp. 37–50 (1996).
198. Oppenheim, A. and Schafer, R. *Digital Signal Processing*, Prentice-Hall (1975).
199. Øksendal, B. *Stochastic Differential equation. An introduction with applications*, Springer (2003).
200. Park, K. and Willinger, W. (eds.) *Self-similar network traffic and performance evaluation*, John Wiley & Sons (2000).
201. Paxson, V. and Floyd, S. *Wide-area Traffic: The Failure of Poisson Modeling*, IEEE/ACM Transactions on Networking, **3**(3), pp. 226–244 (1995).
202. Peng, C. K., Buldyrev, S. V., Simons, M., Stanley, H. E. and Goldberger, A. L. *Mosaic organization of DNA nucleotides*, Physical Review E, **49**, pp. 1685–1689 (1994).
203. Peyriére, J. *Calculs de dimensions de Hausdorff*, Duke Mathematical Journal, **44**(3), pp. 591–601 (1977).
204. Piccoli, B. and Garavello, M. *Traffic Flow on Networks*, AIMS Applied Mathematics, **1** (2006).
205. Pipiras, V. and Taqqu, M. S. *Integration question related to fractional Brownian motion*, Probability Theory and Related Fields, **118**(2), pp. 251–291 (2000).
206. Pipiras, V. and Taqqu, M. S. *The limit of a renewal reward process with heavy-tailed rewards is not a linear fractional stable motion*, Bernoulli, **6**(4), pp. 607–614 (2000).
207. Pipiras, V. and Taqqu, M. S. *Fractional calculus and its connection to fractional Brownian Motion*, In: Doukhan, P., Oppenheim, G. and Taqqu, M. (eds.) Long-Range Dependence: Theory and Application, Birkhäuser, pp. 165–201 (2003).
208. Protter, P. *Stochastic Integration and Differential Equations*, Springer (1990).
209. Rao, M. M. *Probability Theory with Applications*, Academic Press (1984).
210. Reich, E. *On the integrodifferential equation of Takács. Part I*, Annals of Mathematical Statistics, **29**(2), pp. 563–570, (1958).
211. Resnick, S. I. *Adventures in Stochastic Processes*, Birkhäuser (1992).
212. Resnick, S. I. *Heavy tail modelling and teletraffic data*, Annals of Statistics, **25**(5), pp. 1805–1869 (1997).
213. Ribeiro, V.J., Riedi, R. H. and Baraniuk, R. G. *Wavelets and Multifractals for Network Traffic Modelling and Inference*, Proc. IEEE International Conference on Acoustics, Speech and Signal Processing, Salt Lake City, pp. 3429–3432 (2001).
214. Ribeiro, V. J., Riedi, R. H., Crouse, M. S. and Baraniuk, R. G. *Multiscale Queuing Analysis of Long-Range-Dependent Network Traffic*, Proc. IEEE INFOCOM 2000, Tel Aviv, pp. 1026–1035 (2000).
215. Ribeiro, V. J., Riedi, R. H., and Baraniuk, R. G. *Multiscale Queuing Analysis*, IEEE/ACM Transactions on Networking, **14**(5), pp. 1005–1018 (2006).
216. Ribeiro, V. J., Coates, M. J., Riedi, R. H., Sarvotham, S., Hendricks, B. and Baraniuk, R. G. *Multifractal Cross-Traffic Estimation*, Proc. ITC Specialist Seminar on IP Traffic Measurement, Modeling and Management, Monterey (2000).
217. Riedi, R. H., Crouse, M. S., Ribeiro, V. J. and Baraniuk, R. G. *A Multifractal Wavelet Model with Application to TCP Network Traffic*, IEEE Transactions on Information Theory (Special Issue on Multiscale Statistical Signal Analysis and its Applications), **45**(4), pp. 992–1018 (1999).

218. Riedi, R. H. *Multifractal Processes*, In: Doukhan, P., Oppenheim, G. and Taqqu, M. S. (eds.) Long-Range Dependence: Theory and Application, Birkhäuser, pp. 625–716 (2003).
219. Riedi, R. H. *On the multiplicative structure of network traffic*, IMA Conference on Mathematics in Signal Processing, Warwick (2000).
220. Riedi, R. H. and Lévy Véhel, J. *Multifractal properties of TCP traffic: A numerical study*, Technical Report, **RR-3129**, INRIA (1997).
221. Robinson, P. M. *Gaussian semiparametric estimation of long range dependence*, Annals of Statistics, **23**(5), pp. 1630–1661 (1995).
222. Royden, H. L. *Real Analysis*, Macmillan (1963).
223. Rudin, W. *Functional analysis*, McGraw-Hill (1991).
224. Ryu, B. K. and Elwalid, A. *The importance of long-range dependence of VBR video traffic in ATM traffic engineering: Myths and realities*, Proc. ACM SIGCOMM 96, Palo Alto, pp. 3–14 (1996).
225. Salvador, P, Valadas, R. and Pacheco, A. *Multiscale fitting procedure using Markov-modulated Poisson processes*, Telecommunication Systems, **23**(1–2), pp. 123–148 (2003).
226. Samorodnitsky, G. and Taqqu, M. S. *Stable non-Gaussian Random Processes. Stochastic Models with Infinite Variance*, Chapman & Hall/CRC (1994).
227. Sarvotham, S., Riedi, R. H. and Baraniuk, R. *Connection-level analysis and modeling of network traffic*, ACM SIGCOMM Internet Measurement Workshop 2001, San Francisco (2001).
228. Sarvotham, S., Riedi, R. H. and Baraniuk, R. *Network and user driven alpha-beta on-off source model for network traffic*, Computer Networks, **48**(3), pp. 335–350 (2005).
229. Schenker, S. *Fundamental design issues for the future Internet*, IEEE Journal of Selected Areas Communication, **13**(7), pp. 1176–1188 (1995).
230. Sennewald, K. *Controlled stochastic differential equations under Poisson uncertainty and with unbounded utility*, Journal of Economic Dynamics and Control, **31**(4), pp. 1106–1131 (2007).
231. Sennewald, K. und Wälde, K. *'Ito's lemma' and the Bellman equation for Poission processes: and applied view*, Journal of Economics, **89**(1), pp. 1–36 (2006).
232. Schürger, K. *Wahrscheinlichkeitstheorie*, Oldenbourg (1998).
233. Sigman, K. *Appendix: A primer on heavy-tailed distributions*, Queueing Systems, **33**(1–3), pp. 261–275 (1999).
234. Stoev, S. and Taqqu, M. S. *Simulation methods for linear fractional stable motion and FARIMA using the Fast Fourier Transform*, Fractals, **21**(1), pp. 95–121 (2004).
235. Stoev, S., Pipiras, V. and Taqqu, M. S. *Estimation of the self-similarity parameter in linear fractional stable motion*, Signal Processing, **82**(12), pp. 1873–1901 (2002).
236. Stratonovich, R. L. *A new representation for stochastic integrals and equations*, SIAM Journal on Control, **4**(2), pp. 362–371 (1966).
237. Strook, D. W. *Probability Theory. An Analytic View*, Cambridge University Press (1993).
238. Syski, R. *Introduction to Congestion Theory in Telephone Systems*, North-Holland (1986).
239. Takagi, H. *Queueing Analysis: A Foundation of Performance Evaluation. Volume 1: Vacation and Priority Systems, Part 1*, North-Holland (1991).

240. Taqqu, M. S. *Self-similar processes and related ultraviolet and infrared catastrophes*, In: Random Fields: Rigorous Results in Statistical Mechanics and Quantum Field Theory, Colloquia Mathematica Societatis Janos Bolyai, **27**(2), pp. 1057–1096 (1981).
241. Taqqu, M. S. *Fractional Brownian Motion and Long-Range Dependence*, In: Doukhan, P., Oppenheim, G. and Taqqu, M. S. (eds.) Long-Range Dependence: Theory and Application, Birkhäuser, pp. 5–38 (2003).
242. Taqqu, M. S. *A Bibliographical Guide to Self-Similar Processes and Long-Range Dependence*, In: Eberlein, E. and Taqqu, M. S. (eds.) Dependence in Probability and Statistics: A survey of recent results, Birkhäuser, pp. 137–162 (1986).
243. Taqqu, M. S. *Weak converence to Fractional Brownian Motion and to the Rosenblatt Process*, Zeitschrift für Wahrscheinlichkeitstheorie und verwandte Gebiete, **31**, pp. 287–302 (1975).
244. Taqqu, M. S. *Convergence of integrated processes of arbitrary Hermite rank*, Zeitschrift für Wahrscheinlichkeitstheorie und verwandte Gebiete, **50**, pp. 53–83 (1979).
245. Taqqu, M. S. *The modeling of Ethernet data and of signals that are heavy-tailed with infinite variance*, Scandinavian Journal of Statistics, **29**, pp. 273–295 (2001).
246. Taqqu, M. S. and Levy, J. *Using Renewal Processes to Generate Long-Range Dependence and High Variability*, In: Eberlein, E. and Taqqu, M. S. (eds.) Dependence in Probability and Statistics, Birkhäuser, pp. 73–89 (1986).
247. Taqqu, M. S. and Teverovsky, V. *On estimating the Intensity of Long-Range Dependence in Finite and Infinite Variance Time Series*, In: Alder, R. J., Feldmann, A. E. and Taqqu, M. S. (eds.) A Practical Guide to Heavy Tails: Statistical Techniques and Applications, Birkhäuser, pp. 177–217 (1998).
248. Taqqu, M. S., Teverovsky, V. and Willinger, W. *Estimators for long-range dependence: An empirical study*, Fractals, **3**(4), pp. 785–798 (1995).
249. Taqqu, M. S., Teverovsky, V. and Willinger, W. *Is network traffic self-similar or multifractal?*, Fractals, **5**(1), pp. 63–73 (1996).
250. Taqqu, M. S., Willinger, W. and Sherman, R. *Proof of a fundamental result in self-similar traffic modelling*, Computer Communication Review, **27**, pp. 5–23 (1997).
251. Thurner, S., Lowen, S. B., Heneghan, C., Feurstein, M. C., Feichtinger, H. G. and Teich, M. *Analysis, synthesis, and estimation of fractal-rate stochastic point processes*, Fractals, **5**(4), pp. 565–595 (1997).
252. Tjims, H. *Stochastic Models: An algorithmic Approach*, John Wiley & Sons, (1994).
253. Tran-Gia, P. *Analytische Leistungsbewertung verteilter Systeme*, Springer (1996).
254. Tran-Gia, P. *Einführung in die Leistungsbewertung und Verkehrstheorie*, Oldenbourg (2005).
255. Tran-Gia, P., *Discrete-Time analysis for the interdeparture distribution of GI/G/1 queues*, Proc. International Seminar on Teletraffic Analysis and Computer Performance Evaluation, Amsterdam (1986).
256. Tran-Gia, P. *Discrete-time analysis technique and application to usage parameter control modelling in ATM systems*, Proc. 8th Australian Teletraffic Research Seminar, Melbourne (1993).

257. Tran-Gia, P. and Mandjes, M. *Modeling of customer retrial phenomenom in cellular mobile networks*, IEEE Journal on Selected Areas in Communications, **15**(8), pp. 1406–1414 (1997).
258. Tricot, C. *Two definitions of fractal dimension*, Mathematical Proceedings of the Cambridge Philosophical Society, **91**, pp. 57–74 (1982).
259. Veitch, D., Taqqu, M. S. and Abry, P. *Meaningful initialisation for discrete time series*, Signal Processing, **80**(9), pp. 1971–1983 (2000).
260. Verdu, S. *Multiuser Detection*, Cambridge University Press, Cambridge (1998).
261. Wilks, S. S. *Mathematical Statistics*, John Wiley & Sons (1962).
262. Willinger, W. and Paxson, V. *Where Mathematics meets the Internet*, Notices of the American Mathematical Society, **45**(8), pp. 961–970 (1998).
263. Willinger, W., Paxson, V., Riedi, R. H. and Taqqu, M. S. *Long-range Dependence and Data Network Traffic*, In: Doukhan, P., Oppenheim, G. and Taqqu, M. S. (eds.) Long-Range Dependence: Theory and Application, Birkhäuser, pp. 373–407 (2003).
264. Willinger, W., Taqqu, M. S., Sherman, R. and Wilson, D. V. *Self-similarity through high-variability: Statistical analysis of Ethernet LAN traffic at the source level*, IEEE/ACM Transactions on Networking, **5**(1), pp. 71–86 (1997).
265. Willinger, W., Taqqu, M. S., Leland, W. E. and Wilson, V. *Self-Similarity in High-Speed Packet traffic: Analysis and Modeling of Ethernet Traffic Measurement*, Statistical Sciences, **10**, pp. 67–85 (1995).
266. Wischik, D. *The output of a switch, or, effective bandwidths for networks*, Queueing Systems, **32**(4), pp. 383–396 (1999).
267. Whittle, P. *Hypothesis testing in time series analysis*, Almqvist och Wicksel (1951).
268. Whittle, P. *Estimation and information in stationary time series*, Arkiv för Matematik, **2**, pp. 423–434 (1953).
269. Wolpert, R. L. and Taqqu, M. S. *Fractional Ornstein-Uhlenbeck Lévy Processes and the Telecom Prozess: Upstairs and Downstairs*, Signal Processing, **85**(8), pp. 1523–1545 (2005).
270. Yor, M. *Some Aspects of Brownian Motion: Part I, some special functionals*, Lectures in Mathematics, ETH Zürich, Birkhäuser (1992).
271. Yor, M. *Some Aspects of Brownian Motion: Part II, some recent martingale problems*, Lectures in Mathematics, ETH Zürich, Birkhäuser (1997).
272. Yor, M. *On some exponential functionals of Brownian motion*, Advances in Applied Probability, **24**, pp. 509–531 (1992).

Index

α-stable, 309
   distribution, 194
   Lévy process, 243
   motion, 187, 195
   process, 187
      symmetric, 195
σ-algebra, 197

absolute value method, 349
absorbing state, 149
access level, 11
adjusted target region, 356
admission control, 119
aggregated process, 71
Aloha, 55
analytical cycle, 116
application
   best effort, 8
   real time, 8
application layer, 8
application level, 11, 235
ARMA, 215
   model, 211
arrival process
   batch Markovian, 159
   MAP, 162
   Poisson, 244
   time discrete Markov, 159
arrival rates
   fitting, 175
arrival time, 61
asymptotic
   linear, 220

   lognormal, 299
   wavelet self-similar, 313
asymptotic unbiased, 326
autocorrelation function, 191
autocovariance, 206
   function, 198
autoregressive moving average, 215
AV_X, 350
AVM, 349, 350

backward recurrence time, 67
Banach space, 251
batch Markovian arrival process, 159
Bellcore measurements, 95
best effort application, 8
beta distribution, 83, 276
   symmetric, 276
bias, 322
birth rate, 38
Blackwell renewal theorem, 69
blocking, 118
   probability, 42
BMAP, 159, 174
bounded variation, 140
Brownian motion, 186, 242, 311
   complex-valued version, 192
   fractional, 186
buffer
   capacity, 384
   size, 396
burst, 12
burst level, 12

calling attempts, 60

capacity, 425
cascade
　beta-binomial, 300
　binomial, 296, 314
　c-adic, 302
　lognormal, 300
　multinomial, 302
Cauchy
　distribution, 194, 325
　problem, 140
cellular mobile system, 30
central limit theorem, 226, 242
　general, 228
　renewal counting processes, 231
Chapman-Kolmogorov
　equation, 35
　system
　　Jackson network, 124
characteristic function, 195
characteristic moment, 194
classes of priority, 249
Cobb-Douglas production function, 444, 463
completeness relation, 35
confidence interval, 348
connection duration
　heavy-tail distributed, 244
connection level, 12
consistent estimator, 325
control process, 427
correlation function, 313
counting process, 231
covariance function, 186, 188, 189, 241
Cramér-Rao inequality, 324
critical dyadic time scale, 413
　queue, 413
critical time scale, 413
CSMA/CD, 55

D-MAP, 159
D-PH, 155
data link layer, 4
data traffic, see traffic
death rate, 38
deterministic envelope, 288, 309
dialogue level, 11
differentiability of paths, 191
differential equation, stochastic, 239, 443

differentiated services, 21
DiffServ, 21
distribution
　α-stable, 194
　empirical, 329
　Erlang-k, 148
　heavy-tail, 84, 121
　hyperexponential, 153
　integrated
　　complementary, 65
　PH, 152
　phase, 148
　symmetric beta, 418
　time discrete phase, 155
　waiting time, 105
distribution function
　empirical, 246
DNA, 228
Donsker invariance principle, 229
DWT, 370
dyadic tree, 273, 415

ELA, 254
embedded independence, 298
embedded time, 75
end system, 1, 16
entropy admissible, 139
entropy flux, 139
envelope, 302
　deterministic, 288, 289, 309, 311
envelope process
　deterministic, 406
Erlang loss formula, 42
Erlang queueing formula, 47
Erlang-k distribution, 148
Erlang-B formula, 42
Erlang-C formula, 47
estimator
　fractional exponential, 361
　graphical, 353
　parametric, 349, 364
　semiparametric, 349, 360, 379
estimator function, 321
Ethernet, 4
exact self-similar, 184
excess distribution, 92
exponent of the increase
　coarse, 284
　local, 285

exponential decay, 207
exponential distribution, 82

factorization theorem, 327
FARIMA time series, 211, 314, 376, 380
FARIMA$[0, d, 0]$, 212
FARIMA$[p, d, q]$, 212, 217
Fast Fourier transform, 364
FBM, 308, 310, 311
   antipersistent, 191
   chaotic, 191
   persistent, 191
FCFS, 78
FESM, 361
FGN, 208, 308
   process, 380
FIFO, 50, 78, 79, 105
first passage time, 166
flow, 14
   homogenous, 408, 410
   model, 222
flux, 135
forward recurrence time, 67
   distribution, 67
Fourier transform, 192
fractional $s\alpha s$-noise, 200
fractional Brownian motion, 186, 189, 190, 192, 198, 244
   Fourier transform, 248
   multiscaling, 245
fractional Brownian-Poissonian motion
   geometric, 449
fractional Clark-Haussmann-Ocone theorem, 447
fractional Gaussian noise, 24
fractional Girsanov theorem, 446
fractional Itō formula, 448
fractional white noise, 190, 208
free capacity, 385
function, slowly varying, 207, 227
fundamental cycle, 169
fundamental period, 166

G_α-attraction, 227
Gamma function, 213
Gaussian distribution, 83, 194
Gaussian estimator, 360
Gaussian noise, fractional, 208
Gaussian process, 250

Gaussian sequence, linear, 211
Gaussian traffic, 218
generator, 62, 98
Geom(m)/Geom(m)/1 system, 114
geometric approach, 104
GI/G/$n - S$ system, 33
GI/M/$n - S$ system, 33
grade of confidence, 348
grain-based spectrum, 286, 287
   deterministic, 291
grid computing, 23
growth rate
   stochastic, 437, 445
GSM, 31

H-sssi process, 185, 189, 191
   without finite variance, 191
Hölder continuity, 191
   local degree, 281
Hölder exponent, 280
   coarse, 282
   function, 409
Haar wavelet, 283, 303
harmonic representation, 192
Hausdorff dimension, 286
Hausdorff spectrum, 285
heavy-tail distribution, 81, 85
heavy-tail exponent
   estimator, 335
hidden Markov model, 174
Hilbert space, 253
Hill estimator, 336
host-to-host, 15
Hurst exponent, 185, 241, 277
   estimator, 349
Hurst parameter, 185, 186, 242

IETF, 3
increments, stationary, 185, 302
initial value problem
   matrix-valued, 162
inner product, 195
integral equation, 63, 110
   Lindley, 112
   Wiener-Hopf, 110
integral representation, 199
integral, stochastic, 198
integrated distribution, 88
   complementary, 88, 91

integrated services, 21
interarrival time, 32, 61, 72, 232
Internet Protocol, 5
interrupted Poisson process, 154, 174
interval estimation, 348
IntServ, 21
IP, 5
IP traffic
  N-state MMPP
    fitting, 176
  BMAP process, 174
  D-MMPP model, 173, 174, 177
  interrupted Poisson process, 174
  MMPP model, 173
Itō integral, 192

Jackson serving network, 126
jitter, 20
Joseph effect, 199
jump process, 187

Kendall notation, 32, 33
Kolmogorov forward equation, 36
Kruzkov
  entropy, 137

Lévy Process, 195
laddar index, 98
Lagrange multiplicator, 438
Landau symbol, 284
Laplace transform, 63, 86
Laplace-Stieltjes transform, 164
least square estimation, 332
Lebesgue measure, 197
Legendre transform, 278, 293, 294, 302, 311
Leland group, 219
life time, 61
likelihood function, 345, 346
  sample size N, 348
limit theorem
  functional, 229
line speed
  limiting, 177
linear fractional stable motion, 200
linear model, 332
link approximation
  coarse empty, 254
  full, 255

Lipschitz continuity, 281
Little formula, 48, 101
live video streaming, 23
loading coefficient, 45
local Whittle estimator, 364
log periodogram, 359
  estimator
    global, 360
    local, 360
log-fractional stable motion, 200
log-log diagram, 352, 359
logarithmic scale diagram, 373
loglikelihood function, 347
lognormal distribution, 95, 276
long-range dependence, 187, 204, 205, 240, 312
  Allan variance (LRD-SAV), 205
  estimator, 375
  general, 204
  process, 198
loss system
  finite source number, 54
  M/M/∞, 58
LRD, 187
LRD and SRD, 207
LRD-SAV, 205
Lyaponov function, 439

M_K-FBM, 247
M/G/$n-S$ system, 33
M/M/$n-S$ system, 33
Malliavin derivative, 445
MAP, 156
marginal distribution
  Gaussian, 251
  multiscale, 417
Markov chain, 73, 130
  embedded, 73
  fitting, 175
  irreducible, 149
Markov control, 451
Markov modulated Poisson process, 158
Markov process, 32
Markov property, 73
Markovian arrival process, 156
matrix-analytic method, 148
maximum likelihood
  estimator, 345, 363
  function, 364

maximum transmission unit, 5
mean excess function, 92
mean queueing length, 117
measure space, 197
method
  matrix-analytic, 148
  of least squares, 331
  of moments, 344
MFA, 285
mixed traffic, 423
MMPP, 158
model
  multiplicative, 272
  on-off, 240
  TCP/IP, 2
mother wavelet, 282
MPLS, 21
MTU, 5
multifractal
  bothsided, 293
  leftsided, 293
  rightsided, 293
multifractal analysis, 285
multifractal flow
  aggregation, 407
multifractal formalism, 293
  central, 294
multifractal wavelet model, 275
multiplexing, 404
  gain, 42, 404, 409
multiplier, 297
multiprotocol label switching, 21
multiscale FBM, 247
multiscale queueing formula, 415
multiscaled trees, 273
MWM, 275
  model, 418

Net, 16
net-to-net, 16
network
  capacity, 384
  circuit switched, 2
  conservation laws model, 133
  Jackson, 122
  layer, 5
  packet switched, 2
  peer to peer, 23
  topology, 18

new start vector, 150
Noah effect, 200
nodes, 123
non-Gaussian processes, 191
normal attraction, 228
normal distribution, 194
normal equation, 333
Norros approach, 218

on-off models, 235
on-off process, cyclic, 160
optimal selection, 427
optimization
  economic equilibrium, 435
  network flow, 424
    utility approach, 449
  stochastic perturbation, 443
order statistic, 326
  second, 276
overflow probability, 384
overprovisioning, 20

packet level, 12
packet loss, 20, 134
packet switching, 1
PAR protocols, 29
parameter
  state dependent, 126
Pareto distribution, 93, 229
partition function, 287, 301
  deterministic, 278, 288, 295, 302, 310, 311, 313
PASTA, 171
path vector, 253
  most probable, 252, 253
path, càdlàg, 195
PDU, 1
periodogram, 365
phase distribution, 148
phase representation, 148
physical layer, 3
point estimator, 321
Poisson process
  interrupted, 154
  Markov modulated, 158
  memoryless, 68
Pollaczek-Khintchine formula
  queueing state probability, 80
  queueing time, 97

484   Index

state probabilities, 78, 79
polynomial, 281
port-to-port, 14
preserving the mass, 297
preserving the mean, 298
principle of largest deviation, 253, 287
prioritization, 21
process
   α-stable, 187
   α-stable Lévy, 309
   arrival, 32
   H-sssi, 185, 308, 311
   homogeneous, 34
   Lévy stable, 310, 311
   LRD, 370
   Markov renewal, 157
   $s\alpha s$, 310
   self-similar, 370
   serving, 32
   stable, 195
   state, 32
   stationary, 187
processor sharing model, 255
processor sharing system, 254
proportional fair, 437
proportional fairness, 436
protocol data unit, 1
PSTN, 22
public switched telephone networks, 22

QoS, 19
quadratic variation, 379
quality of service, 19
quasi-martingale, 447
queue
   tail probability, 416
queue length
   D-MMPP, 176
   maximal
      time, 406
queueing
   FIFO, 394
   fractional Lévy motion, 395
   LRD processes, 383
   multifractal, 405, 411
   multifractal tree, 411
   multiscale FBM, 392
queueing distribution, 50
   admissible, 132

queueing distribution function, 78
queueing formula, multiscale, 417
queueing length, 96, 101
   mean, 48, 132
queueing loss system, 59
queueing probability, 47, 80
queueing process, 250
queueing space, infinite, 45
queueing system
   discrete
      G/G/1, 109
   Engset, 59
   GI/G/1, 98, 107
   M/G/1, 96
   MAP/G/1, 161
queueing time
   mean, 48, 80
   Pollaczek-Khintchine formula, 172
   virtual, 171
queueing time distribution
   discrete, 113
   storage, 390

R/S estimator, 356
random sum, 71, 230
random vector, 196
random walk, 62, 98
   discrete, 62
   symmetric, 62
Rankine-Hugoniot, 138
rate
   outgoing data, 384
rate control, 436
real time application, 8
real time requirements, 20
recurrence time, 67
regression line, 329, 336, 352, 354
regression method, 335
renewal counting process, 61
renewal density, 63
renewal equation, 63
renewal function, 62
   boundary, 65
   Erlang-k distribution, 64
   exponential distribution, 64
   normal distribution, 65
renewal process, 60, 61, 231
   arithmetic, 69
   asymptotic behavior, 68

Index   485

delayed, 61
embedded Markovian, 163
modified, 61
non-arithmetic, 69
ordinary, 61, 63
simple, 61
stationary, 71
renewal theorem
    Blackwell, 69
    elementary, 68
    fundamental, 69
renewal theory, paradox, 70
renewal time, 61
residual time, 32
residual work, 96
reward process, 235
Riemann problem, 139
Riemann solver, 141
robust estimator, 328
roundtrip time, 6
RTT, 6

sample space, 321
$s\alpha s$ Lévy motion, 200
$s\alpha s$ motion, 196
$s\alpha s$ process, 196
$s\alpha s$-random measure, 197
scaling equation, 310
scaling estimator, 337
scaling property
    fractional Lévy motion, 396
SDH, 4
self-similar, 188
self-similarity, 184, 240
    asymptotic
        Pareto distribution, 233
    second degree, 313
semi-Markov property, 157
sequence
    causal, 211
series, divergence, 206
service duration, 32
service level agreements, 119
service rate, 388, 395, 412, 425, 433
serving network, closed, 127, 128
serving time, 72, 229
    subexponential, 96
set
    closed, 253

open, 253
shadow price, 436, 438
short-range dependence, 207, 220
singularity, 206
    first order, 86
Skorokhod metric, 230
SLA, 119
slowly varying functions, 84
sojourn time, 79
solution
    distributive sense, 138
SONET, 4
spacer, 119
    GI/D/1 system, 119
spectral density, 206, 349
spectral representation, 349
spectrum of the exponent, 285
spectrum, deterministic, 301
stability criteria, 45
stable process, 195
standard process, 186
state
    absorbing, 149
    transient, 149
state equations, 76
    stationary, 37
state probability, 36, 104
    Kolmogorov forward equation, 36
    MAP, 170
    stationary, 166
state transition, 76
stationary increments, 185
stationary process, 187
stationary sequence, 203
Stirling formula, 213
stochastic calculus
    fractional Brownian motion, 445
stochastic continuous, 184
stochastic matrix, 73
storage area networks, 23
storage process, 239, 384
strong law of large numbers, 231
subexponential distributions, 81
sufficient statistic, 327
sum process, 71
Synchronous Digital Hierarchy, 4
Synchronous Optical NETwork, 4
system
    demand, 32

load, 117
M/G/∞
   lognormal distribution, 234
   residual work, 33
systematic error, 322

Taylor polynomial, 281
TCP, 6
   acknowledgment, 6
   congestion control, 7
   connection-oriented, 7
   control cycle, 133
   flow control, 7
   four way close, 6
   segment, 6
   slow start, 7
   three way handshake, 6
   timeouts, 6
TCP influence, 394
TCP/IP
   protocol architecture, 3
theorem of Adler, 191
theorem of Karamata, 224
threshold scale, 416
time discrete Markov arrival process, 159
total variation, 140
traffic
   asymmetry, 17
   circuit switched, 22
   composite, 14
   elastic, 19, 134, 436
   end system, 16
   flow, 14
   Gaussian, 218
   heterogeneous, 258
   host-to-host, 15
   intensity, 101
   load, 110, 388
   management, 119
   model
      GI/G/∞, 232
   models
      special, 58
   net, 16
   net-to-net, 16
   optimization, 423
   packet switched, 22, 219
   port-to-port, 14

priority, 130, 144
relations, 13
TCP, 304
temporal behavior, 17
time sensitive, 20, 245
value, 42
WAN, 279
transient state, 149
transient transition, 157
transition diagram, 46
transition matrix, 73, 103
transition probability, 34
   exponential distributed serving time, 38
   Poisson process, 37
transmission control protocol, 6
transport layer, 5

UDP, 5
   connectionless, 7
UMTS, 31
unbiased estimator, 322, 323
UNI, 119
user datagram protocol, 5
user network interface, 119
utility function, 435, 436

vanishing moment, 369
variance method, 352
variance of residuals, 354
velocity functions, 134
VLA, 255
Voice over IP, 23

waiting model, 232
waiting queue distribution, 102
waiting time
   distribution, 96, 105, 106
      stationary, 110
      subexponential, 97
      Weibull, 391
   mean, 48
   probability, 117
Wald identity, 68, 72
wavelet
   analysis, 307, 368
   coefficient, 303, 368, 379
   differentiable, 306
   energy, 315

exponent, 282
  coarse, 284, 305
  local, 284
  transform, 379
wavelet-domain independent Gaussian, 275
weak solution, 138

Weibull distribution, 83
white noise, fractional, 208
Whittle estimator, 363, 364
  local, 365
WIG model, 274, 417
workload, 100

Printing: Krips bv, Meppel, The Netherlands
Binding: Stürtz, Würzburg, Germany